AF882523

SOTHIS PRESS

Sothis Press aims to completely reimagine the public domain book with an emphasis on universal access to captivatingly designed print editions. We believe that technology will enable every single book ever published to be perpetually accessible in beautiful print form. Sothis is beauty in ubiquity.

www.sothis.press
1 Waterhouse Square, London, EC1N 2ST

The Ancient Volcanoes of Great Britain

Archibald Geikie

SOTHIS PRESS

ISBN 978-1-0365-2292-6

The Ancient Volcanoes of Great Britain
is in the public domain.

This edition is published by
Sothis Press © 2023. The company
is registered in England and Wales.
The Company Number is 14900389.

THE ANCIENT VOLCANOES OF GREAT BRITAIN

G-VI/63

BOEK: —

THE

ANCIENT VOLCANOES

OF

GREAT BRITAIN

GEOLOGISCH-MINERALOGISCH MUSEUM LEIDEN

BY

SIR ARCHIBALD GEIKIE, F.R.S.

D.C.L. Oxf., D.Sc. Camb., Dubl.; LL.D. St. And., Edinb.

DIRECTOR-GENERAL OF THE GEOLOGICAL SURVEY OF GREAT BRITAIN AND IRELAND;
CORRESPONDENT OF THE INSTITUTE OF FRANCE;
OF THE ACADEMIES OF BERLIN, VIENNA, MUNICH, TURIN, BELGIUM, STOCKHOLM, GÖTTINGEN, NEW YORK; OF THE IMPERIAL MINERALOGICAL SOCIETY AND SOCIETY OF NATURALISTS, ST. PETERSBURG; NATURAL HISTORY SOCIETY, MOSCOW; SCIENTIFIC SOCIETY, CHRISTIANIA; AMERICAN PHILOSOPHICAL SOCIETY; OF THE GEOLOGICAL SOCIETIES OF LONDON, FRANCE, BELGIUM, STOCKHOLM, ETC.

WITH SEVEN MAPS AND NUMEROUS ILLUSTRATIONS

IN TWO VOLUMES

VOL. I

London

MACMILLAN AND CO., Limited.

NEW YORK: THE MACMILLAN COMPANY.

1897

All rights reserved

TO

M. Ferdinand Fouqué

MEMBER OF THE INSTITUTE

PROFESSOR OF THE NATURAL HISTORY OF INORGANIC BODIES

IN THE COLLÈGE DE FRANCE

AND

M. Auguste Michel-Lévy

MEMBER OF THE INSTITUTE

DIRECTOR OF THE GEOLOGICAL SURVEY OF FRANCE

DISTINGUISHED REPRESENTATIVES

OF THAT FRENCH SCHOOL OF GEOLOGY

WHICH BY THE HANDS OF DESMAREST FOUNDED THE

STUDY OF ANCIENT VOLCANOES

AND HAS SINCE DONE SO MUCH TO

PROMOTE ITS PROGRESS

THESE VOLUMES ARE INSCRIBED

WITH THE HIGHEST ADMIRATION AND

ESTEEM

PREFACE

In no department of science is the slow and chequered progress of investigation more conspicuous than in that branch of Geology which treats of volcanoes. Although from the earliest dawn of history, men had been familiar with the stupendous events of volcanic eruptions, they were singularly slow in recognizing these phenomena as definite and important parts of the natural history of the earth. Even within the present century, the dominant geological school in Europe taught that volcanoes were mere accidents, due to the combustion of subterranean beds of coal casually set on fire by lightning, or by the decomposition of pyrites. Burning mountains, as they were called, were believed to be only local and fortuitous appearances, depending on the position of the coal-fields, and having no essential connection with the internal structure and past condition of our planet. So long as such fantastic conceptions prevailed, it was impossible that any solid progress could be made in this branch of science. A juster appreciation of the nature of the earth's interior was needed before men could recognize that volcanic action had once been vigorous and prolonged in many countries, where no remains of volcanoes can now be seen.

To France, which has led the way in so many departments of human inquiry, belongs the merit of having laid the foundations of the systematic study of ancient volcanoes. Her groups of Puys furnished the earliest inspiration in this subject, and have ever since been classic ground to which the geological pilgrim has made his way from all parts of the world. As far back as the year 1752, Guettard recognised that these marvellous hills were volcanic cones that had poured forth streams of lava. But it was reserved for Desmarest twelve years later to examine the question in detail, and to establish the investigation of former volcanic action upon a broad and firm basis of careful observation and sagacious inference. His method of research was as well conceived as the region of Auvergne was admirably fitted to be the field of exploration. He soon discovered that the volcanoes of Central France were not all of one age, but had made their appearance in a long series, whereof the individual members became less perfect and distinct in proportion to their antiquity. Beginning with the cones, craters, and lava-streams which stand out so fresh that they might almost be supposed to have been erupted only a few generations ago, Desmarest traced the volcanic series backward in time, through successive stages of the decay and degradation wrought upon them by the influence of the atmosphere, rain and running water. He was thus able, as it were, to watch the gradual obliteration of the cones, the removal of the ashes and scoriæ, and the erosion of the lava-streams, until he could point to mere isolated remnants of lava, perched upon the hills, and overlooking the valleys which had been excavated through them. He showed how every step in this process of denudation could be illustrated by examples of its occurrence in Auvergne, and how, in this way, the various eruptions could be grouped according to their place in the chronological sequence. To this illustrious Frenchman geology is thus indebted, not only for the foundation of the scientific study of former volcanic action, but for the first carefully worked out example of the potency of subærial erosion in the excavation of valleys and the transformation of the scenery of the land.

While these fruitful researches were in progress in France, others of hardly less moment were advancing in Scotland. There likewise Nature had provided ample material to arrest the attention of all who cared to make themselves acquainted with the past history of our globe. Hutton, as a part of his immortal *Theory of the Earth*, had conceived the idea that much molten material had been injected from below into the terrestrial crust, and he had found many proofs of such intrusion among the rocks alike of the Lowlands and Highlands of his native country. His observations, confirmed and extended by Playfair and Hall, and subsequently by Macculloch, opened up the investigation of the subterranean phases of ancient volcanic action.

Under the influence of these great pioneers, volcanic geology would have made steady and perhaps rapid progress in the later decades of last century, and the earlier years of the present, but for the theoretical views unfortunately adopted by Werner. That illustrious teacher, to whom volcanoes seemed to be a blot on the system of nature which he had devised, did all in his power to depreciate their importance. Adopting the old and absurd notion that they were caused by the combustion of coal under ground, he laboured to show that they were mere modern accidents, and had no connection with his universal formations. He proclaimed, as an obvious axiom in science, that the basalts, so widely spread over Central and Western Europe, and which the observations of Desmarest had shown to mark the sites of old volcanoes, were really chemical precipitates from a primeval universal ocean. Yet he had actually before him in Saxony examples of basalt hills which entirely disprove his assertions.

Fortunately for the progress of natural knowledge, Werner disliked the manual labour of penmanship. Consequently he wrote little. But his wide range of acquirement, not in mineralogy only, his precision of statement, his absolute certainty about the truth of his own opinions, and his hardly disguised contempt for opinions that differed from them, combined with his enthusiasm, eloquence and personal charm, fired his pupils with emulation of his zeal and turned them into veritable propagandists. Misled as to the structure of the country in which their master taught, and undisciplined to investigate nature with an impartial mind, they travelled into other lands for the purpose of applying there the artificial system which they had learnt at Freiberg. The methodical but cumbrous terminology in which Werner had trained them was translated by them into their own languages, where it looked still more uncouth than in its native German. Besides imbibing their teacher's system, they acquired and even improved upon his somewhat disdainful manner towards all conclusions different from those of the Saxon Mining School.

Such was the spirit in which the pupils of Werner proceeded to set the "geognosy" of Europe to rights. The views, announced by Desmarest, that various rocks, far removed from any active volcano, were yet of volcanic origin, had been slowly gaining ground when the militant students from Saxony spread themselves over the Continent. These views, however, being irreconcilable with the tenets enunciated from the Freiberg Chair, were now either ignored or contemptuously rejected. Werner's disciples loved to call themselves by their teacher's term "geognosts," and claimed that they confined themselves to the strict investigation of fact with regard to the structure of the earth, in apparent unconsciousness that their terminology and methods were founded on baseless assumptions and almost puerile hypotheses.

With such elements ready for controversy, it was no wonder that before long a battle arose over the origin of basalt and the part played by volcanoes in the past history of the globe. The disciples of Werner, champions of a universal ocean and the deposition of everything from water, were dubbed Neptunists, while their opponents, equally stubborn in defence of the potency of volcanic fire, were known as Vulcanists or Plutonists. For more than a generation this futile warfare was waged with extraordinary bitterness—dogmatism and authority doing their best to stop the progress of impartial observation and honest opinion.

One of the most notable incidents in the campaign is to be found in the way in which the tide of battle was at last turned against the Wernerians.

Cuvier tells us that when some of the ardent upholders of the Freiberg faith came to consult Desmarest, the old man, who took no part in the fray, would only answer, "Go and see." He felt that in his memoir and maps he had demonstrated the truth of his conclusions, and that an unprejudiced observer had only to visit Auvergne to be convinced.

By a curious irony of fate it was from that very Auvergne that the light broke which finally chased away the Wernerian darkness, and it was by two of Werner's most distinguished disciples that the reaction was begun.

Daubuisson, a favourite pupil of the Freiberg professor, had written and published at Paris in 1803 a volume on the Basalts of Saxony, conceived in the true Wernerian spirit, and treating these rocks, as he had been taught to regard then, as chemical precipitates from a former universal ocean. In the following year the young and accomplished Frenchman went to Auvergne and the Vivarais that he might see with his own eyes the alleged proofs of the volcanic origin of basalt. Greatly no doubt to his own surprise, he found these proofs to be irrefragable. With praiseworthy frankness he lost no time in publicly announcing his recantation of the Wernerian doctrine on the subject, and ever afterwards he did good service in making the cause of truth and progress prevail.

Still more sensational was the conversion of a yet more illustrious prophet of the Freiberg school—the great Leopold von Buch. He too had been educated in the strictest Wernerian faith. But eventually, after a journey to Italy, he made his way to Auvergne in 1802, and there, in presence of the astonishing volcanic records of that region, the scales seem to have fallen from his eyes also. With evident reluctance he began to doubt his master's teaching in regard to basalt and volcanoes. He went into raptures over the clear presentation of volcanic phenomena to be found in Central France, traced each detail among the puys, as in the examination of a series of vast models, and remarked that while we may infer what takes place at Vesuvius, we can actually see what has transpired at the Puy de Pariou. With the enthusiasm of a convert he rushed into the discussion of the phenomena, but somehow omitted to make any mention of Desmarest, who had taught the truth so many years before.

Impressed by the example of such men as Daubuisson and Von Buch, the Wernerian disciples gradually slackened in zeal for their master's tenets. They clung to their errors longer perhaps in Scotland than anywhere else out of Germany—a singular paradox only explicable by another personal influence. Jameson, trained at Freiberg, carried thence to the University of Edinburgh the most implicit acceptance of the tenets of the Saxon school, and continued to maintain the aqueous origin of basalt for many years after the notion had been abandoned by some of his most distinguished contemporaries. But the error, though it died hard, was confessed at last even by Jameson.

After the close of this protracted and animated controversy the study of former volcanic action resumed its place among the accepted subjects of geological research. From the peculiarly favourable structure of the country, Britain has been enabled to make many important contributions to the investigation of the subject. De la Beche, Murchison and Sedgwick led the way in recognizing, even among the most ancient stratified formations of England and Wales, the records of contemporaneous volcanoes and of their subterranean intrusions. Scrope threw himself with ardour into the study of the volcanoes of Italy and of Central France. Maclaren made known the structure of some of the volcanic groups of the lowlands of Scotland. Ramsay, Selwyn, and Jukes, following these pioneers, were the first to map out a Palæozoic volcanic region in ample detail. Sorby, applying to the study of rocks the method of microscopic examination by thin slices, devised by William Nicol of Edinburgh for the study of fossil plants, opened up a new and vast field in the domain of observational geology, and furnished the geologist with a key to solve many of the problems of volcanism. Thus, alike from the stratigraphical and petrographical sides, the igneous rocks of this country have received constantly increasing attention.

The present work is intended to offer a summary of what has now been ascertained regarding the former volcanoes of the British Isles. The subject has occupied much of my time and thought all through life. Born among the crags that mark the sites of some of these volcanoes, I was led in my boyhood to interest myself in their structure and history. The fascination which they then exercised has lasted till now, impelling me to make myself acquainted with the volcanic records all over our islands, and to travel into the volcanic regions of Europe and Western America for the purpose of gaining clearer conceptions of the phenomena.

From time to time during a period of almost forty years I have communicated chiefly to the Geological Society of London and the Royal Society of Edinburgh the results of my researches. As materials accumulated, the desire arose to combine them into a general narrative of the whole progress of volcanic action from the remotest geological periods down to the time when the latest eruptions ceased. An opportunity of partially putting this design into execution occurred when, as President of the Geological Society, the duty devolved upon me of giving the Annual Addresses in 1891 and 1892. Within the limits permissible to such essays, it was not possible to present more than a full summary of the subject. Since that time I have continued my researches in the field, especially among the Tertiary volcanic areas, and have now expanded the two Addresses by the incorporation of a large amount of new matter and of portions of my published papers.

In the onward march of science a book which is abreast of our knowledge to-day begins to be left behind to-morrow. Nevertheless it may serve a useful purpose if it does no more than make a definite presentation of the condition of that knowledge at a particular time. Such a statement becomes a kind of landmark by which subsequent progress may be measured. It may also be of service in indicating the gaps that have to be filled up, and the fields where fresh research may most hopefully be undertaken.

I have to thank the Councils of the Royal Society of Edinburgh and the Geological Society for their permission to use a number of the illustrations which have accompanied my papers published in their *Transactions* and *Journal*. To Colonel Evans and Miss Thom of Canna I am indebted for the photographs which they have kindly taken for me. To those of my colleagues in the Geological Survey who have furnished me with information my best thanks are due. Their contributions are acknowledged where they have been made use of in the text.

The illustrations of these volumes are chiefly from my own note-books and sketch-books. But besides the photographs just referred to, I have availed myself of a series taken by Mr. Robert Lunn for the Geological Survey among the volcanic districts of Central Scotland.

GEOLOGICAL SURVEY OFFICE,
28 JERMYN STREET, LONDON,
1st *January* 1897.

CONTENTS

BOOK I

GENERAL PRINCIPLES AND METHODS OF INVESTIGATION

CHAPTER I

CHAPTER II

CHAPTER III

CHAPTER IV

CHAPTER V

CHAPTER VI

CHAPTER VII

BOOK II

VOLCANIC ACTION IN PRE-CAMBRIAN TIME

CHAPTER VIII

Pre-Cambrian Volcanoes

BOOK III

THE CAMBRIAN VOLCANOES

CHAPTER IX

Characteristics of the Cambrian System in Britain

CHAPTER X

CHAPTER XI

BOOK IV

THE SILURIAN VOLCANOES

CHAPTER XII

CHARACTERS OF THE SILURIAN SYSTEM IN BRITAIN. THE ARENIG VOLCANOES

CHAPTER XIII

THE ERUPTIONS OF LLANDEILO AND BALA AGE

CHAPTER XIV

BOOK V

THE VOLCANOES OF DEVONIAN AND OLD RED SANDSTONE TIME

CHAPTER XV

CHAPTER XVI

THE VOLCANOES OF THE OLD RED SANDSTONE

CHAPTER XXIV

Carboniferous Volcanic Plateaux of Scotland

CHAPTER XXV

Geological Structure of the Carboniferous Volcanic Plateaux of Scotland

CHAPTER XXVI

The Carboniferous Puys of Scotland

CHAPTER XXVII

Geological Structure of the Carboniferous Puys of Scotland

CHAPTER XXVIII

Illustrative Examples of the Carboniferous Puys of Scotland

LIST OF ILLUSTRATIONS

MAPS

BOOK I

GENERAL PRINCIPLES AND METHODS OF INVESTIGATION

CHAPTER I

Earliest Knowledge of Volcanoes—Their Influence on Mythology and Superstition—Part taken by Volcanic Rocks in Scenery—Progress of the Denudation of Volcanoes—Value of the Records of former Volcanoes as illustrating Modern Volcanic Action—Favourable Position of Britain for the Study of this Subject.

Among the influences which affected the infancy of mankind, the most potent were those of environment. Whatever in outer nature stimulated or repressed courage, inventiveness, endurance, whatever tended to harden or to weaken the bodily faculties, whatever appealed to the imagination or excited the fancy, became a powerful factor in human development.

Thus, in the dawn of civilization, the frequent recurrence of earthquakes and volcanic eruptions throughout the basin of the Mediterranean could not but have a marked effect on the peoples that dwelt by the borders of that sea. While every part of the region was from time to time shaken by underground commotion, there were certain places that became specially noteworthy for the wonder and terror of their catastrophes. When, after successive convulsions, vast clouds of black smoke rose from a mountain and overspread the sky, when the brightness of noon was rapidly replaced by the darkness of midnight, when the air grew thick with stifling dust and a rain of stones and ashes fell from it on all the surrounding country, when streams of what looked like liquid fire poured forth and desolated gardens, vineyards, fields and villages—then did men feel sure that the gods were angry. The contrast between the peacefulness and beauty of the ordinary landscape and the hideous warfare of the elements at these times of volcanic fury could not but powerfully impress the imagination and give a colour to early human conceptions of nature and religion.

It was not only in one limited district that these manifestations of underground convulsion showed themselves. The islands of the Ægean had their volcanoes, and the Greeks who dwelt among them watched their glowing fires by night and their clouds of steam by day, culminating now and then in a stupendous explosion, like that which, in prehistoric time, destroyed the island of Santorin. As the islanders voyaged eastward they would see, on the coast of Asia Minor, the black bristling lavas of the "Burnt Country," perhaps even then flowing from their rugged heaps of cinders. Or when, more adventurously still, they sailed westward into the Tyrrhenian waters, they beheld the snowy cone of Etna, with its dark canopy of smoke and the lurid nocturnal gleam of its fires; while from time to time they witnessed there on a still more stupendous scale the horrors of a great volcanic eruption.

From all sides, therefore, the early Greek voyagers would carry back to the mother-country marvellous tales of convulsion and disaster. They would tell how the sky rapidly darkened even in the blaze of mid-day, how the land was smothered with dust and stones, how over the sea there spread such a covering of ashes that the oarsmen could hardly drive their vessels onward, how red-hot stones, whirling high overhead, rained down on sails and deck, and crushed or burnt whatever they fell upon, and how, as the earth shook and the sea rose in sudden waves and the mountain gave forth an appalling din of constant explosion, it verily seemed that the end of the world had come.

To the actual horrors of such scenes there could hardly fail to be added the usual embellishments of travellers' tales. Thus, in the end, the volcanoes of the Mediterranean basin came to play a not unimportant part in Hellenic mythology. They seemed to stand up as everlasting memorials of the victory of Zeus over the giants and monsters of an earlier time. And as the lively Greek beheld Mount Etna in eruption, his imagination readily pictured the imprisoned Titan buried under the burning roots of the mountain, breathing forth fire and smoke, and convulsing the country far and near, as he turned himself on his uneasy pallet.

When in later centuries the scientific spirit began to displace the popular and mythological interpretation of natural phenomena, the existence of volcanoes and their extraordinary phenomena offered a fruitful field for speculation and conjecture. As men journeyed outward from the Mediterranean cradle of civilization, they met with volcanic manifestations in many other parts of the world. When they eventually penetrated into the Far East, they encountered volcanoes on a colossal scale and in astonishing abundance. When they had discovered the New World they learnt that, in that hemisphere also, "burning mountains" were numerous and of gigantic dimensions. Gradually it was ascertained that vast lines of volcanic activity encircle the globe. By slow degrees the volcano was recognized to be as normal a part of the mechanism of our planet as the rivers that flow on the terrestrial surface. And now at last men devote themselves to the task of critically watching the operations of volcanoes with as much enthusiasm as they display in the investigation of any other department of nature. They feel that their knowledge of the earth extends to little beyond its mere outer skin, and that the mystery which still hangs over the vast interior of the planet can only, if ever, be dispelled by the patient study of these vents of communication between the interior and the surface.

If, however, we desire to form some adequate idea of the part which volcanic action has played in the past history of the earth, we should be misled were we to confine our attention to the phenomena of the eruptions of the present day. An attentive examination of any modern volcano will convince us that of some of the most startling features of an eruption no enduring memorial remains. The convulsive earthquakes that accompany a great volcanic paroxysm, unless where they actually fissure the ground, leave little or no trace behind them. Lamentably destructive as they are to human life and property, the havoc which they work is mostly superficial. In a year or two the ruins have been cleared away, the earth-falls have been healed over, the prostrated trees have been removed, and, save in the memories and chronicles of the inhabitants, no record of the catastrophe may survive. The clouds of dust and showers of ashes which destroyed the crops and crushed in the roofs of houses soon disappear from the air, and the covering which they leave over the surface of a district gradually mingles with the soil. Vegetation eventually regains its place, and the landscape becomes again as smiling as before.

Even where the materials thrown out from the crater accumulate in much greater mass, where thick deposits of ashes or solid sheets of lava bury the old land-surface, the look of barren desolation, though in some cases it may endure for long centuries, may in others vanish in a few years. The surface-features of the district are altered indeed, but the new topography soon ceases to look new. Another generation of inhabitants loses recollection of the old landmarks, and can hardly realize that what has become so familiar to itself differs so much from what was familiar to its fathers.

But even when the volcanic covering, thus thrown athwart a wide tract of country, has been concealed under a new growth of soil and vegetation, it still remains a prey to the ceaseless processes of decay and degradation which everywhere affect the surface of the land. No feature of a modern volcano is more impressive than the lesson which it conveys of the reality and potency of this continual waste. The northern slopes of Vesuvius, for example, are trenched with deep ravines, which in the course of centuries have been dug out of the lavas and tuffs of Monte Somma by rain and melted snow. Year by year these chasms are growing deeper and wider, while the ridges between them are becoming narrower. In some cases, indeed, the intervening ridges have been reduced to sharp crests which are split up and lowered by the unceasing influence of the weather. The slopes of such a volcanic cone have been aptly compared to a half-opened umbrella. It requires little effort of imagination to picture a time, by no means remote in a geological sense, when, unless renovated by the effects of fresh eruptions, the cone will have been so levelled with the surrounding country that the peasants of the future will trail their vines and build their cots over the site of the old volcano, in happy ignorance of what has been the history of the ground beneath their feet.

What is here predicted as probable or certain in the future has undoubtedly happened again and again in the past. Over many districts of Europe and Western America extinct volcanoes may be seen in every stage of decay. The youngest may still show, perfect and bare of vegetation, their cones and their craters, with the streams of lava that escaped from them. Those of older date have been worn down into mere low rounded hills, or the whole cone has been cleared away, and there is only left the hard core of material that solidified in the funnel below the surface. The lava-sheets have been cut through by streams, and now remain in mere scattered patches capping detached hills, which only a trained eye can recognize as relics of a once continuous level sheet of solid rock.

By this resistless degradation, a volcanic district is step by step stripped of every trace of its original surface. All that the eruptions did to change the face of the landscape may be entirely obliterated. Cones and craters, ashes and lavas, may be gradually effaced. And yet enough may be left to enable a geologist to make sure that volcanic action was once rife there. As the volcano marks a channel of direct communication between the interior of the earth and the atmosphere outside, there are subterranean as well as superficial manifestations of its activity, and while the latter are removed by denudation, the former are one by one brought into light. The progress of denudation is a process of dissection, whereby every detail in the structure of a volcano is successively cut down and laid bare. But for this process, our knowledge of the mechanism and history of volcanic action would be much less full and definite than happily it is. In active volcanoes the internal and subterranean structure can only be conjectured; in those of ancient date, which have been deeply eroded, this underground structure is open to the closest examination.

By gathering together evidence of this nature over the surface of the globe, we learn that abundantly as still active volcanoes are distributed on that surface, they form but a small fraction of the total number of vents which have at various times been in eruption. In Italy, for example, while Vesuvius is active on the mainland, and Etna, Stromboli and Volcano display their vigour among the islands, there are scores of old volcanoes that have been silent and cold ever since the beginning of history, yet show by their cones of cinders and streams of bristling lava that they were energetic enough in their day. But the Italian volcanic region is only one of many to be found on the European Continent. If we travel eastward into Hungary, or northward into the Eifel, or into the heart of France, we encounter abundant cones and craters, many of them so fresh that, though there is no historical record of their activity, they look as if they had been in eruption only a few generations ago.

But when the geologist begins to search among rocks of still older date than these comparatively recent volcanic memorials, he meets with abundant relics of far earlier eruptions. And as he arranges the chronicles of the earth's history, he discovers that each section of the long cycle of geological ages has preserved its records of former volcanoes. In a research of this kind he can best realize how much he owes to the process of denudation. The volcanic remains of former geological periods have in most cases been buried under younger deposits, and have sunk sometimes thousands of feet below the level of the sea. They have been dislocated and upheaved again during successive commotions of the terrestrial crust, and have at last been revealed by the gradual removal of the pile of material under which they had lain.

Hence we learn that the active volcanoes of the present time, which really embrace but a small part of the volcanic history of our planet, are the descendants of a long line of ancestors. Their distribution and activity should be considered not merely from the evidence they themselves supply, but in the light derived from a study of that ancestry. It is only when we take this broad view of the subject that we can be in a position to form some adequate conception of the nature and history of volcanoes in the geological evolution of the globe.

In this research it is obvious that the presently active volcano must be the basis and starting-point of inquiry. At that channel of communication between the unknown inside and the familiar outside of our globe, we can watch what takes place in times of quiescence or of activity. We can there study each successive phase of an eruption, measure temperatures, photograph passing phenomena, collect gases and vapours, register the fall of ashes or the flow of lavas, and gather a vast body of facts regarding the materials that are ejected from the interior, and the manner of their emission.

Indispensable as this information is for the comprehension of volcanic action, it obviously affords after all but a superficial glimpse of that action. We cannot see beyond the bottom of the crater. We cannot tell anything about the subterranean ducts, or how the molten and fragmental materials behave in them. All the underground mechanism of volcanoes is necessarily hidden from our eyes. But much of this concealed structure has been revealed in the case of ancient volcanic masses, which have been buried and afterwards upraised and laid bare by denudation.

In yet another important aspect modern volcanoes do not permit us to obtain full knowledge of the subject. The terrestrial vents, from which we derive our information, by no means represent all the existing points of direct connection between the interior and the exterior of the planet. We know that some volcanic eruptions occur under the sea, and doubtless vast numbers more take place there of which we know nothing. But the conditions under which these submarine discharges are effected, the behaviour of the out-flowing lava under a body of oceanic water, and the part played by fragmentary materials in the explosions, can only be surmised. Now and then a submarine volcano pushes its summit above the sea-level, and allows its operations to be seen, but in so doing it becomes practically a terrestrial volcano, and the peculiar submarine phenomena are still effectually concealed from observation.

The volcanic records of former geological periods, however, are in large measure those of eruptions under the sea. In studying them we are permitted, as it were, to explore the sea-bottom. We can trace how sheets of coral and groves of crinoids were buried under showers of ashes and stones, and how the ooze and silt of the sea-floor were overspread with streams of lava. We are thus, in some degree, enabled to realize what must now happen over many parts of the bed of the existing ocean.

The geologist who undertakes an investigation into the history of volcanic action within the area of the British Isles during past time, with a view to the better comprehension of this department of terrestrial physics, finds himself in a situation of peculiar advantage. Probably no region on the face of the globe is better fitted than these islands to furnish a large and varied body of evidence regarding the progress of volcanic energy in former ages. This special fitness may be traced to four causes—1st, The remarkable completeness of the geological record in Britain; 2nd, The geographical position of the region on the oceanic border of a continent; 3rd, The singularly ample development to be found there of volcanic rocks belonging to a long succession of geological ages; and 4th, The extent to which this full chronicle of volcanic activity has been laid bare by denudation.

1. In the first place, the geological record of Britain is singularly complete. It has often been remarked how largely all the great periods of geological time are represented within the narrow confines of these islands. The gaps in the chronicle are comparatively few, and for the most part are not of great moment.

Thanks to the restricted area of the country and to the large number of observers, this remarkably full record of geological history has been studied with a minute care which has hardly been equalled in any other country. The detailed succession of all the formations has been so fully determined in Britain that the very names first applied here to them and to their subdivisions have in large measure passed into the familiar language of geology all over the globe. Every definite platform in the stratigraphical series has been more or less fully worked out. A basis has thus been laid for referring each incident in the geological history of the region to its proper relative date.

2. In the second place, the geographical position of Britain gives it a notable advantage in regard to the manifestations of volcanic energy. Rising from the margin of a great ocean-basin and extending along the edge of a continent, these islands have lain on that critical border-zone of the terrestrial surface, where volcanic action is apt to be most vigorous and continuous. It has long been remarked that volcanoes are generally placed not far from the sea. From the earliest geological periods the site of Britain, even when submerged below the sea, has never lain far from the land which supplied the vast accumulations of sediment that went to form the Palæozoic and later formations, while, on the other hand, it frequently formed part of the land of former geological periods. It was thus most favourably situated as a theatre for both terrestrial and submarine volcanic activity.

3. In the third place, this advantageous geographical position is found to have been attended with an altogether remarkable abundance and persistence of volcanic eruptions. No tract of equal size yet known on the face of the globe furnishes so ample a record of volcanic activity from the earliest geological periods down into Tertiary time. Every degree of energy may be signalized in that record, from colossal eruptions which piled up thousands of feet of rock down to the feeblest discharge of dust and stones. Every known type of volcano is represented—great central cones like Etna or Vesuvius, scattered groups of small cones like the *puys* of France, and fissure- or dyke-eruptions like those of recent times in Iceland.

Moreover, the accurate manner in which the stratigraphy of the country has been established permits each successive era in the long volcanic history to be precisely determined, and allows us to follow the whole progress of that history stage by stage, from the beginning to the end.

These characteristics may be instructively represented on a map, such as that which accompanies the present volume (Map I.). The reader will there observe how repeatedly volcanic eruptions have taken place, not merely within the general area of the British Isles, but even within the same limited region of that area. The broad midland valley of Scotland has been especially the theatre for their display. From the early part of the Lower Silurian period, through the ages of the Old Red Sandstone, Carboniferous and Permian systems, hundreds of volcanic vents were active in that region, while in long subsequent time there came the fissure-eruptions of the Tertiary series.

4. In the fourth place, the geological revolutions of successive ages have made this long volcanic chronicle fully accessible to observation. Had the lavas and ashes of one period remained buried under the sedimentary accumulations of the next, their story would have been lost to us. We should only have been able to decipher the latest records which might happen to lie on the surface. Fortunately for the progress of geology, the endless vicissitudes of a continental border have brought up the very oldest rocks once more to the surface. All the later formations of the earth's crust have likewise been upraised and exposed to denudation during long cycles of time. In this manner, the rocky framework of the country has been laid bare, and each successive chapter of its geological history may be satisfactorily deciphered. The singularly complete volcanic chronicle, after being entombed under younger deposits, has been broken up and raised once more into view. The active vents of former periods have been dissected, submarine streams of lava have been uncovered, sheets of ashes that fell over the sea-bottom have been laid bare. The progress of denudation is specially favoured in such a variable and moist climate as that of Britain, and thus by the co-operation of underground and meteoric causes the marvellous volcanic records of this country have been laid open in minutest detail.

There is yet another respect in which the volcanic geology of Britain possesses a special value. Popular imagination has long been prone to see signs of volcanic action in the more prominent rocky features of landscape. A bold crag, a deep and precipitous ravine, a chasm in the side of a mountain, have been unhesitatingly set down as proof of volcanic disturbance. Many a cauldron-shaped recess, like the corries of Scotland or the cwms of Wales, has been cited as an actual crater, with its encircling walls still standing almost complete.

The relics of former volcanoes in this country furnish ample proofs to dispel these common misconceptions. They show that not a single crater anywhere remains, save where it has been buried under lava; that no trace of the original cones has survived, except in a few doubtful cases where they may have been preserved under subsequent accumulations of material; that in the rugged tracts, where volcanic action has been thought to have been most rife, there may be not a vestige of it, while, on the other hand, where the uneducated eye would never suspect the presence of any remnant of volcanic energy, lavas and ashes may abound. We are thus presented with some of the most impressive contrasts in geological history, while, at the same time, this momentous lesson is borne in upon the mind, that the existing inequalities in the configuration of a landscape are generally due far less to the influence of subterranean force than to the action of the superficial agents which are ceaselessly carving the face of the land. Those rocks which from their hardness or structure are best able to withstand that destruction rise into prominence, while the softer material around them is worn away. Volcanic rocks are no exception to this rule, as the geological structure of Britain amply proves.

In the following chapters, forming Book I. of this work, I propose to begin by offering some general remarks regarding the nature and causes of volcanic action, so far as these are known to us. I shall then proceed to consider the character of the evidence that may be expected to be met with respecting the former prevalence of that action at any particular locality where volcanic disturbances have long since ceased. The most telling evidence of old volcanoes is naturally to be found in the materials which they have left behind them, and the reader's attention will be asked to the special characteristics of these materials, in so far as they give evidence of former volcanic activity.

As has been already remarked, many of the most prominent phenomena of a modern volcano are only of transient importance. The earthquakes and tremors, and the constant disengagement of steam and gases, that play so conspicuous a part in an eruption, may leave no sensible record behind them. But even the cones of ashes and lava, which are piled up into mountainous masses, have no true permanence: they are liable to ceaseless erosion by the meteoric agencies of waste, and every stage in their degradation may be traced. In successive examples we can follow them as they are cut down to the very core, until in the end they are entirely effaced.

We may well, therefore, ask at the outset by what more enduring records we may hope to detect the traces of former volcanic action. The following introductory chapters will be devoted to an attempt to answer this question. I shall try to show the nature and relative importance of the records of ancient volcanoes; how these records, generally so fragmentary, may be pieced together so as to be made to furnish the history which they contain; how their relative chronology may be established; how their testimony may be supplemented in such wise that the position of long vanished seas, lands, rivers, and lakes may be ascertained; and how, after ages of geological revolution, volcanic rocks that have lain long buried under the surface now influence the scenery of the regions where they have once more been exposed to view.

From this groundwork of ascertained fact and reasonable inference, we shall enter in Book II. upon the story of the old volcanoes of the British Isles. It is usual to treat geological history in chronological order, beginning with the earliest ages. And this method, as on the whole the most convenient, will be adopted in the present work. At the same time, the plan so persistently followed by Lyell, of working backward from the present into the past, has some distinct advantages. The volcanic records of the later ages are much simpler and clearer than those of older times, and the student may, in some respects, profitably study the history of the Tertiary eruptions before he proceeds to make himself acquainted with the scantier chronicles of the eruptions of the Palæozoic periods. But as I wish to follow the gradual evolution of volcanic phenomena, and to show how volcanic energy has varied, waxing and waning through successive vast intervals of time, I will adhere to the chronological sequence.

CHAPTER II

The Nature and Causes of Volcanic Action—Modern Volcanoes.

A VOLCANO is a conical or dome-shaped hill or mountain, consisting of materials which have been erupted from an orifice leading down from the surface into the heated interior of the earth. Among modern and recent volcanoes three types may be recognized. In the first and most familiar of these, the lavas and ashes ejected from the central vent have gathered around it by successive eruptions, until they have built up a central cone like those of Etna and Vesuvius. As this cone grows in height and diameter, lateral or parasitic cones are formed on its flanks, and may become themselves the chief actively erupting vents. This type of volcano, which has been so long well known from its Mediterranean examples, was until recently believed by geologists to be the normal, or indeed the only, phase of volcanic energy on the face of the earth.

A modification of this type is to be found in a few regions where fragmentary discharges are small in amount and where the eruptions are almost wholly confined to the emission of tolerably liquid lava. A vast dome with gently sloping declivities may in this way be formed, as in the Sandwich Islands and in certain parts of Iceland.

The second type of volcano is at the present day extensively developed only in Iceland, but in Tertiary time it appears to have had a wide range over the globe, for stupendous memorials of it are preserved in North-Western Europe, in Western America, and in India. It is distinguished by the formation of numerous parallel fissures from which the lava gushes forth, either with or without the formation of small cinder-cones along the lines of the chasms.

The third type is distinguished by the formation of groups of cinder-cones or lava-domes, which from their admirable development in Central France have received the name of *Puys*. From these vents considerable streams of lava have sometimes been discharged.

Without entering here into a detailed inquiry regarding the nature and causes of Volcanic Action, we may with advantage consider briefly the two main factors on which this action appears to depend.

1. Much uncertainty still exists as to the condition and composition of the earth's interior. The wide distribution of volcanoes over the globe, together with the general similarity of materials brought by them up to the

surface, formerly led to the belief that our planet consists of a central mass of molten rock enclosed within a comparatively thin solid crust. Physical arguments, however, have since demonstrated that the earth, with such a structure, would have undergone great tidal deformation, but that in actual fact it has a greater rigidity than if it were made of solid glass or steel.

From all the evidence obtainable it is certain that the temperature of the earth's interior must be high. The rate of increase of this temperature downward from the surface differs from place to place; but an increase is always observed. At a depth of a few miles, every known substance must be much hotter than its melting point at the surface. But at the great pressures within the earth, actual liquefaction is no doubt prevented, and the nucleus remains solid, though at a temperature at which, but for the pressure, it would be like so much molten iron.

Any cause which will diminish the pressure may allow the intensely hot material within the globe to pass into the liquid state. There is one known cause which will bring about this result. The downward increment of temperature proves that our planet is continually losing heat. As the outer crust is comparatively cool, and does not become sensibly hotter by the uprise of heat from within, the hot nucleus must cool faster than the crust is doing. Now cooling involves contraction. The hot interior is contracting faster than the cooler shell which encloses it, and that shell is thus forced to subside. In its descent it has to adjust itself to a constantly diminishing diameter. It can do so only by plication or by rupture.

When the terrestrial crust, under the strain of contraction, is compressed into folds, the relief thus obtained is not distributed uniformly over the whole surface of the planet. From an early geological period it appears to have followed certain lines. How these came to be at first determined we cannot tell. But it is certain that they have served again and again, during successive periods of terrestrial readjustment. These lines of relief coincide, on the whole, with the axes of our continents. The land-areas of the globe may be regarded as owing their existence above sea-level to this result of terrestrial contraction. The crust underneath them has been repeatedly wrinkled, fractured and thrust upward by the vast oceanic subsidence around them. The long mountain-chains are thus, so to speak, the crests of the waves into which the crust has from time to time been thrown.

Again, the great lines of fracture in the crust of the earth probably lie in large measure within the land-areas, or at least parallel with their axes and close to their borders. Where the disposition of the chief ruptures and of the predominant plications can be examined, these leading structural features are found to be, on the whole, coincident. In the British Islands, for instance, the prevalent trend of the axes of folding from early Palæozoic to Tertiary time has been from south-west to north-east. How profoundly this direction of earth-movement has affected the structure of the region is shown by any ordinary map, in the long hill-ranges of the land and in the long inlets of the sea. A geological map makes the dependence of the scenery upon the building of the rocks still more striking. Not only have these rocks been plicated into endless foldings, the axes of which traverse the British Islands with a north-easterly trend: they have likewise been dislocated by many gigantic ruptures, which tend on the whole to follow the same direction. The line of the Great Glen, the southern front of the Highlands, and the northern boundary of the Southern Uplands of Scotland, are conspicuous examples of the position and effect of some of the greater fractures in the structure of this country.

The ridging up of any part of the terrestrial crust will afford some relief from pressure to the parts of the interior immediately underneath. If, as is probable, the material of the earth's interior is at the melting point proper for the pressure at each depth, then any diminution of the pressure may allow the intensely heated substance to pass into the liquid state. It would be along the lines of terrestrial uplift that this relief would be given. It is there that active volcanoes are found. The molten material is forced upward under these upraised ridges by the subsidence of the surrounding regions. And where by rupture of the crust this material can make its way to the surface, we may conceive that it will be ejected as lava or as stones and ashes.

Viewed in a broad way, such appears to be the mechanism involved in the formation and distribution of volcanoes over the surface of the earth. But obviously this explanation only carries us so far in the elucidation of volcanic action. If the molten magma flowed out merely in virtue of the influence of terrestrial contraction, it might do so for the most part tranquilly, though it would probably be affected by occasional sudden snaps, as the crust yielded to accumulations of pressure. Human experience has no record of the actual elevation of a mountain-chain. We may believe that if such an event were to happen suddenly or rapidly, it would be attended with gigantic catastrophes over the surface of the globe. We can hardly conceive what would be the scale of a volcanic eruption attending upon so colossal a disturbance of the terrestrial crust. But the eruptions which have taken place within the memory of man have been the accompaniments of no such disturbance. Although they have been many in number and sometimes powerful in effect, they have seldom been attended with any marked displacement of the surrounding parts of the terrestrial crust. Contraction is, of course, continuously and regularly in progress, and we may suppose that the consequent subsidence, though it results in intermittent wrinkling and uplifting of the terrestrial ridges, may also be more or less persistent in the regions lying outside these ridges. There will thus be a constant pressure of the molten magma into the roots of volcanoes, and a persistent tendency for the magma to issue at the surface at every available rent or orifice. The energy and duration of outflow, if they depended wholly upon the effects of contraction, would thus vary with the rate of subsidence of the sinking areas, probably assuming generally a feeble development, but sometimes bursting into fountains of molten rock hundreds of feet in height, like those observed from time to time in Hawaii.

2. The actual phenomena of volcanic eruptions, however, show that a source of explosive energy is almost always associated with them, and that while the transference of the subterranean molten magma towards the volcanic vents may be referred to the results of terrestrial contraction, the violent discharge of materials from those vents must be assigned to some kind of energy stored up in the substance of the earth's interior.

The deep-seated magma from which lavas ascend contains various vapours and gases which, under the enormous pressure within and beneath the terrestrial crust, are absorbed or dissolved in it. So great is the tension of these gaseous constituents, that when from any cause the pressure on the magma is suddenly relieved, they are liberated with explosive violence.

A volcanic paroxysm is thus immediately the effect of the rapid escape of these imprisoned gases and vapours. With such energy does the explosion sometimes take place, that the ascending column of molten lava is blown into the finest impalpable dust, which may load the air around a volcano for many days before it falls to the ground, or may be borne in the upper regions of the atmosphere round the globe.

The proportion of dissolved gases varies in different lavas, while the lavas themselves differ in the degree of their liquidity. Some flow out tranquilly like molten iron, others issue in a pasty condition and rapidly congeal into scoriae and clinkers. Thus within the magma itself the amount of explosive energy is far from being always the same.

It is to the co-operation of these two causes—terrestrial contraction and its effects on the one hand, and the tension of absorbed gases and vapours on the other—that the phenomena of volcanoes appear to be mainly due. There is no reason to believe that modern volcanoes differ in any essential respect from those of past ages in the earth's history. It might, indeed, have been anticipated that the general energy of the planet would manifest itself in far more stupendous volcanic eruptions in early times than those of the modern period. But there is certainly no geological evidence in favour of such a difference. One of the objects of the present work is to trace the continuity of volcanic phenomena back to the very earliest epochs, and to show that, so far as the geological records go, the interior of the planet has reacted on its exterior in the same way and with the same results.

We may now proceed to inquire how far volcanoes leave behind them evidence of their existence. I shall devote the next two or three chapters to a consideration of the proofs of volcanic action furnished by the very nature of the materials brought up from the interior of the earth, by the arrangement of these materials at the surface, by the existence of the actual funnels or ducts from which they were discharged above ground, and by the disposition of the masses of rock which, at various depths below the surface, have been injected into and have solidified within the terrestrial crust.

CHAPTER III

Ancient Volcanoes: Proofs of their existence derived from the Nature of the Rocks erupted from the Earth's Interior. A. Materials erupted at the Surface—Extrusive Series. i. Lavas, their General Characters. Volcanic Cycles. ii. Volcanic Agglomerates, Breccias and Tuffs.

THE materials brought by volcanic action from the earth's interior have certain common characters which distinguish them from other constituents of the terrestrial crust. Hence the occurrence of these materials on any part of the earth's surface affords convincing proofs of former volcanic eruptions, even where all outward trace of actual volcanoes may have been effaced from the topographical features of the ground.

Volcanic products may be classed in two divisions—1st, Those which have been ejected at the surface of the earth, or the Extrusive series; and 2nd, Those which have been injected into the terrestrial crust at a greater or less distance below the surface, and which are known as the Intrusive series. Extrusive rocks may be further classified in two great groups—(i.) The Lavas, or those which have been poured out in a molten condition at the surface; and (ii.) The Fragmental Materials, including all kinds of pyroclastic detritus discharged from volcanic vents.

Taking first the Extrusive volcanic rocks, we may in the present chapter consider those characters in them which are of most practical value in the investigation of the volcanic phenomena of former geological periods.

i. LAVAS

The term Lava is a convenient and comprehensive designation for all those volcanic products which have flowed out in a molten condition. They differ from each other in composition and structure, but their variations are comprised within tolerably definite limits.

As regards their composition they are commonly classed in three divisions—1st, The Acid lavas, in which the proportion of silicic acid ranges from a little below 70 per cent upwards; 2nd, The Intermediate lavas, wherein the percentage of silica may vary from 55 to near 70; and 3rd, The Basic lavas, where the acid constituent ranges from 55 per cent downwards. Sometimes the most basic kinds are distinguished as a fourth group under the name of Ultrabasic, in which the percentage of silica may fall below 40.

The structures of lavas, however, furnish their most easily appreciated characteristics. Four of these structures deserve more particular attention: 1st, Cellular, vesicular or pumiceous structure; 2nd, The presence of glass, or some result of the devitrification of an original glass; 3rd, Flow-structure; and 4th, The arrangement of the rocks in sheets or beds, with columnar and other structures.

1. The CELLULAR, VESICULAR, SCORIACEOUS OR PUMICEOUS STRUCTURE of volcanic rocks (Fig. 1) could only have arisen in molten masses from the

FIG. 1.—Vesicular structure, Lava from Ascension Island, slightly less than natural size.

expansion of imprisoned vapours or gases, and is thus of crucial importance in deciding the once liquid condition of the rocks which display it. The vesicles may be of microscopic minuteness, but are generally quite visible to the naked eye, and are often large and conspicuous. Sometimes these cavities have been subsequently filled up with calcite, quartz, agate, zeolites or other mineral deposition. As the kernels thus produced are frequently flattened or almond-shaped (*amygdales*), owing to elongation of the steam-holes by movement of the lava before its consolidation, the rocks containing them are said to be *amygdaloidal*.

This structure, though eminently characteristic of superficial lavas, is not always by itself sufficient to distinguish them from the intrusive rocks. Examples will be given in later chapters where dykes, sills and other masses of injected igneous material are conspicuously cellular in some parts. But, in such cases, the cavities are generally comparatively small, usually spherical or approximately so, tolerably uniform in size and distribution, and, especially when they occur in dykes, distributed more particularly along certain lines or bands, sometimes with considerable regularity (see Figs. 90, 91, and 236).

Among the superficial lavas, however, such regularity is rarely to be seen. Now and then, indeed, a lava, which is not on the whole cellular, may be found to have rows of vesicles arranged parallel to its under or upper surface, or it may have acquired a peculiar banded structure from the arrangement of its vesicles in parallel layers along the direction of flow. The last-named peculiarity is widely distributed among the Tertiary lavas of North-Western Europe, and gives to their weathered surfaces a deceptive resemblance to tuffs or other stratified rocks (see Figs. 260, 310 and 311). It will be more particularly referred to a few pages further on. In general, however, we may say that the steam-cavities of lavas are quite irregular in size, shape and distribution, sometimes increasing to such relative proportions as to occupy most of the bulk of the rock, and in other places disappearing, so as to leave the lava tolerably compact. When a lava presents an irregularly vesicular character, like that of the slags of an iron-furnace, it is said to be *slaggy*. When its upper surface is rugged and full of steam-vesicles of all sizes up to large cavernous spaces, it is said to be *scoriaceous*, and fragments of such a rock ejected from a volcanic vent are spoken of as *scoriae*.

Attention to the flattening of the steam-vesicles in cellular lavas, which has just been alluded to as the result of the onward movement of the still molten mass, may show, by the trend and grouping of these elongated cavities, the probable direction of the flow of the lava before it came to rest. Sometimes the vesicles have been drawn out and flattened to such a degree that the rock has acquired in consequence a fissile structure. In other instances, the vesicles have been originally formed as long parallel and even branching tubes, like the burrows of Annelids or the borings of *Teredo*. Some remarkable examples of this exceptional structure have been obtained from the Tertiary plateau-basalts of the Western Isles, of which an example is represented in Fig. 2.

In many cases the vesicles extend through the whole thickness of a lava. Frequently they may be found most developed towards the top and bottom; the central portion of the sheet being compact, while the top and bottom are rugged, cavernous or scoriaceous.

Though originally the vesicles and cavernous spaces, blown open by the expansion of the vapours dissolved in molten lava, remained empty on the consolidation of the rock, they have generally been subsequently filled up by the deposit within them of mineral substances carried in aqueous solution. The minerals thus introduced are such as might have been derived from the removal of their constituent ingredients by the solvent action of water on the surrounding rock. And as amygdaloids are generally more decayed than the non-vesicular lavas, it has been generally believed that the abstraction of mineral material and its re-deposit within the steam-vesicles have been due to the influence of meteoric water, which at atmospheric temperatures and pressures has slowly percolated from the surface through the cellular lava, long after the latter had consolidated and cooled, and even after volcanic energy at the locality had entirely ceased.

Examples, however, are now accumulating which certainly prove that, in some cases, the vesicles were filled up during the volcanic period. Among the Tertiary basalt-plateaux of the Inner Hebrides, for instance, it can be shown that the lavas were already amygdaloidal before the protrusion of the

FIG. 2.—Elongation and branching of steam-vesicles in a lava, Kilninian, Isle of Mull, a little less than natural size.

gabbros and granophyres which mark later stages of the same continuous volcanic history, and even before the outpouring of much of the basalt of these plateaux. Not improbably the mineral secretions were largely due to the influence of hot volcanic vapours during the eruption of the basalts. This subject will be again referred to in the description of the Tertiary volcanic series.

Vesicular structure is more commonly and perfectly developed among the lavas which are basic and intermediate in composition than among those which are acid.

While the existence of a highly vesicular or scoriaceous structure may generally be taken as proof that the rock displaying it flowed out at the surface as a lava, other evidence pointing to the same conclusion may often be gathered from the rocks with which the supposed lava is associated.

Where, for example, a scoriaceous lava is covered with stratified deposits which contain pieces of that lava, we may be confident that the rock is an interstratified or contemporaneous sheet. It has been erupted after the deposition of the strata on which it rests, and before that of the strata which cover it and contain pieces of it. In such a case, the geological date of the eruption could be precisely defined. Illustrations of this reasoning will be given in Chapter iv., and in the account of the volcanic series of Carboniferous age in Central Scotland, where a basic lava can sometimes be proved to be a true flow and not an intrusive sill by the fact that portions of its upper slaggy surface are enclosed in overlying sandstone, shale or limestone.

2. The presence of GLASS, or of some result of the devitrification of an original glass, is an indication that the rock which exhibits it has once been in a state of fusion. Even where no trace of the original vitreous condition may remain, stages in its devitrification, that is, in its conversion into a stony or lithoid condition, may be traceable. Thus what are called spherulitic and perlitic structures (which will be immediately described), either visible to the naked eye or only observable with the aid of the microscope, afford evidence of the consolidation and conversion of a glassy into a lithoid substance.

Striking evidence of the former glassy, and therefore molten, condition of many rocks now lithoid is to be gained by the examination of thin slices of them under the microscope. Not only are vestiges of the original glass recognizable, but the whole progress of devitrification may be followed into a crystalline structure. The primitive crystallites or microlites of different minerals may be seen to have grouped themselves together into more or less perfect crystals, while scattered crystals of earlier consolidation have been partially dissolved in and corroded by the molten glass. These and other characteristics of once fused rocks have to a considerable extent been imitated artificially by MM. Fouqué and Michel Lévy, who have fused the constituent minerals in the proper proportions.

Since traces of glass or of its representative devitrified structures are so abundantly discoverable in lavas, we may infer the original condition of most lavas to have been vitreous. Where, for instance, the outer selvages of a basic dyke or sill are coated with a layer of black glass which rapidly passes into a fine-grained crystalline basalt, and then again into a more largely crystalline or doleritic texture in the centre, there can be no hesitation in believing that glassy coating to be due to the sudden chilling and consolidation of the lava injected between the cool rocks that enclose it. The part that solidified first may be regarded as probably representing the condition of the whole body of lava at the time of intrusion. The lithoid or crystalline portion between the two vitreous outer layers shows the condition which the molten rock finally assumed as it cooled more slowly.

Some lavas, such as obsidians and pitchstones, have consolidated in the glassy form. More usually, however, a lithoid structure has been developed, the original glass being only discoverable by the microscope, and often not even by its aid. Two varieties of devitrification may be observed among lavas, which, though not marked off from each other by any sharp lines, are on the whole distinctive of the two great groups of acid and basic rocks.

(1) Among the acid rocks, what is called the Felsitic type of devitrification is characteristic. Thus, obsidians pass by intermediate stages from a clear transparent or translucent glass into a dull flinty or horny mass. When thin slices of these transitional forms are examined under the microscope, minute hairs and fibres or trichites, which may be observed even in the most perfectly glassy rocks, are seen to increase in number until they entirely take the place of the glass. Microlites of definite minerals may likewise be observed, together with indefinite granules, and the rock finally becomes a rhyolite, felsite or allied variety (Fig. 3).

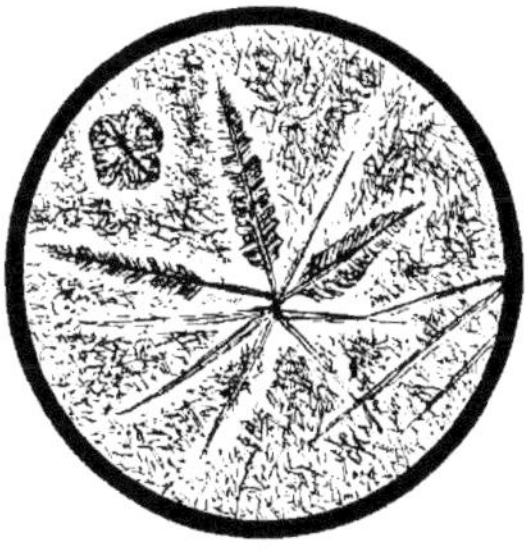

FIG. 3.—Microlites of the Pitchstone of Arran (magnified 70 diameters).

At the same time it should be observed that, even in the vitreous condition of a lava, definite crystals of an early consolidation were generally already present. Felspars and quartz, usually in large porphyritic forms, may be seen in the glass, often so corroded as to indicate that they were in course of being dissolved in the magma at the time of the cooling and solidification of the mass. In obsidians and pitchstones such relics of an earlier or derived series of crystallized minerals may often be recognized, while in felsites and quartz-porphyries they are equally prominent. Where large dispersed crystals form a prominent characteristic in a rock they give rise to what is termed the *Porphyritic* structure.

Accompanying the passage of glass into stone, various structures make

FIG. 4.—Perlitic structure in Felsitic Glass, Isle of Mull (magnified).

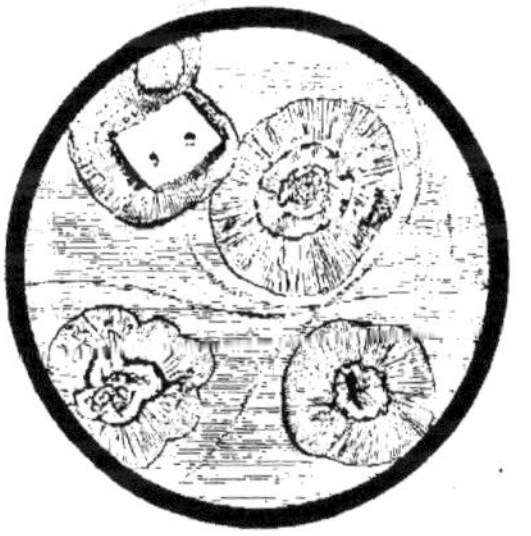

FIG. 5.—Spherulitic structure (magnified).

their appearance, sometimes distinctly visible to the naked eye, at other times only perceptible with the aid of the microscope. One of these structures, known as *Perlitic* (Fig. 4), consists in the formation of minute curved or straight cracks between which the vitreous or felsitic substance, during its contraction in cooling, assumed a finely globular form.

Another structure, termed *Spherulitic* (Fig. 5), shows the development of globules or spherules which may range from grains of microscopic minuteness up to balls two inches or more in diameter. These not infrequently present a well-formed internal fibrous radiation, which gives a black cross between crossed Nicol prisms. Spherulites are more especially developed along the margins of intrusive rocks, and may be found in dykes, sills and bosses (see Figs. 375 and 377). Where the injected mass is not thick it may be spherulitic to the very centre, as can be seen among the felsitic and granophyric dykes of Skye.

Some felsitic lavas possess a peculiar nodular structure, which was developed during the process of consolidation. So marked does this arrangement sometimes become that the rocks which display it have actually been mistaken for conglomerates. It is well exhibited among the Lower Silurian lavas of Snowdon, the Upper Silurian lavas of Dingle, and the Lower Old Red Sandstone lavas near Killarney.

A marked structure among some intrusive rocks, especially of an acid composition, is that called *Micropegmatitic* or *Granophyric*. It consists in a minute intergrowth of two component minerals, especially quartz and felspar,

FIG. 6.—Micropegmatitic or Granophyric structure in Granophyre, Mull (magnified).

FIG. 7.—Ophitic structure in Dolerite, Gortacloghan, Co. Derry (magnified).

and is more especially characteristic of certain granitic or granitoid rocks which have consolidated at some distance from the surface and occur as bosses, sills and dykes. It is also met with, however, in some basic sills. Examples of all these and other structures will occur in the course of the following description of British volcanic rocks.

(2) The second type of devitrification, conspicuous in rocks of more basic composition, is marked by a more complete development of crystallization. Among basic, as among acid rocks, there are proofs of the consolidation of definite minerals at more than one period. Where the molten material has suddenly cooled into a black glass, porphyritic felspars or other minerals are often to be seen which were already floating in the

magma in its molten condition. During devitrification, however, other felspars of a later period of generation made their appearance, but they are generally distinguishable from their predecessors. Probably most basic and intermediate rocks, when poured out at the surface as lavas, were no longer mere vitreous material, but had already advanced to various stages of progress towards a stony condition. These stages are still to some extent traceable by the aid of the microscope.

Microlites of the component minerals are first developed, which, if the process of aggregation is not arrested, build up more or less perfect crystals or crystalline grains of the minerals. Eventually the glass may be so completely devitrified by the development of its constitutent minerals as to be wholly used up, the rock then becoming entirely crystalline, or to survive only in scanty interstitial spaces. In the family of the basalts and dolerites the gradual transition from a true glass into a holocrystalline compound may be followed with admirable clearness. The component minerals have sometimes crystallized in their own distinct crystallographic forms (idiomorphic); in other cases, though thoroughly crystalline, they have assumed externally different irregular shapes, fitting into each other without their proper geometric boundaries (allotriomorphic).

A specially characteristic feature of many basic rocks is the presence of what is termed an *Ophitic* structure (Fig. 7). Thus the component crystals of pyroxene occur as large plates separated and penetrated by small needles and crystals of felspar. The portions of pyroxene, divided by the enclosed felspar, are seen under the microscope to be in optical continuity, and to have crystallized round the already formed felspar. This structure is never found in metamorphic crystalline rocks. It has been reproduced artificially from fusion by Messrs. Fouqué and Michel Lévy.

The name *Variolitic* is applied to another structure of basic rocks (Fig. 8), in which, especially towards the margin of eruptive masses, abundant spheroidal aggregates have been developed from the size of a millet-seed to that of a walnut, imbedded in a fine-grained or compact greenish matrix into which the kernels seem to shade off. These kernels consist of silicates arranged either radially or in concentric zones.

3. FLOW-STRUCTURE is an arrangement of the crystals, vesicles, spherulites, or devitrification-streaks in bands or lines, which sweep round any enclosed object, such as a porphyritic crystal or detached spherulite, and represent the curving flow of a mobile or viscous mass. Admirable examples of this structure may often be observed in old lavas, as well as in dykes and sills, the streaky lines of flow being marked as distinctly as the lines of foam that curve round the boulders projecting from the surface of a mountain-brook.

Flow-structure is most perfectly developed among the obsidians, rhyolites, felsites and other acid rocks, of which it may be said to be a frequently conspicuous character (Fig. 9). In these rocks it is revealed by the parallel arrangement of the minute hair-like bodies and crystals, or by alternate layers of glassy and lithoid material. The streaky lines thus developed are sometimes almost as thin and parallel as the leaves of a book. But they generally show interruptions and curvatures, and may be seen to bend round larger enclosed crystals, or to gather into eddy-like curves, in such a manner as vividly to portray the flow of a viscous substance. These lines represent on a minute scale the same flow-structure which may be traced in large sheets among the lavas. The porphyritic crystals and the spherulites are also drawn out in rows in the same general direction. Sometimes, indeed, the spherulites have been so symmetrically grouped in parallel rows that they appear as rod-like aggregates which extend along the margin of a dyke.

Among lavas of more basic composition flow-structure is not so often well displayed. It most frequently shows itself by the orientation of porphyritic felspars or of lines of steam-vesicles. Occasionally, however, sheets of basalt may be found in which a distinct streakiness has been developed owing to variations in the differentiation of the original molten

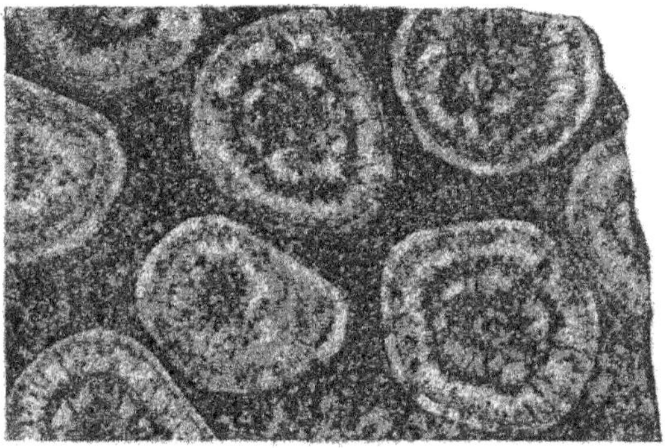

FIG. 8.—Variolitic or Orbicular structure, Napoleonite, Corsica (nat. size).

magma. A remarkable and wide-spread occurrence of such a structure is met with among the Tertiary basalt-plateaux of the Inner Hebrides and the Faroe Islands. In the lower parts of these thick accumulations of successive lava-sheets, a banded character is so marked as to give the rocks the aspect of truly stratified deposits. The observer, indeed, can hardly undeceive himself as to their real nature until he examines them closely. As a full description of this structure will be given in a later chapter, it may suffice to state here that the banding arises from two causes. In some cellular lavas, the vesicles are arranged in layers which lie parallel with the upper and under surfaces of the sheets. These layers either project as ribs or recede into depressions along the outcrop, and thus impart a distinctly stratified aspect to the rock. More frequently, however, the banded structure is produced by the alternation of different varieties of texture, and even of composition, in the same sheet of basalt. Lenticular seams of olivine-basalt may be found intercalated in a more largely crystalline dolerite. These differences appear to point to considerable variations in the constitution of the magma from which the lavas issued—variations which already existed before the discharge of these lavas, and which showed themselves in the successive outflow of basaltic and doleritic material during the eruption of what was really, as regards its appearance at the surface, one continuous stream of molten rock. It is impossible to account for such variations in the same sheet of lava by any process of differentiation in the melted material during its outflow and cooling. Analogous variations occur among the basic sills and bosses of the Tertiary volcanic series of Britain. These, as will be more fully discussed in later chapters, indicate a considerable amount of heterogeneity in the deep-seated magma from which the intrusive sheets and bosses were supplied (see vol. ii. pp. 329, 342).

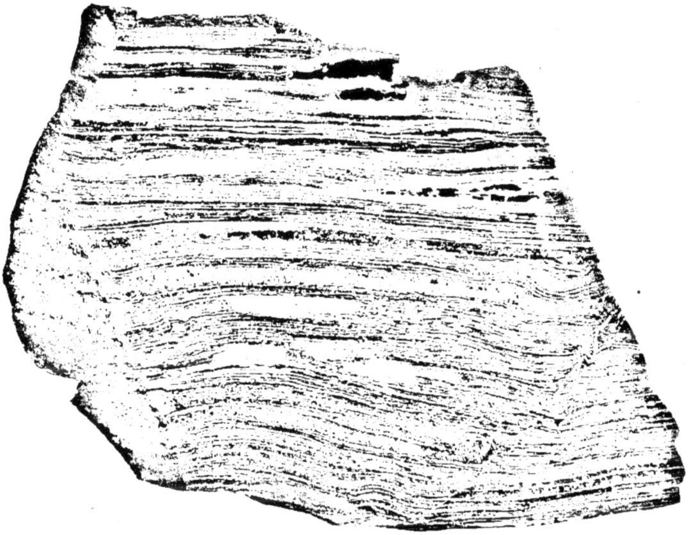

FIG. 9.—Flow-structure in Rhyolite, Antrim, slightly reduced.

It is a common error to assume that flow-structure is a distinctive character of lavas that have flowed out at the surface. In reality some of the most perfect examples of the structure occur in dykes and sills, both among acid and basic rocks. Innumerable instances might be quoted from the British Isles in support of this statement.

Although, in the vast majority of cases, the presence of flow-structure may be confidently assumed to indicate a former molten condition of the rock in which it occurs, it is not an absolutely reliable test for an igneous rock. Experiment has shown that under enormous pressure even solid metals may be made to flow into cavities prepared for their reception.

Under the vast compression to which the earth's crust is subjected during terrestrial contraction, the most obdurate rocks are crushed into fragments varying from large blocks to the finest powder. This comminuted material is driven along in the direction of thrust, and when it comes to rest presents a streakiness, with curving lines of flow round the larger fragments, closely simulating the structure of many rhyolites and obsidians. It is only by attention to the local surroundings that such deceptive resemblances can be assigned to their true cause.

4. The DISPOSITION OF LAVAS IN SHEETS OR BEDS is the result of successive outflows of molten rock. Such sheets may range from only a yard or two to several hundred feet in thickness. As a rule, though with many exceptions, the basic lavas, such as the basalts, appear in thinner beds than the acid forms. This difference is well brought out if we compare, for instance, the massive rhyolites or felsites of North Wales with the thin sheets of basalt in Antrim and the Inner Hebrides. The regularity of the bedded character is likewise more definite among the basic than among the acid rocks, and this contrast also is strikingly illustrated by the two series of rocks just referred to. The rhyolites and felsites, sometimes also

FIG. 10.—Lumpy, irregular trachytic Lava-streams (Carboniferous), East Linton, Haddingtonshire.

the trachytes and andesites, assume lumpy, irregular forms, and some little care may be required to trace their upper and under surfaces, and to ascertain that they really do form continuous sheets, though varying much in thickness from place to place (Fig. 10). Like modern acid lavas, they seem to have flowed out in a pasty condition, and to have been heaped up round the vents in the form of domes, or with an irregular hummocky or mounded surface. The basalts, and dolerites, and sometimes the andesites, have issued in a more fluid condition, and have spread out in sheets of more uniform thickness, as may be instructively seen in the sea-cliffs of Antrim, Mull, Skye, and the Faroe Islands, where the horizontal or gently-inclined flows of basalt lie upon each other in even parallel beds traceable for considerable distances along the face of the precipices (Figs. 11, 265, and 286). The andesites of the Old Red Sandstone (Figs. 99, 100) and Carboniferous series (Figs. 107, 108, 111, 112, 113, 123) in Scotland likewise form terraced hills.

The length of a lava-stream may vary within wide limits. Sometimes an outflow of lava has not reached the base of the cone from the side of which it issued, like the obsidian stream on the flanks of the little cone of the island of Volcano. In other cases, the molten rock has flowed for forty or fifty miles, like the copious Icelandic lava-floods of 1783. In the basalt-plateaux of the Inner Hebrides a single sheet may sometimes be traced for several miles.

Some lavas, more especially among the basic series, assume in cooling a *Columnar structure*, of which two types may be noticed. In one of these the columns pass with regularity and parallelism from the top to the bottom of a bed (Figs. 171, 225). The basalt in which Fingal's Cave, in the isle of Staffa, has been hollowed out may be taken as a characteristic example (Fig. 266a). Not infrequently the columns are curved, as at the well-known Clam-shell Cave of Staffa. In the other type, the columns or prisms are not persistent, but die out into each other and have a wavy, irregular shape, somewhat like prisms of starch. These two types may occur in successive sheets of basalt, or may even pass into each other. At Staffa the regularly

FIG. 11.—View at the entrance of the Svinofjord, Faroe Islands, illustrating the terraced forms assumed by basic lavas.

The island on the left is Borö, that in the centre Viderö, and that on the right Svinö.

columnar bed is immediately overlain with one of the starch-like character. The columnar structure in either case is a contraction phenomenon, produced during the cooling and shrinking of the lava. But it is difficult to say what special conditions in the lava were required for its production, or why it should sometimes have assumed the regular, at others the irregular form. It may be found not only in superficial lavas but in equal perfection in some dykes and intrusive sills or injections, as among the Tertiary volcanic rocks of the island of Canna (Figs. 307 and 308).

The precipitation of a lava-stream into a lake or the sea may cause the outer crust of the rock to break up with violence, so that the still molten material inside may rush into the water. Some basic lavas on flowing into water or into a watery silt have assumed a remarkable spheroidal sack-like or pillow-like structure, the spheroids being sometimes pressed into shapes like piles of sacks. A good instance of this structure occurs in a basalt at Acicastello in Sicily.[1] A similar appearance will be described in a later chapter as peculiarly characteristic of certain Lower Silurian lavas associated with radiolarian cherts in Britain and in other countries (Fig. 12).

It probably seldom happens that a solitary sheet of lava occurs among non-volcanic sedimentary strata, with no other indication around it of former volcanic activity. Such an isolated record does not seem to have been met with in the remarkably ample volcanic register of the British Isles. The outpouring of molten rock has generally been accompanied with the ejection

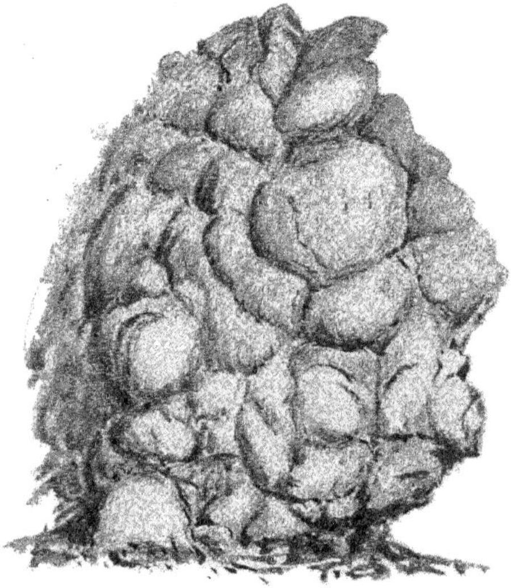

FIG. 12.—Sack-like or pillow-form structure of basic lavas (Lower Silurian), Bennan Head, Ballantrae, Ayrshire.

of fragmentary materials. Hence among the memorials of volcanic eruptions, while intercalated lavas are generally associated with sheets of tuff, bands of tuff may not infrequently be encountered in a sedimentary series without any lava. Instances in illustration of these statements may be culled from the British Palæozoic formations back even into the Cambrian system.

A characteristic feature of some interest in connection with the flow of lava is the effect produced by it on the underlying rocks. If these are not firmly compacted they may be ploughed up or even dislocated. Thus the tuffs of the Velay have sometimes been plicated, inverted, and fractured by

[1] See Prof. G. Platania in Dr. Johnston-Lavis' *South Italian Volcanoes*, Naples (1891), p. 41 and plate xii.

the movement of a flowing current of basalt.[1] The great heat of the lava has frequently induced considerable alteration upon the underlying rocks. Induration is the most common result, often accompanied with a reddening of the altered substance. Occasionally a beautifully prismatic structure has been developed in the soft material immediately beneath a basalt, as in ferruginous clay near the village of Esplot in the Velay, in which the close-set columns are 50 centimetres long and 4 to 5 centimetres in diameter.[2] Changes of this nature, however, are more frequent among sills than among superficial lavas. Many examples of them may be gathered from the Scottish Carboniferous districts.

VARIATIONS OF STRUCTURE IN SINGLE LAVA-SHEETS.—From what has been said above in regard to certain kinds of flow-structure among basic rocks, it will be evident that some considerable range of chemical, but more particularly of mineralogical, composition may be sometimes observed even within the same sheet of lava. Such differences, it is true, are more frequent among intrusive rocks, especially thick sills and large bosses. But they have been met with in so many instances among superficial lavas as to show that they are the results of some general law, which probably has a wide application among the surface-products of volcanic action. Scrope expressed the opinion that in the focus of a volcano there may be a kind of filtration of the constituents of a molten mass, the heavier minerals sinking through the lighter, so that the upper portions of the mass will become more felspathic and the lower parts more augitic and ferruginous.[3]

Leopold von Buch found that in some of the highly glassy lavas of the Canary Islands the felspar increases towards the bottom of the mass, becoming so abundant as almost to exclude the matrix, and giving rise to a compound that might be mistaken for a primitive rock.[4]

Darwin observed that in a grey basalt filling up the hollow of an old crater in James Island, one of the Galapagos group, the felspar crystals became much more abundant in the lower scoriaceous part, and he discussed the question of the descent of crystals by virtue of their specific gravity through a still molten lava.[5]

Mr. Clarence King during a visit to Hawaii found that in every case where he broke newly-congealed streamlets of lava, "the bottom of the flow was thickly crowded with triclinic felspars and augites, while the whole upper part of the stream was of nearly pure isotropic and acid glass."[6] This subject will be again referred to when we come to discuss the characters of intrusive sills and bosses, for it is among them that the most marked petrographical variations may be observed. Examples will be cited both from the intrusive and extrusive volcanic groups of Britain.

VOLCANIC CYCLES.—Closely related to the problem of the range of struc-

[1] M. Boule, *Bull. Cart. Géol. France*, No. 28, tom. iv. (1892), p. 235.

[2] M. Boule. *Op. cit.* p. 234.

[3] *Volcanoes*, p. 125.

[4] *Description Physique des Isles Canaries* (1836), p. 190.

[5] *Geological Observations on Volcanic Islands* (1844), p. 117.

[6] *U.S. Geol. Exploration of the Fortieth Parallel*, vol. i. (1878), p. 716.

ture and composition in a single mass of lava is another problem presented by the remarkable sequence of different types of lava which are erupted within a given district during a single volcanic period. Nearly thirty years ago Baron von Richthofen drew attention to the sequence of volcanic materials erupted within the same geographical area. He showed, more especially from observations in Western America, that a definite order of appearance in the successive species of lava could be established, the earliest eruptions consisting of materials of an intermediate or average composition, and those of subsequent outflows becoming on the whole progressively more acid, but finishing by an abrupt transition to a basic type. His sequence was as follows: 1. Propylite; 2. Andesite; 3. Trachyte; 4. Rhyolite; 5. Basalt.[1] This generalisation has been found to hold good over wide regions of the Old World as well as the New. It is not, however, of universal application.[2] Examples are not uncommon of an actual alternation of acid and basic lavas from the same, or at least from adjacent vents. Such an alternation occurs among the Tertiary eruptions of Central France and among those of Old Red Sandstone age in Scotland.

The range of variation in the nature of the eruptive rocks during the whole of a volcanic period in any district may be termed "a volcanic cycle." In Britain, where the records of many volcanic periods have been preserved, a number of such cycles may be studied. In this way the evolution of the subterranean magma during one geological age may be compared with that of another. It will be one of the objects of the following chapters to trace out this evolution in each period where the requisite materials for the purpose are available. We shall find that back to Archæan time a number of distinct cycles may be observed, differing in many respects from each other, but agreeing in the general order of development of the successive eruptions. Leaving these British examples for future consideration, it may be useful to cite here a few from the large series now collected from the European continent and North America.[3]

Among the older rocks of the European continent, Prof. Brögger has shown that in the Christiania district the eruptive rocks which traverse the Cambrian and Silurian formations began with the outburst of basic material

[1] *Trans. Acad. California*, 1868. Prof. Iddings' *Journ. Geol.*, vol. i. (1893), p. 606.

[2] See Prof. Brögger, "Die Eruptivgesteine des Kristianiagebietes," part ii. (1895), p. 175; *Zeitsch. Kryst. und Mineral*, vol. xvi. (1890) p. 83. This author would, from this point of view, draw a distinction between rocks which have consolidated deep within the earth and those which have flowed out at the surface, since he thinks that we are not justified in applying our experience of the order of sequence in the one series to the other. Yet there can be no doubt that in many old volcanic districts the masses that may be presumed to have consolidated at a great depth have been in unbroken connection with masses that reached the surface. These latter, as Prof. Iddings has urged, furnish a much larger body of evidence than the intrusive sheets and bosses.

[3] Prof. M. Bertrand in a suggestive paper published in 1888 dealt with the general order of appearance of eruptive rocks in different provinces of Europe. But the materials then at his command probably did not warrant him in offering more than a sketch of the subject, *Bull. Soc. Geol.*, France, xvi. p. 573. In the same volume there is a paper by M. Le Verrier, who announces his opinion that the eruption of the basic rocks takes place in times of terrestrial calm, while that of the acid rocks occurs in periods of great disturbance, *op. cit.* p. 498. Compare also Prof. Brögger, *Die Eruptivgesteine des Kristianiagebietes*, ii. p. 169.

such as melaphyre, augite-porphyrite, and gabbro-diabase, having from about 44 to about 52 per cent of silica. These were followed by rocks with a silica-percentage ranging from about 50 to 61, including some characteristic Norwegian rocks, like the rhomben-porphyry. The acidity continued to increase, for in the next series of eruptions the silica-percentage rose to between 60 and 67, the characteristic rock being a quartz-syenite. Then came deep-seated protrusions of highly acid rocks, varieties of granite, containing from 68 to 75 per cent of silica. The youngest eruptive masses in the district show a complete change of character. They are basic dykes (proterobase, diabase, etc.).[1]

The same author institutes a comparison between the post-Silurian eruptive series of Christiania and that of the Triassic system in the Tyrol, and believes that the two cycles closely agree.[2]

During Tertiary time in Central France more than one cycle may be made out in distinct districts. Thus in the Velay, during the Miocene period, volcanic activity began with the outpouring of basalts, followed successively by trachytes, labradorites and augitic andesites, phonolites and basalts. The Pliocene eruptions showed a reversion to the intermediate types of augitic andesites and trachytes, followed by abundant basalts, which continued to be poured forth in Pleistocene time.[3]

Further north, in Auvergne, where the eruptions come down to a later period, the volcanic sequence appears to have been first a somewhat acid group of lavas (trachytes or domites), followed by a series of basalts, then by andesites and labradorites, the latest outflows again consisting of basalts.[4]

Not less striking is the succession of lavas in the Yellowstone region, as described by Mr. Iddings. The first eruptions consisted of andesites. These were followed by abundant discharges of basalt, succeeded by later outflows of andesite, and of basalt like that previously erupted. After a period of extensive erosion, occupying a prolonged interval of time, volcanic energy was renewed by the eruption of a vast flood of rhyolite, after which came a feebler outflow of basalt that brought the cycle to a close, though geysers and fumaroles show that the volcanic fires are not yet entirely extinguished below.[5]

But not only is there evidence of a remarkable evolution or succession of erupted material within the volcanic cycle of a single geological period. One of the objects of the present work is to bring forward proofs that such

[1] *Eruptivgest. Kristianiageb.*, 1895.

[2] *Op. cit.* He supposes in each case the pre-existence of a parent magma from which the eruptive series started and which had a silica-percentage of about 64 or 65. In this difficult subject it is of the utmost importance to accumulate fact before proceeding to speculation.

[3] M. Boule, "Description Géologique du Velay," *Bull. Carte. Géol. France*, 1892. This author draws special attention to the evidence for the alternation of basic and more acid material in the Tertiary eruptions of Central France.

[4] M. Michel Lévy, *Bull. Soc. Géol. France*, 1890, p. 704.

[5] *Journal of Geology*, Chicago, i. (1893) p. 606. See also this author's excellent monograph on "Electric Peak and Sepulchre Mountain," *12th Ann. Rep. U.S. Geol. Survey* (1890-91), and Mr. H. W. Turner on "The Succession of Tertiary Volcanic Rocks in the Sierra Nevada of North America," *14th Ann. Rep. U.S. Geol. Survey* (1892-93), p. 493.

cycles have succeeded each other again and again, at widely separated intervals, within the same region. After the completion of a cycle and the relapse of volcanic energy into repose, there has been a renewal of the previous condition of the subterranean magma, giving rise ultimately to a similar succession of erupted materials.

If we are at a loss to account for the changes in the sequence of lavas during a single volcanic cycle, our difficulties are increased when we find that in some way the magma is restored each time to somewhat the same initial condition. Analogies may be traced between the differentiation which has taken place within a plutonic intrusive boss or sill and the sequence of lavas in volcanic cycles. It can be shown that though the original magma that supplied the intrusive mass may be supposed to have had a fairly uniform composition deep down in its reservoir, differentiation set in long before the intrusive mass consolidated, the more basic constitutents travelling outwards to the margin and leaving the central parts more acid. If some such process takes place within a lava-reservoir, it may account for a sequence of variations in composition. But this would not meet all the difficulties of the case, nor explain the determining cause of the separation of the constituents within the reservoir of molten rock, whether arising from temperature, specific gravity, or other influence. This subject will be further considered in connection with intrusive Bosses.

Another fact which may be regarded as now well established is the persistence of composition and structure in the lavas of all ages. Notwithstanding the oft-repeated cycles in the character of the magma, the materials erupted to the surface, whether acid or basic, have retained with wonderful uniformity the same fundamental characteristics. No part of the contribution of British geology to the elucidation of the history of volcanic action is of more importance than the evidence which it furnishes for this persistent sameness of the subterranean magma. An artificial line has sometimes been drawn between the volcanic products of Tertiary time and those of earlier ages. But a careful study of the eruptive rocks of Britain shows that no such line of division is based upon any fundamental differences.

The lavas of Palæozoic time have of course been far longer exposed to alterations of every kind than those of the Tertiary periods, and certain superficial distinctions may be made between them. But when these accidental differences are eliminated, we find that the oldest known lavas exhibit the same types of structure and composition that are familiar in those of Tertiary and recent volcanoes. Many illustrations of this statement will be furnished in later chapters. As a particularly striking instance, I may cite here the most ancient and most modern lavas of the Grand Cañon of the Colorado. Mr. Walcott and Mr. Iddings have shown that in the Lower Cambrian, or possibly pre-Cambrian, formations of that great gorge, certain basic lavas were contemporaneously interstratified, which, in spite of their vast antiquity, are only slightly different from the modern basalts that have been poured over the surrounding plateau.[1]

[1] *14th Annual Report U.S. Geol. Survey* (1892-93).

THE CHIEF LAVAS FOUND IN BRITAIN.—Of the lavas which have been poured out at the surface within the region of the British Isles, the following varieties are of most frequent occurrence. In the acid series are Rhyolites and Felsites, but the vitreous forms are probably all intrusive. The intermediate series is represented by Trachytes and Andesites (Porphyrites), which range from a glassy to a holocrystalline structure. The basic series includes Dolerites, Diabases, Basalts, Limburgites (or Magma-basalts, containing little or no felspar), and Picrites or other varieties of Peridotites. The intrusive rocks display a greater variety of composition and structure.

ii. VOLCANIC AGGLOMERATES, BRECCIAS AND TUFFS

The coarser fragmentary materials thrown from volcanic vents are known as Agglomerates where they show no definite arrangement, and especially where they actually fill up the old funnels of discharge. When they have accumulated in sheets or strata of angular detritus outside an active vent they are termed Breccias, or if their component stones have been water-worn, Conglomerates. The finer ejected materials may be comprehended under the general name of Tuffs.

Although these various forms of pyroclastic detritus consist as a rule of thoroughly volcanic material, they may include fragments of non-volcanic rocks. This is especially the case among those which represent the earliest explosions of a volcano. The first efforts to establish an eruptive vent lead to the shattering of the terrestrial crust, and the consequent discharge of abundant debris of that crust. The breccias or agglomerates thus produced may contain, indeed, little or no truly volcanic material, but may be made up of fragments of granite, gneiss, sandstone, limestone, shale, or whatever may happen to be the rocks through which the eruptive orifice has been drilled. If the first explosions exhausted the energy of the vent, it may happen that the only discharges from it consisted merely of non-volcanic debris. Examples of this kind have been cited from various old volcanic districts. A striking case occurs at Sepulchre Mountain in the Yellowstone Park, where the lower breccias, the product of the earliest explosions of the Electric Peak volcano, and attaining a thickness of 500 feet, are full of pieces of the Archæan rocks which underlie the younger formations of that district.[1] These non-volcanic stones do not occur among the breccias higher up. Obviously, however, though most abundant at first, pieces of the underlying rocks may reappear in subsequent discharges, wherever by the energy of explosion, fragments are broken from the walls of a volcanic chimney and hurled out of the crater. Illustrations of these features will be given in the account of the British Carboniferous, Permian and Tertiary volcanic rocks.

It will be obvious that where the component materials of such fragmentary accumulations consist entirely of non-volcanic rocks, great caution must be exercised in attributing them to volcanic agency. Two sources of

[1] Prof. J. P. Iddings, *12th Ann. Rep. U.S. Geol. Survey* (1890-91), p. 634.

error in such cases may here be pointed out. In the first place, where angular detritus has been precipitated into still water, as, for instance, from a crag or rocky declivity into a lake, a very coarse and tumultuous kind of breccia may be formed. It is conceivable that, in course of time, such a breccia may be buried under ordinary sediments, and may thereby be preserved, while all trace of its parent precipice may have disappeared. The breccia might resemble some true volcanic agglomerates, but the resemblance would be entirely deceptive.

In the second place, notice must be taken of the frequent results of movements within the terrestrial crust, whereby rocks have not only been ruptured but, as already pointed out, have been crushed into fragments. In this way, important masses of breccia or conglomerate have been formed, sometimes running for a number of miles and attaining a breadth of several hundred feet. The stones, often in huge blocks, have been derived from the surrounding rocks, and while sometimes angular, are sometimes well-rounded. They are imbedded in a finer matrix of the same material, and may be scattered promiscuously through the mass, in such a way as to present the closest resemblance to true volcanic breccia. Where the crushed material has included ancient igneous rocks it might deceive even an experienced geologist. Indeed, some rocks which have been mapped and described as volcanic tuffs or agglomerates are now known to be only examples of "crush-conglomerates." [1]

Not only have vast quantities of detritus of non-volcanic rocks been shot forth from volcanic vents, but sometimes enormous solid masses of rock have been brought up by ascending lavas or have been ejected by explosive vapours. Every visitor to the puys of Auvergne will remember the great cliff-like prominence of granite and mica-schist which, as described long ago by Scrope, has been carried up by the trachyte of the Puy Chopine, and forms one of the summits of the dome (Fig. 344). The same phenomenon is observable at the Puy de Montchar, where large blocks of granite have been transported from the underlying platform. Abich has described some remarkable examples in the region of Erzeroum. The huge crater of Palandokän, 9687 feet above the sea contains, in cliff-like projections from its walls as well as scattered over its uneven bottom, great masses of marmorised limestone and alabaster, associated with pieces of the green chloritic schists, serpentines and gabbros of the underlying non-volcanic platform. These rocks, which form an integral part of the structure of the crater, have been carried up by masses of trachydoleritic, andesitic and quartz-trachytic lavas.[2] Examples will be given in a later chapter showing how gigantic blocks of mica-schist and other rocks have been carried many hundred feet upwards and buried among sheets of lava or masses of agglomerates during the Tertiary volcanic period in Britain (Fig. 262).

[1] For an account of "crush-conglomerates," see Mr. Lamplugh's paper on those of the Isle of Man, *Quart. Journ. Geol. Soc.*, li. (1895), p. 563. Mr. M'Henry has pointed to probable cases of mistake of this kind in Ireland, *Nature*, 5th March 1896. A. Geikie, *Geol. Mag.* November 1896.

[2] Abich, *Geologie des Armenischen Hochlandes* (Part ii., western half), 1882, p. 76.

In the vast majority of cases, the fragmentary substances ejected by ancient volcanic explosions, like those of the present day, have consisted wholly or mainly of material which existed in a molten condition within the earth, and which has been violently expelled to the surface. Such ejected detritus varies from the finest impalpable dust or powder up to huge masses of solid rock. These various materials may come from more than one source. Where a volcanic orifice is blown out through already solidified lavas belonging to previous eruptions, the fragments of these lavas may accumulate within or around the vent, and be gradually consolidated into agglomerate or breccia. Again, explosions within the funnel may break up lava-crusts that have there formed over the cooling upper surface of the column of molten rock. Or the uprising lava in the chimney may be spurted out in lumps of slag and bombs, or may be violently blown out in the form of minute lapilli, or of extremely fine dust and ashes.

Although in theory these several varieties of origin may be discriminated, it is hardly possible always to distinguish them among the products of ancient volcanic action. In the great majority of cases old tuffs, having been originally deposited in water, have undergone a good deal of decomposition, and such early alteration has been aggravated by the subsequent influence of percolating meteoric water.

Where disintegration has not proceeded too far, the finer particles of tuffs may be seen to consist of minute angular pieces of altered glass, or of microlites or crystals, or of some vitreous or semi-vitreous substance, in which such microlites and crystals are enclosed. It has already been stated that the occurrence of glass, or of any substance which has resulted from the devitrification of glass, may be taken as good evidence of former volcanic activity.

Most commonly, especially in the case of tuffs of high antiquity, like those associated with the Palæozoic formations, the fresh glassy and microlitic constituents, so conspicuous in modern volcanic ashes, are hardly to be recognised. The finer dust which no doubt contained these characteristic substances has generally passed into dull, earthy, granular, or structureless material, though here and there, among basic tuffs, remaining as palagonite. In the midst of this decayed matrix, the lapilli of disrupted lavas may endure, but it may be difficult or impossible to decide whether they were derived from the breaking up of older lavas by explosion, or from the blowing out of the lava-crusts within the funnel.

The cellular structure, which we have seen to be a markedly volcanic peculiarity among the lavas, is not less so in their fragments among the agglomerates, breccias and tuffs; indeed it may be said to be eminently characteristic of them. The vesicles in the lapilli, bombs, and blocks are sometimes of large size, as in masses of ejected slag, but they range down to microscopic minuteness. The lapilli of many old tuffs are sometimes so crowded with such minute pores, as to show that they were originally true pumice.

The composition of tuffs must obviously depend upon that of the lavas from which they were derived. But their frequently decayed condition makes it less easy, in their case, to draw definite boundary-lines between varieties. In a group of acid lavas, the tuffs may be expected to be also acid, while among intermediate or basic lavas, the tuffs will generally be found to correspond. There are, however, exceptions to this general rule. As will be afterwards described in detail, abundant felsitic tuffs may be seen among the andesitic lavas of Lower Old Red Sandstone age in Scotland, and rhyolitic tuffs occur also among the Tertiary basalts of Antrim.

As a rule, basic and intermediate tuffs, like the lavas from which they have been derived, are rather more prone to decomposition than the acid varieties. One of their most characteristic features is the presence in them of lapilli of a minutely vesicular pumice, which will be more particularly described in connection with volcanic necks, of which it is a characteristic constituent. Occasional detached crystals of volcanic minerals, either entire or broken, may be detected in them, though perhaps less frequently than in the agglomerates. The earthy matrix is generally greenish in colour, varying into shades of brick-red, purple and brown.

The acid tuffs are, on the whole, paler in colour than the others, sometimes indeed they are white or pale grey, but graduate into tones of hæmatitic red or brown, the varying ferruginous tints being indicative of stages in the oxidation of the iron-bearing constituent minerals. Small rounded lapilli or angular fragments of felsite or rhyolite may be noticed among them, sometimes exhibiting the most perfect flow-structure. As typical examples of such tuffs, I may refer to those of the Pentland Hills, near Edinburgh, and those that lie between the two groups of basalt in Antrim.

Thrown out promiscuously from active vents, the materials that form tuffs arrange themselves, on the whole, according to relative size over the surface on which they come to rest, the largest being generally grouped nearest to the focus of discharge, and the finest extending farthest from it. As the volcanoes of which records have been preserved among the geological formations were chiefly subaqueous, the fragmentary substances, as they fell into water, would naturally be to some extent spread out by the action of currents or waves. They would thus tend to take a more or less distinctly stratified arrangement.

FIG. 13.—Alternations of coarser and finer Tuff.

Moreover, as during an eruption there might be successive paroxysms of violence in the discharges, coarser and finer detritus would successively fall over the same spot. In this way, rapid alternations of texture would arise (Fig. 13). A little experience will enable the observer to distinguish between such truly volcanic variations and those of ordinary sedimentation, where, for instance, layers of gravel and sand repeatedly alternate. Besides the volcanic nature of the fragments and their non-water-worn forms, he will notice that here and there the larger blocks may be placed on end—a position the reverse of that usual in the disposal of aqueous sediments, but one which is not infrequently assumed by ejected stones, even when they fall through some little depth of water. Further, the occurrence of large pieces of lava, scattered at random through deposits of fine tuff, would lead him to recognize the tumultuous discharges of a volcanic focus, rather than the sorting and sifting action of moving water.

Admirable illustrations of these various characteristics may be gathered in endless number from the Palæozoic volcanic chronicles of Britain. I may especially cite the basin of the Firth of Forth as a classical region for the study of Carboniferous examples.

When the conditions of modern volcanic eruptions are considered, it will be seen that where ejected ashes and stones fall into water, they will there mingle with any ordinary sediment that may be in course of deposition at the time. There will thus be a blending of volcanic and non-volcanic detritus, and the transition between the two may be so insensible that no hard line of demarcation can be drawn. Such intermingling has continually taken place during past ages. One of the first lessons learnt by the geologist, who begins the study of ancient volcanic records, is the necessity of recognizing this gradation of material, and likewise the frequently recurring alternations of true tuff with shale, sandstone, limestone or other entirely non-volcanic detritus (Fig. 14). He soon perceives that such facts as these furnish him with some of the most striking proofs of the reality and progress of former eruptions. The intermingling of much ordinary detritus with the volcanic material may be regarded as indicative either of comparatively feeble activity, or at least of considerable distance from the focus of discharge. It is sometimes possible to trace such intermixtures through gradually augmenting proportions of volcanic dust and stones, until the deposit becomes wholly volcanic in composition, and so coarse in texture as to indicate the proximity of the eruptive vent. On the other hand, the gradual decrease of the volcanic ejections can be followed in the upward sequence of a series of stratified deposits, until the whole material becomes entirely non-volcanic.

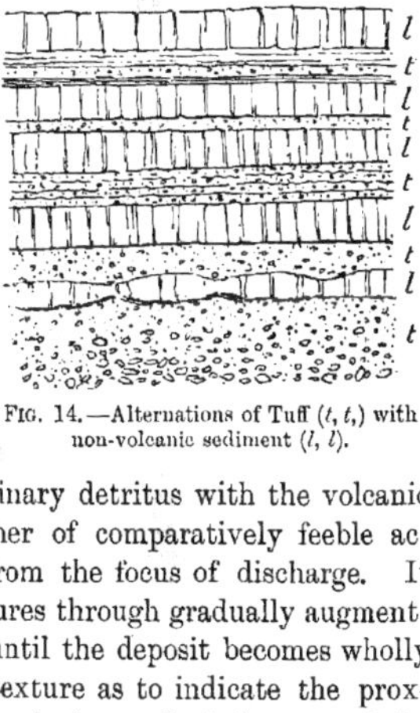

FIG. 14.—Alternations of Tuff (*t*, *t*,) with non-volcanic sediment (*l*, *l*).

The occurrence of thin partings of tuff between ordinary sedimentary strata points to occasional intermittent eruptions of ashes or stones, the vigour and duration of each eruptive interval being roughly indicated by the thickness and coarseness of the volcanic detritus. The pauses in the volcanic activity allowed the deposit of ordinary sediment to proceed unchecked. The nature of such non-volcanic intercalations gives a clue to the physical conditions of sedimentation at the time, while their thickness affords some indication of the relative duration of the periods of volcanic repose.

A little reflection will convince the observer that in such a section as that represented in Fig. 14 the volcanic intercalations must be regarded as a mere local accident. Evidently the normal conditions of sedimentation at the time these strata were accumulated are indicated by the limestone bands (*l*, *l*). Had there been no volcanic eruptions, a continuous mass of limestone would have been deposited, but this continuity was from time to time interrupted by the explosions that gave rise to the intercalated bands of tuff (*t*, *t*).

The application of these rules of geological evidence will be best understood from actual examples of their use. Many illustrations of them will be subsequently given, more especially from the volcanic records of the Carboniferous period.

One of the most interesting peculiarities of interstratified tuffs is the not infrequent occurrence of the remains of plants and animals imbedded in them. Such remains would have been entombed in the ordinary sediment had there been no volcanic eruptions, and their presence in the tuffs is another convincing proof of contemporaneous volcanic action during the deposition of a sedimentary series. But they may be made to furnish much more information as to the chronology of the eruptions and the physical geography of the localities where the volcanoes were active, as will be set forth farther on.

Tuffs, as already remarked, frequently occur without any accompaniment of lava, although lavas seldom appear without some tuff. We thus learn that in the past, as at present, discharges of fragmentary materials alone were more common than the outflow of lava by itself. The relative proportions of the lavas and tuffs in a volcanic series vary indefinitely. In the Tertiary basalt-plateaux of Britain, the lavas succeed each other, sheet above sheet, for hundreds of feet, with few and trifling fragmental intercalations. Among the Carboniferous volcanic ejections, on the other hand, many solitary or successive bands of tuff may be observed without any visible sheets of lava. Viewed broadly, however, in their general distribution during geological time, the two great groups of volcanic material may be regarded as having generally appeared together. In all the great volcanic series, from the base of the Palæozoic systems up to the Tertiary plateaux, lavas and tuffs are found associated, much as they are among the ejections of modern volcanoes. They often alternate, and thus furnish evidence as to oscillations of energy at the eruptive vents.

Now and then, by the explosions from a volcano at the present day, a single stone may be ejected at such an angle and with such force as to fall to the ground at a long distance from the vent. In like manner, among the volcanic records of former periods, we may occasionally come upon a single block of lava imbedded among tuffs or even in non-volcanic strata. Where such a stone has fallen upon soft sediment, it can be seen to have sunk into it, pressing down the layers beneath it, and having the subsequently deposited layers heaped over it. An ejected block of this nature is represented among the tuffs shown in Fig. 13. Another instance from a group of non-volcanic sediments is given in Fig. 15, and is selected from a number of illustrations of this interesting feature which have been observed among the Lower Carboniferous formations of the basin of the Firth of Forth. A solitary block, imbedded in a series of strata, would not, of course, by itself afford a demonstration of volcanic activity. There are various ways in which such stones may be transported and dropped over a muddy water-bottom. They may, for example, be floated off attached to sea-weeds, or wrapped round by the roots of trees. But where a block of basalt or other volcanic rock has obviously descended with such force as to crush down the deposits on

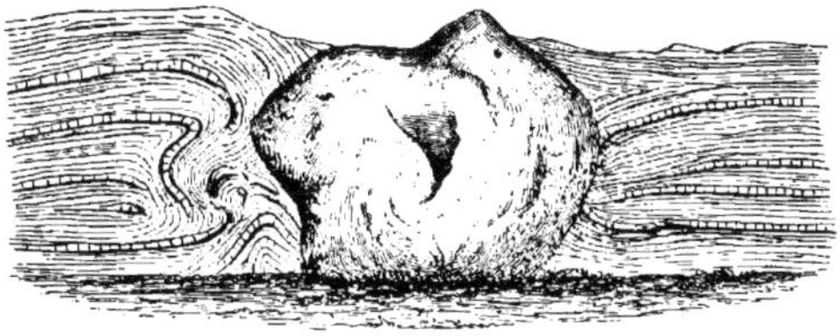

FIG. 15.—Ejected block of Basalt which has fallen among Carboniferous shales and limestones, shore, Pettycur, Fife.

which it fell, and when lavas and tuffs are known to exist in the vicinity, there can be little hesitation in regarding such a block as having been ejected from a neighbouring vent, either during an explosion of exceptional violence or with an unusually low angle of projection.

In conclusion, reference may conveniently be made here to another variety of fragmental volcanic materials which cannot always be satisfactorily distinguished from true tuffs, although arising from a thoroughly non-volcanic agency. Where a mass of lava has been exposed to denudation, as, for instance, when a volcanic island has been formed in a lake or in the sea, the detritus worn away from it may be spread out like any other kind of sediment. Though derived from the degradation of lava, such a mechanical deposit is not properly a tuff, nor can it even be included among strictly volcanic formations. It may be called a volcanic conglomerate, rhyolitic conglomerate, diabase sandstone, felsitic shale, or by any other name that will adequately denote its composition and texture. But it may not afford proof of strictly contemporaneous volcanic activity. All that we are entitled to infer from such a deposit is that, at the time when it was laid down, volcanic rocks of a certain kind were exposed at the surface and were undergoing degradation. But the date of their original eruption may have been long previous to that of the formation of the detrital deposit from their waste.

Nevertheless, it is sometimes possible to make sure that the conglomerate or sandstone, though wholly due to the mechanical destruction of already erupted lavas, was in a general sense contemporaneous with the volcanoes that gave forth these lavas. The detrital formation may be traced perhaps up to the lavas from which it was derived, and may be found to be intercalated in the same sedimentary series with which they are associated. Or it may contain large bombs and slags, such as most probably came either directly from explosions or from the washing down of cinder-cones or other contemporaneously existing volcanic heaps. Examples of such intercalated conglomerates will be given from the Lower Old Red Sandstone of Central Scotland and from the Tertiary volcanic plateaux of the Inner Hebrides.

CHAPTER IV

Materials erupted at the Surface—Extrusive Series—*continued.* iii. Types of old Volcanoes —1. The Vesuvian Type; 2. The Plateau or Fissure Type; 3. The Puy Type. iv. Determination of the relative Geological Dates of ancient Volcanoes. v. How the Physical Geography associated with ancient Volcanoes is ascertained.

HAVING now taken note of the various materials ejected to the surface from volcanic orifices, we may pass to the consideration of these orifices themselves, with the view of ascertaining under what various conditions volcanic action has taken place in the geological past. We have seen that modern and not long extinct volcanoes may be grouped into three types, and a study of the records of ancient volcanoes shows that the same types may be recognized in the eruptions of former ages. The following chapters will supply many illustrations of each type from the geological history of the British Isles. In dealing with these illustrations, however, we must ever bear in mind the all-powerful influence of denudation. We ought not to expect to meet with the original forms of the volcanoes. Some little reflection and experience may be required before we can realize under what aspect we may hope to recognize ancient and much-denuded volcanoes. It may therefore be of advantage to consider here, in a broad way, which of the original characters are most permanent, and should be looked for as mementoes of ancient volcanoes after long ages of denudation.

iii. TYPES OF OLD VOLCANOES

The three forms of ancient volcanoes now to be discussed are—1st, the Vesuvian type; 2nd, the Plateau or Fissure type; and 3rd, the Puy type.

1. *The Vesuvian Type.*—In this kind of volcano, lavas and fragmental ejections are discharged from a central vent, which is gradually built up by successive eruptions of these materials. As the cone increases in size, parasitic cones appear on its sides, and in the energy and completeness of their phenomena become true volcanoes, almost rivalling their parent mountain. Streams of lava descend upon the lower grounds, while showers of dust and ashes are spread far and wide over the surrounding country.

If a transverse section could be made of a modern Vesuvian cone, the volcanic pile would be found to consist of alternations of lavas and tuffs, thickest at the centre, and thinning away in all directions. At some distance from the crater, these volcanic materials might be seen to include layers of soil and remains of land-vegetation, marking pauses between the eruptions, during which soil accumulated and plants sprang up upon it. Where the lavas and ashes had made their way into sheets of fresh water or into the sea, they would probably be found interstratified with layers of ordinary sediment containing remains of the animal or vegetable life of the time.

Conceive now the effects of prolonged denudation upon such a pile of volcanic rocks. The cone will eventually be worn down, the crater will disappear, and the only relics of the eruptive orifice may be the top of the central lava-column and of any fragmental materials that solidified within the vent (Fig. 16). The waste will, on the whole, be greater at the cone than on the more level areas beyond. It might, in course of time, reach the original surface of the ground on which the volcano built up its heap of ejected material. The central lava-plug might thus be left as an isolated

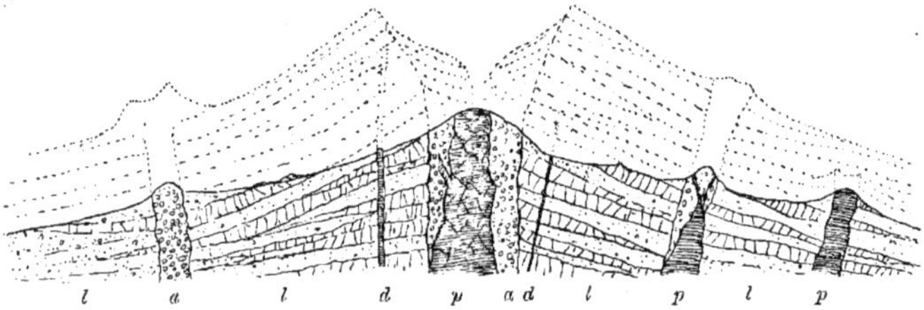

FIG. 16.—Effects of denudation on a Vesuvian cone.

eminence rising from a platform of older non-volcanic rocks, and the distance between it and the dwindling sheets of lava and tuff which came out of it would then be continually increased as their outcrop receded under constant degradation.

This piece of volcanic history is diagrammatically illustrated in Fig. 16. The original forms of the central volcano and of its parasitic cones are suggested by the dotted lines in the upper half of the Figure. All this upper portion has been removed by denudation, and the present surface of the ground is shown by the uppermost continuous line. The general structure of the volcanic pile is indicated underneath that line—the lenticular sheets of lava and tuff (*l*, *l*), the dykes (*d*, *d*), and the lavas (*p*, *p*) and agglomerates (*a*, *a*) of the central vent and of the subordinate cones.

The waste, though greatest on the higher ground of the great cone, would not stop there. It would extend over the flatter area around the volcano. Streams flowing over the plain would cut their way down through the lavas and tuffs, eroding ravines in them, and leaving them in detached and ever diminishing outliers on the crests of the intervening ridges. We can easily picture a time when the last of these relics would have been worn away, and when every vestige of the volcanic ejections would have been removed, save the lava-column marking the site of the former vent.

Every stage in this process of effacement may be recognized in actual progress among the extinct volcanoes of the earth's surface. Probably nowhere may the phenomena be more conveniently and impressively studied than among the volcanic districts of Central France. On the one hand, we meet there with cinder-cones so perfect that it is hard to believe them to have been silent ever since the beginnings of history. On the other hand, we see solitary cones of agglomerate or of lava, which have been left isolated, while their once overlying and encircling sheets of ejected material have been so extensively worn away as to remain merely in scattered patches capping the neighbouring hills. Valleys many hundreds of feet in depth have been cut by the rivers through the volcanic sheets and the underlying Tertiary strata and granite since the older eruptions ceased. And yet these eruptions belong to a period which, in a geological sense, is quite recent. It is not difficult to contemplate a future time, geologically not very remote, when in the valley of the Loire not a trace will remain of the wonderfully varied and interesting volcanic chronicle of that district, save the plugs that will mark the positions of the former active vents.

In the British Islands, ancient volcanoes of the Vesuvian type are well represented among the Palæozoic systems of strata. Their preservation has been largely due to the fact that they made their appearance in areas that were undergoing slow subsidence. Their piles of erupted lava and ashes were chiefly heaped up over the sea-floor, and were buried under the sand, silt and ooze that gathered there. Thus covered up, they were protected from denudation. It is only in much later geological ages that, owing to upheaval, gradual degradation of the surface, and removal of their overlying cover of stratified formations, they have at last been exposed to waste. The process of disinterment may be observed in many different stages of progress. In some localities, only the tops of the sheets of lava and tuff have begun to show themselves; in others, everything is gone except the indestructible lava-plug.

These inequalities of denudation arise not only from variations in the durability of volcanic rocks, but still more from the relative position of these rocks in the terrestrial crust, and the geological period at which, in the course of the general lowering of the surface, they have been laid bare. Not only are volcanic rocks of many different ages, and lie, therefore, on many successive platforms within the crust of the earth: their places have been still further dependent upon changes in the arrangement of that crust. Fracture, upheaval, depression, curvature, unconformable deposition of strata, have contributed to protect some portions, while leaving others exposed to attack. Hence it happens that the volcanic record varies greatly in its fulness of detail from one geological system or one district to another. Some chapters have been recorded with the most surprising minuteness, so that the events which they reveal can be realized as vividly as those of a modern volcano. Others, again, are meagre and fragmentary, because the chronicle is still for the most part buried underground, or because it has been so long exposed at the surface that only fragments of it now remain there.

In the descriptions which will subsequently be given of ancient British volcanoes of the Vesuvian type, it will be shown that at many successive periods during Palæozoic time, and at many distinct centres, lavas and tuffs have been piled up to a depth of frequently more than 5000 feet—that is to say, higher than the height of Vesuvius. Sometimes the vent from which these materials were ejected can be recognized. In other places, it is still buried under later formations, or has been so denuded as to be represented now merely by the column of molten or fragmental rock that finally solidified in it. Examples will be quoted of such ancient vents, measuring not less than two miles in diameter, with subsidiary "necks" on their flanks, like the parasitic cones on Etna.

I shall show that while the ejected volcanic products have accumulated in greatest depth close to the vent that discharged them, they die away as they recede from it, sometimes so rapidly that a volcanic pile which is 7000 feet thick around its source may entirely thin out and disappear in a distance of not more than ten or twelve miles. I shall point out how, as the lavas and tuffs are followed outwards from their centre, they not only get thinner, but are increasingly interstratified among the sedimentary deposits with which they were coeval, and that in this way their limits, their age, and the geographical conditions under which they were accumulated can be satisfactorily fixed.

As illustrations of the Vesuvian type in the volcanic history of Britain, I may refer to the great Lower Silurian volcanoes of Cader Idris, Arenig, Snowdon and the Lake District, and to the Old Red Sandstone volcanoes of Central Scotland.

2. *The Plateau* or *Fissure type* is, among modern volcanoes, best developed in Iceland, as will be more fully detailed in Chapter xl. In that island, during a volcanic eruption, the ground is rent open into long parallel fissures, only a few feet or yards in width, but traceable sometimes for many miles, and descending to an unknown depth into the interior. From these fissures lava issues—in some cases flowing out tranquilly in broad streams from either side, in other cases issuing with the discharge of slags and blocks of lava which are piled up into small cones set closely together along the line of the rent. It was from a fissure of this kind that the great eruption of 1783 took place—the most stupendous outpouring of lava within historic time.

By successive discharges of lava from fissures, or from vents opening on lines of fissure, wide plains may be covered with a floor of rock hundreds or thousands of feet in thickness, made up of horizontal beds. The original topography, which might have been undulating and varied, is completely buried under a vast level lava-desert.

The rivers which drained the country before the beginning of the volcanic history will have their channels filled up, and will be driven to seek new courses across the lava-fields. Again and again, as fresh eruptions take place, these streams will be compelled to shift their line of flow, each river-bed being in turn sealed up in lava, with all its gravels, silts and drift-wood. But the rain will continue to fall, and the drainage to seek its way seaward. When the last eruption ceases, and the rivers are at length left undisturbed at their task of erosion, they will carve that lava-floor into deep gorges or open valleys. Where they flow between the lavas and the slopes against which these ended, they will cut back the volcanic pile, until in course of time the lavas will present a bold mural escarpment to the land that once formed their limit. The volcanic plain will become a plateau, ending off in this vertical wall and deeply trenched by the streams that wind across it. And if the denudation is continued long enough, the plateau will be reduced to detached hills, separated by deep and wide valleys.

This geological history is illustrated by the diagram in Fig. 17. The stippled ground underneath (*x*, *x*) represents the original undulating surface of the country on which the plateau eruptions were poured out. The lavas of these eruptions are shown by the horizontal lines to have entirely buried the heights and hollows of the old land, and to have risen up to the upper dotted line, which may be taken to mark the limit reached by the accumulation of volcanic material. The dark lines (*d*, *d*) which come up

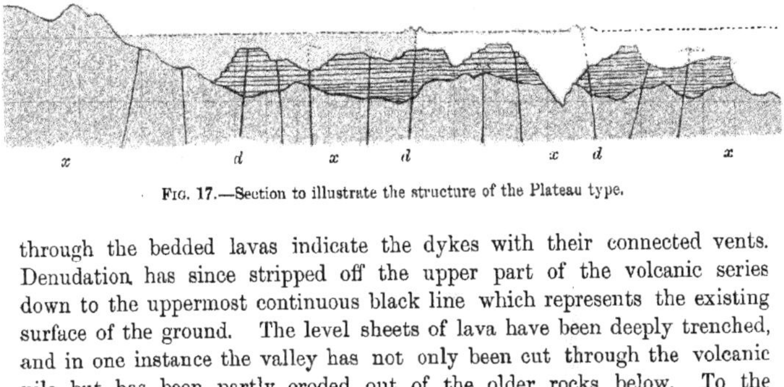

FIG. 17.—Section to illustrate the structure of the Plateau type.

through the bedded lavas indicate the dykes with their connected vents. Denudation has since stripped off the upper part of the volcanic series down to the uppermost continuous black line which represents the existing surface of the ground. The level sheets of lava have been deeply trenched, and in one instance the valley has not only been cut through the volcanic pile, but has been partly eroded out of the older rocks below. To the right and left, the lavas end off abruptly in great escarpments.

The succession of events here depicted has occurred more than once in Britain. The Plateau type is chiefly developed in this country among the great Tertiary basalt districts of Antrim and the Inner Hebrides, which reappear in the Faroe Islands, and again still farther north in Iceland. But it also occurs among the volcanic rocks of the Old Red Sandstone and Carboniferous periods.

As compared with the other volcanic types, that of the Plateaux is distinguished by the wide extent of surface which its rocks cover, by the great preponderance of lavas over tuffs, and by the regularity and persistence of the individual sheets of rock. In Britain, the plateau-lavas are even still often approximately horizontal, and lie piled on each other in tolerably regular beds to a thickness of 1000, and in one place to more than 3000 feet.

They form wide level or gently undulating tablelands, which rise in bold escarpments from the surrounding country and have been deeply carved into valleys. The sides of their cliffs and slopes are marked by parallel lines of terrace, arising from the outcrop of successive sheets of lava (Figs. 11, 265).

With the Tertiary basalt-plateaux are connected thousands of dykes, that follow each other along nearly parallel lines in a general north-westerly direction, and mark the position of fissures up which the molten lava ascended. Occasional necks of agglomerate or basalt indicate the sites of some of the eruptive vents.

The Carboniferous volcanic plateaux have been more extensively denuded than those of Tertiary age, so that a large number of their vents have been laid bare. In general these vents are of comparatively small size, though larger than those of the Carboniferous Puys. In some districts, abundant dykes traverse the rocks on which the plateaux rest, though the fissures seem to have been less numerous than in Tertiary time.

3. *The Puy type*, as before remarked, takes its name from the well-known *puys*, or volcanic cones, of Central France. Volcanoes of this type form conical hills, generally of small size, consisting usually of fragmental materials, sometimes of lava. Where a cone is partially effaced by a second, and even by a third, successive slight shiftings of the vent are to be inferred (see Figs. 29 and 214). In many cases, no lava has issued from such cones, nor were the ashes and cinders dispersed far from the vent. Hence, in the progress of denudation, cones of this kind are easily effaced.

From the uniformity of composition of their materials, the simplicity and regularity of their forms, and their small size, it may be inferred that many of these cones were the products of single eruptions. They may conceivably have been thrown up in a few days, or even in a single day. The history of Monte Nuovo, in the Bay of Naples, which was formed within twenty-four hours in the year 1538, is a memorable example of the rapidity with which a cone more than 400 feet high may be thrown up at some distance from a central vent.

The smallest independent volcanoes are included in the Puy type. In many instances the diameter of the funnel has not exceeded a few yards; the largest examples of the type seldom exceed 1000 feet in width.

Where lavas have been discharged, as well as ashes and stones, a more vigorous activity is indicated than where merely cones of tuff were formed. The lavas may come from more than one side of a cone, and may flow in opposite directions for a distance of several miles. It is observable that considerable streams of lava have issued from the base of a cinder-cone without disturbing it. The molten rock has found a passage between the loose materials and the surface on which they rest,[1] though, in some cases, the cone may have been thrown up after the emission of the lava.

In the history of a puy there is commonly a first discharge of fragmentary material; then lava may flow out, followed by a final discharge of loose stones

[1] M. Boule, *Bull. Carte Géol. France*, No. 28, tome iv. p. 232.

and ashes. Hence the products of such a vent group themselves into three layers—two of breccia separated by an intervening sheet of lava.[1]

Great changes are wrought on puys and their connected lavas and tuffs during the progress of denudation. The cones are eventually destroyed, and only a stump of agglomerate or lava is left to mark its place. The connection of a lava-stream with its parent vent may likewise be effaced, and the lava itself may be reduced to merely a few separate patches, perhaps capping a ridge, while the surrounding ground has been hollowed into valleys. If the waste continues long enough, even these outliers will disappear, and nothing but the neck or stump of the little volcano will remain.

The accompanying diagram (Fig. 18) may help to make these changes more intelligible. The upper dotted lines show the original forms of three puys with the covering of loose materials discharged by them over the surrounding ground. The lower shaded portion represents the surface as left by denudation, and a section of the three vents beneath that surface. The whole of the cones and craters has here been swept away, and only the volcanic "neck" is in each case left. In the vent to the right, the material that

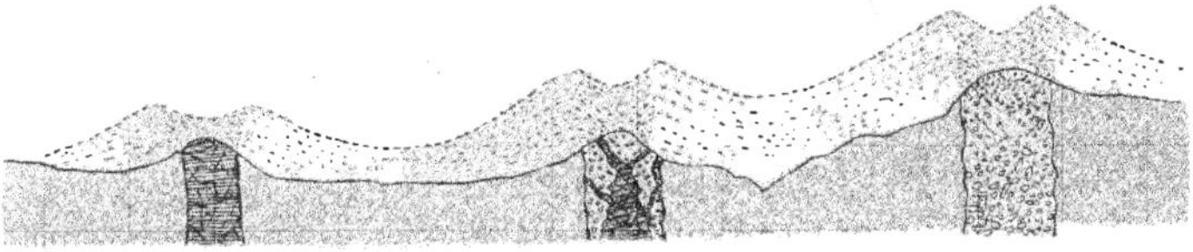

FIG. 18.—Diagram illustrating the structure and denudation of Puys.

fills it up is a coarse agglomerate, which projects as a rounded dome above the surrounding country. The central pipe is filled with fragmentary materials, through which molten rock has risen, giving off dykes and veins. In the vent to the left hand, only lava is seen to occupy the orifice, representing the column of molten rock which solidified there and brought the activity of this little volcano to an end. It will be observed that in each of these volcanic hills the present outlines are very far from being those of the original volcano, and that the eminence projects because of its greater resistance to the forces of denudation that have not only removed the superficial volcanic material, but have made some progress in lowering the level of the ground on which that material was accumulated.

The typical area for the study of Puys is the extraordinarily interesting volcanic region of Central France. There the volcanic cones are clustered in irregular groups, sometimes so close as to be touching each other; elsewhere separated by intervals of several miles. They may be traced in all stages of decay, from the most perfect cones and craters to the isolated eminence that marks the site of a once active chimney. Their lavas, too, may be seen as detached fragments of plateaux, many hundred feet above the valleys that have been excavated since they flowed.[2]

[1] M. Boule, *Bull. Carte Géol. France*, No. 28, tome iv.

[2] See Desmarest's classic map and his papers in *Mem. Acad. Roy. Sciences*, Paris, 1774, 1779; *Journ. de Physique*, 1779; Scrope's *Geology of Central France*, 1827, and *Extinct Volcanoes of*

Another well-known region of modern Puys is that of the Eifel, where the cones and craters are often so fresh that it is difficult to believe them to be prehistoric.[1] One of the most remarkable denuded puy-regions in Europe covers a wide territory in the Swabian Alps of Würtemberg. No fewer than 125 denuded necks filled with tuff, agglomerate and basalt have there been mapped and described. They are of higher antiquity than the Upper Miocene strata, and have thus probably been exposed to prolonged denudation. In external aspect and internal structure they present the closest parallel to the Carboniferous and Permian "necks" of Britain described in Books VI. and VII. of the present work.[2]

Among the Palæozoic volcanoes of Britain many admirable illustrations of the Puy type are to be found. Their cones are almost always entirely gone, though traces of them occasionally appear. The "necks" that show the position of the vents are in some districts crowded together as thickly as those of Auvergne. During the Carboniferous and Permian periods in Central Scotland, clusters of such little volcanoes must have risen among shallow lagoons and inland sheets of water, casting out their ashes and pouring forth their little streams of lava over the water-bottom around them and then dying out. As these eruptions took place in a region that was gradually subsiding, the cones and their ejected ashes and lavas were one by one submerged, the looser materials being washed down and spread out among the silt, sand or mud, and enveloping the remains of any plants or animals that might be strewn over the floor of the lake or sea. Hence the Puys of Palæozoic time in Britain have been preserved with extraordinary fulness of detail. They have been dissected by denudation, both among the hills of the interior and along the margin of the sea. Their structure can thus be in some respects made out even more satisfactorily than that of the much younger and more perfect cones of Central France.

iv. DETERMINATION OF THE RELATIVE GEOLOGICAL DATES OF ANCIENT VOLCANOES

In themselves, accumulations of volcanic materials do not furnish any exact or reliable evidence of the geological period in which they were erupted. The lavas of the early Palæozoic ages may, indeed, on careful examination, be distinguished from those of Tertiary date, but, as we have seen, the difference is rather due to the effects of age and gradual alteration than to any inherent fundamental distinction between them. In all essential particulars of composition and internal structure, the lavas of the Cambrian or Silurian period resemble those of Tertiary and modern volcanoes. The

Central France, 1858; Lecoq's *Époques Géologiques de l'Auvergne*, 1867; M. Michel Lévy, *Bull. Soc. Géol. France*, 1890, p. 688; M. Boule, *Bull. Carte Géol. France*, No. 28, tome iv. 1892.

[1] The Eifel district has been fully described by Hibbert, Von Dechen, and other writers. Von Dechen's little handbooks to the Eifel and Siebengebirge are useful guides.

[2] These Würtemberg vents have been elaborately described and discussed by Professor W. Branco of Tübingen in his *Schwabens 125 Vulkan-Embryonen und deren tufferfüllte Ausbruchsröhren, das grösste Gebiet ehemaliger Maare auf der Erde*, Stuttgart, 1894.

igneous magmas which supply volcanic vents thus appear to have been very much what they are now from early geological epochs. At least no important difference, according to relative age, has yet been satisfactorily established among them.

But although the rocks themselves afford no precise or trustworthy clue to their date, yet where they have been intercalated contemporaneously among fossiliferous stratified formations, of which the geological horizon can be determined from included organic remains, it is easy to assign them to their exact place in geological chronology. A determination of this kind is only an application of the general principle on which the sequence of the geological record is defined. A few illustrations will suffice to make this point quite obvious.

Among the volcanic tuffs in the upper part of Snowdon various fossils occur, which are identical with those found in the well-known Bala Limestone. As the accepted reading of such evidence, we conclude that these tuffs must therefore be of the same geological age as that limestone. Now the position of this seam of rock has been well established as a definite horizon in the series of Lower Silurian formations. And we consequently without hesitation place the eruptions of the Snowdon volcano on that same platform, and speak of them as belonging to the Bala division of the Lower Silurian period.

Again, in West Lothian the tuffs and lavas ejected from many scattered puys were interstratified among shales and limestones in which the characteristic fossils of the Carboniferous Limestone are abundant. There cannot, therefore, be any doubt that these eruptions were much younger than those of Snowdon, and that they took place at the time when the Carboniferous Limestone was being deposited. We thus speak of them as belonging to volcanoes which were active in that early part of the Carboniferous period to which the thick Mountain Limestone of Ireland and Derbyshire belongs.

As yet another illustration of the determination of geological age, an example from the plateau-type of eruption may be given. The great basalt-plateaux of Antrim and the Inner Hebrides are built up of lavas that lie unconformably on the Chalk. They are thus proved to be later than the Cretaceous system, and this deduction would hold true even if no organic remains were found associated with the volcanic rocks. But here and there, intercalated between the basalts, lie layers of shale, limestone and tuff containing well-preserved remains of plants which are recognizable as older Tertiary forms of vegetation. This fossil evidence definitely places the date of the eruptions in older Tertiary time.

It is clear that, proceeding on this basis of reasoning, we may arrange the successive volcanic eruptions of any given district, make out their order of sequence in time, and thus obtain materials for a consecutive history of them. Or, proceeding from that district into other regions, we may compare its volcanic phenomena with theirs, determine the relative dates of their respective eruptions, and in this way compile a wider history of volcanic action in past time. It is on these principles that the general and detailed chronology of the volcanic rocks of the British Isles has been worked out, and that the following chapters have been arranged.

V. HOW THE PHYSICAL GEOGRAPHY ASSOCIATED WITH ANCIENT VOLCANOES IS ASCERTAINED

While the materials erupted from old volcanic vents tell plainly enough their subterranean origin, they may leave us quite in the dark as to the conditions under which they were thrown out at the surface. Yet a careful examination of the strata associated with them may throw much light on the circumstances in which the eruptions took place. Many of the results of such examination will be given in subsequent chapters. I will here submit illustrations of how four different phases of physical geography during former volcanic eruptions may be satisfactorily determined.

1. *Submarine Eruptions.*—As by far the largest accessible part of the crust of the earth consists of old marine sediments, it is natural that the volcanic records preserved in that crust should be mainly those of submarine eruptions. That many lavas during the geological past were poured out upon the sea-bottom is plainly shown by the thick beds of marine organisms which they have overspread and which lie above them (Fig. 19). In Central Scotland, for example, sheets of basalt have flowed over a sea-bottom on which thick groves of crinoids, bunches of coral and crowds of sea-shells were living. Not less striking is the evidence supplied by bands of tuff. Around Limerick, for instance, the thick Carboniferous Limestone is interrupted by many thin layers of tuff marking intervals when showers of volcanic dust fell over the sea-bottom, killing off the organisms that lived there. But the limestone that overlies these volcanic intercalations is again crowded with fossils, proving that the crinoids, corals and shells once more spread over the place and flourished as abundantly as ever above the tuff.

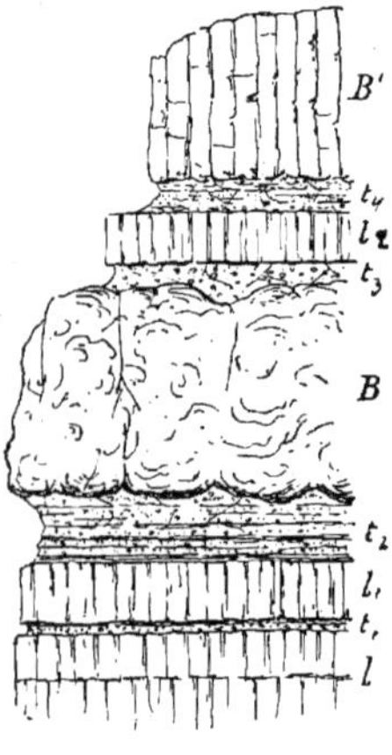

FIG. 19.—Section illustrating submarine eruptions; alternations of lavas and tuffs with limestones and shales full of marine organisms.

The accompanying diagram (Fig. 19) illustrates these statements. At the bottom a thick mass of limestone (*l*) full of crinoids, corals, brachiopods and other marine organisms bears witness to a long time of repose, when the clear sea-water teemed with life. At last a volcanic explosion took place, which threw out the first seam of tuff (*t*). But this was only a transient interruption, for the accumulation of calcareous sediment was immediately resumed, and the next band of limestone was laid down. Thereafter a more prolonged or vigorous eruption ejected a larger mass of dust and stones, which fell over the bottom and prevented the continuation of the limestone. But that the sea still abounded in life is shown by the numerous organisms imbedded in the second stratified band of tuff. At last an access of volcanic vigour gave vent to a stream of slaggy lava, which rolled over the sea-bottom and solidified in the thick sheet of amydaloidal basalt marked B. This outflow was followed by a further discharge of ashes and stones, which, from their absence of stratification, may be supposed to have been the result of a single explosion, or at least to have fallen too rapidly for the marine currents to rearrange them in layers. When the water cleared, the abundant sea-creatures returned, and from their crowded remains limestone once more gathered over the bottom. Yet the volcanic history had not then reached its close, for again there came a discharge of ashes, followed by the outpouring of a second lava, which consolidated as a sheet of columnar basalt (B′).

It is not necessary, in order to prove the eruptions to have been submarine, that organic remains should be found in the tuffs or between them. If the volcanic ejections are intercalated among strata which elsewhere can be proved to be marine, their discharge must obviously have taken place under the sea. The vent that discharged them may have raised its head above the sea-level, but its lavas and tuffs were spread out over the adjoining sea-floor.

2. *Lacustrine Eruptions.*—The same line of evidence furnishes proof that some volcanoes arose in inland sheets of water. If their products are interstratified among sandstones, gravels and shell-marls, wherein the remains of land-plants, insects and lacustrine shells, are preserved, we may be confident that the eruptions took place in or near to some lake-basin. The older lavas and tuffs of Central France supply an instructive example of such an association. In Britain, the abundant and extensive outpouring of lavas and tuffs during the time of the Lower Old Red Sandstone probably occurred in large lakes. Among the sediments of these bodies of water, interstratified between the volcanic sheets, remains of land-plants are abundant, together with, here and there, those of myriapods washed down from the woodlands, and of many forms of ganoid fishes.

3. *Fluviatile Eruptions.*—Volcanoes have sometimes arisen on river-plains or on the edges of valleys and gorges, and have poured out their lavas and discharged their ashes over the channels or alluvial lands of the streams. Volcanic materials, usurping the water-channels, bury or are interstratified with fluviatile sand or shingle, containing perhaps remains of the vegetation or animal life of the surrounding land. There may thus be a constant shifting of the river-courses, and a consequent deposit of fluviatile sediment at many successive levels among the lavas and tuffs. In Fig. 20 some

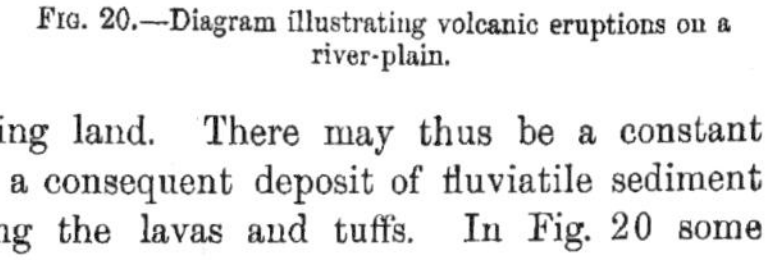

FIG. 20.—Diagram illustrating volcanic eruptions on a river-plain.

of these changes are indicated in a series of bedded lavas (*l*). The lower part of the diagram shows the dying out of a bed of river gravel (*g*) against the sloping end of a lava-stream, and the sealing up of this intercalation by a fresh outpouring of lava. Higher up in the diagram a section is shown of a gully or ravine which has been cut out of the lavas by a stream, and has become choked up with water-worn detritus. Subsequent outflows of lava have rolled over this channel and sealed it up. Examples of such intercalations of lava with old river deposits, and of the burying of water-courses, will be cited in the account of the Tertiary volcanic plateaux of Britain in Chapter xxxviii.

4. *Terrestrial Eruptions.*—That volcanoes in former times broke out on land as well as in water may readily be expected. But it is obvious that the proofs of a terrestrial origin may not be always easy to obtain, for every land-surface is exposed to denudation; and thus the relics of the eruptions of one age may be effaced by the winds, rains, frosts and rivers of the next. In assigning any volcanic group to a terrestrial origin, we may be guided partly by negative evidence, such as the absence of all trace of marine organisms in any of the sedimentary layers associated with the group. But such evidence standing by itself would not be satisfactory or sufficient. If, however, between the sheets of lava there occur occasional depressions, filled with hardened sediment full of land-plants, with possibly traces of insects and other terrestrial organisms, we may with some confidence infer that these silted-up hollows represent pools or lakes that gathered on the surface of the lava-sheets, and into which the vegetation of the surrounding ground was blown or washed. Rain falling on the rugged surface of a lava-field would naturally gather into pools and lakes, as the bottoms of the hollows became "puddled" by the gradual decay of the rock and the washing of fine silt into the crevices of the lava.

Again, it may be expected that prolonged exposure to the air would give rise to disintegration of the lava and to the consequent formation of soil. Terrestrial vegetation would naturally spring up on such soil; trees might take root upon it. Hence, if another lava-flood deluged the surface, the soil and its vegetable mantle would be entombed under the molten rock.

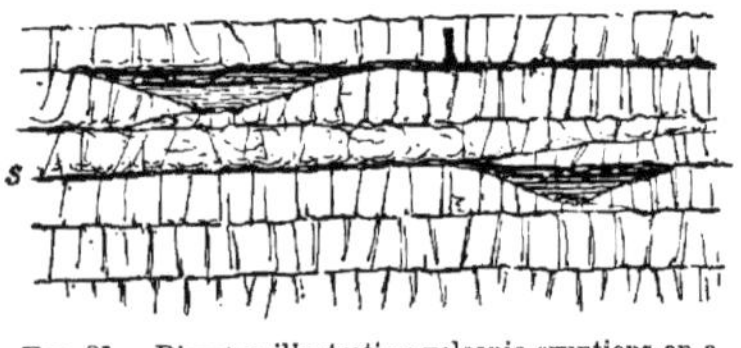

FIG. 21.—Diagram illustrating volcanic eruptions on a land-surface.

These geological changes are represented diagrammatically in Fig. 21. Two hollows among the lavas are there shown to have been filled with silt, including successive layers of vegetation now converted into coal. One of the soils (*s*) is marked between the lavas, and the charred stump of a tree with its roots still in another layer of soil higher up is seen to have been engulphed in the overlying sheet of melted rock.

Admirable illustrations of this succession of events are to be encountered among the great Tertiary basaltic plateaux which cover so large an area in the north-west of Europe. Not only has no trace of any marine organism been found among their interstratified sedimentary layers, but they have yielded a terrestrial flora which is preserved in hollows between the successive sheets of basalt. A full account of these rocks will be given in Book VIII.

CHAPTER V

Underground Phases of Volcanic Action. B. Materials injected or consolidated beneath the Surface—Intrusive Series: I. Vents of Eruption—i. Necks of Fragmentary Materials; ii. Necks of Lava-form Materials; iii. Distribution of Vents in relation to Geological Structure-Lines; iv. Metamorphism in and around Volcanic Cones, Solfataric Action; v. Inward Dip of Rocks towards Necks; vi. Influence of contemporaneous Denudation upon Volcanic Cones; vii. Stages in the History of old Volcanic Vents.

In our profound ignorance of the nature of the earth's interior, we know as yet nothing certain regarding the condition and distribution there of those molten materials which form the prime visible source of volcanic energy. By the study of volcanoes and their products we learn that the fused substances are not everywhere precisely the same and do not remain absolutely uniform, even in the same volcanic region. But in what manner and from what causes these variations arise is still unknown. We are further aware that the molten magma, under a centre of volcanic disturbance, manifests from time to time energetic movements which culminate in eruptions at the surface. But what may be the exciting cause of these movements, to what depth they descend, and over what extent of superficies they spread, are matters regarding which nothing better than conjecture can as yet be offered. It is true that, in some cases, a magma of fairly uniform composition has been erupted over a vast tract of the earth's surface, and must have had a correspondingly wide extent within the terrestrial crust. Thus in the case of the older Tertiary volcanic eruptions of North-Western Europe, basalt of practically the same composition was discharged from thousands of fissures and vents distributed from the south of Antrim northward beyond the Inner Hebrides, through the chain of the Faroe Islands and over the whole breadth of Iceland. Under the British Isles alone, the subterranean reservoirs of molten lavas must have been at least 40,000 square miles in united area. If they stretched continuously northwards below the Faroe Islands and Iceland, as is highly probable, that is, for 600 miles further, their total extent may have been comparable to such a region as Scandinavia.

Was this vast underground body of lava part of a universal liquid mass within the globe, or was it rather of the nature of one or more lakes or large vesicles within the crust? We can only offer speculation for answer. On the other hand, there seems to be good proof that in some districts, both now and in former geological periods, such differences exist between the materials ejected from vents not far distant from each other as to show the existence of more limited distinct reservoirs of liquid rock underneath.

Some of the questions here asked will be further dealt with in later pages in connection with such geological evidence as can be produced regarding them. But it will be found that at every step in the endeavour to ascertain the origin of volcanic phenomena difficulties present themselves which are now and may long remain insoluble.

I. Vents of Eruption

It is a general belief that the first stage in the formation of a volcano of the Vesuvian type by the efforts of subterranean energy is the rending of the terrestrial crust in a line of fissure. Some of the most remarkable groups of active volcanoes on the face of the globe are certainly placed in rows, as if they had risen along some such great rents. The actual fissure, however, is not there seen, and its existence is only a matter of probable inference. Undoubtedly the effect of successive eruptions must be to conceal the fissure, even if it ever revealed itself at the surface.

What is supposed to have marked the initial step in the formation of a great volcano is occasionally repeated in the subsequent history of the mountain. During the convulsive shocks that precede and accompany an eruption, the sides of the cone, and even sometimes part of the ground beyond, are rent open, occasionally for a distance of several miles, and on the fissures thus formed minor volcanoes are built up.

It is in Iceland, as already stated, that the phenomena of fissures are best displayed. There the great deserts of lava are from time to time dislocated by new lines of rent, which ascend up to the surface and stretch for horizontal distances of many miles. From these long narrow chasms lava flows out to either side; while cones of slag and scoriæ usually form upon them. This interesting eruptive phase will be more fully described in the chapters dealing with the Tertiary volcanic rocks of Britain.

There can be no doubt, however, that in a vast number of volcanic vents of all geological periods no trace can be discovered of their connection with any fissure in the earth's crust. Such fissures may indeed exist underneath, and may have served as passages for the ascent of lava to within a greater or less distance from the surface. But it is certain that volcanic energy has the power of blowing out an opening for itself through the upper part of the crust without the existence of any visible fissure there. What may be the limits of depth at which this mode of communication with the outer air is possible we do not yet know. They must obviously vary greatly according to the structure of the terrestrial crust on the one hand, and the amount and persistence of volcanic energy on the other. We may suppose that where a fissure terminates upward under a great depth of overlying rock, the internal magma may rise up to the end of the rent, and even be injected laterally into the surrounding parts of the crust, but may be unable to complete the formation of a volcano by opening a passage to the surface. But where the thickness of rock above the end of the fissure is not too great, the expansive energy of the vapours absorbed in the magma may overcome the resistance of that cover, and blow out an orifice by which the volcanic materials can reach the surface. In the formation of new cones within the historic period at a distance from any central volcano, the existence of an open fissure at the surface has not been generally observed. When, for example, Monte Nuovo was formed, it rose close to the shore among fields and gardens, but without the appearance of any rent from which its materials were discharged.

That in innumerable instances during the geological past, similar vents have been opened without the aid of fissures that reached the surface, will be made clear from the evidence to be drawn from the volcanic history of the British Isles. So abundant, indeed, are these instances that they may be taken as proving that, at least in the Puy type of volcanoes, the actual vents have generally been blown out by explosions rather than by the ascent of fissures to the open air.

In cases where, as in Iceland, fissures open at the surface and discharge lava there, the channel of ascent is the open space between the severed walls of the rent. Within this space the lava will eventually cool and solidify as a *dyke*. It is obvious that a comparatively small amount of denudation will suffice to remove all trace of the connection of such a dyke with the stream of lava that issued from it. Among the thousands of dykes belonging to the Tertiary period in the British Islands, it is probable that many may have served as lines of escape for the basalt at the surface. But it is now apparently impossible to distinguish between those which had such a communication with the outer air and those that ended upward within the crust of the earth. The structure of dykes will be subsequently discussed among the subterranean intrusions of volcanic material.

In an ordinary volcanic orifice the ground-plan is usually irregularly circular or elliptical. If that portion of the crust of the earth through which the vent is drilled should be of uniform structure, and would thus yield equally to the effects of the volcanic energy, we might anticipate that the ascent and explosion of successive globular masses of highly heated vapours would give rise to a cylindrical pipe. But in truth the rocks of the terrestrial crust vary greatly in structure; while the direction and force of volcanic explosions are liable to change. Hence considerable irregularities of ground-plan are to be looked for among vents.

Some of these irregularities are depicted in Fig. 22, which represents the ground plan of some vents from the Carboniferous volcanic districts of Scotland. They are all drawn on the same scale. Other examples will be cited in later chapters from the same and other parts of the British Isles.

Some of the most marked departures from the normal and simple type of vent occur where two orifices have been opened close to each other, or where the same vent has shifted its position (Figs. 29, 125, 205, and 214). Curiously irregular or elongated forms may thus arise in the resultant "necks" now visible at the surface. Many striking examples of these features may be seen among the Carboniferous and Permian volcanoes to be afterwards described. Occasionally where an open fissure has served as a vent it has given rise to a long dyke-like mass (No. 1 in Fig. 22).

The size of a volcanic vent may vary indefinitely from a diameter of not more than a yard or two up to one or two or more miles. As a rule, the smaller the vents the more numerously are they crowded together. In the case of large central volcanoes like Etna, where many subsidiary vents, some of them forming not inconsiderable hills, may spring up along the sides of the parent cone, denudation will ultimately remove all the material that was heaped up on the surface, and leave the stumps or necks of the parasitic vents in groups around the central funnel.

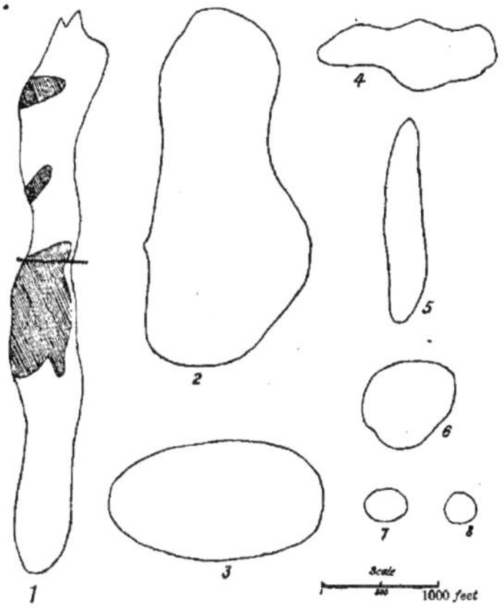

FIG. 22.—Ground-plans of some volcanic vents from the Carboniferous districts of Scotland.

1. Linhope Burn, near Mosspaul, Roxburghshire; the shaded parts are intrusions of trachytic material. 2. Hazelside Hill, two miles W. from Newcastleton, Roxburghshire. 3. St. Magdalen's, Linlithgow. 4. South-west side of Coom's Fell (see Fig. 174). 5. Neck on Greatmoor, Roxburghshire. 6. Pester Hill, Tarras Water. 7. Head of Routing Burn, S.E. side of Hartsgarth Fell, Liddesdale. 8. Hartsgarth Flow, Liddesdale.

Each volcanic chimney, by which vapours, ashes or lava are discharged at the surface, may be conceived to descend in a more or less nearly vertical direction until it reaches the surface of the lava whence the eruptions proceed. After the cessation of volcanic activity, this pipe will be left filled up with the last material discharged, which will usually take the form of a rudely cylindrical column reaching from the bottom of the crater down to the lava-reservoir. It will be obvious that no matter how great may be the denudation of the volcano, or how extensive may be the removal of the various materials discharged over the surrounding ground, the pipe or funnel with its column of solid rock must still remain. No amount of waste of the surface of the land can efface that column. Successively lower and yet lower levels may be laid bare in it, but the column itself goes still further down. It will continue to make its appearance at the surface until its roots are laid bare in the lava of the subterranean magma. Hence, of all the relics of volcanic action, the filled-up chimney of the eruptive vent is the most enduring. Save where it may have been of the less deep-seated nature of a "hornito" upon a lava-stream, we may regard it as practically permanent. The full meaning of these statements will be best understood from a consideration of the numerous illustrations to be afterwards given.

The stumps of volcanic columns of this nature, after prolonged denudation, generally project above the surrounding ground as rounded or conical eminences known as "Necks" (Fig. 23. See also Figs. 52, 82, 102, 109, 123, 133, 144, 178, 192, 195, 203, 204, 209, 294, 298, 306 and 310). Their outlines, however, vary with the nature of their component materials. The softer rocks, such as tuffs and agglomerates, are apt to assume the form of smooth domes or cones, while the harder and especially the crystalline rocks rise into irregular, craggy hills. Occasionally, indeed, it may happen that a neck makes no prominence on the surface of the ground, and its existence may only be discoverable by a careful examination of the geological structure of the locality. Now and then an old vent will be found not to

FIG. 23.—View of an old volcanic "Neck" (The Knock, Largs, Ayrshire, a vent of Lower Carboniferous age).

form a hill, but to sink into a hollow. Such variations, however, have little or no reference to original volcanic contours in the history of the localities which display them. They arise mainly from the differing hardness and structure of the materials that have filled the vents, and the consequent diversity in the amount of resistance which they have offered to the progress of denudation.

The materials now found in volcanic funnels are of two kinds: 1st, Fragmentary, derived from volcanic explosions; and 2nd, Lava-form, arising from the ascent and consolidation of molten rock within the funnel.

i. *Necks of Fragmentary Materials*

By far the most satisfactory evidence of a former volcanic orifice is furnished by a neck of fragmentary materials. Where "bosses" of crystalline rock rise to the surface and assume the outward form of necks, we cannot always be certain that they may not have been produced by subterranean intrusions that never effected any connection with the surface. In other words, such bosses may not mark volcanic orifices at all, though they may have been part of the underground protrusions of volcanoes in their neighbourhood. But where the chimney has been filled with debris, there can be no doubt that it truly marks the site of a once active volcano. The fragmentary material is an eloquent memorial of the volcanic explosions that drilled the vent, kept it open, and finally filled it up. These explosions could not have taken place unless the elastic vapours which caused them had found an escape from the pressure under which they lay within the crust of the earth. Now and then, indeed, where the outpouring of lava or some other cause has left cavernous spaces within the crust, there may conceivably be some feeble explosion there, and some trifling accumulation of fragmentary materials. But we may regard it as practically certain that the mass of tumultuous detritus now found in volcanic necks could not have been formed unless where a free passage had been opened from the molten magma underneath to the outer surface of the planet.

Considerable diversity may be observed in the nature and arrangement of the fragmentary materials in volcanic necks. The chief varieties may be arranged in four groups: (1) Necks of non-volcanic detritus; (2) Necks of volcanic agglomerate or tuff; (3) Necks of agglomerate or tuff with a central plug of lava; and (4) Necks of agglomerate or tuff with veins, dykes or some lateral irregular mass of lava.

(1) *Necks of non-volcanic Detritus.*—During the first convulsive efforts of a volcanic focus to find a vent at the surface, the explosions that eventually form the orifice do so by blowing out in fragments the solid rocks of the exterior of the terrestrial crust. Of the detritus thus produced, shot up the funnel and discharged into the air, part may gather round the mouth of the opening and build up there a cone with an enclosed crater, while part will fall back into the chimney, either to accumulate there, should the explosions cease, or to be thrown out again, should they continue. In the feeblest or most transient kinds of volcanic energy, the explosive vapours may escape without any accompanying ascent of the molten magma to the surface, and even without any sensible discharge of volcanic "ashes" from that magma. In such cases, as I have already pointed out, the detritus of the non-volcanic rocks, whatever they may be, through which volcanic energy has made an opening, accumulate in the pipe and eventually consolidate there. Examples of this nature will be adduced in later chapters from the volcanic districts of Britain.

Where only non-volcanic materials fill up a vent we may reasonably infer that the eruptions were comparatively feeble, never advancing beyond the initial stage when elastic vapours made their escape with explosive violence, but did not lead to the outflow of lava or the discharge of ashes. In the great majority of necks, however, traces of the earliest eruptions have been destroyed by subsequent explosions, and the uprise of thoroughly volcanic fragments. Yet even among these fragments, occasional blocks may be detected which have been detached from the rocks forming the walls of the funnel.

The general name of Agglomerate, as already stated, is given to all accumulations of coarse, usually unstratified, detritus in volcanic funnels, irrespective of the lithological nature of the materials. For further and more precise designation, when an agglomerate is mainly made up of fragments of one particular rock, the name of that rock may be prefixed as sandstone-agglomerate, granite-agglomerate, basalt-agglomerate, trachyte-agglomerate. Volcanic agglomerate is a useful general term that may include all the coarser detritus ejected by volcanic action.

Where volcanic explosions have been of sufficient violence or long continuance, the upper part of the funnel may be left empty, and on the cessation of volcanic activity, may be filled with water and become a lake. The ejected detritus left round the edge of the orifice sometimes hardly forms any wall, the crater-bottom being but little below the level of the surrounding ground. Explosion-lakes are not infrequent in Central France and the Eifel (Maare). A more gigantic illustration is afforded by the perfectly circular crater of Coon Butte in Arizona, about 4000 feet in diameter and 600 feet deep. It has been blown out in limestone, the debris of which forms a rampart 200 feet high around it. Examples will afterwards be cited from the Tertiary volcanic plateaux of North-Western Europe. Vents may also be formed by an engulphment or subsidence of the material, like that which has taken place at the great lava cauldron of Hawaii, still an active volcano. The picturesque Crater Lake of Oregon is an admirable instance of this structure.

(2) *Necks of Agglomerate or Tuff.*—In the vast majority of cases, the explosions that clear out a funnel through the rocks of the upper part of the crust do not end by merely blowing out these rocks in fragments. The elastic vapours that escape from the molten lava underneath are usually followed by an uprise of the lava within the pipe. Relieved from the enormous pressure under which it had before lain, the lava as it ascends is kept in ebullition, or may be torn into bombs which are sent whirling up into the air, or may even be blown into the finest dust by the sudden expansion of the imprisoned steam. If its ascent is arrested within the vent, and a crust is formed on the upper surface of the lava-column, this congealed crust may be disrupted and thrown out in scattered pieces by successive explosions, but may re-form again and again.

In many vents, both in recent and in ancient times, volcanic progress has never advanced beyond this early stage of the ejection of stones and dust. The column of lava, though rising near enough to the surface to supply by its ebullition abundant pyroclastic detritus, coarse and fine, has not flowed out above ground, nor even ascended to the top of the funnel. It may have formed, at the surface, cones of stones and cinders with enclosed craters. But thereafter the eruptions have ceased. The vents, filled up with the fragmentary ejected material, have given passage only to hot vapours and gases. As these gradually ceased, the volcanoes have become finally extinct. Denudation has attacked their sides

Fig. 24.—Section of neck of agglomerate, rising through sandstones and shales.

and crests. If submerged in the sea or a lake, the cones have been washed down, and their materials have been strewn over the bottom of the water. If standing on the land, they have been gradually levelled, until perhaps only the projecting knob or neck of solidified rubbish in each funnel has remained to mark its site. The buried column of compacted fragmentary material will survive as the only memorial of the eruptions (Fig. 24. For views of necks formed of agglomerate or tuff see Figs. 23, 82, 102, 123, 144, 178, 192, 203, 204, 209, 210, 212, 216).

The volcanic agglomerates of such vents sometimes include, among their non-volcanic materials, pieces of rock which bear evidence of having been subjected to considerable heat (see vol. ii. p. 78). Carbonaceous shales, for instance, have had their volatile constituents driven off, limestones have been converted into marble, and a general induration or "baking" may be perceptible. In other cases, however, the fragments exhibit no sensible alteration. Fossiliferous limestones and shales often retain their organic remains so unchanged that specimens taken out of the agglomerate cannot be distinguished from those gathered from the strata lying *in situ* outside. Some stones have evidently been derived from a deeper part of the chimney, where they have been exposed to a higher temperature than others, or they may have been lain longer within the influence of hot ascending vapours.

The volcanic materials in agglomerate range in size from the finest dust to blocks several yards in length, with occasionally even much larger masses. The proportions of dust to stones vary indefinitely, the finer material sometimes merely filling in the interstices between the stones, at other times forming a considerable part of the whole mass.

The stones of an agglomerate may be angular or subangular, but are more usually somewhat rounded. Many of them are obviously pieces that have been broken from already solid rock and have had their edges rounded by attrition, probably by knocking against each other and the walls of the chimney as they were hurled up and fell back again. Their frequently angular shapes negative the supposition that they could have been produced by the discharge of spurts of still liquid lava. As already stated, they have probably been in large measure derived from the violent disruption of the solidified cake or crust on the top of the column of lava in the pipe. Many of them may have been broken off from the layer of congealed lava that partially coated the rough walls of the funnel after successive uprises of the molten material. Among them may be observed many large and small blocks that appear to have been derived from the disruption of true lava-streams, as if beds of lava had been pierced in the formation of the vent, or as if those that congealed on the slopes of the cone had been broken up by subsequent explosions. These fragments of lava are sometimes strongly amygdaloidal. A characteristic feature, indeed, of the blocks of volcanic material in the agglomerates is their frequent cellular structure. Many of them may be described as rough slags or scoriae. These have generally come from the spongy crust or upper part of the lava where the imprisoned steam, relieved from pressure, is able to expand and gather into vesicles.

Less frequently evidence is obtainable that the blocks were partially or wholly molten at the time of expulsion. Sometimes, for example, a mass which presents on one side such a broken face as to indicate that it came from already solidified material, will show on the other that its steam-vesicles have been pulled out in such a way as to conform to the rounded surface of the block. This elongation could only take place in lava that was not yet wholly consolidated. It seems to indicate that such blocks were derived from a thin hardened crust lying upon still molten material, and that they carried up parts of that material with them. As each stone went whirling up the funnel into the open air, its melted part would be drawn round the gyrating mass, and would rapidly cool there.

In other cases, we encounter true volcanic bombs, that is, rounded or bomb-shaped blocks of lava, with their vesicles elongated all round them and conforming to their spherical shape. Sometimes such blocks are singularly vesicular in the centre, with a more close-grained crust on the outside. Their rapid centrifugal motion during flight would allow of the greater expansion of the dissolved steam in the central part of each mass, while the outer parts would be quickly chilled, and would assume a more compact texture. Bombs of this kind are met with among ancient volcanic products, and, like those of modern volcanoes, have obviously been produced by the ejection of spurts or gobbets of lava from the surface of a mass in a state of violent ebullition. Occasionally they are hollow inside, the rotation in these cases having probably been exceptionally rapid.

Passing from the larger blocks to the smaller fragments, we notice the great abundance of nut-like subangular or rounded pieces of lava in the agglomerates. These include lumps of fine grain not specially vesicular, and probably derived from the disruption of solidified rock. But in many agglomerates, especially those associated with the outpouring of basalts or other basic lavas (as those of Carboniferous and Tertiary age described in later chapters), they comprise also vast numbers of very finely cellular material or pumice. These pumiceous lapilli have been already alluded to as ingredients of the stratified tuffs. But they are still more characteristic of the necks, and reach there a larger size, ranging from the finest grains up to lumps as large as a hen's egg, or even larger.

The peculiar distinctions of this ejected pumice are the extreme minuteness of its vesicles, their remarkable abundance, their prevalent spherical forms, and the thinness of the walls which separate them. In these respects they present a marked contrast to the large irregularly-shaped steam-cavities of the outflowing lavas, or even of the scoriæ in the agglomerates.

This characteristic minutely vesicular pumice is basic in composition. Where not too much decayed, it may be recognized as a basic glass. Thus among the remarkable agglomerates which fill up the Pliocene or Pleistocene vents of the Velay, the fragments consist of a dark very basic glass, which encloses such a multitude of minute steam-cavities that, when seen under the microscope, they are found to be separated from each other by walls so thin that the slice looks like a pattern of delicate lace.[1] In necks of earlier date, such as those of older Tertiary, and still more of Palæozoic, time, the glass has generally been altered into some palagonitic material.

This finely pumiceous substance appears to be peculiar to the vents and to the deposits of tuff immediately derived from them. It is not found, so far as I know, among any of the superficial lavas, and, of course, would not be looked for among intrusive rocks. It was evidently a special product of the volcanic chimney, as distinguished from the mass of the magma below. We may perhaps regards it as in some way due to a process of quiet simmering within the vent, when the continual passage of ascending vapours kept the molten lava there in ebullition, and gave it its special frothy or finely pumiceous character.

The compacted dust, sand or gravelly detritus found in necks, and comprised under the general name of Tuff, consists partly of the finer particles produced during the violent disruption of already solidified rocks, partly of the detritus arising from the friction and impact of stones ascending and descending above an active vent during times of eruption, and partly of the extremely light dust or ash into which molten lava may be blown by violent volcanic explosions. In old volcanic necks, where the rocks have long been subjected to the influence of percolating meteoric water, it is not perhaps possible to discriminate, except in a rough way, the products from these three sources. The more minutely comminuted material has generally undergone considerable alteration, so that under the microscope it seldom reveals any distinctive structures. Here and there in a slide, traces may occasionally be detected of loose volcanic microlites, though more usually these can only be found in lapilli of altered glass or finely pumiceous lava.

The composition of the detritus in a neck of agglomerate or tuff has almost always a close relation to that of any lavas which may have been emitted from that vent. If the lavas have been of an acid character, such as rhyolites, felsites or obsidians, the pyroclastic materials will almost always be found to be also acid. Where, on the other hand, the lavas have been intermediate or basic, so also will be the tuffs and agglomerates. Occasionally, however, as has already been pointed out, from the same or closely adjoining vents lavas of very different chemical composition have been successively erupted. Felsites or rhyolites have alternated with diabases, basalts or andesites. In such cases, a commingling of acid and basic detritus may be observed, as, for example, among the volcanoes of the Old Red Sandstone. It has even happened sometimes that such a mixture of material has taken place when only one class of lavas has been poured out at the surface, as in the agglomerates that fill vents among the basalts of the Inner Hebrides. But we may be sure that, though not discharged at the surface, the lavas of which pieces are found in the tuffs must have risen high enough in the vents to be actually blown out in a fragmentary form. The occurrence of felsitic fragments among the otherwise basic

[1] M. Boule, *Bull. Cart. Géol. France*, No. 28, tome iv. (1892) p. 193.

agglomerates of Mull and Skye will be described in subsequent pages, likewise the intercalation of rhyolitic detritus between the basalts of Antrim. A similar association occurs among the modern vents of Iceland.

Among the contents of the tuffs and agglomerates that occupy old volcanic vents, some are occasionally to be observed of which the source is not easily conjectured. Detached crystals of various minerals sometimes occur abundantly which were certainly not formed *in situ*, but must have been ejected as loose lapilli with the other volcanic detritus. Where these crystals belong to minerals that enter into the composition of the lavas of the district in which they are found, they may be regarded as having probably been derived from the explosion of such lavas in the vents, the molten magma being blown into dust, and its already formed crystals being liberated and expelled as separate grains. But it seems to be extremely rare to find any neighbouring lava in which the minerals in question are so largely and so perfectly crystallized as they are in these loose crystals of the neck. The beautifully complete crystals of augite found in the old tuffs of Vesuvius and on the flanks of Stromboli may be paralleled among Palæozoic tuffs and agglomerates in Britain. Thus the necks belonging to the Arenig and Llandeilo volcanoes of southern Scotland are sometimes crowded with augite, varying from minute seed-like grains up to perfectly formed crystals as large as hazel nuts. The conditions under which such well-shaped idiomorphic minerals were formed were probably different from those that governed the cooling and consolidation of the ordinary lavas.

But besides the minerals that may be claimed as belonging to the volcanic series of a district, others occur not infrequently in some tuff-necks, the origin of which is extremely puzzling. Such are the large felspars, micas, garnets and the various gems that have been obtained from necks. The large size of some of these crystals and their frequently perfect crystallographic forms negative the idea that they can, as a rule, be derived from the destruction of any known rocks, though they may sometimes be conceivably the residue left after the solution of the other constituents of a rock by the underground magma, like the large residual felspars enclosed in some dykes. The crystals in question, however, seem rather to point to some chemical processes still unknown, which, in the depths of a volcanic focus, under conditions of pressure and temperature which we may speculate about but can perhaps hardly ever imitate in our laboratories, lead to the elaboration of the diamond, garnet, sahlite, smaragdite, zircon and other minerals.[1] Examples of such foreign or deep-seated crystals will be described from the probably Permian necks of Central Scotland.

Whatsoever may be the source and nature of the fragmentary materials that fill old volcanic vents, they present, as a general rule, no definite arrangement in the necks. Blocks of all sizes are scattered promiscuously through

[1] For lists of the minerals found in the diamond-bearing necks of Kimberley, see M. Boutan in Frémy's *Encyclopédie Chimique* (1886), vol. ii. p. 168; Dr. M. Bauer's *Edelsteinkunde* (1895), p. 223.

the agglomerate, just as they fell back into the chimney and came to rest there. The larger masses are placed at all angles, or stand on end, and are sometimes especially conspicuous in the centre of a neck, though more usually dispersed through the whole. Such a thoroughly tumultuous accumulation is precisely what might be expected where explosions have taken place in still liquid and in already consolidated lavas, and where the materials, violently discharged to the surface, have fallen back and come finally to rest in the chimney of the volcano.

Nevertheless, this absence of arrangement sometimes gives place to a stratification which becomes more distinct in proportion as the material of the vent passes from coarse agglomerate into fine tuff. It is possible that the existence and development of this structure depend on the depth at which the materials accumulate in the funnel. We may conceive, for instance, that in the lower parts of the chimney, the stones and dust, tumultuously falling and rebounding from projections of the rugged walls, will hardly be likely to show much trace of arrangement, though even there, if the explosions continue to keep an open though diminishing passage in the vent, alternations of coarser and finer layers, marking varying phases of eruptivity, may be formed in the gradually heightening pile of agglomerate. Rude indications of some such alternations may sometimes be detected in what are otherwise quite unstratified necks.

In the upper part of a volcanic funnel, however, close to and even within the crater, the conditions are not so unfavourable to the production of a stratified arrangement. As the pipe is filled up, and the activity of eruption lessens, explosions may occur only from the very middle of the orifice. The debris that falls back into the vent will gather most thickly round the walls, whence it will slide down to the central, still eruptive hole. It will thus assume a stratified arrangement, the successive layers lying at the steepest angles of repose, or from 30° to 35°, and dipping down in an inverted conical disposition towards the centre. If the process should continue long enough, the crater itself may be partially or completely filled up with detritus (Fig. 25).

Of this gradual infilling of a volcanic chimney with stratified agglomerate and tuff, examples belonging to different geological periods will be cited in subsequent chapters. I may here especially allude to one of the most recently observed and best marked illustrations, which occurs on the west side of Stromö, in the Faroe Islands (see Figs. 310, 311, 312). A neck has there been filled up with coarse agglomerate, which is rudely stratified, the layers dipping steeply into the centre, where the tumultuous assemblage of large blocks no doubt points to the final choking up of the diminished orifice of explosion. The walls of the neck are nearly vertical, and consist of the bedded basaltic lavas through which the vent has been opened. They terminate upward in a conical expansion, evidently the old crater, which has subsequently been filled up by the inroads of several lava-streams from adjacent vents. It is here manifest that the bedded agglomerate belongs to the uppermost part of the volcanic funnel.

Where vents have been filled up with tuff rather than with agglomerate, the stratified structure is best developed. Alternations of coarser and finer detritus give rise to more or less definite layers, which, though inconstant and irregular, serve to impart a distinctly stratified character to the mass. Where there has been no subsequent disturbance within a vent, these layers show the same inward dip towards the centre just referred to, at the ordinary angles of repose. Now and then, where a neck with

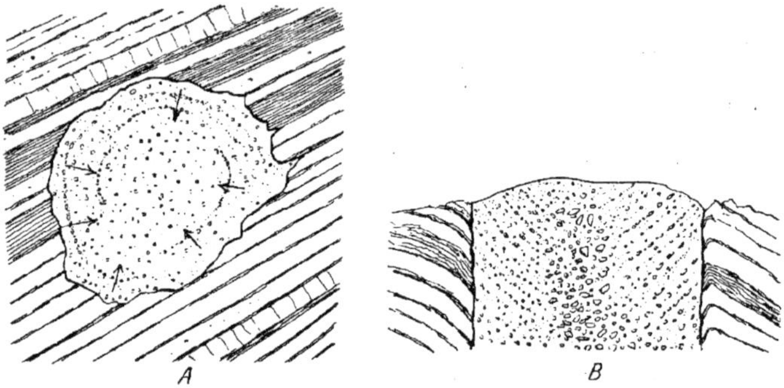

FIG. 25.—Neck filled with stratified tuff. A. ground plan; B. transverse section.

this structure has been laid bare on a beach, its denuded cross-section presents a series of concentric rings of strata from the walls towards the centre. Good illustrations of these features are supplied by the probably Permian necks of eastern Fife (Figs. 25 A and 217).[1]

It has frequently happened, however, that, owing to subsidence of the materials filling up the vents or to later volcanic disturbances, the compacted tuffs have been broken up and thrown into various positions, large masses being even placed on end. Among the Carboniferous and Permian necks of Central Scotland such dislocated and vertical tuffs are of common occurrence (see Figs. 145, 218). If, as is probable, we are justified in regarding the stratified parts of necks as indicative of the uppermost parts of volcanic funnels, not far from the surface, the importance of this inference will be best understood when the Carboniferous and Permian volcanoes are described.

(3) *Necks with a central Lava-plug.*—Some vents of agglomerate or tuff are pierced by a plug of lava, as may be instructively seen in many of the Carboniferous and Permian necks of the centre and south of Scotland (Fig. 26; compare also Figs. 148, 174, 207, and 226). Where this structure shows itself, the contrast in hardness and durability between the more destructible fragmentary material and the solid resisting lava leads to a topographical distinction in the outer forms of necks. The smooth declivities

[1] See also the sections of vents on the west coast of Stromö Faroes, above referred to.

of the friable tuffs are crowned or interrupted by more craggy features, which mark the position of the harder intrusive rock.

The plug, like the pipe up which it has risen, is in general irregularly circular in ground-plan. It may be conceived to be a column of rock, descending to an unknown depth into the interior, with a casing of pyroclastic debris surrounding it. It may vary considerably in the proportion which its cross-section bears to that of the surrounding fragmental material. Sometimes it does not occupy more than a small part of the whole, often appearing in the centre. In other cases, it more than equals all the rest of the material in the vent, while instances may be noted where only occasional patches of tuff or agglomerate are visible between the lava-plug and the wall of the pipe. From these we naturally pass to the second type of vent, where no fragmentary material is to be seen, but where the chimney is now entirely filled with some massive once-molten rock.

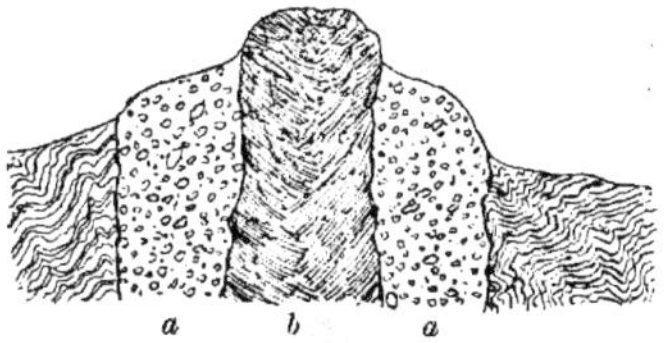

FIG. 26.—Section of neck of agglomerate (*a a*) with plug of lava (*b*).

A neck with a lava-plug probably contains the records of two stages in volcanic progress, the first of which, indicated by the tuff or agglomerate, was confined to the discharge of fragmentary materials; while the second, shown by the lava-plug, belonged to the time when, after the earlier explosions, lava ascended in the vent and solidified there, thus bringing the eruptions from that particular orifice to an end. Where a small central column of lava rises through the tuff, we may suppose that the funnel had been mainly choked up by the accumulation in it of ejected detritus, which was compacted to a solid mass adhering to the wall of the funnel, but leaving a central orifice to be kept open by the gradually waning energy of the volcano. By a final effort that impelled molten rock up that duct and allowed it to consolidate there, the operations of the vent were brought to a close.

Where, on the other hand, only occasional strips of tuff or agglomerate are to be found between the lava-plug and the wall of the pipe, the last uprise of lava may be supposed to have been preceded by more vigorous explosions which cleared the throat of the volcano, driving out the accumulated detritus and leaving only scattered patches adhering to the sides of the funnel.

There is, no doubt, some downward limit to the production of fragmentary material, and if we could lay bare successive levels in the chimney of a volcano we should find the agglomerate eventually replaced entirely by lava.

The materials of the lava-plugs vary widely in composition. Sometimes they are remarkably basic, and present rocks of the picrite or limburgite type; in other cases they are thoroughly acid rocks such as felsite and granophyre. Many intermediate varieties may be found between these extremes. It is noteworthy that, in districts where the lavas erupted to the surface have been andesitic or basaltic, the material which has finally solidified in the vents is often more acid in composition, trachytic rocks being specially frequent.

(4) *Necks with Dykes, Veins, or irregular intrusions of Lava.*—While the presence of a central plug of lava in a neck of fragmental material may indicate that the vent was still to some extent open, there is another structure which seems to point to the ascent of lava after the funnel has been choked up. Numerous instances have been observed where lava has been forced upward through rents in a mass of tuff or agglomerate, and has solidified there in the form of dykes or veins (Fig. 27). Illustrations of this structure abound among the Carboniferous and Permian necks of Britain. Here, again, though on a less marked scale, the contrast in the amount and character of the weathering of the two groups of rock gives rise to corresponding topographical features, which are especially observable in cliffs and coast-sections, where the dykes and veins project out of the tuffs as dark prominent walls (Figs. 135, 149, 166, 168, 219, 221, 222).

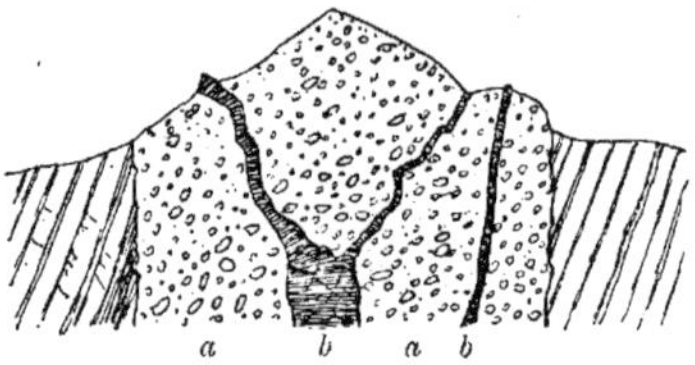

FIG. 27.—Section of agglomerate neck (*a a*) with dykes and veins (*b b*).

These intrusive injections are generally irregular in their forms, the lava having evidently been driven through a mass of material which, not having yet consolidated sufficiently to acquire a jointed structure, afforded few dominant lines of division along which it could ascend. Now and then, however, sharply defined dykes or veins, which at a distance look like dark ribbons, may be seen running vertically or at a high angle, and with a straight or wavy course, through the fine compacted tuff of a vent. Frequently the injected material has found its readiest line of ascent along the walls of the funnel, between the tuff and the surrounding rocks. Occasionally it has made its way into rents in these rocks, as well as into the body of the neck.

It is worthy of remark in passing that complete consolidation of the fragmentary material does not appear to be always requisite in order to allow of the formation of such fissures as are needed for the production of dykes. A singularly interesting illustration of this fact may be seen on the northern crest of the outer crater of the Puy Pariou in Auvergne. A dyke of andesite 8 or 10 feet broad may there be traced running for a distance of about 300 yards through the loose material of the cone. The rock is highly vesicular, and the vesicles have been elongated in the direction of the course of the dyke so as to impart a somewhat fissile structure to the mass.

There can be little doubt that the dykes and veins which traverse necks of agglomerate belong to one of the closing phases in the history of the vents in which they occur. They could only have been injected after the pipes had been so choked up that explosions had almost or entirely ceased, and eruptions had consequently become nearly or quite impossible. They show, however, that volcanic energy still continued to manifest itself by impelling the molten magma into these extinct funnels, while at the same time it may have been actively discharging materials from other still open vents in the same neighbourhood.

With regard to the composition of these dykes and veins, it may be remarked that in a district of acid lavas they may be expected to be felsitic or rhyolitic, sometimes granophyric. Where, on the other hand, the lavas poured out at the surface have been intermediate or basic, the veins in the necks may be andesites, basalts or other still more basic compounds. But it is observable, as in the case of the lava-plugs, that the injections into the necks may be much more acid than any of the superficial lavas. The advent of acid material in the later part of a volcano's history has been already alluded to, and many examples of it will be given in this work.

After all explosions and eruptions have ceased, heated vapours may still for a long period continue to make their way upward through the loose spongy detritus filling up the vent. The ascent of such vapours, and more particularly of steam, may induce considerable metamorphism of the agglomerate, as is more particularly noticed at p. 71.

ii. *Necks of Lava-form Material*

The second type of neck is that in which the volcanic pipe has been entirely filled up with some massive or crystalline rock. As already remarked, it is not always possible to be certain that bosses of rock, having the external form of necks of this kind, mark the sites of actual volcanic orifices. Eruptive material that has never reached the surface, but has been injected into the crust of the earth, has sometimes solidified there in forms which, when subsequently exposed by denudation, present a deceptive resemblance to true volcanic necks. Each example must be examined by itself, and its probable origin must be determined by a consideration of all the circumstances connected with it. Where other evidence exists of volcanic activity, such, for instance, as the presence of bedded tuffs or intercalated sheets of lava, the occurrence of neck like eminences or bosses of felsite, andesite, dolerite, basalt or other eruptive rock, would furnish a presumption that these marked the sites of some of the active vents of the period to which the tuffs and lavas belonged.

If a neck-like eminence of this kind were found to possess a circular or elliptical ground-plan, and to descend vertically like a huge pillar into the crust of the earth; if the surrounding rocks were bent down towards it and altered in the manner which I shall afterwards describe in detail; if, moreover, the material composing the eminence were ascertained to be closely related petrographically to some parts of the surrounding volcanic series, it might with some confidence be set down as marking the place of one of the active vents from which that series was ejected.

The chief contrast in external form between this type of neck and that formed of fragmentary material arises from differences in the relative durability of their component substance. The various kinds of lava-form rock found in necks are, as a whole, much harder and more indestructible than agglomerates and tuffs. Consequently bosses of them are apt to stand out more prominently. They mount into higher points, present steeper declivities, and are scarped into more rugged crags. But essentially they are characterized by similar conical outlines, and by rising in the same solitary and abrupt way from lower ground around them (see Figs. 109, 133, and 195, 294).

Various joint-structures may be observed in these necks. In some cases there is a tendency to separate into joints parallel to the bounding walls, and occasionally this arrangement goes so far that the rock has acquired a fissile structure as if it were composed of vertical strata. In other instances, the rock shows a columnar structure, the columns diverging from the outer margin, or curving inwards, or displaying various irregular groupings. More usually, however, this jointing is so indefinite that no satisfactory connection can be traced between it and the walls of the orifice in which the rock has solidified.

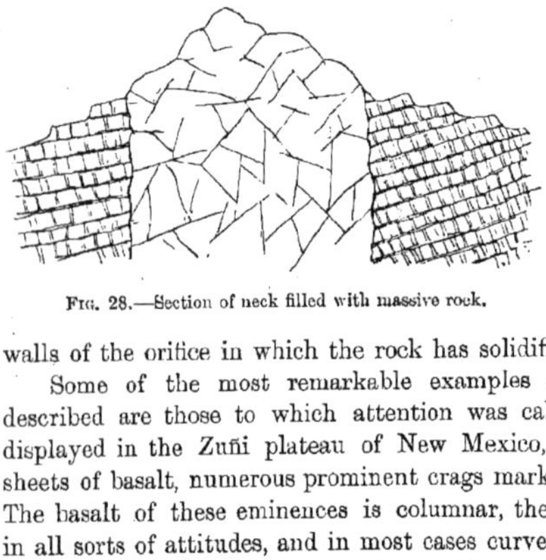

FIG. 28.—Section of neck filled with massive rock.

Some of the most remarkable examples of necks ever figured and described are those to which attention was called by Captain Dutton as displayed in the Zuñi plateau of New Mexico, where, amid wide denuded sheets of basalt, numerous prominent crags mark the sites of eruptive vents. The basalt of these eminences is columnar, the columns standing or lying in all sorts of attitudes, and in most cases curved.[1] In the Upper Velay, in Central France, numerous conspicuous domes and cones of phonolite rise amidst the much-worn basalt-plateau of that region (Fig. 345). Many instances will be cited in later chapters from the British Isles.

iii. *Distribution of Vents in Relation to Geological Structure-lines*

Where the positions of true volcanic necks can be accurately determined, it is interesting to study their distribution and their relation to the main lines of geological structure around them. Sometimes a distinct linear arrangement can be detected in their grouping. Those of the Lower Old Red Sandstone of Central Scotland, for instance, can be followed in lines for

[1] *U.S. Geol. Survey, 6th Annual Report*, 1884-85, p. 172.

distances of many miles (Map No. III). Yet when we try to trace the connection of such an arrangement with any known great lines of dislocation in the terrestrial crust, we can seldom establish it satisfactorily. In the case of the Scottish Old Red Sandstone just cited, it is obvious that the vents were opened along a broad belt of subsidence between the mountains of crystalline schist on the north, and those of convoluted Silurian strata on the south, either margin of that belt being subsequently, if not then, defined by lines of powerful fault. No vents have risen along these faults, nor has any relation been detected between the sites of the volcanic foci and dislocations in the area of ancient depression.

Indeed, it may be asserted of the vents of Britain that they are usually entirely independent of any faults that traverse at least the upper visible part of the earth's crust. They sometimes rise close to such lines of fracture without touching them, but they are equally well developed where no fractures are to be found. Now and then one of them may be observed rising along a line of fault, but such a coincidence could hardly fail occasionally to happen. From the evidence in the British Isles, it is quite certain that if volcanic vents have, as is possible, risen preferably along lines of fissure in the terrestrial crust, these lines are seldom those of the visible superficial faults, but must lie much deeper, and are not generally prolonged upward to the surface. The frequent recurrence of volcanic outbursts at successive geological periods from the same or adjacent vents seems to point to the existence of lines or points of weakness deep down in the crust, within reach of the internal molten magma, but far beneath the horizon of the stratified formations at the surface, with their more superficial displacements.

While sometimes running in lines, old volcanic vents of the Vesuvian and Puy types often occur also in scattered groups. Two or three may be found together within an area of a few hundred yards. Then may come an interval where none, or possibly only a solitary individual, may appear. And beyond that space may rise another sporadic group. These features are well exhibited by the Carboniferous and Permian sèries of Scotland, to the account of which the reader is referred.

A large neck may have a number of smaller ones placed around it, just as a modern Vesuvian cone has smaller parasitic cones upon its flanks. An instructive example of this arrangement is to be seen at the great vent of the Braid Hills belonging to the Lower Old Red Sandstone and described in Chapter xx. Other instances may be cited from the Carboniferous and Permian volcanic series (see Figs. 90, 148, 213).

Not infrequently the irregularities in the ground-plan of a neck, as already remarked, may be accounted for on the supposition that they mark the site of more than one vent. Sometimes, indeed, it is possible to demonstrate the existence of two or even more vents which have been successively opened nearly on the same spot. The first orifice having become choked up, another has broken out a little to one side, which in turn ceasing to be effective from the same or some other cause, has been succeeded by a third (Fig. 29). The three cones and craters of the little island of Volcanello supply a singularly perfect recent instance of this structure (Fig. 214). Here the funnel has twice shifted its position, each cone becoming successively smaller and partially effacing that which preceded it. In Auvergne, the Puy de Pariou has long been celebrated as an example of a fresh cinder-cone partially effacing an earlier one. In the much denuded Palæozoic volcanic tracts of Britain, where the cones have long since disappeared and only the stumps of the volcanic cylinders are left, many illustrations occur of a similar displacement of the funnel, especially among the volcanoes of the Carboniferous system.

Among the irregularities of necks that may indicate a connection with lines of fissure, reference may be made here to dykes or dyke-like masses of agglomerate which are sometimes to be seen among the volcanic districts of Britain. In these cases the fragmentary materials, instead of lying in a more or less cylindrical pipe, appear to fill up a long fissure. We may suppose that the explosions which produced them did actually occur in

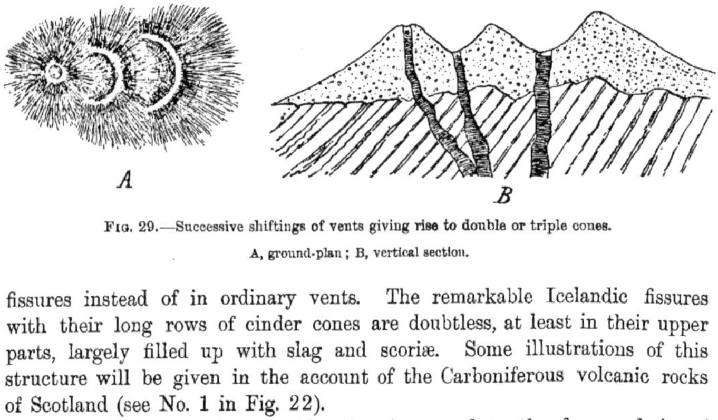

Fig. 29.—Successive shiftings of vents giving rise to double or triple cones.
A, ground-plan; B, vertical section.

fissures instead of in ordinary vents. The remarkable Icelandic fissures with their long rows of cinder cones are doubtless, at least in their upper parts, largely filled up with slag and scoriæ. Some illustrations of this structure will be given in the account of the Carboniferous volcanic rocks of Scotland (see No. 1 in Fig. 22).

There is yet another consideration in regard to the form and size of necks which deserves attention. Where the actual margin of a neck and its line of vertical junction with the rocks through which it has been drilled can be seen, there is no room for dispute as to the diameter of the original funnel, which must have been that of the actual neck. But in many cases it is impossible to observe the boundary; not merely because of superficial soil or drift, but occasionally because the volcanic detritus extends beyond the actual limits of the funnel. In such cases the necks have retained some portion of the original volcanic cone which accumulated on the surface around the eruptive vent. It may even chance that what appears to be a large neck would be considerably reduced in diameter, and might be shown to include more than one pipe if all this outer casing could be removed from it. In Fig. 30, for example, a section is given of a neck (n)

from which on the right-hand side all the cone and surrounding tuffs (*t*) have been removed by denudation, the original form of the volcano being suggested by the dotted lines. On the left side, however, the tuffs which were interstratified with the contemporaneous sediments are still connected with the neck, denudation not having yet severed them from it. The overlying strata (*l*, *l*) which originally overspread the extinct volcano have been

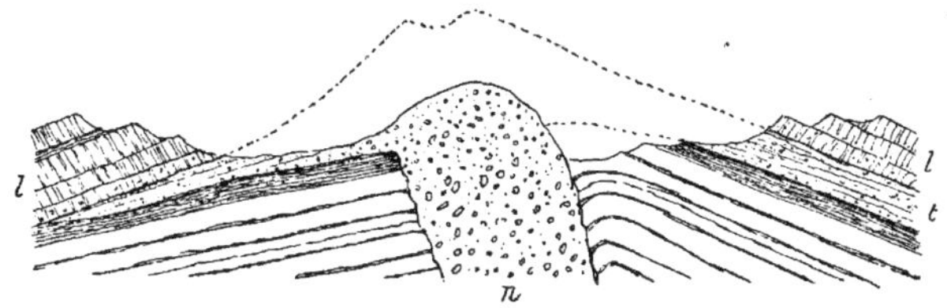

FIG. 30.—Section to show the connection of a neck with a cone and surrounding bedded tuffs.

bent into an anticline, and the neck of the vent has thus been laid bare by the removal of the crest of the arch.

The instances where a structure of this kind is concealed are probably fewer in number in proportion to their antiquity. But among Tertiary cones they may perhaps not be so rare. The possibility of their occurrence should be kept in view during the investigation of extinct volcanoes. The term Neck ought not properly to be applied to such degraded volcanic cones. The true neck still remains preserved in the inside of them. As illustrative of the structure here referred to, I may cite the example of the Saline Hill (Fig. 148) and of Largo Law (Fig. 226), both in Fife.

iv. *Metamorphism in and around Volcanic Vents—Solfataric Action*

The prolonged ascent of hot vapours, stones, dust and lava, in the funnel of a volcano must necessarily affect the rocks through which the funnel has been driven. We may therefore expect some signs of alteration in the material forming the walls of a volcanic neck. The nature of the metamorphism will no doubt depend, in the first place, on the character and duration of the agents producing it, and in the second, on the susceptibility of the rocks to undergo change. Mere heat will indurate rocks, baking sandstone, for instance, into quartzite, and shales into porcellanite. But there will almost invariably be causes of alteration other than mere high temperature. Water-vapour, for instance, has probably always been one of the most abundant and most powerful of them. The copious evolution of steam from volcanoes is one of their most characteristic features at the present day, and that it was equally so in past time seems to be put beyond question by the constantly recurring vesicular structure in ancient lavas and in the lapilli and ejected blocks of old agglomerates and tuffs. Direct experiment has demonstrated, in the hands of various skilful observers, from the time of Sir James Hall to that of Professor Daubrée, how powerfully rocks are acted upon when exposed to superheated vapour of water under great pressure. But the steam of volcanoes often contains other vapours or mineralizing agents dissolved in it, which increase its metamorphic influence. The mineral acids, for instance, must exert a powerful effect in corroding most minerals and rocks. At the Solfatara of Naples and at other volcanic orifices in different parts of Italy, considerable alteration is seen to be due to this cause.

Bearing these well-known facts in mind, we may be prepared to find various proofs of metamorphism around and within old volcanic vents. The surrounding rocks are generally much hardened immediately contiguous to a neck, whether its materials be fragmental or massive. Sandstones, for example, are often markedly bleached, acquire the vitreous lustre and texture of quartzite, lose their usual fissility, break irregularly into angular blocks, and on an exposed surface project above the level of the unaltered parts beyond. Shales are baked into a kind of porcelain-like substance. Coal-seams are entirely destroyed for economic purposes, having been burnt into a kind of cinder or fused into a blistered slag-like mass. Limestones likewise lose their usual bluish-grey tint, become white and hard, and assume the saccaroid texture of marble.

The distance to which this metamorphism extends from the wall is, among the exposed necks in Britain, smaller than might be anticipated. Thus I have seldom been able to trace it among those of Carboniferous or Permian age for more than 15 or 20 yards in ordinary arenaceous and argillaceous strata, even where every detail of a neck and its surroundings has been laid bare in plan upon a beach. The alteration seems to reach furthest in carbonaceous seams, such as coals.

It is evident that the element of time must enter into the question of the amount of metamorphism produced in the terrestrial crust immediately surrounding a volcanic pipe. A volcano, of which the eruptions begin and end within an interval of a few days or hours, cannot be expected to have had much metamorphic influence on the rocks through which its vent was opened. On the other hand, around a funnel which served for many centuries as a channel for the escape of hot vapours, ashes or lava to the surface, there could hardly fail to be a considerable amount of alteration. The absence or comparatively slight development of metamorphism at the Carboniferous and Permian necks of Scotland may perhaps be regarded as some indication that these volcanoes were generally short-lived. On the other hand, more extensive alteration may be taken as pointing to a longer continuance of eruptive vigour.

The same causes which have induced metamorphism in the rocks surrounding a volcanic vent might obviously effect it also among the fragmentary materials by which the vent may have been filled up. When the eruptions ceased and the funnel was left choked with volcanic debris, hot vapours and gases would no doubt still continue for a time to find their way upward through the loose or partially compacted mass. In their ascent they would permeate this material, and in the end produce in it a series of changes similar to, and possibly even more pronounced than, those traceable in the walls of the vent. Instances of this kind of metamorphism will be cited in the following chapters (see in particular p. 404).

v. *Inward Dip of Rocks towards Necks*

One concluding observation requires to be made regarding the relation of old volcanic necks to the rocks which immediately surround them. Where a vent has been opened through massive rocks, such as granite, felsite, andesite or basalt, it is generally difficult or impossible to determine whether there has been any displacement of these rocks, beyond the disruption of them caused by the explosions that blew out the orifice. But where the pipe has been drilled through stratified rocks, especially when these still lie nearly flat, the planes of stratification usually supply a ready test and measure of any such movement. Investigation of the volcanic rocks of Britain has shown me that where any displacement can be detected at a neck, it is almost invariably in a downward direction. The strata immediately around the vent tend to dip towards it, whatever may be their prevalent inclination in the ground beyond (Fig. 24). This is the reverse of the position which might have been expected. It is so frequent, however, that it appears to indicate a general tendency to subsidence at the sites of volcanic vents. After copious eruptions, large cavernous spaces may conceivably be left at the roots of volcanoes, and the materials that have filled the vents, losing support underneath, will tend to gravitate downwards, and if firmly welded to their surrounding walls may drag these irregularly down with them. Examples of such sagging structures are abundantly to be seen among the dissected vents of the Carboniferous and Permian volcanic series of Scotland.

vi. *Influence of Contemporaneous Denudation upon Volcanic Cones*

It must be remembered that former vents, except those of the later geological periods, are revealed at the surface now only after extensive denudation. As a rule, the volcanoes that formed them appeared and continued in eruption during periods of general subsidence, and were one by one submerged and buried beneath subaqueous deposits. We can conceive that, while a volcanic cone was sinking under water, it might be seriously altered in form and height by waves and currents. If it consisted of loose ashes and stones, it might be entirely levelled, and its material might be strewn over the floor of the sea or lake in which it stood. But, as has been already pointed out, the destruction of the cone would still leave the choked-up pipe or funnel from which the materials of that cone had been ejected. Though, during the subsidence, every outward vestige of the actual volcano might disappear, yet the agglomerate or lava that solidified in the funnel underneath would remain. And if these materials had risen some way within the cone or crater, or if they reached at least a higher level in the funnel than the surrounding water-bottom or land-surface, the destruction of the cone might leave a projecting knob or neck to be surrounded and covered by the accumulating sediments of the time. It is thus evident that the levelling of a cone of loose ashes during gradual subsidence, and the deposition of a contemporary series of sedimentary deposits, might give rise to a true neck, which would be coeval with the geological period of the volcano itself.

In practice it is extremely difficult to decide how far any now visible neck may have been reduced to the condition of a mere stump or core of a volcano before being buried under the stratified accumulations of its time. In every case the existence of the neck is a proof of denudation, and perhaps, in most cases, the chief amount of that denudation is to be ascribed not to the era of the original volcano, but to the comparatively recent interval that has elapsed since, in the progress of degradation, the volcanic rocks, after being long buried within the crust, were once more laid bare by the continuous waste and lowering of the level of the land.

vii. *Stages in the History of old Volcanic Vents*

Let us now try to follow the successive stages in the history of a volcano after its fires had quite burnt out, and when, slowly sinking in the waters of the sea or lake wherein it had burst forth, it was buried under an ever-growing accumulation of sedimentary material. The sand, mud, calcareous ooze, shell-banks, or whatever may have been the sediment that was gathering there, gradually crept over the submerged cone or neck, and would no doubt be more or less mixed with any volcanic detritus which waves or currents could stir up. If the cone escaped being levelled, or if it left a projecting neck, this subaqueous feature would be entombed and preserved beneath these detrital deposits. Hundreds or thousands of feet of strata might be laid down over the site of the volcano, which would then remain hidden and preserved for an indefinite period, until in the course of geological revolutions it might once again be brought to the surface.

These successive changes involve no theory or supposition. They must obviously have taken place again and again in past time. That they actually did occur is demonstrated by many examples in the British Isles. I need only refer here to the interesting cases brought to light by mining operations in the Dalry coalfields of Ayrshire, which are more fully described in Chapter xxvii. (p. 433). In that district a number of cones of tuff, one of which is 700 feet in height, have been met with in the course of boring and mining for ironstone and coal. The well-known mineral seams of the coalfield can be followed up to and over these hidden hills of volcanic tuff which in the progress of denudation have not yet been laid bare (Fig. 146).

The subsidence which carried down the water-bottom and allowed the volcanic vents to be entombed in sedimentary deposits may have been in most cases tolerably equable, so that at any given point these deposits would be sensibly horizontal. But subsequent terrestrial disturbances might seriously affect this regularity. The sedimentary formations, piled above each other to a great depth, and acquiring solidity by compression, might be thrown into folds, dislocated, upheaved or depressed. The buried volcanic funnels would, of course, share in the effects of these disturbances, and eventually might be so squeezed and broken as to be with difficulty recognizable. It is possible that some of the extreme stages of such subterranean commotions are revealed among the "Dalradian" rocks of Scotland. Certain green schists which were evidently originally sediments, and probably tuffs, are associated with numerous sills and bosses of eruptive material. The way in which these various rocks are grouped together strikingly suggests a series of volcanic products, some of the crushed bosses recalling the forms of true necks in younger formations. But they have been so enormously compressed and sheared that the very lavas which originally were massive amorphous crystalline rocks have passed into fissile hornblende-schists.

Among the Palæozoic systems of Britain, however, where considerable

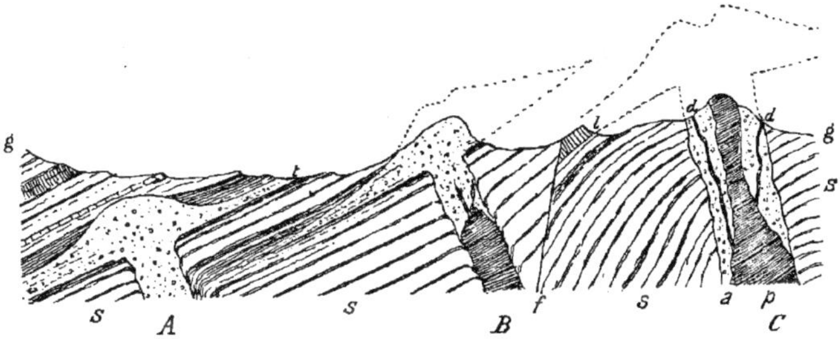

FIG. 31.—Diagram illustrating the gradual emergence of buried volcanic cones through the influence of prolonged denudation.

fracture and displacement have taken place, examples of successive stages in the reappearance of buried volcanic cones and necks may be gathered in abundance. As an illustrative diagram of the process of revelation by the gradual denudation of an upheaved tract of country, Fig. 31 may be referred to (compare also Fig. 147).

Here three volcanic vents are represented in different stages of re-emergence. In the first (A) we see a cone and funnel which, after having been buried under sedimentary deposits (*s, s*,) have been tilted up by subterranean movements. The overlying strata have been brought within the influence of denudation, and their exposed basset edges along the present surface of the land (*g, g*) bear witness to the loss which they have suffered. Already, in the progress of degradation, a portion of the volcanic materials which, ejected from that vent, were interstratified with the contemporaneous sediments of the surrounding sea-floor, has been exposed at *t*. A geologist coming to that volcanic intercalation would be sure that it pointed to the existence of some volcanic vent in the neighbourhood, but without further evidence he would be unable to tell whether it lay to right or left, whether it was now at the surface or lay still buried under cover of the stratified deposits which were laid down upon it.

In the second or central example (B) we have a pipe and cone which have been similarly disturbed. But in this case denudation has proceeded so far as to reveal the cone and even to cut away a portion of it, as shown by the dotted lines to the right hand. Owing, however, to the general inclination of the rocks towards the left, that side of the cone, together with the tuffs or lavas connected with it, still lies buried and protected under cover of the sedimentary formations (*s*, *s*).

The third example (C) shows a much more advanced stage of destruction. Here the whole of the cone has been worn away. All the lavas and tuffs which were ejected from it towards the right have likewise disappeared, and strata older than the eruptions of this vent now come to the surface there. To the left, however, a little portion of its lavas still remains at *l*, though all the intervening volcanic material has been removed. That solitary fragment of the outpourings of this volcano once extended further to the left hand, but the occurrence of the large dislocation (*f*) has carried this extension far down below the surface. The vent in this instance, owing to its position, has suffered more from denudation than the other two. Yet, judged by the size of its neck, it was probably larger than either of them, and threw out a more extensive pile of volcanic material. Its funnel has been filled with agglomerate (*a*), through which a central plug of lava (*p*) has ascended, and into which dykes or veins (*d*, *d*), the last efforts of eruption, have been injected.

This diagram will serve to illustrate the fact already so often insisted on, that although denudation may entirely remove a volcanic cone, and also all the lavas and tuffs which issued from it, the actual filled-up pipe cannot be so effaced, but is practically permanent.

CHAPTER VI

Underground Phases of Volcanic Action—*continued.* II. Subterranean Movements of the Magma: i. Dykes and Veins; ii. Sills and Laccolites; iii. Bosses (Stocks, Culots)—Conditions that govern the Intrusion of Molten Rock within the Terrestrial Crust.

II. Subterranean Movements of the Magma

In the foregoing pages attention has been more specially directed to those aspects of volcanic energy which reveal themselves above ground and in eruptive vents. We have now to consider the various ways in which the molten magma is injected into the crust of the earth.

Such injection must obviously take place during the expulsion of volcanic materials to the surface. If the explosive violence of an eruption, or the concomitant movements of the earth's crust, should lead to ruptures among the subterranean rocks, the molten magma will be forced into these rents. It is evident that this may happen either with or without any discharge of lava at the surface. It may be either entirely a plutonic, that is, a deep-seated phenomenon, or it may be part of a truly volcanic series of events.

It is clear that, by the study of old volcanoes that have had their structure laid bare by denudation, we may hope to obtain fresh light in regard to some of the more deeply-seated features of volcanic energy, which in a modern volcano are entirely concealed from view. A little reflection will convince us that the conditions for consolidation within the crust are so different from those at the surface that we may expect them to make themselves visible in the internal characters of the rocks.

An essential distinction between underground propulsions of molten rock and superficial outflows of the same material lies in the fact that while the latter are free to take any shape which the form and slope of the ground may permit, the subterranean injections, like metal poured into a mould, are always bounded by the walls of the aperture into which they are thrust. According, therefore, to the shape of this aperture a convenient classification of such intrusions may be made. Where the molten material has risen up vertical fissures or irregular cracks, it has solidified as Dykes and Veins. Where it has been thrust between the divisional planes either of stratified or unstratified rocks, so as to form beds, these are conveniently known as Sills, Laccolites or Intrusive Sheets. Where it has taken the form of large cylindrical masses, which, ascending through the crust, appear at the surface in rounded, elliptical or irregularly-shaped eminences, these are called Bosses (Stocks, Culots).

Further contrasts between the superficial and subterranean consolidation of molten material are to be found in the respective textures and minute structures of the rocks. The deep-seated intrusions are commonly characterized by a general and markedly greater coarseness of crystallization than is possessed by lavas poured out at the surface. This difference of texture, obviously in great measure the result of slower cooling, shows itself in acid, intermediate, and basic magmas. A lava which at the surface has cooled as a fine-grained, compact black basalt, in which neither with the naked eye nor with the lens can the constituent minerals be distinctly determined, may conceivably be represented at the roots of its parent volcano by a coarse-textured gabbro, in which the felspars and pyroxenes may have grown into crystals or crystalline aggregates an inch or more in length. Mr. Iddings has pointed out that the various porphyrites which form the dykes and sills of Electric Peak are connected with a central boss of coarsely crystalline diorite.[1] Examples of the same relation from different volcanic centres in Britain will be cited in later chapters.

This greater coarseness of texture is shown by microscopic examination to be accompanied by other notable differences. In particular, the glassy residuum, or its devitrified representatives, which may be so frequently detected among the crystals of outflowing lavas, is less often traceable in the body of subterreanean intrusive rocks, though it may sometimes be noticed at their outer margins where they have been rapidly chilled by contact with the cool upper part of the crust into which they have been impelled. Various minerals, the constituents of which exist in the original magma, but which may be hardly or not all recognisable in the superficial lavas, have had leisure to crystallize out in the deep-seated intrusions and appear sometimes among the components of the general body of the rock, or as well-terminated crystals in its drusy cavities.

Considerable though the variations may be between the petrographical characters of the intrusive and extrusive rocks of a given district and of the same eruptive period, they appear generally to lie within such limits as to suggest a genetic relation between the whole series. Conditions of temperature and pressure, and the retention or escape of the absorbed vapours which play so large a part in volcanic activity, must exercise great influence on the crystallization of constituent minerals, and on the consolidation and ultimate texture of the rocks. Slow cooling under great pressure and with the mineralizing vapours still largely retained seems to be pre-eminently favourable for the production of a holocrystalline texture in deep-seated portions of the magma, while rapid cooling under merely atmospheric pressure and with a continuous disengagement of vapours, appears to be required for the finer grain, more glassy structure, and more vesicular character of lavas poured out at the surface.

Besides these differences, however, there is evidence of a migration of

[1] *12th Ann. Rep. U.S. Geol. Survey* (1890-91), p. 595.

the constituent minerals in the body of large intrusive masses before consolidation. In particular, the heavier and more basic constituents travel towards the cooling margin, leaving the central portions more acid. This subject will be more fully considered in connection with the internal constitution of Bosses, and some British examples will then be cited.

Reference, however, may here be made to one of the most exhaustive and instructive studies of the relations of the subterranean and superficial erupted rocks of an old volcano, which will be found in the monograph by Mr. Iddings on Electric Peak and Sepulchre Mountain in the Yellowstone Park of Western America. From the data there obtainable he draws the deduction that one parent magma, retaining the same chemical composition, may result in the ultimate production of rocks strikingly different from each other in structure and mineralogical constitution, yet chemically identical. Electric Peak includes the central funnel filled up with coarsely crystalline diorite, and having a connected series of sills and dykes of various porphyrites. Sepulchre Mountain, separated from its neighbouring eminence by a fault of 4000 feet, displays some of the superficial discharges from the vent—coarse breccias with andesite-lavas. These rocks are not chemically distinguishable from the intrusive series, but the lavas are, on the whole, more glassy, while the materials of the bosses, sills and dykes are more crystalline. The latter display much more visible quartz and biotite.[1]

By practice in the field, supplemented by investigation with the aid of the microscope, a geologist acquires a power of discriminating with fair accuracy, even in hand specimens, the superficial from the subterranean igneous rocks of an old volcanic district.

Denudation, while laying bare the underground mechanism of an ancient volcano, has not always revealed the evidence of the actual structural relations of the rocks, or has first exposed and then destroyed it. Sometimes a mass of eruptive rock has been worn down and left in such an isolated condition that its connection with the rest of the volcanic network cannot be determined. So far as its position goes, it might perhaps be either a remnant of a lava-stream or the projecting part of some deeper-seated protrusion. But its texture and internal structure will often enable a confident opinion to be expressed regarding the true relations of such a solitary mass.

i. *Dykes and Veins*

For the study of these manifestations of volcanic energy, the British Isles may be regarded as a typical region. It was thence that the word "dyke" passed into geological literature. Thousands of examples of both dykes and veins may be seen from the Outer Hebrides southwards across the length and breadth of the southern half of Scotland, far into the north of England and towards the centre of Ireland. They may be found cut-

[1] *12th Ann. Rep. U.S. Geol. Survey*, 1890-91. As already stated, the eruptions of this volcanic centre became progressively more acid, and this change appears to be exhibited by the extrusive lavas as well as by the intrusive rocks.

ting the crests of the mountains and extending as reefs below the level of the sea. They are thus exposed in every conceivable divergence of position and in endless varieties of enclosing rock. Moreover, they can be shown to represent a vast range of geological time. One system of them belongs to some remote part of the Archæan periods, another is as young as the older Tertiary ages.

Full details regarding these interesting relics of volcanic activity will be given in later chapters, especially in Chapters xxxiv. and xxxv. It may suffice here to note that each of the three types of old volcanoes above described has, in Britain, its accompaniment of dykes and veins. The plateaux, however, present by far the most abundant and varied development of them. The dykes of this series are characterized not only by their prodigious numbers in and around some of the plateaux, but by the long distances to which they may be traced beyond these limits. They are chiefly found in

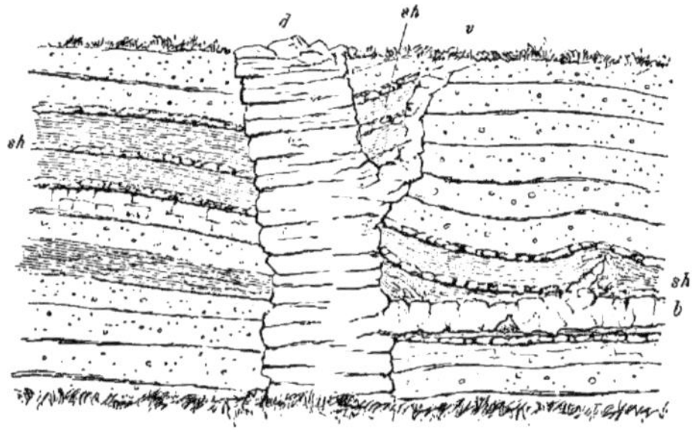

FIG. 32.—Dyke, Vein and Sill.

The dyke (*d*) rises along a small fault among sandstones, shales, and ironstones (*sh*), and gives off a vein (*v*) and an intrusive sheet or sill (*b*).

connection with the Tertiary basalt-plateaux, though the Carboniferous andesite-plateaux present a feebler display of them. The Tertiary dykes are pre-eminently distinguished by their persistent rectilinear lines, sometimes for distances of many miles, and their general north-westerly direction. They form a vast system extending over an area of some 40,000 square miles. Throughout that wide region their persistence of direction and of petrographical characters point to the former existence of one or more reservoirs of an andesitic and basaltic magma underneath the northern half of Britain, and to the rupture of the crust overlying this subterranean reservoir by thousands of parallel fissures. They thus constitute perhaps the most astonishing feature in the volcanic history of Tertiary time.

The dykes and veins connected with the puys are mainly to be found at or close to the vents. Not infrequently they traverse the agglomerates of the necks, and are sometimes to be traced to a central pipe or core of basalt.

The larger cones are likewise intersected with similar vertical, inclined or tortuously irregular walls of intruded lava. Occasionally a radiate arrangement may be observed in such cases, like that noticeable at some modern volcanoes, the dykes diverging from the eruptive centre.

Many dykes exist regarding which there is no evidence to connect them with any actual volcanic rocks. They have been injected into fissures, but whether this took place during volcanic paroxysms, or owing to some subterranean movements which never culminated in any eruption, cannot be decided.

The question of the age of dykes, like that of intrusive masses of all kinds, is often difficult or impossible to decide. A dyke must of course be younger than the rocks which it traverses, and a limit to its antiquity is thus easily fixed. But we cannot always affirm that because a dyke stops short of a particular rock, or series of rocks, it is older than these. The Hett Dyke, in the north of England, rises through the Coal-measures, but stops at the Magnesian Limestone; yet this cessation does not necessarily imply that the dyke was in place before the deposition of that limestone. The structure may have arisen from the dyke-fissure having ended at the bottom of the limestone. Where dykes rise up to the base of an unconformable formation without in any single case entering it, and where fragments of them are enclosed in that formation, they must be of higher antiquity, and must have been laid bare by extensive denudation before the unconformable strata were deposited upon them. The great system of dykes in the Lewisian Gneiss of the north-west of Scotland is in this way proved to be much more ancient than the Torridon Sandstones under which it passes (Figs. 35, 36).

Where two dykes cross each other, it is sometimes not difficult to decide upon their relative antiquity. In intrusive rocks, the finest-grained parts are those which lie nearest the outer margin, where the molten material was rapidly chilled by coming in contact with cool surfaces of rock. Such "chilled margins" of closer grain are common characteristics of dykes. Wherever a dyke carries its chilled margin across another dyke, it must be the younger of the two, and wherever such a margin is interrupted by another dyke, it must belong to the older.

As a rule, the uprise of molten material in a fissure has so effectually sealed it up that in the subsequent disturbances of the terrestrial crust the fissure has not been reopened, though others may have been produced near it, or across it. Sometimes, however, the enormous tension to which the crust was exposed opened the fissure once more, sometimes even splitting a dyke along its centre, and a new ascent of molten rock took place within the rent. Hence double or treble or compound dykes have been produced. The second or later infillings are generally somewhat different from the original dyke. Occasionally, indeed, they present a strong contrast to it. Thus, among the dykes of Skye examples occur where the centre is occupied by an acid granophyre, while the sides are occupied by dykes of basalt. Instances of this compound type of dyke will be given in the account of the Tertiary volcanic rocks of Britain.

It is obvious that in a wide fissure the central portion may remain molten for some time after the sides have consolidated. If the fissure served as a channel for the ascent of lava to the surface, it is conceivable that the central still fluid part might be driven out and be replaced by other material from below, and that this later material might differ considerably in composition from that which first filled the opening. Such, according to Mr. Iddings, has been the probable history of some of the dykes at the old volcano of Electric Peak.[1] But we can hardly suppose that this explanation of compound dykes can have any wide application. It could only hold good of broad fissures having an outlet, and is probably inadmissible in the case of the numerous compound dykes not more than 10 or 15 feet in diameter, where the several bands of rock are sharply marked off from each other. The abrupt demarcation of the materials in these dykes, their closer texture along their mutual boundaries, the indications of solution of the older parts of the group by the younger, and of injection of the latter into the former, show that they belong to separate and unconnected intrusions. These questions will be again referred to in the account of the British Tertiary dykes (Chapter xxxv. vol. ii. p. 159).

Another kind of compound dyke has arisen from the manner in which the original fissure has been produced. While, in general, the dislocation has taken the form of a single rectilinear rent, which on opening has left two clean-cut walls, cases occur where the rupture has followed several parallel lines, and the magma on rising into the rents appears as two or more vertical sheets or dykes, separated by intervening partitions of the surrounding rock. Examples of this structure are not infrequent among the Tertiary dykes of Scotland. One of these may be noticed rising through the cliffs of Lewisian gneiss on the east coast of the island of Lewis, south of Stornoway. One of the most extraordinary instances of the same structure yet observed is that described by Professor A. C. Lawson from the Laurentian rocks at the mouth of White Gravel River, on the N.E. coast of Lake Superior. In a breadth of only about 14 feet no less than 28 vertically intrusive sheets or dykes of diabase, from 1 inch to $6\frac{1}{2}$ inches broad, rise through the granite, which is thus split into 27 thin sheets. The diabase undoubtedly cuts the granite, some of the sheets actually anastomosing and sending veins into the older rock.[2]

From the evidence supplied by the modern eruptions of Iceland, it is evident that gaping fissures, which are filled by ascending lava and thereby converted into dykes, in many instances serve as channels by which molten rock escapes to the surface. It would be interesting if any test could be discovered whereby those dykes could be distinguished which had ever established a connection with the outer air. If the lava continued to ascend in the fissures, and to pour out in superficial streams for a long time, the rocks on either side would be likely to undergo considerably more metamorphism than where there was only one rapid injection of the magma, which would soon cool. Possibly in the much

[1] *12th Ann. Rep. U.S. Geol. Survey* (1890-91), p. 587.
[2] *American Geologist* (1894), p. 293.

greater alteration of the same rocks by some dykes than by others, a sign of such a connection with the surface may survive. This subject will be again referred to in the account of the Tertiary dykes of Britain in Book VIII., where the whole of the phenomena of this phase of volcanic action will be fully discussed (see vol. ii. p. 163).

ii. *Sills and Laccolites*

The word "sill," derived from a remarkable sheet of eruptive rock in the north of England, known as the Great Whin Sill (Chapter xxix.), is now applied as a convenient general term to masses of intrusive material, which have been injected between such divisional planes as those of stratification, and which now appear as sheets or beds (Fig. 33). These masses are likewise called Intrusive Sheets, and where the injected material has accumulated in large blister-like expansions, these are known as Laccolites (Fig. 34).

Sills vary from only an inch or two up to 500 feet or more in thickness. Lying, as they frequently do, parallel with strata above and below them, they resemble in some respects true lava-sheets erupted contemporaneously

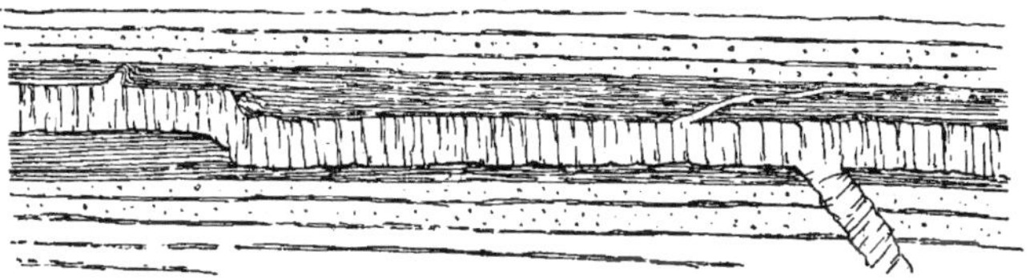

FIG. 33.—Section of Sill or Intrusive Sheet.

with the series of sediments among which they are intercalated. And, indeed, cases occur in which it is hardly possible to decide whether to regard a given mass as a sill or as a superficial lava. In general, however, sills exhibit the coarser texture above referred to as specially characteristic of subterranean eruptive masses. Moreover they are usually, though not always, free from the vesicular and amygdaloidal structures of true surface-lavas. Their under and upper surfaces, unlike the more scoriaceous parts of lavas, are commonly much closer in grain than the general body of the mass; in other words, they possess chilled borders, the result of more rapid consolidation by contact with cooler rock. Again, instead of conforming to the stratification of the formations among which they lie, as truly interstratified lavas do, they may be seen to break across the bedding and pursue their course on a higher or lower platform. The strata that overlie them, instead of enclosing pieces of them and wrapping round irregularities on their surface, as in the case of contemporaneously erupted lava-sheets, are usually indurated, sometimes even considerably altered, while in many cases they are invaded by veins from the eruptive sheet, or portions of them are involved in it, and are then much hardened or metamorphosed.

The petrographical character of the sills in a volcanic district depends primarily on the constitution of the parent magma, whence both they and the outflowing lavas have issued. Where the lavas are rhyolites or felsites the sills are acid, where basalts have been erupted the sills are basic, though there has often been a tendency towards the appearance of more acid material, such as trachyte. As we have seen, considerable differences in petrographical characters may arise between the intrusive and extrusive offshoots from the same parent magma during the course of a volcanic cycle. This question will be more appropriately discussed together with the leading characters of Bosses.

Between the upper and under surface of a thick sill considerable petrographical variation may sometimes be observed, especially where the rock is of basic constitution. Differences both of texture and even to some extent of composition can be detected. Sometimes what have been called "segregation veins" traverse the mass, consisting of the same minerals as the general body of the rock, but in larger crystals and in somewhat different proportions. That these veins belong to the period of original consolidation appears to be shown by the absence of fine-grained, chilled margins, and by the way in which the component crystals of the veins are interlocked with those of the body of the rock. Other veins of finer grain and more acid composition probably belong to a later phase of consolidation, when, after the separation and crystallization of the more basic minerals, the more acid mother liquor that remained was, in consequence of terrestrial movements, injected into cracks in the now solidified, though still highly heated, rock. Examples of these features will be cited from various geological formations in the following chapters.

Reference has already been made to the difference occasionally perceptible between the constitution of the upper and that of the under portions of superficial lavas. A similar variation is sometimes strongly marked among sills, especially those of a basic character, the felspars remaining most abundant above, while the olivines and augites preponderate below. Mr. Iddings has observed some excellent illustrations of this character in the great series of sills connected with the volcanic pipe of Electric Peak in the Yellowstone country.[1] Some examples of the same structure will subsequently be cited from the Carboniferous volcanic series of Central Scotland.

The greatest extreme of difference which I have observed in the petrographical characters of any group of sills is that displayed by the Tertiary gabbros of Skye. These rocks occur as sheets interposed among the bedded basalts, and injected between each other in such a manner as to form thick piles of rudely stratified sills. They possess a remarkable banded structure, due to the aggregation of their component minerals in distinct layers, some of which are dark in colour, from the abundance of their iron-ore, pyroxene and olivine; while others are light-coloured, from the predominance of their felspar. From the manner in which the component minerals of one band interlace with those of the contiguous bands, it is quite certain that the

[1] "Electric Peak and Sepulchre Mountain," *12th Ann. Rep. U.S. Geol. Survey* (1890-91), p. 584.

structure is not due to successive injections of material among already consolidated rocks, but belongs to the original conditions of expulsion of the gabbro as a whole. It seems to indicate that the magma which supplied the sills was at the time of its extrusion heterogeneous in composition, and that the banding arises from the simultaneous or rapidly successive protrusion of different portions of this variously-constituted magma. The details of the structure will be described in the general account to be given of the Tertiary volcanic rocks (Chapters xliii. and xliv.).

Besides such visible differences in the composition of sills, others much less obtrusive may occasionally be detected with the aid of microscopic or chemical research. The outer parts of some sills are thus discovered to be more basic or more acid than the inner portions. Or evidence may be obtained pointing to the probable melting down of surrounding rocks by the erupted magma, with a consequent local change in the chemical and mineralogical constitution of the mass.

In regard to their position in the geological structure of an old volcanic district I may here remark that sills, seldom entirely absent, are more especially developed either among the rocks through which the volcano has driven its vent, or about the base of the erupted lavas and tuffs. Many illustrations of this distribution will be described from the various volcanic areas of Britain belonging to Palæozoic and Tertiary time. At the base of the great Cambrian and Lower Silurian volcanic series of Merionethshire, sills are admirably developed, while among the basaltic eruptions which closed the long volcanic record in the north of Ireland and the Inner Hebrides, they play a notable part.

From the frequent place which sills take at the base of a volcanic series, it may be inferred that they generally belong to a late phase in the history of an eruptive episode or cycle, when the orifices of discharge had become choked up, and when the volcanic energy found an easier passage laterally between the strata underneath the volcanic pile or between the sheets of that pile itself, than upward through the ever-increasing thickness of ejected material.

While there is an obvious relation between most sills and some eruptive centre in their neighbourhood, cases occur in which no trace of any contemporaneous volcano can be found, but where the intrusive sheet remains as the sole evidence of the movements of the subterranean magma. The Great Whin Sill, one of the most extensive intrusive sheets in the British Isles, is an instance of this kind. Though this large mass of injected material can be traced for a distance of about 80 miles, and though the strata beneath and above it are well exposed in innumerable sections, no evidence has yet been detected to show that it was connected with any vent that formed a volcano at the surface (see vol. ii. p. 2). The absence of this evidence may, of course, arise from the failure of denudation to uncover the site of the vent, which may possibly still remain buried under the Carboniferous strata that overlie the sill towards the south-east. But it may be due to the non-existence of any such vent. We can quite conceive that volcanic energy should sometimes have failed to complete the formation of an actual volcano. Aided by subterranean movements, it might have been potent enough to disrupt the lower parts of the terrestrial crust, to propel the molten magma into fissures, even to inject it for many miles between the planes of stratification, which would be lines of least resistance, and yet in default of available rents, might have been unable to force its way through the upper layers and so reach the surface. Examples of such incompleted volcanoes are perhaps to be recognized among solitary sills, which not infrequently present themselves in the geological structure of Britain. But the positive decision of this question is almost always frustrated by the imperfection of the evidence, and the consequent possibility that a connected vent may still lie concealed under overlying strata.

Besides the more usual intrusions of molten material in the form of sheets of which the vertical thickness bears but a small proportion to the horizontal extent, there occur also large and thick cakes of intruded material in which the vertical thickness may approach, or perhaps even surpass, the horizontal diameter. These dome-shaped or irregular expansions form a connecting link between ordinary sills and the bosses to be subsequently described.

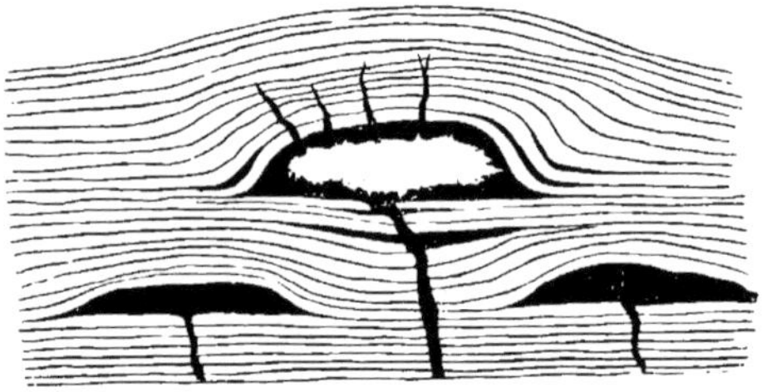

FIG. 34.—Ideal section of three Laccolites. (After Mr. Gilbert.)

They have received the name of *Laccolites* from Mr. G. K. Gilbert, who worked out this peculiar type of structure in the case of the Henry Mountains in southern Utah [1] (Fig. 34). The same type has since been found distributed over Arizona and Colorado, and it has been recognized as essentially that of many eruptive masses or bosses in all parts of the world.

In Western America, owing in large measure to the previously undisturbed condition of the sedimentary formations, the relations of the injected igneous material to these formations can be satisfactorily ascertained. The geological structure of the various isolated laccolites thus clearly presented, helps to explain the structure of other intrusive bodies which, having been injected among plicated and dislocated rocks, do not so readily admit of interpretation.

[1] "Geology of the Henry Mountains," *U.S. Geog. and Geol. Survey of the Rocky Mountain Region*, 1877. For a review of the whole subject of laccolites in Western America see a paper by Mr. Whitman Cross, in the 14*th Annual Report of the Director of the U.S. Geological Survey*, 1892-93 (pub. 1895), p. 157.

In Colorado, Utah and Arizona the eruptive magma, usually a porphyrite, diorite or quartz-porphyry, has risen in one or more pipes, and has then intruded itself laterally between the planes of the sedimentary formations which, over the centre of intrusion, have been pushed upward into a vast dome-shaped or blister-like elevation. The horizon on which this lateral and vertical expansion of the intruded material took place would seem to have lain several thousand feet below the surface. It ranges from the Cambrian to the Tertiary formations. Subsequent denudation has cut down the upraised mantle of sedimentary layers, and has revealed more or less of the igneous rock underneath, which is thus allowed to protrude and to be affected by atmospheric erosion. In this way, wide plains of horizontal or gently undulating Secondary and Tertiary strata have been diversified by the appearance of cones, detached or in groups, which have become more peaked and varied in outline in proportion as their original sedimentary covering has been removed from them. The largest of the laccolitic masses in the Henry Mountains is about 7000 feet deep and about 4 miles in diameter. Less than one-half of the cover of overarching strata has been removed, and denudation has cut deeply into the remaining part.

That the type of structure, so well exhibited among the Henry Mountains, has not been more abundantly recognized elsewhere probably arises from the fact not that it is rare, but that the conditions for its development are seldom so favourable as in Western America. Obviously where stratified rocks have been much disturbed, they cease to furnish definite or regular platforms for the reception of eruptive material, and to afford convenient datum-lines for estimating what was probably the shape of the intruded magma. We may believe that the effect of the propulsion of eruptive material is usually to upheave the overlying crust, and thus to give rise to a laccolitic form of intrusion. The upheaval relatively to the surrounding country will be apt to be practically permanent, the intruded body of rock being welded to the surrounding formations, and forming in this way a solid and resisting core directly united by pipes or funnels with the great magma-reservoir underneath. On the other hand, where the molten rock, instead of consolidating underground, has been copiously discharged at the surface, its emission must tend towards the production of cavernous spaces within the crust. The falling in of the roofs of such caverns will give rise to shocks of earthquakes. Subsequent uprisings of the magma may fill these spaces up, and when the rock has solidified in the form of laccolites or bosses, it may effectually put an end there to further eruptions.

Some contact metamorphism may be observed along the upper and under surfaces of large sills. The rocks over the American laccolites have sometimes been highly altered. But as the change is the same in kind as that attendant upon Bosses, though generally less in degree, it will be considered with these intrusive masses. The problems in terrestrial physics suggested by the intrusion of such thick and persistent masses of eruptive material as those which form sills and laccolites will likewise be discussed in connection with the mechanism of the remaining intrusive masses which have now to be described.

iii. *Bosses* (*Stocks*, *Culots*)

The term Boss has been applied to masses of intrusive rock which form at the surface rounded, craggy or variously-shaped eminences, having a circular, elliptical or irregular ground-plan, and descending into the terrestrial crust with vertical or steeply-inclined sides (Fig. 28). Sometimes they can be seen to have pushed the surrounding rocks aside. In other places they seem to occupy the place of these rocks through which, as it were, an opening has been punched for the reception of the intrusive material.

Occasionally, more especially in the case of large bosses, like those in which granite so frequently appears, the eruptive mass may be observed to rise here and there in detached knobs through the surrounding rocks, or to enclose patches of these, in such a manner as to indicate that the large body of eruptive material terminates upward in a very irregular surface, of which only the more prominent parts project through the cake of overlying rocks. In true bosses, unlike sills or laccolites, we do not get to any bottom on which the eruptive material rests. Laccolites, indeed, may be regarded as intermediate between the typical sill and the typical boss. The difference between a laccolite and a boss lies in the fact that the body of the laccolite does not descend into an unknown depth in the crust, but lies upon a platform on which it has accumulated, the magma having ascended by one or more ducts, which generally bear but a small proportion in area to the mass of the laccolite. The boss, on the other hand, is not known to lie on any horizon, nor to proceed from smaller ducts underneath, but plunges as a great pillar or irregular mass, which may frequently be noticed to widen downwards into the crust. There can be no doubt, however, that many masses of eruptive rock, which, according to the definition here given, should be called bosses, would be found to be truly laccolites if their structure below ground could be ascertained. It is obvious that our failure to find any platform on which the body of a boss lies, may arise merely from denudation having been as yet insufficient to lay such a platform bare. It is hardly probable that a boss several miles in diameter should descend as a column of that magnitude to the magma-reservoir from which its material came. More probably it has been supplied through one or more smaller ducts. The large boss now visible at the surface may thus be really a laccolitic expansion on one or more horizons. M. Michel Lévy lays stress on the general widening of granitic bosses as they descend into the crust.[1] While his observations are supported by many illustrations from all parts of the globe, and are probably true of the deeper-seated masses of granite, it is no less true that numerous examples have been met with where a granite boss is sharply marked off from the rocks which it has invaded and on which it may

[1] M. Michel Lévy, *Bull. Carte Géol. France*, No. 35, tome v. (1893), p. 32. The view stated in the text is also that adopted by Prof. Brögger with reference to the granite of the Christiania district. "Die Eruptivgesteine des Kristianiagebietes."

be seen to lie. Apart from the cases where granite seems to form part of a vast internal, once molten mass, into which its encircling gneisses seem to graduate, there are others in which this rock, as now visible, has been injected into the crust as a boss or as a laccolite. Instances will be described in later chapters where such bosses have risen through Cambrian, Silurian, Devonian and Carboniferous formations. It may be said that between such granitic intrusions and volcanic operations no connection can be traced. But reasons will be brought forward in later chapters to regard some of the granitic bosses as parts of the mechanism of Palæozoic volcanoes. It will also be shown that among the intrusive rocks of the Tertiary volcanic series of Britain there occur bosses of truly granophyric and granitic material. Hence, though mainly what is called a "plutonic" rock, granite has made its appearance among the subterranean protrusions of volcanoes.

It is no doubt true that many intrusive masses, which must be included under the general name of bosses, have probably had no connection whatever with volcanic action properly so called. They are plutonic injections, that is, portions of the subterranean magma which have been intruded into the terrestrial crust during its periods of disturbance, and have not been accompanied with any superficial discharges, which are essential in truly volcanic energy. It has been proposed to draw a distinction between such deep-seated intrusions and those which represent volcanic funnels.[1] If this were always practicable it would certainly be desirable. But the distinction is not one that can in every case be satisfactorily drawn. Even in regard to granitic bosses, which may generally be assumed to be plutonic in origin, the British examples just referred to have in all likelihood been connected with undoubted volcanic outbursts. Without, therefore, attempting here to separate the obviously volcanic necks of eruptive material from the probably plutonic bosses, I propose to describe briefly the general characters of bosses considered as a group of intrusive rocks, together with the phenomena which accompany them, and the conditions under which they may have been injected.

Bosses, whether of plutonic or volcanic origin, are frequently not merely single masses of eruptive rock, but are accompanied with a system of dykes and veins, some of which can be traced directly into the parent-mass, while others traverse it as well as the surrounding rocks. Hence the history of a boss may be considerably more complex than the external form of the mass might suggest.

The petrographical characters of bosses link them with the other underground injections of igneous material, more especially with sills and laccolites. Indeed, on mere lithological grounds no satisfactory line could be drawn between these various forms of intrusive rocks. The larger the mass the more coarsely crystalline it may be expected to be. But the whole range of structure, texture and composition, from those of the narrowest vein to those of the widest boss, constitutes one connected series of gradations.

Acid, intermediate and basic rocks are abundantly displayed among the

[1] M. Michel Lévy, *Bull. Carte Géol. France*, No. 35, tome v. (1893).

bosses. Huge masses of granite, granophyre, quartz-porphyry, felsite or rhyolite, represent the acid series. Intermediate varieties consist of trachyte, phonolite, diorite, andesite or other rock. The basic bosses include varieties of gabbro, dolerite, basalt, picrite, and other compounds.

In a boss of large size, a considerable range of texture, composition and structure may often be observed. The rock is generally much coarser in grain than that of thin sills or dykes. Sometimes it exhibits a finer texture along the margin than in the centre, though this variation is not usually so marked as in sills and dykes. The rapidly-chilled and therefore more close-textured selvage seems to have been developed much more fully in small than in large masses of eruptive material. The latter, cooling more slowly, allowed even their marginal parts to retain their heat, and sometimes perhaps even their molten condition, longer than small injections. Some influence must also have been exercised by the temperature of the rocks into which the eruptive material was intruded. Where this temperature was high, as in deep-seated parts of the crust, it would allow the intrusive magma to cool more slowly, and thus to assume a more coarsely crystalline condition. The absence of a close grain round the margins of granitic bosses may be due to this cause.

But a much more important distinction may be traced between the central and marginal parts of some large bosses and thick sills. I have already alluded to the fact that while the middle of a large intrusive mass may be decidedly acid, taking even the form of granite, the outer borders are sometimes found to be much more basic, passing into such a rock as gabbro, or even into some ultra-basic compound. Between these extremes of composition no sharp division is sometimes discoverable, such as might have been expected had the one rock been intruded into the other. The differences graduate so insensibly into each other as to suggest that originally the whole mass of the rock formed one continuous body of eruptive material. It is possible that in some cases the magma itself was heterogeneous at the time of intrusion.[1] But the frequency of the distribution of the basic ingredients towards the outer margin, and the acid towards the centre, points rather to a process of differentiation among the constituents of the boss before consolidation. In some instances the differentiation would appear to have taken place before crystallization to any great extent had set in, because the minerals ultimately developed in the central parts differ from those at the sides. In other cases, the transference of material would seem to have been in progress after the component minerals had crystallized out of the magma, for they are the same throughout the whole intrusive mass, but differ in relative proportions from centre to circumference.[2]

As illustrations of these features I may cite two good examples, one from Scotland and one from England. The mass of Garabol Hill, in the

[1] The Tertiary gabbros of the Inner Hebrides have already been cited, and will be more fully described in a later chapter as exhibiting the heterogeneousness of an eruptive magma.

[2] See Messrs. Dakyns and Teall, *Quart. Journ. Geol. Soc.* xlviii. (1892), p. 104; Prof. Brögger, *op. cit.* l. (1894), p. 15; Mr. A. Harker, *op. cit.* p. 320; Prof. Iddings, *Journ. Geol. Chicago*, i. (1893), p. 833; *Bull. Phil. Soc. Washington*, ii. (1890), p. 191; 1892, p. 89.

Loch Lomond district, consists mainly of granite, occupying an area of about 12½ square miles. Messrs. Dakyns and Teall have shown that while the central portions consist of granite, the south-eastern margin affords a remarkable series of intermediate rocks, such as hornblende-biotite-granite, tonalite (quartz-mica-diorite), diorite and augite-diorite, which lead us outwards into highly basic compounds, including wehrlites (olivine-diallage rocks), picrites (olivine-augite rocks), serpentine (possibly representing dunites, saxonites, and lherzolites), and a peculiar rock consisting essentially of enstatite, diallage, brown hornblende and biotite. The authors regard the whole of these widely different rocks as the products of one original magma, the more basic marginal area having consolidated first as peridotites, followed by diorites, tonalites and granites in the order of increasing acidity. The most acid rock in the whole series consists of felspar and quartz, is almost devoid of ferro-magnesian minerals, and occurs in narrow veins in the granite and tonalite. It indicates that after the segregation and consolidation of the whole boss, ruptures occurred which were filled in by the ascent of the very latest and most acid remaining portion of still fluid magma.[1]

The case of Carrock Fell in Cumberland has been described by Mr. A. Harker, who has ascertained that the gabbro of this boss has in its central portions a specific gravity of less than 2·85 and a silica-percentage sometimes as high as 59·46, whilst its marginal zone gives a specific gravity above 2·95 and a silica-percentage as low as 32·50. The migration of the heavy iron ores towards the margin is readily apparent to the naked eye, and is well established by chemical analysis, the oxides of iron amounting in the centre to 6·24 (Fe_2O_3 3·60, FeO 2·64), and at the margin to 25·54 (Fe_2O_3 8·44, FeO 17·10).[2] Neither in this instance nor in that of Garabol Hill has any evidence been noticed which would suggest that the basic and acid rocks belong to different periods of intrusion. They pass so insensibly into each other as to form in each case one graduated mass.

From these and other examples which have been observed, it is difficult to escape the conclusion that the differences between the basic margin and the acid centre are due to some process of segregation or differentiation while the mass was still in a liquid condition, and its constituents could pass from one part of the boss to another. According to Professor Brögger, it may be stated as a general law that differentiation sets in during consolidation, and is determined by, and dependent on, the laws of crystallization in a magma, in so far as the compounds which, on given conditions, would first crystallize out, diffuse themselves towards the cooling margin so as to produce in the contact-stratum a peculiar chemical composition in the still liquid material before crystallization takes place.[3]

If during the process of differentiation, and before consolidation, injections of the magma occur, they may be expected to differ in character according

[1] Messrs. Dakyns and Teall, *Quart. Journ. Geol. Soc.* xlviii. (1892), p. 104.

[2] Mr. A. Harker, *op. cit.* p. 320.

[3] This general conclusion is stated by Professor Brögger from his investigation of the rocks of Gran, *Quart. Journ. Geol. Soc.* l. (1894), p. 36.

to the portion of the magma from which they are derived. Professor Brögger believes that among the basic eruptive rocks of Gran in the Christiania district, one and the same magma has in the bosses solidified as olivine-gabbro-diabases, and in the dykes as camptonites, bostonites, pyroxenites, hornblendites, and more acid augite-diorites.[1]

Various opinions have been propounded as to the cause or causes of this so-called differentiation, but none of them are entirely satisfactory. We must await the results of further exploration in the field and of continued research in the laboratory.

What appears to have taken place within a subterranean molten magma which has been propelled into the earth's crust as a boss or laccolite, with or without a connected system of dykes, may possibly be made to throw some light on the remarkable changes in the characters of lavas successively erupted from the same vent during the continuance of a volcanic cycle. Whether or not any such process of differentiation can be proved to take place within a subterranean volcanic reservoir, the sequence of erupted lavas bears a curious resemblance to the order in which the constituents of some large bosses succeed each other from margin to centre. The earliest lavas may be of an intermediate or even basic character, but they generally tend to become more acid. Nevertheless alternations of basic and acid lavas which have been noted in various districts would seem to show that if there be a process of differentiation in the magma-basins, it is not regular and continuous, but liable to interruption and renewal. The return to basic eruptions, which so often marks the close of a volcanic cycle, is likewise not easily explicable on the supposition of continuous differentiation.

Where no sensible evidence of differentiation is traceable in the general body of a large intrusive mass, indications that some such process has there been in progress are perhaps supplied by the more acid dykes or veins, and the so-called "segregation veins," which have been already alluded to as traversing large intrusive masses. Though these portions differ to a greater or less extent in texture and composition from the main substance of the boss, the differences are not such as to prevent us from regarding them as really parts of the same parent magma. The veins, which are more acid than the rock that they traverse, may be regarded as having emanated from some central or deeper-seated part of a boss, which still remained fluid after the marginal or upper portion had consolidated sufficiently far to be capable of being rent open during subterranean disturbance. But that the mass, though coherent enough to be fissured, still remained at a high temperature, may be inferred from the general absence of chilled edges to these veins. The evidence of differentiation supplied by "segregation veins" has been referred to in the case of Sills.

The study of the petrographical variations in the constitution of large eruptive bosses has a twofold interest for the geologist. In the first place, it affords him material for an investigation of the changes which a volcanic magma undergoes during its eruption and consolidation, and thereby pro-

[1] *Quart. Journ. Geol. Soc.* l. (1894), p. 35.

vides him with some data for an elucidation of the cause of the sequence of erupted products during a volcanic cycle. In the second place, it yields to him some interesting analogies with the structures of ancient gneisses, and thus helps towards the comprehension of the origin and history of these profoundly difficult but deeply fascinating rocks.

Bosses, like sills, occur in the midst of volcanic sheets, and also as solitary protrusions. Where they rise amidst interstratified lavas and tuffs they may often be recognized as occupying the position of volcanic vents. They are then necks, and their characters in this connection have already been given. Where, however, as so frequently happens, they appear among rocks in which no trace of any contemporaneous volcanic material is to be detected, their relation to former volcanic activity remains uncertain.

Of this doubtful nature some of the most notable examples are supplied by the great granitic bosses which occur so frequently among the older Palæozoic rocks of Britain. The age of these can sometimes be approximately fixed, and is then found to correspond more or less closely with some volcanic episode. Thus the granite-bosses of Galloway, in the south of Scotland, disrupt Upper Silurian strata, but are older than the Upper Old Sandstone. Hence they probably belong to the period of the Lower Old Red Sandstone, which was eminently characterized by the vigour and long continuance of its volcanoes. The granite of Arran and of the Mourne Mountains can be shown by one line of reasoning to be younger than surrounding Carboniferous formations, by other arguments to be probably later than the Permian period, and by a review of the whole evidence to form almost certainly part of the volcanic history of Tertiary time.

But even where it can be shown that the uprise of a huge boss of eruptive material was geologically contemporaneous with energetic volcanic action, this coincidence may not warrant the conclusion that the boss therefore marks one of the volcanic centres of activity. Each example must be judged by itself. There have, doubtless, been many cases of the intrusion of molten material in bosses, as well as in sills, without the establishment of any connection with the surface. Such incompleted volcanoes have been revealed by denudation after the removal of a great thickness of superincumbent rock. The evidence which would have decided the question to what extent any of them became true volcanic vents has thus been destroyed. We can only reason tentatively from a careful collation of all the facts that are now recoverable. Illustrations of this kind of reasoning will be fully given in subsequent chapters.

It has been supposed that a test for the discrimination of a subterranean protrusion from a true volcanic chimney may be found in the condition of the surrounding rocks, which in the case of the prolonged flow of molten matter up a vent would be likely to undergo far more metamorphism than would be the case in the injection of a single eruptive mass.[1] But, as has been already pointed out, no special or excessive metamorphism of the encircling rocks is noticeable around many vents. There is certainly no

[1] See, for example, Mr. Harker, *Quart. Journ. Geol. Soc.* l. (1894), p. 329.

more alteration contiguous to numerous true necks than around bosses, which there is no reason to suppose ever communicated directly with the surface, and which were probably the result of a single intrusion. We must always remember that the denudation which has revealed these bosses has generally removed the evidence of their upward termination and of their possible connection with any volcanic ejections. Many of them may mark the sites of true vents from which only single eruptions took place. The opening of a volcanic vent does not necessarily imply a prolonged ascent of volcanic material. In a vast number of cases the original eruption was the first and last effort of the volcano, so that in such circumstances there seems no more reason for much alteration of the walls of the chimney than for the metamorphism of the rocks round a boss, laccolite, sill or dyke.

The metamorphism produced by intrusions of molten material upon the rocks with which they have come in contact has long been studied. Its amount varies so greatly in different cases that the conditions on which it has specially depended are not easily determined. Three factors have obviously been of great importance—first, the bulk of the intruded material; secondly, the chemical composition and lithological texture and structure of the rocks affected; and thirdly, the constitution and temperature of the invading magma.

1. It is clear that a huge boss of eruptive material will be likely to effect much more alteration of the surrounding rocks than a small boss, sill or dyke. Its initial temperature will probably be higher at the time of its assuming its final place than that of the same material after it has found its way into the narrower space of a thin sill or dyke. It will likewise take much longer to cool. Hence the influence of its heat and its vapours will continue to act long after those of the dyke or sill have ceased to manifest themselves.

2. It is equally evident that much of the resultant metamorphism will depend on the susceptibility of the rocks to change. An obdurate material such as pure quartz-sand, for example, will resist further alteration than mere hardening into quartzite. Shales and mudstones may be indurated into cherty substances of various textures. Limestones and dolomites, on the other hand, may become entirely crystalline, and may even have new minerals, such as garnet, tremolite, pyroxene, etc., developed in them. Hence in comparing the amount of metamorphism attendant on two separate bosses we must always take into account the nature of the rocks in which it has been induced.

3. But perhaps the most effective cause of variation in the nature and amount of contact metamorphism has been the constitution of the eruptive magma. A broad distinction may be drawn between the alteration produced by basic and by acid rocks. The intrusion of basic material has often produced singularly little change, even when the eruptive mass has been of considerable size. The greatest amount of alteration is to be found where the basic boss has caught up and enveloped portions of the surrounding rocks. Thus where the gabbro of Carrock Fell has invaded the basic

Lower Silurian lavas of the Lake District, the enveloped portions of the latter show considerable modification. Their groundmass becomes darker and more lustrous, the felspars assume a clearer appearance and lose some of their conspicuous inclusions, the pyroxenic constituents are converted into pale amphibole, and the glassy base disappears. At the actual line of contact the felspars of the lavas have become disengaged from their original matrix, which seems to have been dissolved and absorbed in the gabbro-magma. Brown mica has been exceptionally developed in the altered lava. At the same time, a change is noticeable in the character of the gabbro itself near the contact. Brown mica is there to be seen, though not a constituent of the rock elsewhere. The eruptive material has incorporated the basic groundmass of the lavas, leaving the felspars undissolved.[1]

Much more serious are the changes produced by intrusions of acid material, though here again the metamorphism varies within wide limits, being sometimes hardly perceptible, and in other cases advancing so far as to convert mere sedimentary material into thoroughly crystalline rocks. Small sills and dykes of felsite and granophyre may produce very slight change even upon shales and limestones, as may be seen among the eruptive rocks of Skye and Raasay. Large bosses of granophyre, and still more of granite, have been accompanied with the most extensive metamorphism. Round these eruptive masses every gradation may be traced among sandy and argillaceous sediments, until they pass into crystalline mica-schists, which do not appear to be distinguishable from rocks of Archæan age. Admirable examples of this extreme alteration may be observed around the great granite bosses of Galloway.[2] Again, among calcareous rocks a transition may be traced from dull grey ordinary fossiliferous limestones and dolomites into pure white crystalline marbles, full of crystals of tremolite, zoisite, garnet and other minerals. The alteration of the fossiliferous Cambrian limestones of Strath in Skye by the intrusive bosses of Tertiary granite well illustrates this change.[3]

Without entering further here into the wide subject of contact metamorphism, to which a large literature has now been devoted, we may note the effects which have been produced in the eruptive material itself by its contact with the surrounding rocks. Not only have these rocks been altered, but very considerable modifications have likewise taken place in the active agent of the change.

Sometimes the alteration of the invading material has been effected without any sensible absorption of the mineral constituents of the rocks invaded. This appears to be the case in those instances where sheets of basalt, intruded among coals or highly carbonaceous shales, have lost their compact crystalline character and have become mere clays. In the coal-

[1] Mr. Harker, *Quart. Journ. Geol. Soc.* vol. l. (1894), p. 331.

[2] See Explanation to Sheet 9 of the *Geological Survey of Scotland*, p. 22; Prof. Bonney and Mr. Allport, *Proc. Roy. Soc.* xvi. (1889); Miss Gardiner, *Quart. Journ. Geol. Soc.* vol. xlvi. (1890), p. 569.

[3] Macculloch, *Trans. Geol. Soc.* vol. iii. (1816), p. 1; *Description of the Western Isles*, vol. i. p. 322. See also *Quart. Journ. Geol. Soc.* vol. xiv. (1857), p. 1; and vol. xliv. (1888), p. 62.

fields of Britain, where many examples of this change have been noted, the igneous material is known as "white trap." The iron oxides have been in great part removed, or, together with the lime of the component minerals, have been converted into carbonates. Traces of the original felspar crystals may still be detected, but the groundmass has been changed into a dull, earthy, friable and decomposed substance.

Nearly always, however, the alteration of the intrusive magma has resulted from the incorporation of portions of the surrounding rocks. Reference has been made above to the alteration of the Carrock Fell gabbro by the absorption of some of the basic lavas around it. But still more remarkable is the change produced in some acid rocks by the incorporation of basic material into their substance. Professor Sollas has described in great detail a remarkable instance of this effect in the probably Tertiary eruptive rocks of the Carlingford district in the north-east of Ireland. He has ascertained that the eruptive gabbro of that district is older than the granite, for it is traversed by granophyre dykes which enclose pieces of it. The granophyre dykes, on the other hand, often show a lithoidal or chilled margin, which is not visible in the gabbro. He believes that the gabbro is not only older than the acid protrusions, but was already completely solid, traversed by contraction-joints, and probably fractured by earth-movements, before the injection of the granophyric material, which at the time of its intrusion was in a state of extreme fluidity, for it has found its way into the minutest cracks and crevices. He has especially studied the alteration produced by the granophyre upon the enclosed pieces of basic rock. The diallage, isolated from the other constituents of the gabbro, may commonly be seen to have broken up into numerous granules, like the augite grains of basalt, while in some cases biotite and hornblende have been developed with the concomitant excretion of magnetite. The acid rock itself has undergone considerable modification owing to the incorporation of basic material into its substance. Professor Sollas distinguishes the following varieties of the rock:—Biotite-granophyre, biotite-amphibole-granophyre, augite-granophyre, diallage-amphibole-augite-granophyre.[1]

Similar phenomena have been described by Mr. Harker as occurring where granophyre has invaded the gabbro of Carrock Fell.[2] The same observer has more recently detected some interesting examples furnished by injections of Tertiary granophyre in the agglomerates of Skye. The acid rock is roughly estimated by him to have taken up about one-fourth of its bulk of gabbro fragments. He has investigated the minute structure of the rock thus constituted, and has been able to recognize the augite of the original gabbro, in various stages of alteration and completely isolated, the other minerals having been dissolved in the acid magma.[3]

1 *Trans. Roy. Irish Acad.* xxx. (1894), part xii. p. 477.

2 *Quart. Journ. Geol. Soc.* li. (1895), p. 183.

3 *Op. cit.* lii. The metamorphism produced upon fragments of different kinds of foreign material enclosed within various igneous rocks has in recent years been studied in great detail by Professor Lacroix—*Les Enclaves des Roches Volcaniques*, Macon, 1893.

It is not easy to comprehend the conditions under which large masses of molten material have been injected into the crust of the earth. The two main factors in volcanic action—terrestrial contraction and the energy of the vapours in the magma—have no doubt played the chief part in the process. But the relative share of each and the way in which the enormous load of overlying rock has been overcome are not readily intelligible.

Let us first consider for a moment the pressure of the superincumbent crust under which the injection in many cases took place. The Whin Sill of England may serve as a good illustration of the difficulties of the problem. This notable mass of intrusive rock has been forced between the stratification planes of the Carboniferous Limestone series in one, or sometimes more than one, sheet. It stretches for a horizontal distance of not less than 80 miles with an average thickness of between 80 and 100 feet. From the area over which it can be traced its total extent underground must be at least 400 square miles (see Chapter xxix.).

In any single section the Whin Sill might be supposed to be a truly interstratified sheet, so evenly does it seem to be intercalated between the sedimentary strata. But here and there it diverges upward or downward in such a way as to prove it to be really a vast injected sheet. The age of the injection cannot be precisely fixed. It must be later than the Carboniferous Limestone. There is no trace of any stratigraphical break in the Carboniferous system of the region traversed by the sill. If the injection took place during the Carboniferous period, it does not appear to have been attended with any local disturbance, such as we might suppose would have been likely to accompany the extravasation of so enormous a mass of igneous material. If the date of injection be assigned to the next volcanic episode in the geological history of Britain—that of the Permian period—it will follow that the Whin Sill was intruded into its present position under the superincumbent weight of the whole of the Carboniferous system higher than the platform followed by the injected rock. The overlying body of strata would thus exceed 5000 feet in thickness, or in round numbers would amount at least to an English mile. The pressure of this mass of superincumbent material, at the depth at which the injected magma was forced between the strata, must have been so gigantic that it is difficult to believe that the energy of the magma would have been able to achieve of itself so stupendous a task as the formation of the Great Whin Sill.

The volume of injected material is likewise deserving of special attention. Many sills exceed 300 or 400 feet in thickness; and some laccolites must enormously surpass these limits. The intrusion of so vast a body of new material into the terrestrial crust will necessitate either a corresponding elevation of that part of the crust overlying the injected magma or a subsidence of that part underlying it, or some combination of both movements. It is conceivable that, where the body of protruded magma was large and the thickness of overlying crust was small, the expansive force of the vapours under high tension in the molten rock may have sufficed for the uplift. This result will be most likely to be effected around a volcanic chimney where the magma has the least amount of overlying load, and encounters that relief from pressure which enables it to become a powerful agent in terrestrial physics.

But in the case of the larger bodies of injected rock, especially where they do not seem to have been accompanied by the opening of any volcanic vents, the propulsion of the igneous material into the crust has probably been effected as a consequence of disturbance of the terrestrial crust. When the strain of contraction leads to the pushing upward of the terrestrial areas intervening between wide regions of subsidence, even though the differential movement may be slight, the isogeotherms undergo deformation. The intensely hot nucleus is squeezed upward, and if in the process of compression ruptures take place in the crust, and cavities in it are consequently opened, the magma will at once be forced into them. Such ruptures may be expected to take place along lines of weakness. Rocks will split along their stratification-planes, and the tendency to separation along these lines may be aided by the readiness of the energetic magma to find its way into and to enlarge every available opening. Hence we may expect that, besides vertical fractures, leading to the production of dykes and bosses, there will often be horizontal thrusts and ruptures, which will give rise to the formation of sills.

There is still another feature of terrestrial contraction which may help us to follow the behaviour of the magma within the crust. Plication of the crust is one of the most characteristic results of the contracting strain. Where a great series of sedimentary formations has been violently compressed so that its component strata have been thrown into rapid folds and squeezed into a vertical position, the portion of the crust thus treated may possibly be on the whole strengthened against the uprise of molten material through it. But the folding is often accompanied with dislocation. Not only are the rocks thrown into endless plications, but portions of them are ruptured and even driven horizontally over other parts. Such greatly disturbed areas of the crust are not infrequently found to have been plentifully injected with igneous rocks in the form of dykes, veins, sills, laccolites and bosses.

The elevation of a mountain-chain is known to be accompanied with a diminution of density in the crust underneath. Mr. O. Fisher has suggested that along such lines of terrestrial uplift there may be a double bulge in the crust, one portion rising to form the upheaved land and the other sinking down into the hot nucleus. If the lighter descending crust were there melted it might form a magma ready to be poured out as lava on the opening of any vent. The lava thus ejected would be of the lighter kinds. It has been remarked as certainly a curious fact that the lavas which issue from high mountain ranges are generally much more acid than the heavy basic lavas which are so characteristic of volcanoes close to the level of the sea.

But even where no actual mountain-chain is formed, there are gentle undulations of the crust which no doubt also affect the isogeotherms. If any series of disturbances should give rise to a double system of such undulations, one crossing the other, there would be limited dome-shaped elevations at the intersections of these waves, and if at the same time actual rupture of the crust should take place, the magma might find its way upward under such domes and give rise to the formation of laccolitic intrusions. Cessation of the earth-movements might allow the intruded material slowly to solidify without ever making an opening to the surface and forming a volcano. Doubtless many sills, laccolites and bosses represent such early or arrested stages in volcanic history.

Propelled into the crust at a high temperature, and endowed with great energy from the tension of its absorbed vapours and gases, the magma will avail itself of every rent which may be opened in the surrounding crust, and where it has succeeded in reaching the surface, its own explosive violence may enable it to rupture the crust still further, and open for itself many new passages. Thus an eruptive laccolite or boss is often fringed with veins, dykes and sills which proceed from its mass into the rocks around.

The question how far an ascending mass of magma can melt down its walls is one to which no definite answer can yet be given. Recent observations show that where the difference in the silica percentage between the magma and the rock attacked is great, there may be considerable dissolution of material from this cause. Allusion has already been made to Mr. Harker's computation that some of the acid granophyres of Skye have melted down about a fourth of their bulk of the basic gabbros. If such a reaction should take place between the magma of a boss, sill or laccolite and the rocks among which it has been intruded, great changes might result in the composition of the intruded rock. We are not yet, however, in possession of evidence to indicate that absorption of this kind really takes place on an extensive scale within the earth's crust. If it did occur to a large extent, we should expect much greater varieties in the composition of eruptive rocks than usually occur, and also some observable relation between the composition of the igneous material and that of the rocks into which it has been injected. But enough is not yet known of this subject to warrant any decided opinion regarding it.

CHAPTER VII

Influence of Volcanic Rocks on the Scenery of the Land—Effects of Denudation.

As considerable popular misapprehension exists respecting the part which volcanism has played in the evolution of the existing topography of the earth's surface, and as the British Isles, from their varied geological structure, offer special facilities for the discussion of this subject, it may not be out of place to devote a final section of the present Introduction to a consideration of the real topographical influence of volcanic action.

With modern, and especially with active, volcanoes we need not here concern ourselves. Their topographical forms are well known, and give rise to no difficulty. The lofty cones of the Vesuvian type, with their wide-spread lavas and ashes, their vast craters and their abundant parasitic volcanoes; the crowded, but generally diminutive, cones and domes of the puy type, so well displayed in Auvergne, the Eifel and the Bay of Naples; and the vast lava deserts of the plateaux, so characteristically developed in Iceland and Western America, illustrate the various ways in which volcanic energy directly changes the contours of a terrestrial surface.

But the circumstances are altered when we deal with the topographical influence of long extinct volcanoes. Other agencies then come into play, and some caution may be needed in the effort to disentangle the elements of the complicated problem, and to assign to each contributing cause its own proper effect.

Reference has already been made to the continuous denudation of volcanic hills from the time that they are first erupted. But the comparative rapidity of the waste and the remarkable topographical changes which it involves can hardly be adequately realized without the inspection of an actual example. A visit to the back of Monte Somma, already alluded to, will teach the observer, far more vividly than books can do, how a volcanic cone is affected by daily meteoric changes. The sides of such a cone may remain tolerably uniform slopes so long as they are always being renewed by deposits from fresh eruptions. But when the volcanic activity ceases, and the declivities undergo no such reparation, they are rapidly channelled by the descent of rain-water, until the furrows grow by degrees wide and deep ravines, with only narrow and continually-diminishing crests between them. If unchecked by any fresh discharge of volcanic material, the degradation will at last have removed the whole cone.

It is thus obvious that purely volcanic topography, that is, the terrestrial scenery due directly to the eruption of materials from within the earth, can never become in a geological sense very old. It can only endure so long as it is continually renewed by fresh eruptions, or where it is carried down by subsidence under water and is there buried under a cover of protecting sediments. When, therefore, we meet with volcanic rocks of ancient date exposed at the surface, we may be quite certain that their present contours are not those of the original volcano, but have been brought about by the processes of denudation.

It is true that, in the general erosion of the surface of the land, volcanic rocks of ancient date sometimes rise into wonderfully craggy heights, including, perhaps, cones and deep crater-like hollows, which to popular imagination betoken contours left by now extinguished volcanic fires. Examples of such scenery are familiar in various parts of Britain; but the resemblance to recent volcanic topography is deceptive. There are, indeed, a few hills wherein the progress of denudation seems not as yet to have entirely removed the lavas and tuffs that gathered round the original vents. Some of the tuff-cones of eastern Fife, for example, present cases of this kind. Again, the great granophyre domes and cones of the Tertiary volcanic series of the Inner Hebrides, though they have undoubtedly been extensively denuded, may possibly retain contours that do not greatly differ from those which these protruded bosses originally assumed under the mass of rock which has been removed from them. Nevertheless, putting such doubtful exceptions aside, we may confidently affirm that hills composed of ancient volcanic material give no clue to the forms of the original volcanoes.

It can hardly be too often repeated that the fundamental law in the universal decay and sculpture of the land is that the waste is proportioned to the resistance offered to it: the softer rocks are worn down with comparative rapidity, while the harder varieties are left projecting above them. As a general rule, volcanic rocks are more durable than those among which they are interstratified, and hence project above them, but this is not always the case. No universal rule can, indeed, be laid down with regard to the relative durability of any rocks. While, therefore, topographic contours afford a valuable indication of the nature and disposition of the rocks below the surface, they cannot be relied upon as in all circumstances an infallible guide in this respect. No better proof can be offered of the caution that is needed in tracing such contours back to their origin than is furnished by the old volcanic rocks of Britain. These eruptive masses, consisting usually of durable materials and ranging through a vast cycle of geological time, usually rise into prominent features and thus support the general law. But they include also many easily eroded members, which, instead of forming eminences, are worn into hollows. They include, in short, every type of scenery, from featureless plains and rolling lowlands to craggy and spiry mountains.

The first point, then, which is established in an investigation of the topographical influence of old volcanic rocks is that their prevailing prominence arises from relative durability amidst universal degradation. When we proceed further to inquire why they vary so much from each other in different places, and how their complicated details of feature have been elaborated, we soon learn that such local peculiarities have arisen mainly from variations in the internal structure and grouping of the rocks themselves.

Here again the general law of sculpture comes into play. The local features have depended upon the comparative resistance offered to the sculpturing agents by the different portions of a volcanic series. Each distinct variety of rock possesses its own characteristic internal structure. The lines along which atmospheric disintegration will most effectually carry on its carving work are thus already traced in the very substance and architecture of the rock itself. Each rock consequently yields in its own way to the processes of disintegration, and thus contributes its own distinctive share to topographical feature.

Among the massive rocks abundant examples of such special types of weathering may be cited, from the acid and basic series, and from superficial lavas as well as from intrusive bosses and sills. Acid bosses, such as those of granite, granophyre and quartz-porphyry, tend to weather into blocks and finally into sand, and as this tendency is somewhat uniformly distributed through the rocks, they are apt to assume rounded, dome-shaped or conical forms which, at a distance, may seem to have smooth declivities, but on examination are generally found to be covered with a slowly-descending sheet of disintegrated blocks and debris (Fig. 346). When less prone to decay, and especially where traversed by a strongly-defined system of vertical joints, they may shoot up into tower-like heights, with prominent spires and obelisks. Basic bosses, when their materials decay somewhat rapidly, give rise to analogous topographical forms, though the more fertile soils which they produce generally lead to their being clothed with vegetation. Where they consist of an obdurate rock, much jointed and fissured, like the gabbro of the Inner Hebrides, they form exceedingly rugged mountains, terminating upward in serrated crests and groups of aiguilles (Figs. 331, 333).

Acid lavas that have been superficially erupted weather into irregularly craggy hills, like the flanks of Snowdon. Those of intermediate composition, where they have accumulated in thick masses, are apt to weather into conical forms, as may be seen among the Cheviot, Pentland and Garleton Hills (Figs. 109, 110, 133); but where they have been poured out in successive thin sheets they have built up undulating plateaux with terraced sides, as among the Ayrshire and Campsie Fells and the hills of Lorne (Figs. 99, 107). Basic lavas have issued in comparatively thin sheets, frequently columnar or slaggy, forming flat-topped hills and terraced escarpments, such as are typically developed among the Tertiary basalt-plateaux of the Inner Hebrides and the Faroe Islands (Figs. 11, 265, 283, 284, 286).

One of the most frequent causes of local peculiarities of topography among old volcanic rocks is the intercalation of very distinct varieties of material in the same volcanic series. Where, for instance, lavas and tuffs alternate, great inequalities of surface may be produced. The tuffs, being generally more friable, decay faster and give rise to hollows, while the lavas, being more durable, project in bold ridges or rise into mural escarpments (Fig. 265). Again, where dykes weather more readily than the rocks which they traverse, they originate deep narrow clefts, while where they weather more slowly than the rocks around them, they project as dark ribs. Thus in Skye some dykes which rise through the obdurate gabbro are marked by chasms which reach up even to the highest crests of the mountains (Fig. 333), while of those which run in the pale crumbling granophyre, some stand up as black walls that can be followed with the eye across the ridges even from a long distance.

Many further illustrations of these principles might be cited here from the old volcanic districts of Britain. But they will present themselves successively in later chapters. For my present purpose it is enough to show that the scenery of these districts is not directly due to volcanic action, but is the immediate result of denudation acting upon volcanic rocks, modified and directed by their geological structure.

It may, however, be useful, in concluding the discussion of this subject, to cite some typical volcanic regions in the British Isles as illustrations of the relations between geology and topography, which, besides impressing the main lesson here enforced, may serve also to show some of the striking contrasts which geology reveals between the present and former conditions of the surface of the globe. Among these contrasts none are more singular than those offered by tracts where volcanic action has once been rife, and where the picture of ancient geography presented in the rocks differs so widely from the scenery of the same places to-day as to appeal vividly to the imagination.

The first district to which I may refer where ancient volcanic rocks are well developed is that of Devonshire. The story of the Devonian volcanoes will be told in some detail in later chapters, when it will be shown that the eruptions were again and again renewed during a long course of ages. Yet, abundant as the intercalated lavas and tuffs are, they can hardly be said to have had any marked effect on the scenery, though here and there a harder or larger mass of diabase rises into a prominent knoll or isolated hill. When the amount of volcanic material in this region is considered, we may feel some surprise at the trifling influence which it has exerted in the general denudation of the surface.

To one who wanders over the rich champaign of southern Devonshire, and surveys from some higher prominence the undulating tree-crowned ridges that slope down into orchard-filled hollows, and the green uplands that sweep in successive waves of verdure to the distant blue tors of Dartmoor, the scene appears as a type of all that is most peaceful, varied and fertile in English landscape. In the trim luxuriance that meets the eye on every side, the hand of man is apparent, though from many a point of vantage no sound may be heard for a time to show that he himself is anywhere near us. Yet ever and anon from the deep lanes, hidden out of sight under their canopy of foliage, there will come the creak of the groaning waggon and the crack of the waggoner's whip, as evidence that there are roads and human traffic through this bosky silent country.

Amid so much quiet beauty, where every feature seems to be eloquent of long generations of undisturbed repose, it must surely stir the imagination to be told that underneath these orchards, meadows and woodlands lie the mouldering remnants of once active and long-lived volcanoes. Yet we have only to descend into one of the deep lanes to find the crumbling lavas and ashes of the old eruptions. The landscape has, in truth, been carved out of these volcanic rocks, and their decomposition has furnished the rich loam that nourishes so luxuriant a vegetation.

Not less impressive is the contrast presented between the present and former condition of the broad pastoral uplands of the south of Scotland. Nowhere in the British Islands can the feeling of mere loneliness be more perfectly experienced than among these elevated tracts of bare moorland. They have nothing of the grandeur of outline peculiar to mountain tracts. Sometimes, for miles around one of their conspicuous summits, we may see no projecting knob or pinnacle. The rocks have been gently rounded off into broad featureless hills, which sink into winding valleys, each with its thread of streamlet and its farms along the bottom, and its scattered remnants of birch-wood or alder-copse along its slopes and dingles. Across miles of heathy pasture and moorland, on the summits of this great tableland, we may perchance see no sign of man or his handiwork, though the bleating of the sheep and the far-off barking of the collie tell that we are here within the quiet domain of the south-country shepherd.

In this pastoral territory, also, though they hardly affect the scenery, volcanic rocks come to the surface where the foldings of the earth's crust have brought up the oldest formations. Their appearance extends over so wide an area as to show that a large part of these uplands lies on a deeply-buried volcanic floor. A whole series of submarine volcanoes, extending over an area of many hundreds of square miles, and still in great part overlain with the accumulated sands and silts of the sea-bottom, now hardened into stone, underlies these quiet hills and lonely valleys.

A contrast of another type meets us in the broad midland valley of Scotland. Around the city of Edinburgh, for instance, the landscape is diversified by many hills and crags which show where harder rocks project from amidst the sediments of the Carboniferous system. On some of these crags the forts of the early races, the towers of Celt and Saxon, and the feudal castles of the middle ages were successively planted, and round their base clustered for protection the cots of the peasants and the earliest homesteads of the future city. Beneath these crags many of the most notable events in the stormy annals of the country were transacted. Under their shadow, and not without inspiration from their local form and colour, literature, art and science have arisen and flourished. Nowhere, in short, within the compass of the British Isles has the political and intellectual progress of the people been more plainly affected by the environment than in this central district of Scotland.

When now we inquire into the origin and history of the topography which has so influenced the population around it, we find that its prominences are relics of ancient volcanoes. The feudal towers are based on sills and dykes and necks. The fields and gardens, monuments and roadways, overlie sheets of lava or beds of volcanic ashes. Not only is every conspicuous eminence immediately around of volcanic origin, but even the ranges of blue hills that close in the distant view to south and north and east and west are mainly built up of lavas and tuffs. The eruptions of which these heights are memorials belong to a vast range of geological ages, the latest of them having passed away long before the advent of man. But they have left their traces deeply engraven in the rocky framework of the landscape. While human history, stormy or peaceful, has been slowly evolving itself during the progress of the centuries in these fertile lowlands, the crags and heights have remained as memorials of an earlier history when Central Scotland continued for many ages to be the theatre of vigorous volcanic activity.

As a final illustration of the influence of volcanic rocks in scenery, and of the contrast between their origin and their present condition, I may cite the more prominent groups of hills in the Inner Hebrides. In the singularly varied landscapes of that region three distinct types of topography attract the eye of the traveller. These are best combined and most fully developed in the island of Skye. Throughout the northern half of that picturesque island, the ground rises into a rolling tableland, deeply penetrated by arms of the sea, into which it slopes in green declivities, while along its outer borders it plunges in ranges of precipice into the Atlantic. Everywhere, alike on the cliffs and the inland slopes, long parallel lines of rock-terrace meet the eye. These mount one above another from the shores up to the flat tops of the highest hills, presenting level or gently-inclined bars of dark crag that rise above slopes of debris, green sward and bracken. It is these parallel, sharply-defined bars of rock, with their intervening strips of verdure, that give its distinctive character to the scenery of northern Skye. On hillside after hillside and in valley after valley, they reappear with the same almost artificial monotony. And far beyond the limits of Skye they are repeated in one island after another, all down the chain of the Inner Hebrides.

In striking contrast to this scenery, and abruptly bounding it on the south, rise the Red Hills of Skye—a singular group of connected cones. Alike in form and in colour, these hills stand apart from everything around them. The verdure of the northern terraced tableland here entirely disappears. The slopes are sheets of angular debris,—huge blocks of naked stone and trails of sand, amidst which hardly any vegetation finds a footing. The decay of the rock gives it a pale yellowish-grey hue, which after rain deepens into russet, so that in favourable lights these strange cones gleam with a warm glow as if they, in some special way, could catch and reflect the radiance of the sky.

Immediately to the west of these pale smooth-sloped cones, the dark mass of the Cuillin Hills completes the interruption of the northern tableland. In almost every topographical feature these hills present a contrast to the other two kinds of scenery. Their forms are more rugged than those of any other hill-group in Britain (Fig. 331). Every declivity among them is an irregular pile of crags, every crest is notched like a saw, every peak is sharpened into a pinnacle. Instead of being buried under vast sheets of their own debris, these hills show everywhere their naked rock, which seems to brave the elements as few other rocks can do. Unlike the pale Red Hills, they are dark, almost black in tone, though when canopied with cloud they assume a hue of deepest violet.

Each of these three distinct types of topography owes its existence to the way in which a special kind of volcanic rock yields to the influences of denudation. The terraced tableland of the north is built up of hundreds of sheets of basaltic lava, each of the long level ledges of brown rock marking the outcrop of one or more of these once molten streams. The black rugged mass of the Cuillin Hills consists of a vast protruded body of eruptive material, which, in the form of endless sills and bosses of gabbro and dolerite, has invaded the basalt-plateau, and has now been revealed by the gradual removal of the portion of that plateau which it upraised. The pale cones and domes of the Red Hills mark the place of one of the last protrusions in the volcanic history of Britain—that of large masses of an acid magma, which broke through the basalt-plateau and also disrupted the earlier gabbro.

In no part of North-Western Europe has volcanic activity left more varied and abundant records of its operations than in these three contiguous tracts of Skye. It is interesting therefore to note the striking contrast between the former and the present landscapes of the region. The lavas of the basaltic tableland crumble into a rich loam, that in the mild moist climate of the Hebrides supports a greener verdure than any of the other rocks around will yield. The uplands have accordingly become pasture-grounds for herds of sheep and cattle. The strips of lowland along the valleys and in the recesses of the coast-line furnish the chief tracts of arable land in the island, and are thus the main centres of the crofter population. The bays and creeks of the much-indented shores form natural harbours, which in former days attracted the Norse sea-rovers, and supplied them with sites for their settlements. Norse names still linger on headland and inlet, but the spirit of adventure has passed away, and a few poor fishing-boats, here and there drawn up on the beach, are usually the only token that the islanders make any attempt to gather the harvest of the sea.

The mountain groups which so abruptly bound the basalt-plateau on the south, and present in their topographical features such distinctive scenery, comprise a region too lofty, too rugged and too barren for human occupation. The black Cuillins and the pale Red Hills are solitudes left to the few wild creatures that have not yet been exterminated. The corries are the home of the red deer. The gabbro cliffs are haunts of the eagle and the raven. Where patches of soil have gathered in the crannies of the gabbro, alpine plants find their home. In the chasms left by the decay of the dykes between the vertical walls of their fissures, the winter snows linger into summer, and conceal with their thick drifts the mouldering surface of the once molten rock beneath them. On every side and at every turn a mute appeal is made to the imagination by the strange contrasts between the quiet restfulness of to-day, when the sculpture-tools of nature are each busily carving the features of the landscape, and the tumult of the time when the rocks, now so silent, were erupted.

The general discussion of the subject of Volcanism in this Introduction will, I hope, have prepared the reader who has no special geological training for entering upon the more detailed descriptions in the rest of this treatise. As already stated, the chronological order of arrangement will be followed. Beginning with the records of the earliest ages, we shall follow the story of volcanic action down to the end of the latest eruptions.

Each great geological system will be taken as a whole, representing a long period of time, and its volcanic evolution will be traced from the beginning of the period to the close. Some variety of treatment is necessarily entailed by the wide range in the nature and amount of the evidence for the volcanic history of different ages. But where practicable, an outline will first be given of what can be gathered respecting the physical geography of each geological period in Britain. In the description which will then follow of the volcanic phenomena, an account of the general characters of the erupted rocks will precede the more detailed narrative of the history of the volcanic eruptions in the several regions where they took place. References to the published literature of each formation will be given in the first part of each section, or will be introduced in subsequent pages, as may be found most convenient.

BOOK II

VOLCANIC ACTION IN PRE-CAMBRIAN TIME

CHAPTER VIII

PRE-CAMBRIAN VOLCANOES

The Beginnings of Geological History—Difficulties in fixing on a generally-applicable Terminology—i. The Lewisian (Archaean) Gneiss; ii. The Dalradian or Younger Schists of Scotland; iii. The Gneisses and Schists of Anglesey; iv. The Uriconian Volcanoes; v. The Malvern Volcano; vi. The Charnwood Forest Volcano.

THE early geological history of this globe, like the early history of mankind, must be drawn from records at once scanty and hardly decipherable. Exposed to the long series of revolutions which the surface of the planet has undergone, these records, never perhaps complete at the first, have been in large measure obliterated. Even where they still exist, their meaning is often so doubtful that, in trying to interpret it, we find little solid footing, and feel ourselves to be groping, as it were, in the dimness of mythological legend, rather than working in the light of trustworthy and intelligible chronicles. These primeval records have been more particularly the objects of sedulous study during the last twenty years all over Europe and in North America. A certain amount of progress in their decipherment has been made. But the problems they still present for solution are numerous and obscure. Fortunately, with many of these problems the subject of the present treatise is not immediately connected. We need only concern ourselves with those which are related to the history of primeval volcanic activity.

To the earliest and least definite division of the geological annals various names have been applied. Some writers, believing that this period preceded the first appearance of plants or animals upon the globe, have named it Azoic—the lifeless age of geological history. But the absence of any hitherto detected trace of organic existence among the oldest known rocks cannot be held to prove that these rocks were formed before the advent of living things on the surface of the earth. The chance discovery of a single fossil, which might at any moment be made, would show the name "Azoic" to be a misnomer. Other geologists, believing that, as a matter of fact, organic structures of low types do actually occur in them, have called these old rocks "Eozoic," to denote that they were deposited during the dawn of life upon our planet. But the supposed organisms have not been everywhere accepted as evidence of former life. By many able observers they are regarded as mere mineral aggregates. Another term, "Archaean," has been proposed for the primeval ages of geological history, which are recorded in rocks that carry us as far as may ever be possible towards the beginnings of that history.

In choosing some general term to include the oldest known parts of the earth's crust, geologists are apt unconsciously to assume that the rocks thus classed together represent a definite section of geological time, comparable, for instance, to that denoted by one of the Palæozoic systems. Yet it is obvious that, under one of these general terms of convenient classification, a most multifarious series of rocks may be included, representing not one but possibly many, and widely separated, periods of geological history.

In many countries the oldest sedimentary accumulations, whether fossiliferous or not, are underlain by a series of crystalline rocks, which consist in great part of coarse massive gneisses and other schists. All over the world these rocks present a singular sameness of structure and composition. What might be found below them no man can say. They are in each country the oldest rocks of which anything is yet known, and whatsoever may be our theory of their origin, we must, at least for the present, start from them as the fundamental platform of the terrestrial crust.

But though crystalline rocks of this persistent character are widely distributed, both in the Old World and in the New, they in themselves furnish no means of determining their precise geological age. No method has yet been devised whereby the oldest gneiss of one country can be shown to be the true stratigraphical equivalent of the oldest gneiss of another. Palæontology is here of no avail, and Petrology has not yet provided us with such a genetic scheme as will enable us to make use of minerals and rock-structures, as we do of fossils, in the determination of geological horizons. All that can be positively affirmed regarding the stratigraphical relations of the rocks in question is that they are vastly more ancient than the oldest sedimentary and fossiliferous formations in each country where they are found. The "Lewisian" gneiss of the north-west of Scotland, the "Urgneiss" of Central Europe, and the "Laurentian" gneiss of Canada occupy similar stratigraphical positions, and present a close resemblance in lithological characters. We may conveniently class them under one common name to denote this general relationship. But we have, as yet, no means of determining how far they belong to one continuous period of geological history. They may really be of vastly different degrees of antiquity.

From the very nature of the case, any name by which we may choose to designate such ancient rocks cannot possess the precise stratigraphical value of the terms applied to the fossiliferous formations. Yet the convenience of possessing such a general descriptive epithet is obvious.

Until much more knowledge of the subject has been gained, any terminology which may be proposed must be regarded as more or less provisional. The comprehensive term "pre-Cambrian" may be usefully adopted as a general designation for all rocks older than the base of the Cambrian system, irrespective of their nature and origin. Already it is well known that under this term a vast series of rocks, igneous and sedimentary, is included. In some regions several successive formations, or systems of formations, may be recognized in this series. But until some method has been devised for determining the stratigraphical relations of these formations in different regions, it would seem safest not to attempt to introduce general names for universal adoption, but to let the sequence of rocks in each distinct geological province be expressed by a local terminology. This caution is more especially desirable in the case of sedimentary deposits. We may surmise as to the equivalence of the rocks called Huronian, Torridonian and Longmyndian, but whilst so much is mere conjecture, it is certainly injudicious to transfer the local names of one province to the rocks of another.

The only relaxation of this general precaution which I think may at present be made is the adoption of a common name for the oldest type of gneisses. The term "Archaean" has been applied to these rocks, and if it is used simply to express a common petrographical type, occupying the lowest horizon in the stratigraphical series of a country, it has obvious advantages. But I would still retain the local names as subordinate terms to mark the local characteristics of the Archaean rocks of each province. Thus the "Laurentian" rocks of Canada and the "Lewisian" rocks of Scotland are widely-separated representatives of the peculiar stratigraphical series which is known as Archaean.

The pre-Cambrian rocks of Britain include several distinct systems or groups. How far those of even one part of this comparatively limited region are the proper equivalents of those of another and distant part is a problem still unsolved. Hence each distinct area, with its own type of rocks, will here be treated by itself. The following rock-types will be described: I. The Lewisian (Archaean) Gneiss; II. The Younger (Dalradian) Schists of Scotland; III. The Gneisses and Schists of Anglesey; IV. The Uriconian Group; V. The Malvern Group; VI. The Charnwood Forest Group (see Map I.).

i. THE LEWISIAN (ARCHAEAN) GNEISS

The British Isles are singularly fortunate in possessing an admirable development of pre-Cambrian rocks. These ancient masses rise up in various parts of the islands, but the region where they are most extensively displayed, and where their stratigraphical position and sequence are most clearly shown, lies in the north-west of Scotland.[1] In that territory they

[1] These rocks have been the subject of much discussion, but geologists are now agreed as to their succession and structure. A full summary of the literature of the controversy regarding them will be found in the *Quarterly Journal of the Geological Society*, vol. xliv. (1888), p. 378.

form the whole chain of the Outer Hebrides, and likewise extend as an irregular selvage along the western margin of the counties of Sutherland and Ross. The lowest known platform of the fossiliferous formations has there been discovered and has been traced for a distance of more than 100 miles. From this definite horizon, the high antiquity of all that lies below it is impressively demonstrated. The accompanying diagram (Fig. 35) will explain the general relations of the various geological formations of the region.

In certain dark shales (*b*) which occupy a well-defined and readily-traceable position among the rocks of Sutherland and Ross, numerous specimens of the trilobite genus *Olenellus*, together with other fossils, have been found. By common consent among geologists, the zone of rock in which this genus appears is taken as the lowest stage of the Cambrian system. In Britain it marks the oldest known group of fossiliferous strata—the platform on which the whole of the Palæozoic systems rest.

From the definite geological epoch indicated by this platform, we can go

FIG. 35.—Diagram illustrating the stratigraphical relations of the pre-Cambrian and Cambrian rocks of the North-west Highlands of Scotland.

c, Durness Limestones, with Upper Cambrian and perhaps Lower Silurian fossils, 1500 feet, top nowhere seen. *b*, Serpulite grit and "fucoid" shales, 70 to 80 feet, containing the *Olenellus*-zone. *a*, Quartzite, with abundant annelid tubes, about 600 feet. II. Red Sandstones and Conglomerates, sometimes 8000 feet or more (Torridonian). I. Gneiss with dykes, etc. (Lewisian).

backward into pre-Cambrian time, and realize in some measure how prodigious must be the antiquity of the successive groups of rock which emerge from beneath the base of the Palæozoic systems. Nowhere is this antiquity more impressively proclaimed than in the north-west of Scotland. From below the *Olenellus*-zone with its underlying sheets of quartzite (*a*), a thick group of dull red sandstones and conglomerates (II.) rises into a series of detached conical or pyramidal mountains, which form one of the most characteristic features in the scenery of that region. As this detrital formation is well developed around Loch Torridon, it has been termed Torridonian. It attains a thickness of at least 8000 or 10,000 feet, and is traceable all the way from the extreme northern headlands of Sutherland to the southern cliffs of the island of Rum.

In judging of the chronological significance of the geological structure of the north-west of Scotland, we are first impressed by the stratigraphical break between the base of the Cambrian system and the Torridonian deposits below. This break is so complete that here and there the thick intervening mass of sandstones and conglomerates has been nearly or wholly removed by denudation before the lowest Cambrian strata were laid down. Such a discordance marks the passage of a protracted interval of time.

Again, when the composition of the Torridonian rocks is considered, further striking evidence is obtained of the lapse of long periods. The sandstones, conglomerates and shales of this pre-Cambrian system present no evidence of cataclysmal action. On the contrary, they bear testimony that they were accumulated much in the same way and at the same rate as the subsequent Palæozoic systems. In that primeval period, as now, sand and silt were spread out under lakes and seas, were ripple-marked by the agitation of the water, and were gradually buried under other layers of similar sediment. The accumulation of 10,000 feet of such gradually-assorted detritus must have demanded a long series of ages. Here, then, in the internal structure of the Torridonian rocks, there is proof that in passing across them, from their summit to their base, we make another vast stride backward into the early past of geological history.

But when attention is directed to the relations of the Torridonian strata to the rocks beneath them, a still more striking proof of an enormously protracted period of time is obtained. Between the two series of formations lies one of the most marked stratigraphical breaks in the geological structure of the British Isles. There is absolutely nothing in common between them, save that the conglomerates and sandstones have been largely made out of the waste of the underlying gneiss. The denudation of the crystalline rocks before the deposition of any of the Torridonian sediments must have been prolonged and gigantic. The more, indeed, we study the gneiss, the more do we feel impressed by the evidence for the lapse of a vast interval of time, here unrecorded in rock, between the last terrestrial movements indicated by the gneiss and the earliest of the Torridonian sediments.

In this manner, reasoning backward from the horizon of the *Olenellus*-zone, we are enabled to form some conception of the vastness of the antiquity of the fundamental rocks of the North-west Highlands. The nature and origin of these rocks acquire a special interest from a consideration of their age. They contain the chronicles of the very beginnings of geological history, in so far as this history is contained in the crust of the earth. No part of the geological record is so obscure as this earliest chapter, but we need not here enter further into its difficulties than may be necessary for the purpose of understanding what light it can be made to throw on the earliest manifestations of volcanic action.

Under the term Lewisian Gneiss (I. in Fig. 35) a series of rocks is comprised which differ from each other in composition, structure and age, though most of them possess such crystalline and generally foliated characters as may be conveniently included under the designation of gneiss. The complexity of these ancient crystalline masses was not recognized at the time when Murchison called them the "Fundamental" or "Lewisian" gneiss. It is only since the Geological Survey began to study and map them in full detail that their true nature and history have begun to be understood.[1]

The researches of the Survey have shown the so-called Lewisian gneiss to comprise the following five groups of rock: 1. A group of various more or less banded and foliated rocks which form together the oldest and chief part of the gneiss (Fundamental complex); 2. Highly basic dykes cutting the first group; 3. Dykes and sills of dolerite, epidiorite and hornblende-schist; 4. A few dykes of peculiar composition; 5. Gneissose granite and pegmatite.

The first of these groups, forming the main body of the gneiss, has been critically studied on the mainland from Cape Wrath to Skye. But its development in the Outer Hebrides has not yet been worked out, although the name "Lewisian" was actually taken from that chain of islands. So far as at present known, however, the gneiss of the Hebrides repeats the essential characters of that of the mainland.

Mr. Teall, as the result of a careful investigation in the field and with the microscope, has ascertained that on the mainland between Skye and Cape Wrath the rocks of the "fundamental complex" are essentially composed of olivine, hypersthene, augite (including diallage), hornblende, biotite, plagioclase, orthoclase, microcline and quartz. He has further observed that these minerals are associated together in the same manner as in peridotites, gabbros, diorites and granites. Treating the rocks in accordance with their composition and partly with their structure, but excluding theoretical considerations, he has arranged them in the following five subdivisions:—

1. Rocks composed of ferro-magnesian minerals, without felspar or quartz—Pyroxenites, Hornblendites.
2. Rocks in which pyroxenes are the dominating ferro-magnesian constituents, felspar always being present, sometimes quartz: A, Without quartz, Hypersthene-augite-rocks (pyroxene granulites; rocks of the Baltimore-gabbro type) and augite-rocks (gabbros); B, With quartz, Augite-gneiss.
3. Rocks in which hornblende is the prevalent ferro-magnesian constituent: A, Without quartz, or containing it only in small quantity; rocks basic in composition: (*a*) massive or only slightly foliated (Amphibolites, as epidote-amphibolite, zoisite-amphibolite, garnet-amphibolite); (*b*) foliated (Hornblende-schist). B, With quartz; rocks intermediate or acid in composition: (*a*) with compact hornblende and a granular structure (Hornblende-gneiss proper); (*b*) with hornblende occurring in fibrous or other aggregates; (*c*) with compact hornblende and a more or less granulitic structure (Granulitic hornblende-gneiss).
4. Rocks in which biotite is the predominant ferro-magnesian constituent; felspar and quartz both present: (*a*) Biotite occurring as independent plates or in aggregates of two or three large individuals (Biotite-gneiss); (*b*) Biotite occurring in aggregates of numerous small individuals (rare type); (*c*) Biotite occurring as independent plates in a granulitic structure.
5. Rocks in which muscovite and biotite are present, together with felspar and quartz —Muscovite-biotite-gneiss. These, though not forming a well-defined natural

[1] See the Report of this Survey work by Messrs. Peach, Horne, Gunn, Clough, Cadell and Hinxman, *Quart. Journ. Geol. Soc.* vol. xliv. (1888), pp. 378-441; and Annual Reports of Director-General of the Geological Survey in the *Report of The Science and Art Department for* 1894, p. 279, and 1895, p. 17 of reprint. The general area of the gneiss is shown in Map I.

group, are placed together for purposes of description. They are all foliated, some having the aspect of mica-schists, others being typical augen-gneisses, or light grey gneisses with abundant oligoclase and inclusions of microlitic epidote.

The rocks of each of these types are usually restricted to relatively small areas, and they succeed each other with much irregularity all the way from Skye to Cape Wrath. Their chemical and mineralogical composition proves them to have decided affinities with the plutonic igneous masses of the earth's crust.

The only exceptions to this prevalent igneous type occur in the districts of Gairloch and Loch Carron, where the gneiss appears to be associated with a group of mica-schists, graphitic-schists, quartzites and siliceous granulites, limestones, dolomites, chlorite-schists and other schists. That these are altered sedimentary formations can hardly be doubted. What their precise relations to the fundamental complex of the gneiss may be has not yet been satisfactorily determined. They are certainly far older than the Torridon sandstone which covers them unconformably. Possibly they may represent a sedimentary formation still more ancient than the gneiss.

Save these obscure relics of a pre-Torridonian system of strata, the gneiss never presents any structure which suggests the alteration of clastic constituents. Everywhere its mineral composition points to a connection with the subterranean intrusions of different igneous magmas, while the manner in which its different rock-groups are associated together, and the internal structure of some of them, still further link it with phenomena which will be described in succeeding chapters as parts of the records of volcanic action.

An interesting feature of the fundamental complex, as bearing on the origin of the gneiss, is to be found in the occurrence of bosses and bands which are either non-foliated or foliated only in a slight degree. These comparatively structureless portions present much of the character of bosses or sills of true eruptive rocks. They occur in various parts of Sutherland and Ross. Their external margins are not well defined, and they pass insensibly into the ordinary gneiss, the dark basic massive rocks shading off into coarse basic gneisses, and the pegmatites of quartz and felspar which traverse them merging into bands of grey quartzose gneiss.

So far, therefore, as present knowledge goes, the main body or fundamental complex of the Lewisian gneiss in the North-west Highlands of Scotland consists of what may have been originally a mass of various eruptive rocks. It has subsequently undergone a succession of deformations from enormous stresses within the terrestrial crust, which have been investigated with great care by the Geological Survey. But it presents structures which, in spite of the abundant proofs of great mechanical deformation, are yet, I venture to think, original, or at least belong to the time of igneous protrusion before deformation took place. The alternation of rocks of different petrographical constitution suggests a succession of extravasations of eruptive materials, though it may not be always possible now to determine the order in which these followed each other. In the feebly foliated or massive bands and bosses there is a parallel arrangement of their constituent minerals or of fine and coarse crystalline layers which recalls sometimes very strikingly the flow-structure of rhyolites and other lavas. This resemblance was strongly insisted on by Poulett Scrope, who believed that the laminar structure of such rocks as gneiss and mica-schist was best explained by the supposition of the flow of a granitic magma under great pressure within the earth's crust.[1]

The conviction that these parallel structures do, in some cases, really represent traces of movements in the original unconsolidated igneous masses, not yet wholly effaced by later mechanical stresses, has been greatly strengthened in my mind by a recent study of the structures of various eruptive bosses, especially those of gabbro in the Tertiary volcanic series of the Inner Hebrides. The banded structure, the separation of the constituent minerals into distinct layers or zones, the alternation of markedly basic with more acid layers, and the puckering and plication of those bands, can be seen as perfectly among the Tertiary gabbro bosses of Skye as in the Lewisian gneiss (see Figs. 336 and 337). It cannot be contended that such structures in the gabbro are due to any subsequent terrestrial disturbance and consequent deformation. They must be accepted as part of the original structure of the molten magma.[2] It seems to me, therefore, highly probable that the parallel banding in the uncrushed cores of the Lewisian gneiss reveals to us some of the movements of the original magma at the time of its extrusion and before it underwent those great mechanical stresses which have so largely contributed to the production of many of its most characteristic structures.

While the material of the oldest gneiss presents many affinities to plutonic rocks of much younger date, a wide region of mere speculation opens out when we try to picture the conditions under which this material was accumulated. Some geologists have boldly advanced the doctrine that the Archaean gneisses represent the earliest crust that consolidated upon the surface of the globe. But these rocks offer no points of resemblance to the ordinary aspect of superficial volcanic ejections. On the contrary, the coarsely-crystalline condition even of those portions of the gneiss which seem most nearly to represent original structure, the absence of anything like scoriae or fragmental bands of any kind, and the resemblances which may be traced between parts of the gneiss and intrusive bosses of igneous rock compel us to seek the nearest analogies to the original gneiss in deep-seated masses of eruptive material. It is difficult to conceive that any rocks approaching in character to the gabbros, picrites, granulites and other coarsely-crystalline portions of the old gneiss could have consolidated at or near the surface.

When the larger area of gneiss forming the chain of the Outer

[1] *Volcanoes*, pp. 140, 283, 299.

[2] See A. Geikie and J. J. H. Teall, *Quart. Journ. Geol. Soc.* vol. l. (1894), p. 645.

Hebrides is studied, we may obtain additional information regarding the probable origin and the earliest structures of the fundamental complex of the Lewisian gneiss. In particular, we may look for some unfoliated cores of a more acid character, and perhaps for evidence which will show that both acid and basic materials were successively protruded. We may even entertain a faint hope that some trace may be discovered of superficial or truly volcanic products connected with the bosses which recall those of later date and obviously eruptive nature. But up to the present time no indication of any such superficial accompaniments has been detected. If any portions of the old gneiss represent the deeper parts of columns of molten rock that flowed out at the surface as lava, with discharges of fragmentary materials, all this superincumbent material, at least in the regions which have been studied in detail, had disappeared entirely before the deposition of the very oldest part of the Torridonian rocks, unless some trace of it may remain among the pebbles of the Torridonian conglomerates, to which reference will be immediately made.

So far, then, as the evidence now available allows a conclusion to be drawn, the Lewisian gneiss reveals to us a primeval group of eruptive rocks presenting a strong resemblance to some which in later formations are connected, as underground continuations, with bedded lavas and tuffs that were erupted at the surface; and although no proof has yet been obtained of true volcanic ejections associated with the fundamental complex, the rocks seem to be most readily understood if we regard them as having consolidated from igneous fusion at some depth, and we may plausibly infer that they may have been actually connected with the discharge of volcanic materials at the surface. The graphite-schists, mica-schists, and limestones of the Gairloch and Loch Carron may thus be surviving fragments of the stratified crust into which these deep-seated masses were intruded, and through which any volcanic eruptions that were connected with them had to make their way.

The limited areas occupied by the several varieties of rock in the fundamental complex suggests the successive protrusion of different magmas, or of different portions from one gradually changing magma. Mr. Teall has ascertained that whenever in this series of rocks the relative ages of two petrographical types can be clearly ascertained, the more basic is older than the more acid.

But besides all the complexity arising from original diversity of area, structure and composition among the successive intrusions, a further intricacy has been produced by the subsequent terrestrial disturbances, which on a gigantic scale affected the north-west of Europe after the formation of the fundamental complex of the old gneiss, but long before the Torridonian period. By a series of terrestrial stresses that came as precursors of those which in later geological times worked such great changes among the rocks of the Scottish Highlands, the original bosses and sheets of the gneiss were compressed, plicated, fractured and rolled out, acquiring in this process a crumpled, foliated structure. Whether or not these disturbances were accompanied by any manifestations of superficial volcanic action has not yet been determined. But we know that they were followed by a succession of dyke-eruptions, to which, for extent and variety, there is no parallel in the geological structure of Britain, save in the remarkable assemblage of dykes belonging to the Tertiary volcanic period[1] (Fig. 36).

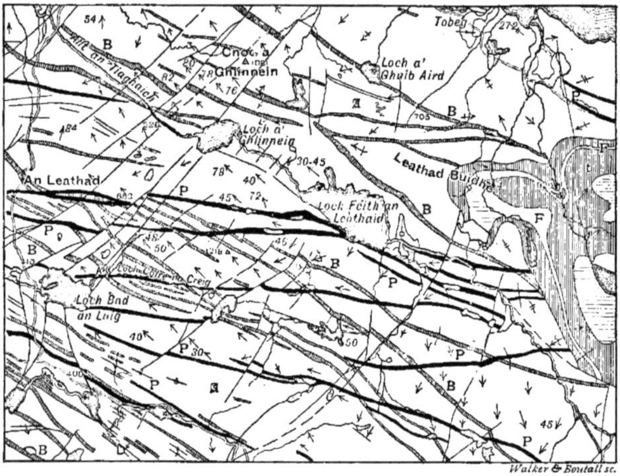

FIG. 36.—Map of a portion of the Lewisian gneiss of Ross-shire.

Taken from Sheet 107 of the Geological Survey of Scotland on the scale of one inch to a mile. The white ground (𝔄) marks the general body of the Lewisian gneiss. This is traversed by dykes of dolerite (B), which are cut by later dykes of highly basic material (peridotite, picrite, etc., P). The gneiss and its system of dykes is overlain unconformably by the nearly horizontal Torridon Sandstone (*t*), which is injected by sheets of oligoclase-porphyry (F).

For the production of these dykes a series of fissures was first opened through the fundamental complex of the gneiss, having a general trend from E.S.E. to W.N.W., running in parallel lines for many miles, and so close together in some places that fifteen or twenty of them occurred within a horizontal space of one mile. The fissures were probably not all formed at the same time; at all events, the molten materials that rose in them exhibit distinct evidence of a succession of upwellings from the igneous magma below.

Considered simply from the petrographical point of view, the materials that have filled the fissures have been arranged by Mr. Teall in the following groups: 1. Ultra-basic dykes, sometimes massive (peridotites), sometimes foliated (talcose schists containing carbonates and sometimes gedrite); 2. Basic dykes which where massive take the forms of dolerite and epidiorite, and where foliated appear as hornblende-schist, the same dyke often presenting the three conditions of dolerite, epidiorite and hornblende-schist; 3. Dykes of peculiar composition, comprising microcline-mica rocks and

[1] *Quart. Journ. Geol. Soc.* vol. xliv. (1888), p. 389 *et seq.*

biotite-diorite with macro poikilitic plagioclase; 4. Granites and gneissose granites (biotite-granite with microcline); 5. Pegmatites (microcline-quartz rocks with a variable amount of oligoclase or albite).[1]

Distinct evidence of a succession of eruptions can be made out among these rocks. By far the largest proportion of the dykes consists of basic materials. The oldest and most abundant of them are of plagioclase-augite rocks, which, where uncrushed, differ in no essential feature of structure or composition from the dolerites and basalts of more modern periods, though they have been plentifully changed into epidiorite and hornblende-schist.[2] They present, too, most of the broad features that characterize the dykes of later times—the central more coarsely-crystalline portion, the marginal band of finer grain, passing occasionally into what was probably a basic glass, and the transverse jointing. They belong to more than one period of emission, for they cross each other. They vary in width up to nearly 200 - feet, and sometimes run with singular persistence completely across the whole breadth of the strip of gneiss in the west of Sutherland and Ross. Dozens of dykes have been followed by the Geological Survey for distances of ten or twelve miles.

Later in time, and much less abundant, are certain highly basic dykes —peridotites with schistose modifications—which cut across the dolerites in a more nearly east-and-west direction. There are likewise occasional dykes of peculiar composition, which, as above stated, have been distinguished by Mr. Teall as microcline-mica rocks and biotite-diorite.

Last of all comes a group of thoroughly acid rocks—varieties of granite and pegmatite—which form intrusive sheets and dykes. The granites contain biotite with microcline, and are sometimes gneissose. The pegmatites are microcline-quartz rocks with a variable amount of oligoclase or albite. These dykes coincide in direction with the basalts and dolerites, but they are apt to run together into belts of granite and pegmatite, sometimes 1500 feet broad.

Up to the present time no evidence has been found of any superficial outpouring of material in connection with this remarkable series of dykes in the Lewisian gneiss. That they may have been concomitant with true volcanic eruptions may be plausibly inferred from the close analogy which, in spite of their antiquity and the metamorphism they have undergone, they still present to the system of dykes that forms a part of the great Tertiary volcanic series of Antrim and the Inner Hebrides. The close-set fissures running in a W.N.W direction, the abundant uprise into these fissures of basic igneous rocks, followed by a later and more feeble extravasation of acid material, are features which in a singular manner anticipate the volcanic phenomena of Tertiary time.

There can be no question as to the high antiquity of these dykes. They were already in place before the advent of those extraordinary vertical lines of shearing which have so greatly affected both the gneiss and the

[1] *Annual Report of Geological Survey for* 1895, p. 18 of reprint.

[2] See Mr. Teall, *Quart. Journ. Geol. Soc.* vol. xli. (1885), p. 133.

dykes; and these movements, in turn, had long been accomplished before the Torridon Sandstone was laid down, for the dykes, with their abundant deformation, run up to and pass beneath the sandstone which buries them and all the rocks with which they are associated. Though later than the original fundamental complex, the dykes have become so integral and essential a part of the gneiss as it now exists that they must be unhesitatingly grouped with it.

With so wide an extension of the subterranean relics of volcanic energy, it is surely not too much to hope that somewhere there may have been preserved, and may still be discovered, proofs that these eruptive rocks opened a connection with the surface, and that we may thus recognize vestiges of the superficial products of actual Archaean volcanoes. Among the pebbles in the conglomerates of the Torridon Sandstone there occur, indeed, fragments of felsites which possess great interest from the perfection with which they retain some of the characteristic features of younger lavas. Mr. Teall has described their minute structure. They are dark, purplish, compact rocks, consisting of a spherulitic micro-pegmatitic, micro-poikilitic or micro-crystalline groundmass, in which are imbedded porphyritic crystals or crystal-groups of felspar, often oligoclase. These spherulitic rocks occasionally show traces of perlitic structure. They bear a striking resemblance to some of the Uriconian felsites of Shropshire, pebbles from which occur in the Longmynd rocks.[1] These fragments suggest the existence of volcanic materials at the surface when the Torridon Sandstone was deposited. Possibly they may represent some vanished Lewisian lavas. But the time between the uprise of the dykes and the formation of the Torridonian series was vast enough for the advent of many successive volcanic episodes. The pebbles may therefore be the relics of eruptions that took place long after the period of the dykes.

Among the Torridonian strata no undoubted trace of any contemporaneous volcanic eruptions has been met with.[2] The only relics of volcanic rocks in this enormous accumulation of sediments are the pebbles just referred to, which may be referable to a time long anterior to the very oldest parts of the Torridonian series.

That Archaean time witnessed volcanic eruptions on a considerable scale, and with great variety of petrographical material, has recently been shown in detail by Mr. Otto Nordenskjöld from a study of the rocks of Småland in Sweden. He has described a series of acid outbursts, including masses of rhyolite and dacite, together with agglomerates and tuffs, likewise basic eruptions, with dioritic rocks, augite-porphyrite and breccia. He refers these rocks to the same age as most of the Scandinavian gneisses, and remarks that though they have undergone much mechanical deformation and metamorphism, they have yet here and there retained some of their distinctive volcanic structures, such as the spherulitic.[3] When the large

[1] *Annual Report of Geological Survey for* 1895, p. 21 of reprint.

[2] The supposed tuff referred to in *Quart. Journ. Geol. Soc.* vol. xlviii. (1892), p. 168, is probably not of truly volcanic origin.

[3] "Über Archæische Ergussgesteine aus Småland," *Sveriges Geol. Undersökn*, No. 135 (1894).

area of Lewisian gneiss forming the chain of the Outer Hebrides is investigated it may possibly supply examples of a similar series of ancient volcanic masses.

ii. THE DALRADIAN OR YOUNGER SCHISTS OF SCOTLAND

We now come to one of the great gaps in the geological record. The Lewisian gneiss affords us glimpses of probable volcanic activity at the very beginning of geological history. An enormous lapse of time, apparently unrepresented in Britain by any geological record, must be marked by the unconformability between the gneiss and the Torridon Sandstone. Another prodigious interval is undoubtedly shown by the Torridonian series. Neither this thick accumulation of sediment nor the Cambrian formations, which to a depth of some 2000 feet overlie the Torridon Sandstone, have yielded any evidence of true superficial eruptions, though they are traversed by numerous dykes, sills and bosses. The age of these intrusive masses cannot be precisely fixed; a large proportion of them is certainly older than the great terrestrial displacements and concurrent metamorphism of the North-West Highlands.

While from the Lewisian gneiss upward to the highest visible Cambrian platform in Sutherland, no vestige of contemporaneous volcanic rocks is to be seen, the continuity of the geological record is abruptly broken at the top of the Durness Limestone. By a series of the most stupendous dislocations, the rocks of the terrestrial crust have there been displaced to such a degree that portions have been thrust westward for a horizontal distance of sometimes as much as ten miles, while they have been so crushed and sheared as to have often lost entirely their original structures, and to have passed into the crystalline and foliated condition of schists. Portions of the floor of Lewisian gneiss, and large masses of the Torridon Sandstone, which had been buried under the Cambrian sediments, have been torn up and driven over the Durness Limestone and quartzite.

Though much care has been bestowed by the officers of the Geological Survey on the investigation of the complicated mass of material which, pushed over the Cambrian strata, forms the mountainous ground that lies to the east of a line drawn from Loch Eribol, in the north of Sutherland, to the south-east of Skye, some uncertainty still exists as to the age and history of the rocks of that region. For the purposes of this work, therefore, the rest of the country eastwards to the line of the Great Glen—that remarkable valley which cuts Scotland in two—may be left out of account.

To the east of the Great Glen the Scottish Highlands display a vast succession of crystalline schists, the true stratigraphical relations of which to the Lewisian gneiss have still to be determined, but which, taken as a whole, no one now seriously doubts must be greatly younger than that ancient rock. Murchison first suggested that the quartzites and limestones found in this newer series are the equivalents of those of the North-West.

This identification may yet be shown to be correct, but must be regarded as still unproved. Traces of fossils (annelid-pipes) have been found in some of the quartzites, but they afford little or no help in determining the horizons of the rocks. In Donegal, where similar quartzites, limestones and schists are well developed, obscure indications of organic remains (corals and graptolites) have likewise been detected, but they also fail to supply any satisfactory basis for stratigraphical comparison.

Essentially the schists of the Scottish Highlands east of the Great Glen consist of altered sedimentary rocks. Besides quartzites and limestones, there occur thick masses of clay-slate and other slates and schists, with bands of graphitic schist, greywacke, pebbly grit, quartzite, boulder-beds and conglomerates. Among rocks that have been so disturbed and foliated it is necessarily difficult to determine the true order of succession. In the Central Highlands, however, a certain definite sequence has been found to continue as far as the ground has yet been mapped. Were the rocks always severely contorted, broken and placed at high angles, this sequence might be deceptive, and leave still uncertain the original order of deposition of the whole series. But over many square miles the angles of inclination are low, and the successive bands may be traced from hill to hill, across strath and glen, forming escarpments along the slopes and outliers on the summits, precisely as gently-undulating beds of sandstone and limestone may be seen to do in the dales of Yorkshire. It is difficult to resist the belief, though it may, perhaps, be premature to conclude, that this obvious and persistent order of succession really marks the original sequence of deposition. In Donegal also a definite arrangement of the rock-groups has been ascertained which, when followed across the country, gives the key to its geological structure.[1]

In the order of succession which has been recognized during the progress of the Geological Survey through the Central and Southern Highlands, it is hard in many places to determine whether the sequence that can be recognized is in an upward or downward direction. Two bands of limestone, which appear to retain their relative positions across Scotland for a distance of some 230 miles, may afford a solution of this difficulty, and if, as is probable, they are to be identified with the similar limestones of Donegal, Mayo and Galway, their assistance will thus be available across a tract of more than 400 miles. What is regarded as the lower zone of limestone is particularly well seen about Loch Tay; what is believed to be the upper is typically displayed in the heart of Perthshire, about Blair-Athol.

From under the Loch Tay Limestone a great thickness of mica-schists, "green schists," schistose grits and conglomerates, slates and greywackes, emerges up to the border of the Highlands. Above that calcareous band thick masses of mica-schist and sericite-schist are succeeded by a well-marked zone of quartzite, which forms the mountains of Ben-y-Glo and Schihallion, and stretches south-westward across Argyllshire into Islay and Jura. The

[1] *Geol. Survey Memoirs: Geology of N.W. Donegal*, 1891.

second or Blair-Athol Limestone lies next to this quartzite. If the limestones are identical with those of Donegal, Mayo and Galway, the quartzites may doubtless be also regarded as continued in those of the same Irish counties, where they form some of the most conspicuous features in the scenery, since they rise into such conspicuous mountains as Erigal, Slieve League, Nephin, and the twelve Bins of Connemara.

The age of this vast system of altered rocks has still to be determined. It is possible that they may include some parts of the Torridonian series, or even here and there a wedge of the Lewisian gneiss driven into position by gigantic disruptions, like those of the North-West Highlands. But there can be no doubt that the schists, quartzites and limestones form an assemblage of metamorphosed sedimentary strata which differs much in variety of petrographical character, as well as in thickness, from the Torridonian sandstone, and which has not been identified as the equivalent of any known Palæozoic system or group of formations in Britain. It may conceivably embrace the Cambrian series of the North-West Highlands, and also the sedimentary deposits that succeeded the Durness Limestone, of which no recognizable vestige remains in Sutherland or Ross.

That the metamorphic rocks east of the line of the Great Glen are at least older than the Arenig formation of the Lower Silurian system may be inferred from an interesting discovery recently made by the officers of the Geological Survey. A narrow strip of rocks has been found which, from their remarkable petrographical characters, their order of sequence and their scanty fossil contents (*Radiolaria*), are with some confidence identified with a peculiar assemblage of rocks on the Arenig horizon of the Silurian system in the Southern Uplands of Scotland, to which fuller reference will be made in Chapter xii. This strip or wedge of probably Lower Silurian strata intervenes between the Highland schists and the Old Red Sandstone in Kincardineshire, Forfarshire and Dumbartonshire. It has been recognized also, occupying a similar position, in Tyrone in Ireland. The schists in some places retain their foliated character up to the abrupt line of junction with the presumably Lower Silurian strata, while in other districts, as at Aberfoyle, they have been so little affected that it is hardly possible to draw a line between the Highland rocks and those of this border-zone, which indeed are there perhaps more metamorphosed than the Highland grits to the north of them. The metamorphism of the schists may have been mainly effected before the final disturbances that wedged in this strip of Silurian strata along the Highland border, though some amount of crushing and schist-making seems to have accompanied these disturbances. No trace of any similar strip of Palæozoic rocks has ever been detected among the folds of the schists further into the Highlands. But some of the Highland rocks in the region of Loch Awe lose their metamorphosed character, and pass into sedimentary strata which, so far as petrographical characters are concerned, might well be Palæozoic.

Until some clue is found to the age of the Younger or Eastern schists, quartzites and limestones of the Highlands, it is desirable to have some short convenient adjective to distinguish them. As a provisional term for them I have proposed the term "Dalradian," from Dalriada, the name of the old Celtic kingdom of the north of Ireland and south-west of Scotland.[1]

The special feature for which this Dalradian series is cited in the present volume is the evidence it furnishes of powerful and extensive volcanic action. In a series of rocks so greatly dislocated, crumpled and metamorphosed, we cannot look for the usual clear proofs of contemporaneous eruptions. Nevertheless all over the Scottish Highlands, from the far coast of Aberdeenshire to the Mull of Cantyre, and across the west of Ireland from the headlands of Donegal into Galway, there occurs abundant evidence of the existence of rocks which, though now forming an integral part of the schists, can be paralleled with masses of undoubtedly volcanic origin.

Intercalated in the vast pile of altered sediments lie numerous sheets of epidiorite and hornblende-schist, which were erupted as molten materials, not improbably as varieties of diabase-lava. Most of these sheets are doubtless intrusive "sills," for they can be observed to break across from one horizon to another. But some of them may possibly be contemporaneous lava-streams. A sheet may sometimes be followed for many miles, occupying

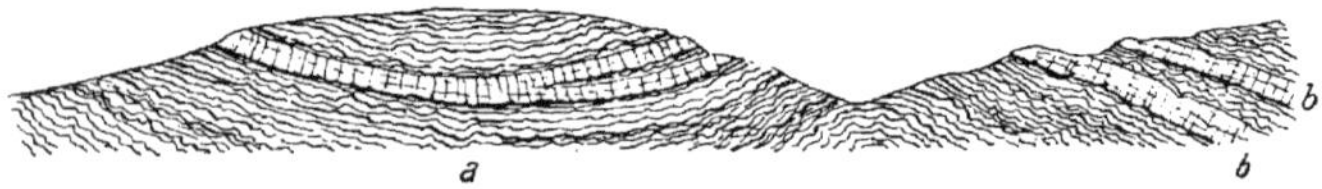

FIG. 37.—Section showing the position of Sills in the mica-schist series between Loch Tay and Amulree. *a*, Mica-schist; *b b*, Sills.

the same stratigraphical platform. Thus a band of sills may be traced from the coast of Banffshire to near Ben Ledi, a distance of more than 100 miles. Among the hornblendic sills of this band some occur on a number of horizons between the group of Ben Voirlich grits and the Ben-y-Glo quartzite. One of the most marked of these is a sheet, sometimes 200 feet thick, which underlies the Loch Tay Limestone. Another interesting group in the same great band has been mapped by the Geological Survey on the hills between Loch Tay and Amulree, some of them being traceable for several miles among the mica-schists with which they alternate (Fig. 37).

In Argyllshire also, between Loch Tarbert and Loch Awe, and along the eastern coasts of the islands of Islay and Jura, an abundant series of sheets of epidiorite, amphibolite and hornblende-schist runs with the prevalent strike of the schists, grits and limestones of that region. Similar rocks reappear in a like position in Donegal, where, as in Scotland, the frequency of the occurrence of these eruptive rocks on the horizons of the limestones is worthy of remark. The persistence, number and aggregate thickness of the sills in this great band mark it out as the most extensive series of intrusive sheets in the British Isles.

[1] *Presidential Address to Geological Society*, 1891, p. 39.

In addition to the sills there occur also bosses of similar material, which in their form and their obvious relation to the sheets recall the structure of volcanic necks. They consist of hornblendic rocks, like the sills, but are usually tolerably massive, and show much less trace of superinduced foliation.

Besides the obviously eruptive masses there is another abundant group of rocks which, I believe, furnishes important evidence as to contemporaneous volcanic action during the accumulation of the Dalradian series. Throughout the Central and South-Western Highlands certain zones of "green schist" have long occupied the attention of the officers of the Geological Survey. They occur more especially on two horizons between the Loch Tay Limestone and a much lower series of grits and fine conglomerates, which run through the Trossachs and form the craggy ridges of Ben Ledi, Ben Voirlich and other mountains near the Highland border. In the lower group of "green schists," thick hornblendic sills begin to make their appearance, increasing in number upwards. The upper group of "green schists" lies between two bands of garnetiferous mica-schist, above the higher of which comes the Loch Tay Limestone. The peculiar greenish tint and corresponding mineral constituents of these schists, however, are likewise found diffused through higher parts of the series.

So much do the "green schists" vary in structure and composition that no single definition of them is always applicable. At one extreme are dull green chlorite-schists, passing into a "potstone," which, like that of Trondhjem, can be cut into blocks for architectural purposes.[1] At the other extreme lie grits and quartzites, with a slight admixture of the same greenish-coloured constituent. Between these limits almost every stage may be met with, the proportion of chlorite or hornblende and of granular or pebbly quartz varying continually, not only vertically, but even in the extension of the same bed. The quartz-pebbles are sometimes opalescent, and occasionally larger than peas. An average specimen from one of the zones of "green schists" is found, on closer examination, to be a thoroughly schistose rock, composed of a matrix of granular quartz, through which acicular hornblende and biotite crystals, or actinolite and chlorite, are ranged along the planes of foliation.

That these rocks are essentially of detrital origin admits of no doubt. They differ, however, from the other sedimentary members of the Dalradian series in the persistence and abundance of the magnesian silicates diffused through them. The idea which they suggested to my mind some years ago was that the green colouring-matter represents fine basic volcanic dust, which was showered out during the accumulation of ordinary quartzose, argillaceous and calcareous sediments, and that, under the influence of the metamorphism which has so greatly affected all the rocks of the region, the original pyroxenes and felspars suffered the usual conversion into hornblendes, chlorites and micas. This view has occurred also to my colleagues on the Survey, and is now generally adopted by them.

[1] From such a rock, which crosses the upper part of Loch Fyne, the Duke of Argyll's residence at Inveraray has been built.

Not only are these "green schists" traceable all through the Central and South-Western Highlands, rocks of similar character, and not improbably on the same horizons, reappear in the north-west of Ireland, and run thence south-westward as far as the Dalradian rocks extend. If we are justified in regarding them as metamorphosed tuffs and ashy sediments, they mark a widespread and long-continued volcanic period during the time when the later half of the Dalradian series was deposited.

Besides the extensive development of basic sills which, though probably in great part later than the "green schists," may belong to the same prolonged period of subterranean activity, numerous acid protrusions are to be observed in the Dalradian series of Scotland and Ireland. That these masses were erupted at several widely-separated intervals is well shown by their relation to the schists among which they occur. Some of the great bosses and sills of granite were undoubtedly injected before the metamorphism of the schists was completed, for they have shared in the foliation of the region. Others have certainly appeared after the metamorphism was complete, for they show no trace of having suffered from its effects. Thus some of the vast tracts of newer granite in the Grampian chain, which cover many square miles of ground, must be among the newest rocks of that area. They have recently been found by Mr. G. Barrow, of the Geological Survey, to send veins into the belt of probably Lower Silurian strata which flanks the Highland schists. They are thus later than the Arenig period. Not impossibly they may be referable to the great granite intrusions which formed so striking a feature in the history of the Lower Old Red Sandstone.

iii. THE GNEISSES AND SCHISTS OF ANGLESEY

In the island of Anglesey an interesting series of schists and quartzites presents many points of resemblance to the Dalradian or younger schists of the Highlands. At present the geologist possesses no means of determining whether these Welsh rocks are the equivalents of the Scottish in stratigraphical position, but their remarkable similarity justifies a brief allusion to them in this place. Much controversy has arisen regarding the geology of Anglesey, but into this dispute it is not necessary for my present purpose to enter.[1] I will content myself with expressing what seems to me, after several traverses, to be the geological structure of the ground.

[1] The literature of Anglesey geology is now somewhat voluminous, but I may refer to the following as the chief authorities. The island is mapped in Sheet 78 of the Geological Survey of England and Wales, and its structure is illustrated in Horizontal Sections, Sheet 40. A full account of its various formations and of their relations to each other is given in vol. iii. of the *Memoirs of the Geological Survey*, "The Geology of North Wales," by Sir A. C. Ramsay, 2nd edit. 1881. The subject has been discussed by Professor Hughes, *Quart. Journ. Geol. Soc.* vols. xxxiv. (1878) p. 137, xxxv. (1879) p. 682, xxxvi. (1880) p. 237, xxxviii. (1882) p. 16; *Brit. Assoc. Rep.* (1881) pp. 643, 644; *Proc. Camb. Phil. Soc.* vol. iii. pp. 67, 89, 341; by Professor Bonney, *Quart. Journ. Geol. Soc.* vol. xxxv. (1879) pp. 300, 321; *Geol. Mag.* (1880) p. 125; by Dr. H. Hicks, *Quart. Journ. Geol. Soc.* vols. xxxiv. (1878) p. 147, xxxv. (1879) p. 295; *Geol. Mag.* (1879) pp. 433, 528 (1893) p. 548; by Dr. C. Callaway, *Quart. Journ. Geol. Soc.* vols. xxxvii. (1881) p. 210, xl. (1884) p. 567; and by the Rev. J. F. Blake, *Quart. Journ. Geol. Soc.*

There are two groups of rocks in Anglesey to which a pre-Cambrian age may with probability be assigned. In the heart of the island lies a core of gneiss which, if petrographical characters may be taken as a guide, must certainly be looked upon as Archaean. In visiting that district with my colleague Mr. Teall I was much astonished to find there so striking a counterpart to portions of the Lewisian gneiss of the north-west of Sutherland and Ross. The very external features of the ground recall the peculiar hummocky surface which so persistently characterizes the areas of this rock throughout the north-west of Scotland. If the geologist could be suddenly transported from the rounded rocky knolls of Sutherland, Ross-shire or the Hebrides to those in the middle of Anglesey, south of Llanerchymedd, he would hardly be aware of the change, save in the greater verdure of the hollows, which has resulted from a more advanced state of decomposition of the rocks at the surface, as well as from a better climate and agriculture.

When we examine these rocky hummocks in detail we find them to consist of coarse gneisses, the foliation of which has a prevalent dip to N.N.W. Some portions abound in dark hornblende and garnets, others are rich in brown mica, the folia being coarsely crystalline and rudely banded, as in the more massive gneisses of Sutherland. Abundant veins of coarse pegmatite may here and there be seen, with pinkish and white felspars and milky quartz. Occasionally the gneiss is traversed by bands of a dark greenish-grey rock, which remind one of the dykes of the north-west of Scotland. There are other rocks, some of them probably intrusive and of later date, to be seen in the same area; but they require more detailed study than they have yet received.

The relation of this core of gneiss and its associated rocks to the second group of pre-Cambrian rocks has not hitherto been satisfactorily ascertained. The core may conceivably be an eruptive boss in that group, and may have acquired its foliation during the movements that produced the foliation of the surrounding schists. But it seems more probable that the gneiss is much older than these schists, though it would undoubtedly participate in the effects of the mechanical movements which gave rise to their deformation, cleavage and foliation.

The second group of rocks occupies a large area in the west and in the centre and south of Anglesey. The schists of which it consists are obviously in the main a clastic series. One of their most conspicuous members is quartzite, which, besides occurring sporadically all over the island, forms the prominent mass of Holyhead Mountain. There are likewise flaggy chloritic schists, green and purple phyllites or slates, and bands of grit, while parts of the so-called "grey gneiss" consist of pebbly sandstones that have acquired a crystalline structure. That some order of sequence among these various strata may yet be worked out is not impossible, but the task will be one of no ordinary difficulty, for the plications

vol. xliv. (1888) p. 463. Further references to the work of these observers in Anglesey are given in Chapter xiii. p. 220 *et seq.* The Pre-Cambrian areas of Anglesey are shown in Map II.

and fractures are numerous, and much of the surface of the ground is obscured by the spread of Palæozoic formations and superficial deposits.

These Anglesey schists are so obviously an altered sedimentary series that it is not surprising that they should have been regarded as metamorphosed Cambrian strata. All that can be positively affirmed regarding their age is that they are not only older than the lowest fossiliferous rocks around them—that is, than Arenig or even Tremadoc strata—but that they had already acquired their present metamorphic character before these strata were laid down unconformably upon them. There is no actual proof that they include no altered Cambrian rocks. But when we consider their distinctly crystalline structure, and the absence of such a structure from any portion of the Cambrian areas of the mainland; when, moreover, we reflect that the metamorphism which has affected them is of the regional type, and can hardly have been restricted to merely the limited area of Anglesey;

FIG. 38.—Sketch of crushed basic igneous rock among the schists, E. side of Porth-tywyn-mawr, E. side of Holyhead Straits.

we must agree with those observers who, in spite of the absence of positive proof of their true geological horizon, have regarded these rocks as of much higher antiquity than the Cambrian strata of the neighbourhood. No one familiar with the Dalradian rocks of Scotland and Ireland can fail to be struck with the close resemblance which these younger Anglesey schists bear to them, down even into the minutest details. Petrographically they are precisely the counterparts of the quartzites and schists of Perthshire and Donegal, and a further connection may be established of a palæontological kind. The upper part of the Holyhead quartzite was found by Mr. B. N. Peach and myself in the autumn of the year 1890 to be at one place crowded with annelid-pipes, and I subsequently found the same to be the case with some of the flaggy quartzites near the South Stack.

For the purpose of the inquiry which forms the theme of this work, the feature of greatest interest about these younger schists of Anglesey is the association of igneous rocks with them. They include bands of dark basic material, the less crushed parts of which resemble the diabases of later formations, while the sheared portions pass into epidiorites and true hornblende-schists. As in other regions where eruptive rocks have been crushed down and changed into the schistose modification, it is frequently possible to see groups of uncrushed cores round which, under severe mechanical stresses, the rock has undergone this conversion. Lines of movement through the body of the rock may be detected by bands of schist, the gradation from the solid core to the hornblende-schist being quite gradual. The accompanying figure (Fig. 38) represents a portion of one of these crushed basic igneous rocks on the east side of Holyhead Straits.

As in the Dalradian series of the Highlands, many, perhaps most, of these igneous bands are probably intrusive sills, but others may be intercalated contemporaneous sheets. They occur across the whole breadth of the island from the Menai Strait to the shores of Holyhead.

Besides these undoubtedly igneous rocks, the green chloritic slates of Anglesey deserve notice. They are well-bedded strata, consisting of alternations of foliated fine grit or sandstone, with layers more largely made up of schistose chlorite. The gritty bands sometimes contain pebbles of blue quartz, and evidently represent original layers of sandy sediment, but with an admixture of chloritic material. The manner in which this green chloritic constituent is diffused through the whole succession of strata, and likewise aggregated into bands with comparatively little quartz-ose sediment, reminds one of the "green schists" of the Central Highlands and Donegal, and suggests a similar explanation. Taken in connection with the associated basic igneous rocks, these chloritic schists seem to me to represent a thick group of volcanic tuffs and interstratified sandy and clayey layers. If this inference is well founded, and if we are justified in grouping these Anglesey rocks with the Dalradian schists of Scotland and Ireland, a striking picture is presented to the mind of the wide extent and persistent activity of the volcanoes of that primeval period in Britain.[1]

iv. THE URICONIAN VOLCANOES

Along the eastern borders of Wales a ridge of ancient rocks, much broken by faults and presenting several striking unconformabilities, has long been classic ground in geology from the descriptions and illustrations of Murchison's *Silurian System*.[2] The main outlines of the structure of that district, first admirably worked out by this great pioneer, were delineated on the maps and sections of the Geological Survey, wherein it was shown that in the Longmynd an enormously thick group of stratified rocks, which, though unfossiliferous, were referred to the Cambrian system, rose in the very heart of the country; that to the east of these rocks lay

[1] Mr. E. Greenly, late of the Geological Survey of Scotland, has recently established himself on the Menai Strait for the purpose of working out in detail the geological structure of this interesting and complicated region. We may therefore hope that some of the still unsolved problems presented by the rocks of Anglesey will before long be satisfactorily explained.

[2] See especially chap. xix. vol. i. p. 225.

strata of Caradoc or Bala age; that by a great hiatus in the stratigraphy the Upper Silurian series transgressively wrapped round everything below it; that yet again the Coal-measures crept over all these various Palæozoic formations, followed once more unconformably by Permian and Triassic deposits.[1] Besides all this evidence of extraordinary and repeated terrestrial movement, it was found that the region was traversed by some of the most powerful dislocations in this country, while to complete the picture of disturbance, many protrusions of igneous rocks were recognized.

In a territory so complicated, though it had been sedulously and skilfully explored, there could hardly fail to remain features of structure which had escaped the notice of the first observers. In particular, the igneous rocks had been dealt with only in a general way, and they consequently offered a favourable field for more detailed study; while by a more searching examination of some of the rocks for fossils, important corrections of the earlier work might yet be made.

A notable step towards a revision of the received opinions regarding the igneous rocks of this region was taken by Mr. Allport, who showed that the so-called "greenstone" included masses of devitrified spherulitic pitchstones and perlites, together with indurated volcanic breccias, agglomerates and ashes.[2] Subsequently Professor Bonney described more fully the petrographical characters of the Wrekin igneous rocks, confirming and extending the observations of Mr. Allport.[3]

But the correction of the prevalent error as to the geological age of these rocks was due to Dr. Callaway, who, after spending much time and labour in ascertaining, by a careful search for fossils, the position of the superincumbent rocks (wherein he discovered Cambrian organisms), and in a detailed investigation of the structure and relationships of the igneous masses themselves, was led to regard them as part of an ancient pre-Cambrian ridge; and he proposed for the volcanic group the name of Uriconian, from the name of the former Roman town which stood not far to the west of them.[4] He has shown how essentially volcanic this ancient series of rocks is, how seldom they present any clearly-marked evidence of stratification, and how small is the proportion of sedimentary material associated with them.[5]

Subsequently Professor Lapworth, by his discovery of the *Olenellus*-fauna, marking the lowest known fossiliferous Cambrian zone in the Wrekin

[1] The area is embraced in Sheet 61 of the Geological Survey, and is illustrated by Nos. 33 and 36 of the sheets of Horizontal Sections. In the early editions of the Survey maps the "felspathic traps" and the "greenstones" of the Wrekin district were distinguished by separate colours, but unfortunately this useful and so far correct discrimination was given up in subsequent editions, where all the acid and basic rocks are merged into one.

[2] *Quart. Journ. Geol. Soc.* vol. xxxiii. (1877) p. 449.

[3] *Op. cit.* vol. xxxv. (1879) p. 662; vol. xxxviii. (1882) p. 124.

[4] *Quart. Journ. Geol. Soc.* vols. xxx. (1874) p. 196, xxxiv. (1878) p. 754, xxxv. (1879) p. 643, xlii. (1886) p. 481. For a criticism of Dr. Callaway's views as to the order of succession among the rocks of this district, see Prof. Blake, *op. cit.* vol. xlvi. (1890) p. 386, and Dr. Callaway's reply, vol. xlvii. (1891) p. 109.

[5] *Op. cit.* vol. xlvii. (1891) p. 123.

district, and his recognition of Cambrian fossils under the Coal-measures of Warwickshire, supplied valuable evidence for the discussion of the geological position of the older rocks of the Midlands. He has mapped in minute detail the rocks of the Wrekin, and has exhausted all the evidence that is at present obtainable on the subject. But unfortunately the publication of his researches is still delayed.[1]

It is now recognized that the core of the ancient ridge, extending from near Wellington through the Wrekin, Caer Caradoc and other hills, until it sinks beneath the Upper Silurian formations, is formed of igneous rocks that consist partly of lavas, partly of volcanic breccias and fine tuffs. The lavas are thoroughly acid rocks of the felsitic or rhyolitic type. One of them, about 100 feet thick, which forms a prominent feature on the flanks and crest of Caer Caradoc, shows abundant finely-banded flow-structure, often curved or on end, while its bottom and upper parts are strongly amygdaloidal, the cavities being occasionally pulled out in the direction of flow and lined with quartz or chalcedony. Some of the detached areas of eruptive rocks show the beautiful spherulitic and perlitic structures first noticed in this region by Mr. Allport. More recently the structures of these acid rocks have been described by Mr. F. Rutley.[2]

The breccias and tuffs appear to consist mainly of felsitic material. In the coarser varieties, fragments of finely-banded felsite may be noticed, while the finer kinds pass into a kind of hornstone (hälleflinta), which in hand-specimens could hardly be distinguished from close-grained felsite. In some places, these pyroclastic rocks are well stratified, but elsewhere no satisfactory bedding can be recognized in them. Various other rocks, which are probably intrusive, occur in the ridge. At either end of the Wrekin there is a mass of pink microgranite, while at Caer Caradoc numerous sheets of "greenstone," intercalated in the fine tuffs, sweep across the hill. Mr. Rutley has published an account of these basic rocks, which he classes as "melaphyres," or altered forms of basalt or andesite.[3] That at least some of them are intrusive is manifest by the way in which they ramify through the surrounding strata. But others are so strongly amygdaloidal and slaggy that they may possibly be true interbedded lavas, though there may be some hesitation in admitting that such basic outflows could be erupted in the midst of thoroughly acid ejections.[4] Leaving these doubtful flows out of account, we have here a group of undoubted volcanic rocks represented by acid lavas and pyroclastic materials, by intrusive bosses of acid rocks, and by younger basic sills. The general lithological characters of these masses and the sequence of their appearance thus strongly resemble those of subsequent Palæozoic volcanic episodes.

[1] *Geol. Mag.* (1882) p. 563, (1886) p. 319, (1887) p. 78, (1888) p. 484; and a joint paper with Mr. W. W. Watts on the Geology of South Shropshire, *Proc. Geol. Assoc.* vol. xiii. (1894) pp. 302, 335.

[2] *Quart. Journ. Geol. Soc.* vol. xlvii. (1891) p. 540. Mr. Rutley more particularly describes those of Caradoc Hill.

[3] *Op. cit.* p. 534.

[4] This difficulty, however, need not be in itself insuperable, as is evident from the remarkable alternation of basic and acid lavas and tuffs in the Cambrian volcanic group of St. David's and in the Old Red Sandstone series of the Pentland Hills.

The geological age of this volcanic group is a question of much interest and importance in regard to the history of volcanism in this country. An inferior limit to the antiquity of the group can at once be fixed by the fact that, as originally pointed out by Dr. Callaway, the quartzite which overlies the volcanic rocks passes under a limestone containing Cambrian fossils in which Professor Lapworth has since recognized *Olenellus, Paradoxides* and other Lower Cambrian forms. The eruptions, therefore, must be at least as old as the earlier part of the Cambrian period. But it is affirmed that the quartzite rests with a complete unconformability on the volcanic rocks. If this be so, then the epoch of eruption must be shifted much farther back.

The evidence adduced in favour of this great break appears to me to be threefold. In the first place, the quartzite contains fragments of the volcanic rocks. I do not think much stress can be laid on this fact. When I visited the ground, what struck me most in the composition of the quartzite was its singularly pure quartzose character, and the comparative scarcity of felsite-pebbles in it. Any deposit laid down conformably upon the top of the breccias and tuffs might obviously contain some of these materials, while, if laid down unconformably, it might reasonably be expected to be full of them.

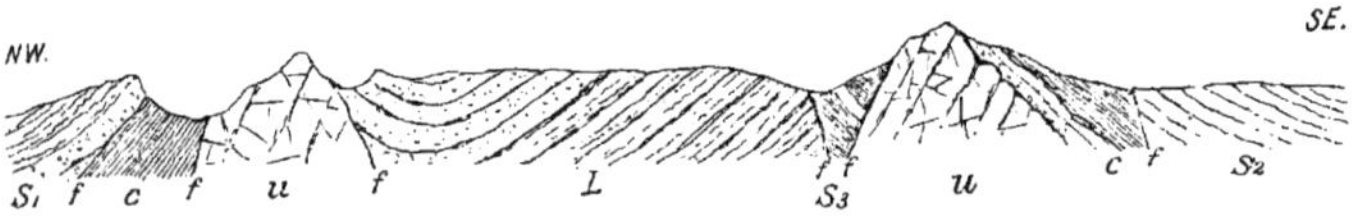

FIG. 39.—Section across the Uriconian series of Caer Caradoc.

S3, Upper Silurian; S2, Bala group; S1, Arenig group; C, Cambrian; L, Longmyndian; *u*, Uriconian; *ff*, faults.

In the second place, this quartzite is alleged to pass transgressively across the edges of successive sheets of the volcanic group, and thus to have a quite discordant dip and strike. I failed to find satisfactory evidence of this unconformability in the northern part of the district. But in the Caer Caradoc area the quartzite does appear to steal across the outcrops of the older rocks, which plunge at nearly right angles in an opposite direction. In the third place, the felsitic volcanic group is believed by Professor Lapworth to pass upwards into the Longmynd rocks. Obviously, if this group lies at the very bottom of the vast Longmynd series, the discordance between it and the quartzite must be enormous, and the date of the volcanic eruptions must be placed vastly farther back in geological antiquity. Though the evidence does not seem to me to amount to clear proof, I am disposed, in the meantime, to accept it as affording the most probable solution of the difficulties presented by the structure of the ground.

The sequence of the rocks around Caer Caradoc is partly concealed by surface accumulations, but if these could be cleared away the structure of the ground would be, according to Messrs. Lapworth and Watts, as shown in Fig. 39.[1]

[1] *Proc. Geol. Assoc.* vol. xiii. (1894), pp. 314, 315.

If, then, this volcanic group underlies the whole of the Longmynd series, and if, as it now appears, that series is older than the *Olenellus*-zone of the Lower Cambrian rocks, we can hardly include the volcanic rocks of the Wrekin and Caer Caradoc in the Cambrian system. They must belong to a still older geological formation, and I think we cannot do better than adopt for them Dr. Callaway's name, Uriconian.

There are still, however, many problems to be solved before the geological history of that region is completely understood. The rocks of the Longmynd must be more fully worked out. It is improbable that strata which look so likely to yield fossils should for ever prove barren. The lower half at least may be hopefully searched, although the upper massive reddish sandstones and conglomerates offer less prospect of success. On the west side of the Longmynd, above Pontesbury, there occurs a small area of volcanic rocks like those of the Wrekin district, including a well-marked nodular felsite and fine tuffs. These rocks have been regarded by Dr. Callaway as another axis of the Uriconian series. It is very difficult, however, by any combination of geological structures, to bring up a portion of the very bottom of the Longmynd series and place it apparently at the top. This is a feat which a detailed study of the region, and the detection of unconformabilities in the Longmynd, may possibly accomplish. In the meantime, however, I would venture to suggest whether it is not more probable that we have here a detached area of much younger volcanic rocks, like those which, in various districts, may be included in the Cambrian system, and which will be referred to in some detail in subsequent pages.

V. THE MALVERN VOLCANO

Regarding the age and origin of the oldest rocks of the Malvern Hills some controversy has arisen, and no general agreement has yet been reached.[1] It is clear that the core of crystalline rocks which is overlain unconformably by the Hollybush Sandstone must be older than the Upper Cambrian rocks. There is no good evidence of any stratigraphical break in the Cambrian system of England or Wales, and it may be reasonably inferred that the break seen at the base of the Hollybush Sandstones indicates that the rocks underneath that horizon are pre-Cambrian. Some portions of these certainly very ancient rocks are gneisses or schists; others have been described as "felsites," and have been regarded as passing into schists, and as the original material from which portions of the foliated series of the range have been produced by mechanical deformation. Not improbably the

[1] There is no room here for a full bibliography of the geological literature devoted to this locality. In the monograph by J. Phillips in vol. ii. part i. of the *Memoirs of the Geological Survey*, a list of writings is given up to the time of its publication in 1848. Since that year many additional papers have appeared. I may especially refer to H. B. Holl, *Quart. Journ. Geol. Soc.* xxi. (1865) p. 72; J. H. Timins, *op. cit.* xxii. (1867); Mr. F. Rutley, *op. cit.* xliii. (1887) p. 481; Dr. Callaway, *op. cit.* xliii. (1887) p. 525, xlv. (1889) p. 475, xlix. (1893) p. 398; Prof. Green, *op. cit.* li. (1895) p. 1; Mr. H. D. Acland, *Geol. Mag.* 1894, p. 48.

whole series of rocks is of igneous origin, but has been subsequently rendered more or less schistose.

There is one area where the rocks have escaped metamorphism, and where they present some of the well-known features of ancient volcanic materials. This tract was first indicated by Dr. H. B. Holl as one occupied by "altered primordial rocks and post-primordial trap." Its evidently igneous materials have been examined and described by different observers, among whom Dr. Callaway has contributed some detailed papers on the subject. More recently Professor Green, who had the advantage of sections exposed in the excavations for the construction of a reservoir for supplying water to Great Malvern, came to the conclusion that the rocks consist mainly of felsites, having many of the characters of rhyolites. With these are associated felsitic tuffs, while bands of dolerite, probably intrusive, form likewise part of the series. So far as the somewhat meagre evidence allows an opinion to be formed, there appears to be an alternation of felsites, lavas and tuffs placed in a more or less vertical position, striking in a northerly direction, and traversed by several sheets of intrusive dolerite.

No junction has been found between these unfoliated volcanic rocks and the schists that form the core of the range. Judging merely from their present relative condition, one would naturally infer that the volcanic rocks must be the younger of the two groups. But, as Professor Green has pointed out, it is conceivable that the latter may have locally escaped crushing, and yet be of the same age as the felsites and epidiorites of the neighbouring Raggedstone Hill, which have been in part considerably affected by mechanical movements.[1]

For our present inquiry it is perhaps sufficient to take note that in the heart of the Malvern Hills there lies a remnant of a volcanic district, probably of pre-Cambrian age, the rocks of which had been raised up into a vertical position so as to form islets or reefs in the sea in which the Upper Cambrian strata (Hollybush Sandstone and Upper Lingula shales) were deposited. Until some more precise evidence is obtained as to the geological age of these rocks it may be convenient to place them provisionally with the volcanic Uriconian series.

vi. THE CHARNWOOD FOREST VOLCANO

In the heart of England the great Triassic plain is diversified by the uprise through it of the peaks and crests of an old Triassic land-surface, which are embraced in the district known as Charnwood Forest. These scattered eminences consist of materials not only immensely older than the Trias, but once doubtless buried under thousands of feet of Palæozoic strata. They had been laid bare by denudation and carved into picturesque crags and pinnacles before the New Red Sandstone was deposited around and above them.

[1] *Op. cit.* p. 7. The metamorphism of the igneous rocks of the Malvern Hills into schists has been especially investigated by Dr. Callaway.

To these vestiges of an early Mesozoic land, still half buried among Triassic strata, a peculiar interest attaches from the obviously high antiquity of their rocks and their uprise in the very centre of the island. Various opinions have been expressed as to the age of their component rocks. When they were mapped by the Geological Survey they were recognized to be as old as any group of rocks then known, and they were accordingly placed in the Cambrian system. More recent research has suggested that they may be still more ancient, and may be regarded as pre-Cambrian.

The rocks of Charnwood Forest have been the subject of an exhaustive research by the Rev. E. Hill and Professor Bonney, to whom most of our knowledge regarding them is due. These observers first pointed out the truly volcanic nature of the coarse clastic rocks of the district. They have traced their relations in the field, and have likewise described their structure and composition as shown by the microscope. Subsequently the district has been re-mapped on the scale of six inches to a mile by Mr. Fox Strangways for the Geological Survey, while Mr. W. W. Watts, another member of the Survey, has studied the petrography of the ground, and has traced the boundaries of the several rock-groups so far as these can be determined. Confirming generally the stratigraphical arrangement sketched by Messrs. Hill and Bonney, Mr. Watts has proposed the following classification of the rocks:—[1]

7. Groby and Swithland slates.	The Brand series.
6. Hanging Rocks conglomerate and Bradgate quartzite.	
5. Woodhouse beds (ashy grits).	The Maplewell series (volcanic tuffs and agglomerates).
4. Slate-agglomerate of Roecliffe.	
3. Hornstone beds of Beacon Hill.	
2. Felsitic agglomerate of Benscliffe.	
1. Quartzose, felspathic and felsitic grits.	The Blackbrook series.

Under any computation or measurement, the total thickness of detrital material in this series of formations must amount to several thousand feet. The chief interest centres in the middle series, which consists largely of fragmental volcanic rocks, with intercalations of slate and grit. As was first shown by Mr. Hill and Professor Bonney, these volcanic materials vary from exceedingly coarse agglomerates to fine, ashy or felspathic slates. In most cases distinct bedding can be recognized in them, but more particularly in the fine-grained material. Yet even among the massive agglomerates a tendency may be seen towards an orientation of the blocks with their long axes parallel. That this arrangement is not entirely due to the effects of cleavage may be inferred from the many exceptions to it, which would hardly have occurred had such powerful cleavage affected the whole district, as would be needed to rearrange the large blocks in the agglomerates. Besides, the coarser parts often inter-

[1] Annual Report of Director-General of the Geological Survey, in the *Report of Science and Art Department for* 1895.

calate with fine felspathic grits, which distinctly mark the stratification of the whole.

The remarkably coarse breccia of Benscliffe is mainly made up of blocks of quartz-porphyry, felsite or rhyolite, with slate fragments. The Roecliffe agglomerate, another extraordinarily coarse rock, consists of slate fragments imbedded in an andesitic matrix, some of the blocks of slate being six feet long. The finer tuffs have been ascertained to consist of felsitic or andesitic detritus, sometimes forming exceedingly compact flinty rocks or hornstones.

In this thick accumulation of detrital rocks we are presented with a series of alternations of coarser and finer pyroclastic material, interstratified among green, grey and purple slates and grits, which probably represent the non-volcanic sediments of the time of eruption. The succession of strata bears witness to a long series of eruptions of varying intensity, but culminating at two distinct periods in the discharge of huge blocks of rock (Benscliffe and Roecliffe agglomerates).

After some search I have been unable to detect a single vesicular fragment among the stones in the breccias and tuffs, and Messrs. Hill and Bonney were not more successful. Not a trace of anything in the least degree scoriaceous is anywhere to be found. The paste in which the blocks lie consists of such fine material as would result from the trituration of felsite and slate. It contains many broken crystals of felspar, with grains of clear quartz. A gradation can be traced from the coarser into the finer bands of volcanic and non-volcanic material, fine slates being also interleaved with highly-felspathic partings of grit.

Having looked with some care for a trace of a true volcanic neck in the district, I have not seen anything that could be unhesitatingly so designated. Even in the north-western part of the district, where the breccias are coarsest, and there is least trace of ordinary sediments, some signs of bedding can usually be detected in the position of the imbedded stones and the partings of finer tuff. Both the coarser and finer detritus suggest the kind of material discharged from vents before the uprise of any lava. The entire absence of scoriaceous fragments is noteworthy, and the abundance of slate blocks rather points to the early eruptions of a volcanic focus. Possibly, while the chief centre of eruption lay towards the north-west, numerous vents may have been opened all over the district, discharging abundant showers of dust and stones, but seldom or never culminating in the actual outpouring of lava.

No indubitable lava-sheet has, in my judgment, been yet recognized in Charnwood Forest. Various opinions have been expressed as to some of the more compact close-grained rocks, and even the verdicts of the same observers have varied from time to time, the rocks once considered as felsites being afterwards regarded as tuffs, and subsequently placed with the felsites or andesites after all. It is not necessary for my present purpose to enter into these questions, which are rather of local interest. I will only say that, in my opinion, the rocks of Sharpley, Peldar, and Bardon Hill are massive rocks, as they have finally been classed by Messrs. Hill and Bonney. But I cannot look upon them as lavas, at least I have seen no evidence to lead me to believe that they were ever erupted at the surface. I have fully considered the arguments of Mr. Hill and Professor Bonney on this point.[1] There can, I think, be no doubt of the close association of these felsitic rocks and the breccias, but the structure of the rocks in the field seems to me to be decidedly in favour of the view expressed above. The microscope affords no assistance in the question.[2] The doubtful rocks seem to me rather to be intrusive masses which have been protruded into the volcanic sedimentary series among which they rise. They are acid, fine-grained, porphyritic rocks, which would formerly have been included under the general name of felsites or quartz-porphyries. Their coarse porphyritic parts rapidly pass into close-grained felsitic material. Many of the blocks in the breccias are precisely like parts of these rocks. It might hence be asserted that these fragmental deposits are later than the eruptive bosses. At least it is obvious that rocks of the same type as those of Sharpley, Peldar, and Bardon Hill must have been disrupted to produce the coarse breccias.

Later eruptive rocks, consisting of masses of syenite and granite, with still younger dykes of dolerite, andesite, diorite and felsite, have successively made their appearance, and add to the diversity of the igneous phenomena of this district.

The question of the age of this isolated volcanic series is one of much interest, but of great perplexity. Though a resemblance may be admitted to exist between some of the slates and parts of the Cambrian system of North Wales, the difference between the Charnwood rocks and the undoubted Cambrian series of Warwickshire, only thirteen miles to the south-west, is such as to indicate that the former are probably older than the latter. While the Charnwood rocks have been intensely cleaved and crushed, those of Warwickshire have undergone no such change. The argillaceous strata in the one region have been converted into slates, in the other they remain mere shales. Though cleavage is sometimes irregularly developed, its rapid disappearance in so short a distance as the interval between Charnwood Forest and Nuneaton seems most explicable if we suppose that the rocks at the more easterly locality were cleaved before those towards the west were deposited. If this inference be well grounded the pre-Cambrian age of the Charnwood volcanoes would be established. But the argument is not conclusive. No fossils of any kind have yet been found in any of the old rocks of Charnwood.[3] Merely lithological resemblances or differences are all that can be used as a guide to the geological age of these masses. Mr. Watts has suggested that possibly the quartzite

[1] *Quart. Journ. Geol. Soc.* xlvii. (1891), pp. 80-88.

[2] See Messrs. Hill and Bonney, *op. cit.* xxxiv. (1878), p. 211.

[3] Since this page was in type, Professor Lapworth has found a worm-burrow low down in the Brand Series, and one or two additional examples have since been obtained by Mr. J. Rhodes of the Geological Survey. These are the first undoubted organisms from the Charnwood Forest rocks. Mr. Watts, *Geol. Mag.* 1896, p. 487.

of Bradgate (No. 6 of the Charnwood groups) may be the equivalent of the quartzite which in Shropshire and Warwickshire forms the base of the sedimentary Cambrian formations. If that correlation could be established, the volcanic series below the quartzite in Charnwood might be regarded as representing the Uriconian volcanic series of Shropshire.

BOOK III

THE CAMBRIAN VOLCANOES

CHAPTER IX

CHARACTERISTICS OF THE CAMBRIAN SYSTEM IN BRITAIN

The Physical Geography of the Cambrian Period—The Pioneers of Palæozoic Geology in Britain—Work of the Geological Survey in Wales—Subdivisions of the Cambrian System in Britain.

In leaving the investigation of the pre-Cambrian rocks and entering upon that of the Palæozoic systems, that is, the great series of sedimentary formations which include the earliest records of organized life upon the surface of the globe, the geologist feels much as the historian when, quitting the domain of legend and tradition, he can tread firmly in the region of documentary evidence. From the bottom of the Cambrian system upward through the long series of geological formations, the chronicle, though often sadly incomplete, is usually clear and legible. As we follow the lowest fossiliferous strata across a territory, we recognize that they bear witness to the same processes of denudation and deposition which have been going on uninterruptedly on the face of the globe ever since. The beds of conglomerate represent the gravels and shingles of old coast-lines and river-beds. The sandstones recall the familiar features of sandy sea-bottoms not far from land. The shales bear witness to the fall of fine sediment in stiller water, such as now takes place in the deeper parts of seas and lakes. Notwithstanding their vast antiquity, the strata themselves exhibit no exceptional peculiarities of origin. They seem to be just such familiar deposits as are gathering under fitting conditions at the present time.

Some writers have speculated on the far greater intensity of all geological activities in the early times of the planet's history. But if we may interpret the record of the stratified formations by the phenomena of today, there is for these speculations no confirmation in the sedimentation of the oldest stratified deposits. It is of course quite intelligible, if not probable, that many geological forces may have been more vigorous in primeval times than they afterwards became. But of the gigantic tides, prodigious denudation and violent huddling together of the waste of the earth's surface, which have been postulated for the early Palæozoic ages, there is assuredly nowhere any indication among the stratified formations. In those vast orderly repositories, layer succeeds layer among thinly-laminated shales, as gently and equably as the fine silt of each tide sinks to-day over the floor of a sheltered estuary. At the primeval period of which these sediments are the memorial, the waters receded from flat shores and left tracts of mud bare to the sky, precisely as they do still. Then as now, the sun shone and dried such mud-flats, covering their surfaces with a network of cracks; the rain fell in heavy drops, that left their imprints on the drying mud; and the next tide rose so gently as to overflow these records of sunshine and shower without effacing them, but spreading over them a fresh film of sediment, to be succeeded by other slowly-accumulating layers, under which they have lain preserved during the long cycles of geological history.

That organized creatures had already appeared upon the earth's surface before the beginning of the Cambrian period cannot be doubted. The animal remains in the lowest Cambrian strata are far from being the simple forms which might be expected to indicate the first start of animal life upon the surface of the earth. On the contrary, though they are comparatively scanty in types, and often rare or absent throughout a thick mass of sedimentary deposits, they show beyond dispute that, when they flourished, invertebrate life had already reached such a stage of advancement and differentiation that various leading types had appeared which have descended, in some cases with generic identity, down to our own day. There must have been a long pedigree to these organisms of the oldest known fossiliferous rocks. And somewhere on the earth's surface we may yet hope to find the remains of their progenitors in pre-Cambrian deposits.

The researches of many explorers in Europe and North America have brought to light an interesting series of organic remains from the Cambrian system. Of the plants of the time hardly any traces have survived, save some markings which have been referred to sea-weeds. The earliest known sponges and corals occur in this system, likewise the ancestors of the graptolites, which played so prominent a part in the life of the next or Silurian period. There were already representatives of crinoids and star-fishes, besides examples of the extinct group of cystideans. Sea-worms crawled over the muddy and sandy sea-bottom, for they have left their trails and burrows in the hardened sediments. Molluscs had by this time appeared in their four great divisions of Brachiopods, Lamellibranchs, Gasteropods and Cephalopods, though the forms yet discovered among Cambrian rocks are comparatively few. The most abundant and characteristic inhabitants of the Cambrian seas were the trilobites, of which many genera have been disinterred from the strata. In the lowest fossiliferous Cambrian group the trilobitic genus *Olenellus*, already referred to, is the characteristic form. Higher up *Paradoxides* is predominant, while towards the top of the system the most characteristic genus is *Olenus*.

From the organic remains which have been preserved, we may legitimately infer the existence of others which have entirely disappeared. There seems no reason to doubt that the leading grades of invertebrate life which are wanting in the known Cambrian fauna were really represented in the Cambrian seas. The chance discovery of a band of limestone may any day entirely alter our knowledge as to the relative proportions of the several divisions of the animal kingdom in the earliest Palæozoic rocks. Sand is rather adverse to the preservation of a varied representation of the organisms of the overlying sea-water. Mud is generally favourable, but calcareous accumulations are greatly more so, and they usually consist almost entirely of organic remains. Thus in the Cambrian series of the north-west of Scotland the quartzites that form the lower group, though sometimes crowded with worm-burrows, contain hardly any other sign of organisms. The overlying shales, besides their abundant worm-castings, have yielded perfect specimens of *Olenellus* and other fossils. But in the uppermost group, consisting of limestones, every particle of the sediment appears to have passed through the intestines of worms, and as it gathered on the sea-bottom it enclosed and has preserved a varied and abundant assemblage of organisms, including trilobites, gasteropods and a number of cephalopods. While in the Cambrian rocks of Europe calcareous bands are comparatively rare, in those of North America they are not infrequent. Hence it is largely from American deposits that our knowledge of the Cambrian fauna has been derived.

Not a vestige of any vertebrate organism has yet been detected among the earlier Palæozoic sediments. So far as we know, there were no fishes in the Cambrian seas. The highest organisms then existing were chambered shells, a once abundant and singularly varied class, of which the living Nautilus is now the sole representative.

In trying to realize the general geographical conditions of Cambrian time, the geologist finds himself entirely without any evidence as to the character of the terrestrial vegetation. We can hardly doubt that the land was clothed with plants, probably including lycopods and ferns, possibly even cycads and conifers. But no remains of this flora have yet been recovered. Nor have any traces of land-animals been detected. All that we yet know of the life of the period has been gleaned from marine sediments, which show that the invertebrate population by which the sea was then tenanted embraced some of the leading types of structure that have survived through all the long vista of geological time down to our own day.

Some of the shore-lines of the Cambrian waters may still be traced, and it is possible to say where the land of the time stood and where lay the sea. In the British area the largest relic of Cambrian land is found in the far north-west of Scotland. Formed partly of the Lewisian Gneiss and partly of the Torridon Sandstone, it takes in the whole chain of the Outer

Hebrides and likewise part of the present western seaboard of Sutherland and Ross. Along the margin of that northern land the white sand was laid down which now gleams in sheets of snow-like quartzite on most of the higher mountains from Cape Wrath to Skye. The sea lay to the east and, so far as we know, may have stretched across the rest of Scotland, and the north and centre of England. Another vestige of the land of this ancient era occurs in Anglesey. There, and likewise over scattered tracts in the Midlands, and in the south-west of England, the geologist seems to descry the last relics of islets that rose out of the Cambrian sea, and are now surrounded with its hardened sediments.

While such was the general aspect of the region of the British Isles during Cambrian time, volcanic action manifested itself at various localities over the area, breaking out on the sea-bottom, and pouring forth sheets of lava and showers of ashes, which mingled with the sand and silt that were settling there at the time. In the northern or Scottish tract no trace of this subterranean activity has been found; but in the English Midlands and over much of Wales abundant evidence has been obtained to show that in those districts the Cambrian period was marked by frequent and prolonged eruptions.

As its name denotes, the Cambrian system is typically developed in Wales. It was there that Sedgwick first worked out the stratigraphical relations of its ancient sediments, and that Murchison demonstrated the succession of organic remains contained in them, applying to them the principles of classification laid down by William Smith in regard to the Secondary formations. It was there too that some of the earliest and most memorable achievements were made in the investigation of ancient volcanic rocks. Sedgwick and Murchison, besides the admirable work which they accomplished in establishing the stratigraphy of the older Palæozoic formations, clearly recognized that among these formations there were preserved the records of contemporaneous submarine eruptions. Sedgwick showed that the mountainous masses of eruptive rock in North Wales were really lavas and ashes, which had been discharged over the sea-floor at the time when the ancient sediments of that region were deposited, while Murchison established the same fact by numerous observations in the east and south of Wales, and in the bordering English counties. De la Beche had found similar evidence among the "grauwacke" rocks of Devonshire.[1]

Following in the track thus opened up by these great masters, the officers of the Geological Survey were enabled to unravel, as had never before been attempted, the complicated structure of the old volcanic regions of Wales.

[1] For early researches on the older Palaeozoic volcanic rocks of Britain, see Sedgwick, *Proc. Geol. Soc.* vols. ii. (1838) pp. 678, 679, iii. (1841) p. 548, iv. (1843) p. 215; *Quart. Journ. Geol. Soc.* vols. i. (1845) pp. 8-17, iii. (1847) p. 134. Murchison, *Proc. Geol. Soc.* vol. ii. (1833-34) p. 85; *Silurian System* (1839) pp. 225, 258, 268, 287, 317, 324, 401; *Siluria*, 4th edit. (1867) p. 76 *et seq.* De la Beche, *Mem. Geol. Survey*, vol. i. (1846) pp. 29-36. A. C. Ramsay in the Maps and Horizontal Sections of Wales published by the Geological Survey; also Descriptive Catalogue of the Rock-Specimens in the Museum of Practical Geology, 1st edit. (1858), 2nd edit. (1859), 3rd edit. (1862); "The Geology of North Wales," forming vol. iii. of *Memoirs of the Geological Survey*, 1st edit. (1866), 2nd edit. (1881).

At the outset of the following discussion I wish to express my admiration of the labours of the early pioneers who thus laid for us the foundation of our knowledge of volcanic action in the Palæozoic periods. To De la Beche and his associates in the Survey a special measure of gratitude is due from all who have followed in their steps and profited by their work. When we consider the condition of geological science, and especially of the department of petrography, in this country at the time when these early and detailed investigations were carried on, when we remember the imperfection of much of the topography on the old one-inch Ordnance maps (which were the only maps then available), when we call to mind the rugged and lofty nature of the ground where some of the most complicated geological structures are displayed, we must admit that at the period when these maps and sections were produced they could not have been better done; nay, that as in some important respects they were distinctly in advance of their time, their publication marked an era in the progress of structural, and especially of volcanic, geology. The separation of lavas and tuffs over hundreds of square miles in a mountainous region, the discrimination of intrusive sheets and eruptive bosses, the determination of successive stratigraphical zones of volcanic activity among some of the oldest fossiliferous formations, were achievements which will ever place the names of Ramsay, Selwyn, Jukes and their associates high in the bede-roll of geological science. No one ever thinks now of making a geological excursion into Wales without carrying with him the sheets of the Geological Survey map. These form his guide and handbook, and furnish him with the basis of information from which he starts in his own researches.

But science does not stand still. The most perfect geological map that can be made to-day will be capable of improvement thirty or forty years hence. The maps of the Geological Survey are no exception to this rule. In criticizing and correcting them, however, let us judge them not by the standard of knowledge which we have now reached, but by that of the time when they were prepared. It is easy to criticize; it is not so easy to recognize how much we owe to the very work which we pronounce to be imperfect.

The ancient volcanoes of Wales, thanks mainly to the admirable labours of my former friend and chief, Sir Andrew C. Ramsay, have taken a familiar place in geological literature. But a good deal has been learnt regarding them since he mapped and wrote. The volcanic history, as he viewed it, began in the Arenig period. The progress of subsequent inquiry, however, has shown that there are volcanic rocks in Wales of much older date. I shall show that the Cambrian period, both in South and North Wales, was eminently volcanic.

Much controversy having arisen as to the respective limits and nomenclature of the older Palæozoic rocks, let me state, at the outset of the inquiry into the volcanic eruptions of Cambrian time, that under the term "Cambrian" I class all the known Palæozoic rocks which lie below the bottom of what is termed the Arenig group. It was maintained by Sir

Andrew Ramsay and his colleagues on the Geological Survey that on the mainland of Wales no base is ever found to the Cambrian system. More recently certain conglomerates have been fixed upon as the true Cambrian base, both in South and North Wales, and endeavours have been made to trace an unconformability at that line, all rocks below it being treated as pre-Cambrian. But conglomerates do not necessarily mark a stratigraphical discordance, and in South Wales there is no trace of any unconformability between the strata above and below the supposed line of break.[1] Professor Bonney has shown that in North Wales several zones of conglomerate have been erroneously identified as the supposed basal platform of the Cambrian series, and more recently Mr. Blake has pointed out that some of these conglomerates are unquestionably Lower Silurian.

My own examination so far confirms the conclusions arrived at by these observers. Like my predecessors in the Geological Survey, however, I have been unable to detect anywhere in Caernarvonshire or Merionethshire a base to the Cambrian system, and I am compelled to agree with them in regarding as Cambrian (partly even as Lower Silurian) all the rocks from Bangor to Llanllyfni, which have more recently been classed as pre-Cambrian. But though thus supporting their general stratigraphy, I am bound to acknowledge that they failed to recognize the existence of a great volcanic series below the Arenig horizon. The existence of this series, noticed by Sedgwick, was first definitely stated by Professor Hughes,[2] and his statements have been confirmed and extended by subsequent observers, notably by Professor Bonney and Mr. Blake. The Cambrian period is thus proved to have been perhaps even more continuously volcanic than the Lower Silurian period was in Wales.

The following table shows the subdivisions of the Cambrian system now recognized in Britain:—

Wales. (Ranging up to 12,000 feet or more.)		Western England. (About 3000 feet.)	N. W. Scotland. (About 2000 feet.)
Upper or *Olenus* Zones.	Tremadoc Slates Lingula Flags (*Lingulella, Olenus*, etc.).	Shineton Shales (*Dictyograptus* or *Dictyonema, Olenus*, etc.).	Limestones, about 1500 feet thick, divisible into seven groups (*Archæocyathus, Maclurea, Ophileta, Murchisonia, Orthoceras* and vast quantities of annelid castings).
Middle or *Paradoxides* Zones.	Menevian group (*Paradoxides*).	Conglomerates and limestones (Comley), with *Paradoxides*, etc.	
Lower or *Olenellus* Zones.	Harlech and Llanberis group with basement volcanic rocks; bottom not seen.	Thin quartzite passing up into green flags, grits, shales and sandstone (Comley Sandstone), containing *Olenellus*.	Shales ("fucoid beds"), with *Olenellus, Saltcrella*, etc. Quartzites with annelid burrows. The base of the series lies unconformably on pre-Cambrian rocks.

[1] See a discussion of this subject in *Quart. Journ. Geol. Soc.* vol. xxxix. (1883), p. 305.

[2] *Proc. Camb. Phil. Soc.* vol. iii. (1877), p. 89. The Cambrian volcanic areas of North Wales are represented in Map II.

CHAPTER X

THE CAMBRIAN VOLCANOES OF SOUTH WALES

In the southern part of the Principality of Wales a remarkably varied display of British Cambrian volcanic rocks has been preserved. The district around St. David's has the distinction of being the first in which volcanic rocks of such high antiquity were recognized. As far back as the year 1842, Ramsay found that "felspathic volcanic ash" was associated with other proofs of igneous action, and this fact was recorded by him on the published Horizontal Sections of the Geological Survey. Unfortunately he afterwards came to regard the rocks as "altered Cambrian," thus following certain hypothetical views which, as will be further alluded to in the sequel, he had adopted in explanation of the phenomena in Caernarvonshire and in Anglesey. The volcanic nature of these ancient materials was subsequently rediscovered by Dr. Hicks, who has devoted much time and labour to their study. Distinguishing the volcanic series of St. David's by the name "Pebidian," he has contended that it forms a pre-Cambrian system separated by an unconformability from the base of the Cambrian formations. He likewise endeavoured to show that an older system of rhyolitic lavas, felsitic breccias and hälleflintas could be distinguished, which he termed "Arvonian"; and more ancient still, a core of granitoid or gneissic rocks, which he separated under the name of "Dimetian." My own investigation of the ground thoroughly convinced me that there are no pre-Cambrian rocks at St. David's; that the "Arvonian" and "Dimetian" series are merely intrusive rocks (quartz-porphyry, granite, etc.) which have invaded the volcanic series; and that the "Pebidian," instead of being a pre-Cambrian formation on which the Cambrian base rests unconformably, is a group of volcanic rocks into which the Cambrian strata pass down conformably, and which in the St. David's district constitutes the lowest group of the Cambrian system.[1]

[1] For Dr. Hicks' views, see especially his papers in the *Quart. Journ. Geol. Soc.* vols. xxxi. xxxiii. xxxiv. xl. My criticism of them will be found in *op. cit.* vol. xxxix. (1883), subsequently in the main confirmed by Prof. Lloyd Morgan, *op. cit.* xlvi. p. 241. See also Prof. Blake, *op. cit.* xl. (1884). Dr. Hicks in his more recent papers has merely reiterated his previously published opinions.

The volcanic geology of St. David's possesses a special interest inasmuch as it embraces a tolerably full development of various features which characterize the volcanic groups of later Palæozoic systems. Though the

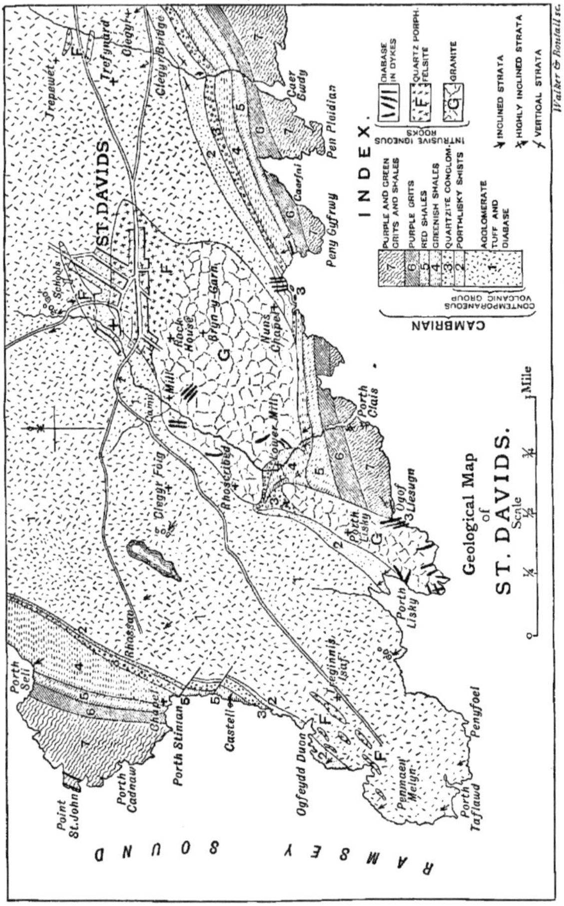

FIG. 40.—Map of the volcanic district of St. David's.

rocks are chiefly tuffs, they include also sheets of lava, as well as sills, dykes and bosses. They show a remarkable range in chemical composition from quite basic to highly acid materials. They present the amplest proofs of having been erupted and spread out over the sea-bottom, and they likewise afford clear evidence of alternation with the ordinary non-volcanic sediment of the time to which they belong. In these respects they are particularly noteworthy, for they prove that in the earliest Palæozoic ages the essential features of volcanic action were already as well developed as in any subsequent epoch of geological history.

The volcanic group of St. David's attains a visible thickness of about 1800 feet. Its upper part graduates upward into purple and green Lower Cambrian sandstones. The base of the group is not seen owing to the plicated structure of the district. Hence the total thickness of volcanic material cannot be determined, neither can we tell on what it rests, whether on a still lower sedimentary series or on some platform of pre-Cambrian rocks.

The structure of the group, notwithstanding all that has been written about it, has never yet been adequately worked out. The unfortunate and barren controversy about supposed pre-Cambrian rocks at St. David's has tended to obscure the real importance of these rocks as the oldest well-preserved record of volcanic action in Britain. They deserve to be carefully surveyed on maps of a large scale, in the same detailed manner as has been so successfully applied to the elucidation of younger volcanic tracts. Until such detailed investigation is made, any account of them which is given can be little more than a general outline of the subject. The following description is the result of my examination of the ground in company with my colleague Mr. B. N. Peach, and afterwards with the late Mr. W. Topley.[1] A few additional observations, from the subsequent exploration of Professor Lloyd Morgan,[2] are incorporated in the narrative.

The geologist who traces these St. David's rocks in the field cannot fail to be struck with their general resemblance to volcanic masses of later Palæozoic date. Many of the lavas and tuffs are in outward characters quite indistinguishable from those of the Lower Old Red Sandstone and Carboniferous systems of Britain. So many points of detail may be observed to be common to the Palæozoic eruptive rocks all over the country from the Cambrian to the Permian periods as to indicate that volcanic phenomena must have recurred under much the same conditions throughout Palæozoic time.

By far the larger part of the Cambrian volcanic group of St. David's consists of bedded tuffs, though a few lavas are interstratified in it, particularly towards the top. The whole has subsequently been invaded by acid protrusions, and lastly by basic dykes.

1. *Bedded Tuffs and Lavas.*—The tuffs, which are the predominant members of the volcanic group, present many varieties of colour, from dark purple, through tints of brick-red and lilac, to pale pink, yellow and creamy white, but not unfrequently assume various shades of dull green. They

[1] *Quart. Journ. Geol. Soc.* vol. xxxix. (1883), p. 294 *et seq.* While the essential parts of the investigation are given in the following pages, I would refer the reader to this paper for details not transferred to the present volume.

[2] *Op. cit.* vol. xlvi. (1890), p. 241.

vary likewise in texture from somewhat coarse breccias or agglomerates, through many gradations, into fine silky schists in which the tuffaceous character is almost lost. Generally they are distinctly granular, presenting to the naked eye abundant angular and subangular lapilli, among which broken crystals of a white, somewhat kaolinized, felspar and fragments of fine-grained felsite are often conspicuous. The greater part of the tuffs, particularly the purple, red and dark-green varieties, which constitute so large a proportion of the whole, has been derived from the explosion of basic rocks similar in character to the diabases now found associated with them. On the other hand, the paler varieties, both in the form of fine tuffs and of breccias, have probably resulted mainly from the destruction of more siliceous lavas, probably felsites (rhyolites) or other acid rocks.

That many of the tuffs are due to the destruction of diabase-lavas may be surmised from their close general external resemblance to these rocks, and from the way in which they are associated with the contemporaneous sheets of diabase. Some of the dull dark-purple tuffs might almost at first sight be mistaken for truly eruptive rocks. The analyses of two typical examples of these basic tuffs (Nos. I. and II.), and one (No. III.) of an intermediate variety containing an admixture of acid fragments, are given in the subjoined table.

	SiO_2	Al_2O_3	Fe_2O_3	FeO	MnO	CaO	MgO	K_2O	Na_2O	H_2O and Loss on Ignition.	Total.	Specific Gravity.
I.	51·25	20·41	3·02	3·91	0·21	4·53	7·22	2·93	1·82	5·02	100·32	2·84
II.	48·11	13·30	3·70	8·10	1·43	8·48	9·51	1·57	1·96	4·21	100·37	2·92
III.	61·54	16·30	4·40	3·66	0·32	3·08	2·99	1·62	2·81	2·99	99·71	...

I. Purplish-red shaly tuff from below olivine-diabase, Crag Rhosson. Analysis by Mr. J. S. Grant Wilson.

II. Dull purple and green tuff from the lowest group of tuffs between Pen-maen-melyn and Pen-y-foel. Analysis by Mr. Wilson.

III. Greenish shaly finely granular tuff, from road-side, north of Board Schools, St. David's. Analysis by Prof. A. Renard of Ghent.

Although the majority of the tuffs are more or less basic, they frequently contain evidence in the form of small felsitic lapilli that acid lavas were present in the eruptive vents, while the pale tuffs show that at the time of their discharge it was these acid lavas and not the diabases that were blown out by the explosions. Appended are three analyses of the acid tuffs (Nos. IV. V. and VI.).

	SiO_2	Al_2O_3	Fe_2O_3	FeO	MnO	CaO	MgO	K_2O	Na_2O	H_2O and Loss on Ignition.	Total	Specific Gravity.
IV.	80·59	11·29	0·28	1·41	trace	0·52	0·95	2·98	0·72	1·96	100·70	2·55
V.	73·42	12·09	0·91	3·13	0·25	2·94	1·12	1·67	3·88	1·28	100·69	2·74
VI.	72·63	16·23	2·70	0·48	...	0·18	1·36	3·35	0·15	3·04	100·12	...

IV. Greenish felsitic breccia, Clegyr Hill; angular fragments of various felsites in a greenish base. Analysis by Mr. J. S. Grant Wilson.

V. Grey granular felsitic tuff, Bridge over Allan River north from St. David's Board Schools. Analysis by Mr. Wilson.

VI. Pale pinkish-white, finely schistose tuff—a characteristic sample of the "Porth-lisky schists." Analysis by Prof. Renard.

Many varieties of texture can be traced among the tuffs, from coarse breccias or agglomerates, with blocks a yard or more in length, to fine schistose mudstones or sericitic schists. One of the most remarkable of the finer kinds, found near Pen-y-foel, is externally dirty-green, compact and tolerably homogeneous, but with distinct evidence of its clastic character. Under the microscope it is found to be composed mainly of lapilli of a peculiar rock, which is characterized by the abundance and freshness of its plagioclase (an unusual feature in the volcanic group of St. David's); by the large, well-defined crystals (one of which measured 0·022 inch by 0·0125 inch) of augite; by large crystals replaced by green decomposition-products, but having the external form of olivine; by the absence or scantiness of any base or groundmass; and, in many of the lapilli, by the abundance of spherical cells, either empty or filled up as amygdales with decomposition-products. These spherical vapour-vesicles, so characteristic of the basic or palagonitic lapilli in many Palæozoic volcanic vents, were found in one fragment, where they were particularly abundant, to range from a minimum of 0·0008 inch to a maximum of 0·0033 inch, with a mean of about 0·0018. The rock from which these lapilli have been derived comes nearest to one of the diabases from the same part of the district (which will afterwards be referred to), but shows a closer approach to basalt rocks.

Another interesting tuff is that of which the analysis (No. II.) has been given. It occurs not far from the horizon of the rock just described. Under a low power, it is seen to be composed mainly of fragments of diabase like the rocks of Rhosson and Clegyr Foig. These fragments are subangular, or irregular in shape, and vary considerably in size. They are sometimes finely cellular—the cavities, as in the case just referred to, being spherical. The plagioclase crystals in the diabase-lapilli are everywhere conspicuous; so also is the augite, which occurs in larger forms than in the rock of Rhosson or Clegyr Foig. Next in abundance to these basic fragments are rounded or subangular pieces of felsite. These weather out in conspicuous grey rough projections on the exposed face of the rock; under the microscope they are seen to consist of fine granular felsite, which shows a groundmass remaining dark between crossed nicols, but with luminous points and filaments, and an occasional spherulite giving the usual cross in polarized light. Lapilli of an older tuff may here and there be detected. A few angular and subangular grains of quartz are scattered through the rock. The lapilli are bound together by a finely-granular dirty-green substance.

As a typical illustration of the minute structure of the felsitic tuffs, I may refer to the rock No. V. of the foregoing analyses. It is composed mainly of fragments of various felsites, many of which show good flow-structure. Large, and usually broken, crystals of orthoclase are dispersed among the other ingredients. Here and there a fragment of diabase may be detected; but I could find no trace of pieces of the peculiar micro-crystalline spherulitic quartz-porphyries of St. David's. There is but little that could be called matrix cementing the lapilli together. The presence of fragments of diabase may possibly reduce the proportion of silica and increase that of magnesia, as compared with what would otherwise have been present in the rock.

Some of the tuffs appear to have been a kind of volcanic mud. A specimen of this nature collected from the road-side section, north of the Board School, presents a finely-granular paste enclosing abundant angular and subangular lapilli of diabase, a smaller proportion of felsite (sometimes displaying perfect flow-structure), broken plagioclase crystals, and a greenish micaceous mineral which has been subsequently developed out of the matrix between the lapilli.

Though they lie in the sedimentary series above the main volcanic group, I may refer here to certain thin bands of tuff at Castell, on account of their interest in relation to the true Cambrian age of the volcanic group. They are not quite so fresh as the tuff that occurs in thicker masses, but their volcanic origin is readily observable. One band appears to be made up of the debris of some basic rock, like the diabase of the district, through which detached plagioclase crystals are scattered. The lapilli are subangular; and around their border a granular deposit of haematite has taken place, giving a red colour to the rock. Another band presents small angular lapilli, almost entirely composed of a substance which to the naked eye, or with a lens, is dull, white and clay-like, easily scratched, and slightly unctuous to the touch. Under the microscope, with a low power, it becomes pale greyish-green and transparent, and is seen to consist in large part of altered felspar crystals, partially kaolinized and partially changed into white mica and calcite. These scattered crystals are true volcanic lapilli, and have not been derived from the mechanical waste of any pre-existing volcanic rock. In the tuffs interstratified with the conglomerate, at the quarry above Porth-clais, though much decomposed, crystals of plagioclase can likewise still be traced. These strata are also true tuffs, and not mere detritus due to mechanical degradation (see Fig. 41).

The general result of the study of the microscopic structure of the Cambrian tuffs of St. David's may be briefly summed up as follows:—

1. These pyroclastic deposits are almost wholly composed of fragments of eruptive rocks, sometimes rounded, but usually angular or subangular. In the more granular varieties very little matrix is present; it consists of fine debris of the same materials. No detached microlites have been noted, such as are common among modern volcanic ashes; but there are abundant ejected crystals. In these respects the Cambrian tuffs resemble those of the other Palæozoic systems. A mingling of grains of quartz-sand may indicate the intermixture of ordinary with volcanic sediment.

2. They may be divided into two groups—one composed mainly of fragments of diabase or other similar basic rocks, the other of felsite. The former group has doubtless been derived from the explosion of such rocks as the diabase-sheets of the district. The felsitic tuffs have not been observed to contain any fragments of the microcrystalline quartz-porphyries of St. David's. They have been derived from true fine-grained felsites or rhyolites. There are various intermediate varieties of tuff, due to the mingling in various proportions of the two kinds of debris.

3. They are marked by the presence of some characteristic features of the volcanic vents of later Palæozoic time, and in particular by presenting the following peculiarities: (*a*) lapilli of a minutely-cellular pumice with spherical cells; (*b*) lapilli with well-developed flow-structure; (*c*) lapilli consisting of a pale green serpentinous substance resembling altered palagonite and probably originally glass; (*d*) lapilli derived from the destruction of older tuffs; and (*e*) lapilli consisting of ejected crystals, especially of felspars, sometimes entire, often broken.

4. They frequently show that they have undergone metamorphism, by the development of a pale greenish micaceous mineral between the lapilli, the change advancing until the fine tuffs occasionally pass into fine silky schists.

In my study of the St. David's district, I was unable to observe any evidence that the basic and siliceous tuffs characterize two distinct periods of volcanicity. From the foregoing analyses it appears that some of the oldest visible tuffs which are seen between Pen-maen-melyn and Pen-y-foel contain only 48·11 per cent of silica; while a specimen from Porth-lisky yielded 72·63 per cent of that ingredient. Specimens taken even from adjacent beds show great differences in the percentage of silica, as may be seen in the analyses Nos. III. and V.

This alternation of basic and siliceous fragmental materials has its parallel in the neighbouring eruptive rocks, some of which are olivine-diabases containing only 45 per cent of silica, while others are highly siliceous quartz-porphyries. But all the siliceous eruptive rocks, so far as I have been able to discover, are intrusive, and belong, I believe, to a later period than that of the volcanic group; in no single instance do they appear to me to be true superficial lava-flows. On the other hand, the basic eruptive rocks occur both as contemporaneous sheets and as intrusive masses. The presence of both siliceous and basic lavas in the Cambrian volcanic reservoirs, however, is proved by the character of the tuffs. It would appear from the evidence at present known, that while the basic lavas were most abundant in the vents during the volcanic period recorded by the rocks of St. David's, furnishing the material for most of the fragmental eruptions, and issuing in occasional superficial streams of molten rock, the siliceous lavas did not flow forth at the surface, though their debris was copiously discharged in the form of dust and lapilli.

The rise of both basic and acid lavas at different periods in the same or adjoining vents, so familiar in recent volcanic phenomena, thus appears to have also characterized some of the oldest examples of volcanic action. An interesting parallel may be traced between the succession of events at St. David's and that which occurred in the volcanic group of the Lower Old Red Sandstone of the Pentland Hills, near Edinburgh, of which a detailed account will be given in Chapter xx. of this volume. It is also worthy of remark that in the latest of the volcanic episodes in British geology a remarkable similarity to the St. David's volcanic group may be observed. Some of the older Tertiary agglomerates are full of pieces of acid rocks (felsites, rhyolites or granophyres), while the lavas poured out at the surface were mainly basalts.

In the volcanic group of St. David's the tuffs contain evidence that ordinary sedimentation was not entirely interrupted by the volcanic discharges. Thus, in the Allan valley, west from the Cathedral, one of the schistose tuffs is full of well-rounded pebbles of white quartz. Occasional shaly bands indicate the deposit of mud with the tuffs.

Excluding the granites and porphyries (which are described at p. 155), two kinds of eruptive rocks are associated with the volcanic group. One of these is certainly intrusive and of late date, viz. dykes and veins of diabase, to be afterwards referred to. The other kind occurs in long parallel sheets, some of which, if not all, are true contemporaneous lava-streams, erupted at intervals during the accumulation of the volcanic group. They form prominent crags to the west of St. David's, such as Clegyr Foig, Rhosson, and the rocky ground rising from the eastern shores of Ramsey Sound. Their dip and strike coincide with those of the tuffs above and below them. It is possible that some of these sheets may be intrusive sills intercalated along the bedding of the tuffs; and in one or two cases I have observed indications of what, on further and more careful exploration, may prove to be disruption across the bedding.

But it is the interbedded sheets that possess the chief interest as superficial lava-streams of such venerable antiquity. They present many of the ordinary features of true lava-flows. In particular a slaggy structure may be detected at the bottom of a sheet, the vesicles being here and there lengthened in the direction of flow. Some of the sheets are in part amygdaloidal. The alternation of these sheets with tuffs, evidently derived from lavas of similar character, is another argument in favour of their contemporaneous date. One of the best localities for studying these features lies between Clegyr Foig and the coast, west of Rhosson.

The eruptive rocks thicken towards the south-west, as if the main vents had lain in that direction. There are doubtless intrusive as well as contemporaneously interbedded masses in the rough ground between Pen-maen-melyn and Treginnis. To separate these out would be a most interesting and beautiful piece of mapping for any competent geologist in possession of a good map on a sufficiently large scale.

The interbedded lavas, so far as I have had an opportunity of studying them, appear to present remarkable uniformity of petrographical characters. Megascopically they are dull, fine-grained to compact, sparingly porphyritic, ranging in colour from an epidote-green to dull blackish-green and dark chocolate-brown. Some of them are finely porphyritic from the presence of small glistening surfaces which present the colour and metallic lustre of haematite and yield its characteristic streak. Obviously basic rocks, they present, as I have said, a close external resemblance to many of the lavas of the Lower Old Red Sandstone and Carboniferous districts of Scotland. From their chemical composition and microscopic structure they may be most appropriately ranged among the diabases. The analyses of two of the most conspicuous diabases of this class in the district, those of Rhosson (VII.) and Clegyr Foig (VIII.), by Mr. J. S. Grant Wilson, are shown in the following table:—

	SiO_2	Al_2O_3	Fe_2O_3	FeO	MnO	CaO	MgO	K_2O	Na_2O	H_2O and Loss on Ignition.	Insoluble Residue.	Total.	Specific Gravity.
VII.	45·92	18·16	1·18	9·27	0·19	7·19	10·07	1·78	2·12	4·22	0·04	100·14	2·96
VIII.	45·38	16·62	4·06	8·63	0·14	8·19	9·41	0·71	2·20	4·34	0·08	99·76	2·99

The two rocks here analyzed, likewise that from the crag south of Castell and that from the cliffs at the southern end of the promontory between Ramsey Sound and Pen-y-foel, show under the microscope a general similarity of composition and structure. They present a variable quantity of a base, which under a $\frac{1}{5}$ objective is resolved into ill-defined coalescent globulites and fibre-like bodies, that remain dark when rotated between crossed nicols. In some varieties, as in part of Rhosson Crag, the base is nearly lost in the crowd of crystalline constituents; in others, as in the crag south of Castell, it forms a large part of the whole mass, and may be seen in distinct spaces free from any crystalline particles. Through this base are diffused, in vast numbers, irregularly shaped grains of augite, seldom showing idiomorphic forms. These grains, or granules, may perhaps average about 0·003 inch in diameter. Plagioclase is generally hardly to be recognized, though here and there a crystal with characteristic twinning may be detected in the base. Magnetite occurs abundantly—its minute octahedra, with their peculiar colour and lustre, being apparent with reflected light on the fresher specimens, though apt to be lost as diffused ferruginous blotches in the more decomposed varieties. But perhaps the most remarkable ingredient is olivine. Red haematitic crystals are visible, even to the naked eye, dispersed through the groundmass of the rocks. With a lens these may be observed to be orthorhombic in form, and to be evidently pseudomorphs after some mineral which has been converted chiefly into haematite. Such red pseudomorphs are common in Carboniferous and Old Red Sandstone lavas, where in some cases they appear to be after hornblende, and in others after augite, but occasionally are suggestive of olivine, though with no trace of the original substance of that mineral. In the lava associated with the tuffs at the south end of the promontory between Ramsey Sound and Pen-y-foel, however, certain large, well-developed pseudomorphs are undoubtedly after olivine. They have the characteristic contour of that mineral and its peculiar transverse curved and irregular fractures. The average length of these pseudomorphs was found, from the measurement of six examples, to be 0·023 inch, the largest being 0·034, and the smallest 0·014. Seen by transmitted light they present a structureless pale-green material nearly inert in polarized light, round the borders and across fissures in which an opaque substance has been developed, as serpentine and magnetite have been grouped in the familiar alteration of olivine. The opaque material is bright brick-red in reflected light, and is evidently now chiefly oxidized into haematite. Every stage may be traced, from orthorhombic forms with the incipient development of transverse fissures filled with iron-oxide, to others of distorted shapes in which the ferruginous matter occupies the whole, or nearly the whole, of the mould of the original crystal.

The rocks now described differ from the Palæozoic andesites or "porphyrites," with which I am acquainted, in their more basic composition, in the less abundance of their microscopic base, in the comparatively inconspicuous development of their felspars of later consolidation, in the absence of large porphyritic felspars of earlier consolidation, in the extraordinary prominence of the granular augite, and in the presence of olivine. In composition and structure they are essentially forms of olivine-diabase.

Above the volcanic group of St. David's lies a band of quartz-conglomerate which has been taken by Dr. Hicks to mark the base of the Cambrian system. This rock, though mainly composed of quartz and quartzite, contains fragments of the underlying volcanic rocks. But that it does not mark any decided break in the sedimentation, much less the violent unconformability and vast interval of time which it has been erroneously supposed to do, is well illustrated by the occurrence of bands of tuff, as well as diffused volcanic dust, in the conglomerate and also in the green and red shales and sandstones which conformably overlie it. An example of this intercalation of volcanic material is represented in Fig. 41. On the left side vertical layers

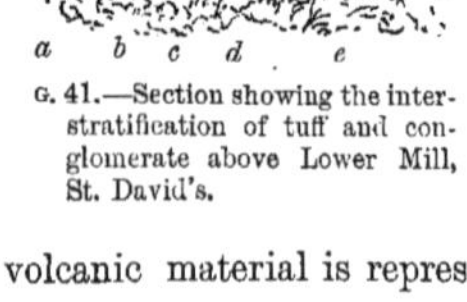

G. 41.—Section showing the interstratification of tuff and conglomerate above Lower Mill, St. David's.

of fine reddish tuff (*a*) are succeeded by a band of quartz conglomerate (*b*) of the usual character. Parallel to this conglomerate comes a band, about six inches thick, of fine tuff (*c*), followed by ashy sandstone (*d*), which graduates into conglomerate (*e*). No more complete evidence could be desired of the perfect inosculation of the conglomerate with the volcanic group. On the coast at Nun's Chapel similar evidence presents itself. The conglomerate there contains some thin seams of tuff, and is intercalated between a series of dull green agglomerates and tuffs and some greenish shales and sandstones with layers of tuff.

Not less striking is the evidence of the contemporaneous eruption of fine volcanic dust in the overlying shales and sandstones.[1] Some of the red shales are full of this material, which here and there is gathered into the thin seams or ribs of which the microscopic characters have already been described. This diffused volcanic detritus marks, no doubt, the enfeebled discharges of fine dust towards the close of the volcanic episode in the Lower Cambrian period at St. David's. It would be difficult to find an instance of a more perfect transition from a series of thoroughly volcanic masses into a series of ordinary mechanical sediments.

2. *Intrusive Granite and Quartz-Porphyry.*—The heart of the volcanic group is pierced by a mass of granite which also cuts the conglomerate and overlying shales and sandstones on the east side. The age of this intrusive boss cannot be more definitely fixed than by saying that it must be later than the volcanic group. This rock has been the subject of a remarkable amount of description, and has been dignified by being actually elevated into a distinct Archaean "formation" composed of "highly crystalline gneissic rocks," with "bands of limestone, hornblende, chlorite, and micaceous schists."[2] Into this somewhat dreary chapter of English geological literature it is fortunately not necessary to enter here. I will only say that the rock is unquestionably a granite, with no essential differences from many other bosses regarding which there has been no controversy. It is a holocrystalline rock with a thoroughly granitic texture, and composed of the ordinary minerals of granite—quartz, orthoclase and plagioclase, among which a green chloritic mineral, more or less plentiful, probably represents original hornblende, biotite or augite. Sometimes the quartz and felspar in the body of the rock show a micropegmatitic arrangement, and the same structure occurs in veins that traverse it. This structure gives the rock some resemblance to the Tertiary granites and granophyres of Ireland and Scotland.

This granite has undergone a good deal of decomposition, for its felspars are turbid, and its original ferro-magnesian constituents are always replaced by green chloritic aggregates, while epidote is also present. The rock tends to become finer in grain towards the margin, and then some-

[1] These are a portion of Dr. Hicks' "Caerfai group" in the Lower Cambrian series. They have yielded Lower Cambrian fossils.

[2] See the papers cited on p. 145 and my discussion of the relations of this granite in *Quart. Journ. Geol. Soc.* vol. xxxix.; also Prof. Lloyd Morgan, *op. cit.* vol. xlvi. (1890).

times assumes a more decidedly pegmatitic structure, like graphic granite. At the northern end of the granite ridge, a gradation can be traced from the ordinary texture through increasingly fine-grained varieties until we pass into a microcrystalline spherulitic porphyry. After a careful examination of the ground I satisfied myself that the spherulitic quartz-porphyries, which form a conspicuous feature in the geology of St. David's, are really offshoots from this granitic core.[1]

These spherulitic rocks have been fully described.[2] They consist of a base composed of a microcrystalline aggregate of quartz and orthoclase, which is distributed between the spherulites. These have been developed in remarkable beauty and perfection. While the microcrystalline structure is everywhere recognizable, the spherulites occasionally disappear. But their absence is merely local, and they may be found both in large dykes and narrow veins. A further porphyritic structure is given to the rocks by the presence in them of abundant quartz, which takes the form of conspicuous rounded blebs or worn crystals sometimes distinctly dihexihedral, but with somewhat blunted angles. Porphyritic plagioclase is also common. Flow-structure is occasionally traceable. Some parts of these rocks where the porphyritic elements are locally absent might be cursorily classed as felsites; but they all possess a microcrystalline and not a felsitic base. They cannot be confounded with the true felsites of which fragments occur in the tuffs.

In addition to the parallelism that may be traced between the earliest Palæozoic agglomerates and those of the youngest volcanic series of Britain, a close analogy may also be noticed between the acid intrusive rocks of the two widely-separated periods. In both cases we have a granitic core sending out apophyses which assume a spherulitic structure and traverse earlier volcanic products of the district.

These spherulitic quartz-porphyries of St. David's occur as bosses, dykes (elvans) or veins, cutting through all horizons of the volcanic group, and in one case apparently, if not actually, reaching the quartz conglomerate. One of the best exposures of this intrusive character may be seen in the cliff below Nun's Chapel, where the elvan runs along the face of the cliff through the uppermost zone of the volcanic group, cutting the strata somewhat irregularly. Apparently in connection with this dyke, a network of intrusions of decomposed quartz-porphyry may be observed in the shales along the face of the cliff immediately below Nun's Chapel. On the whole, the intruded material has forced its way along the bedding-planes of the shales, but has also broken across them, sending out finger-like branches.

3. *Diabase Dykes and Sills.*—The latest rocks of the St. David's district are dykes and intrusive sheets of diabase, which traverse all the other

[1] These apophyses from the granite constitute the "Arvonian" formation of Dr. Hicks' pre-Cambrian series of St. David's.

[2] See, for example, J. Davies, *Quart. Journ. Geol. Soc.* vol. xxiv. p. 164, xxxv. p. 203; also the paper already referred to, *op. cit.* xxxix. p. 315; and Mr. Teall's *British Petrography*, p. 334.

formations. The dykes are specially abundant in the granite. One or two may be detected in almost every artificial opening which has been made in that rock; while on the coast-section they are here and there profusely abundant. They are likewise frequent in the quartz-porphyries, and occur also in the volcanic group as well as in the sandstones and shales above the conglomerate, but become fewer in number as they recede from the granite centre.[1]

In external characters, the rock composing these dykes and sheets may be described as usually a dull dirty-green or yellowish-brown mass, to which the old name of "wacke" might appropriately be given. It exhibits the texture and mode of weathering of the more distinctly crystalline members of the basalt family. It is occasionally amygdaloidal or cellular, the kernels or cavities being arranged parallel with the sides of the dyke. Here and there a rudely prismatic structure extends between the walls.

The microscopic structure of this rock has been described by Professor Judd, Mr. Davies and Mr. Tawney. It is a diabase, but more allied in structure to true basalt than the olivine-diabase of the volcanic group. It especially differs from the older rock in the abundance and freshness of its felspars, in the comparative scarcity of its augite, and in the absence of olivine. The magnesian silicates are very generally replaced by green decomposition-products diffused through the mass. An occasional crystal of hornblende, recognizable by its cleavages and dichroism, may be detected. Some of the diabase dykes present excellent examples of flow-structure. A beautiful instance occurs in a dyke that cuts the shales, in a small cove to the east of Nun's Chapel. The shale and eruptive rock are in contact; and the small acicular prisms of felspar, besides ranging themselves in line parallel to the side of the dyke, stream round the larger felspar crystals.

Some of the dykes or veins are only three inches broad. They send out fingers, and sometimes break abruptly across from one line to another. They appear generally to have followed the lines of joint in the granite, as Mr. Tawney has observed;[2] consequently they must be posterior to the development of the system of jointing in that rock.

Besides the abundant dykes, there has been a more limited extrusion of the same material in sheets parallel (or approximately so) to the bedding of the sandstones and shales. These sheets are well displayed at St. John's Point, where evidence of their being intrusive, and not truly bedded, may be seen along the fine cliffs which have been cut by the waves on this part of the coast-line.

The sedimentary series which overlies the volcanic group of St. David's, and contains the fossils of the lower part of the Cambrian system, gradually loses all trace of volcanic material, as its members are followed upward in

[1] The occurrence of these dykes is paralleled by that of the similar intrusions in the quartz-felsite of Llyn Padarn to be afterwards described.

[2] *Proc. Nat. Hist. Soc. Bristol*, vol. ii. part ii. (1879), p. 115.

stratigraphical order.[1] We thus learn that the eruptions of this district came to an end in an early part of the Cambrian period. But as we shall see in the following pages, volcanic activity was subsequently renewed at no great distance in the next or Silurian period.

[1] Dr. Hicks has noted the occurrence of "volcanic tuff" in the Lower Lingula Flags of Porth-y-Rhaw, a little to the east of St. David's (*Quart. Journ. Geol. Soc.* vol. xx. 1864, p. 240). This intercalation is marked as a "dyke" in the MS. notes of Sir A. C. Ramsay on a copy of the Geological Survey map of the district.

CHAPTER XI

THE CAMBRIAN VOLCANOES OF NORTH WALES, THE MALVERN HILLS AND WARWICKSHIRE

NORTH WALES

THE Cambrian volcanic rocks in the northern part of the Welsh Principality have their main development in Caernarvonshire. Southwards from that tract, though the Lower Cambrian strata form a vast pile of sedimentary material in the Harlech anticline, which is estimated by the Geological Survey to be from 6000 to 7000 feet thick, they have yielded no trace of any contemporaneous volcanic rocks.[1] The purple slates that rise along the centre of the anticline dip below the grits and conglomerates on either side without disclosing a glimpse of the base of the system. This enormous accumulation of sedimentary deposits seems to diminish in thickness as it is traced northwards, for towards the Menai Strait it does not reach more than a fourth part of the depth which it is said to display in the Harlech anticline.[2] In the Pass of Llanberis the series of grits that overlies the purple slates is estimated to be about 1300 feet thick.[3] This gradual thinning away of the Cambrian series towards the north was, in the opinion of Sir Andrew Ramsay, accompanied by an increasing metamorphism of the lower portions of the system. In his view, the long ridge of quartz-porphyry which crosses the lower end of Llyn Padarn represents the result of the extreme alteration of the stratified rocks. He believed that he could trace an insensible passage from the slates, grits and conglomerates into the porphyry, and he was led to the "conviction that the solid porphyry itself is nothing but the result of the alteration of the stratified masses carried a stage further than the stage of porcellanite, into the condition of that kind of absolute fusion that in many other regions seems to have resulted

[1] *Mem. Geol. Surv.* vol. iii. 2nd edit. "Geology of North Wales," p. 21. It is possible that this thickness has been somewhat overestimated. Dr. Hicks (*Geol. Mag.* 1880, p. 519) has referred to certain "highly felsitic rocks, for the most part a metamorphic series of schists, alternating with harder felsitic bands, probably originally felsitic ashes," lying at the bottom of the whole pile, and he has claimed them as pre-Cambrian. But I have not found any evidence of such rocks, nor any trace of igneous materials save dykes and sills, acid and basic, such as are indicated on the Survey map.

[2] *Ibid.* p. 24.

[3] *Ibid.* p. 173.

in the formation of granites, syenites and other rocks, commonly called intrusive."[1] Certain structural lines in the quartz-porphyry he looked upon as indicating "traces of stratification in a rock, the original felspathic and quartzose material of which has been metamorphosed into true porphyry."[2] In conformity with these ideas, the remarkable felspathic strata which lie nearest the porphyry were regarded as metamorphosed Cambrian rocks, and where similar rocks reappear over a large area near Bangor they were coloured on the map with the same tint and lettering as were used for the so-called "altered Cambrian" of Anglesey.

No one who has examined this Caernarvonshire ground can have failed to find the sections which doubtless led my predecessor to form the convictions to which he gave expression in the passages I have just quoted. It is easy to see how these sections, wherein it is certainly difficult to draw a sharp line between the igneous rock and the clastic materials derived from it, would be welcomed as appearing to offer confirmation of the ideas concerning metamorphism which were then in vogue. There cannot, however, be any doubt that my friend was mistaken in his interpretation of the structure of that part of the country. It is to me a subject of keen regret that in his later years, when the subject was revived, he was no longer able to re-examine this ground himself, for no one would have confessed more frankly his error, and done more ample justice to those who, coming after him, have been able in some parts to correct his work.

The quartz-porphyry, felsite or rhyolite of Llyn Padarn, as well as that of Llandeiniolen, is not a metamorphic but an eruptive rock, as has been demonstrated by Professors Hughes and Bonney. There is no true passage of the sedimentary rocks into it; on the contrary, the conglomerates which abut against it are in great part made out of its fragments, so that it was already in existence before these Cambrian strata were deposited upon it. These conclusions must be regarded as wholly indisputable. But most of the critics of the work of the Geological Survey have proceeded to certain further deductions. They have maintained that the presence of fragments of the porphyry in the overlying conglomerate marks an unconformability between the two rocks, that the conglomerate shows the base of the Cambrian system, and that the porphyry is therefore pre-Cambrian.

These assertions and inferences do not seem to me to be warranted. They have, in my judgment, been disproved by Mr. Blake,[3] who shows that there is no break in the Cambrian series, that the various porphyries and their accompaniments are parts of that series, and that there is no certain proof of the existence of any pre-Cambrian rocks in the whole district.[4]

That the igneous rocks of the Llyn Padarn area mark a volcanic period has been recognized by most writers since Professor Bonney pointed out the flow-structure of the quartz-porphyry, and other proofs of active volcanic

[1] *Mem. Geol. Surv.* vol. iii. 2nd edit. p. 173. [2] *Ibid.* p. 174.

[3] In an excellent memoir read before the Geological Society in 1888, with the main conclusions of which I agree.

[4] *Quart. Journ. Geol. Soc.* vol. xliv. p. 271. For subsequent papers by Mr. Blake, see *op. cit.* vols. xlviii. (1892) p. 243, xlix. (1893) p. 441.

eruptions have been traced by him, as well as by Professor Hughes and Mr. Blake, in the stratified rocks which stretch north-eastwards to Bangor. The extent and persistence of these ancient volcanic phenomena, and their probable connection with the remarkable northward attenuation of the Cambrian sedimentary rocks, deserve special attention.

It is generally agreed that the rocks variously termed quartz-porphyries, felsites or rhyolites form the oldest members of this volcanic series.[1] They come to the surface in two long ridges, one running from Caernarvon to near Bangor, the other from Llanllyfni to Ann's Chapel, at the mouth of Nant Francon (Map. II.). Whether the materials of these two ridges are parts of one originally continuous sheet or group of sheets, or, if different protrusions, whether they belong to the same geological horizon, or whether, as Mr. Blake believes, they are distinct masses, separated by a considerable thickness of detrital material, cannot in the present state of our knowledge be positively decided. It seems to me probable that they are connected underground, as a continuous platform beneath the overlying pyroclastic materials.

These acid rocks have been regarded by some observers as intrusive sheets, by others as lava-streams that were poured out at the surface. If account be taken simply of their petrographical characters, they find their nearest analogies among the intrusive quartz-porphyries of older geological periods. The presence of flow-structure in them has been thought to indicate that they were superficial streams. But this structure may be found in dykes and intrusive sheets as perfectly as in lava-flows, so that it cannot by itself be taken as proof of a surface-discharge of lava. It must be confessed that, both in the main mass of quartz-porphyry and in the abundant fragments of it in the overlying conglomerates and breccias, there is an absence of such scoriform portions as one would naturally look for in a superficial lava-stream;[2] while, on the other hand, the rock generally presents the tolerably uniform flinty texture so familiar in intrusive sheets of similar material.[3] My own impression is that these igneous masses were probably erupted to the surface as long banks which rose above the waves; that they were thus exposed to prolonged subaerial and marine denudation; that by this means any upper more cellular portions of the lava which may have existed were broken up and pounded down into detritus, and thus that what is now visible is a part of the eruptive rock which originally lay at some depth within its body. This view is confirmed by a study of other lavas which are found on different platforms in the detrital deposits that overlie the Llyn Padarn quartz-porphyry.

[1] Whether the granitic rock of Twt Hill, Caernarvon, is connected with the porphyry or belongs to an older eruption is immaterial for my present purpose.

[2] But the Llyn Padarn rock, like many acid lavas, may never have possessed a vesicular structure in any portion of its mass. The sheets of felsite which occur among the overlying tuffs are not cellular, but present the closest resemblance to the main mass below.

[3] Mr. Blake brought forward the evidence of a section on the north or under side of the Llyn Padarn ridge to show that the rock has there been intruded into the Cambrian strata (*Quart. Journ. Geol. Soc.* vol. xliv. (1888), p. 283). But the rock supposed by him to be altered slate has been shown to be intrusive "greenstone" (Miss Raisin, *op. cit.* vol. xlvii. (1891), p. 336).

That the material of each of the two main ridges is the result of more than one eruption has been inferred from the supposed intercalation of bands of slate and of breccia in the rock.[1] Considerable lithological differences may be detected in each mass, but they are not greater than may be observed in single sills and bosses. In some parts of the Llyn Padarn porphyry a distinct nodular structure appears which shades off into bands and lenticular streaks, reminding one of the characters of some of the Bala rhyolites. Other portions are markedly brecciated, the separated fragments being surrounded in a matrix of the rock, which shows flow-structure sweeping past them. On Moel Gronw angular fragments of a dark pinkish tint are scattered through the general mass. Again, some parts are crowded with quartz-grains, while others are comparatively free of these, and occasionally a spherulitic structure has been observed.[2]

The microscopic structure of this ancient eruptive rock has been studied by Professor Bonney, who found that the general type was a compact dull grey felsite, with porphyritic crystals of felspar and grains of quartz, closely resembling some modern rhyolites. Though unable to detect any actual glass in the base, he had no doubt that the rock was originally vitreous, and he found abundant and fresh examples of the most perfect flow-structure.[3]

Reference may be made here to the remarkable influence of the intense cleavage of the district upon this rock.[4] Along its southern margin, where it has been exposed to pressure from the south-east, the quartz-porphyry has been so crushed that it passes here and there into a fine unctuous slate or almost a schist. Nowhere can this change be more clearly seen than on the slopes of Mynydd y Cilgwyn. The cleavage planes strike about N. 40° E., with an inclination to dip towards the N.W. Within a space of a few yards a series of specimens may be collected showing at one end an ordinary or only slightly-sheared quartz-porphyry with abundant quartz-blebs, and at the other a fine greenish sericitic slate or phyllite, wherein the quartz has been almost entirely crushed down. Lines of shearing may be detected across the breadth of the porphyry ridge, each of them coinciding with the prevalent trend of the cleavage. Sometimes also certain basic dykes, which traverse the porphyry in some numbers,

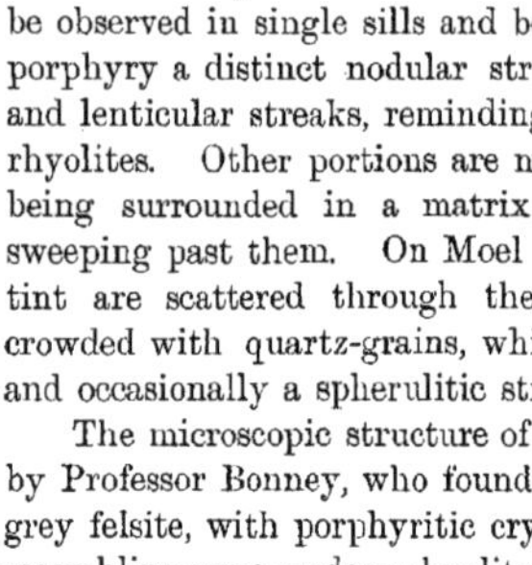

FIG. 42.—Basic dyke traversing quartz-porphyry and converted into a kind of slate by cleavage. West side of Llyn Padarn.

p p, quartz-porphyry; *d d*, dyke and connected veins.

[1] See for example, Prof. Bonney, *Quart. Journ. Geol. Soc.* vol. xxxv. (1879), p. 312; Mr. Blake, *op. cit.* vol. xliv. (1888), pp. 277, 287. But some at least of the supposed "slates," as stated in a previous footnote, have been since shown to be dykes.

[2] Mr. Blake, *ibid.* p. 277.

[3] *Op. cit.* vol. xxxv. p. 312.

[4] The secondary planes due to cleavage must not be confounded with the original flow-structure.

have undergone considerable deformation from the same cause. Their thinner portions are so well cleaved that they have been mistaken for included bands of green slate (Fig. 42). But these cleaved branches may sometimes be traced into a thicker and more solid dyke, whose uncrushed cores still preserve the original character of the rock and prove it to be eruptive.

The rocks which succeed the porphyry in the Valley of Llanberis are of great interest, for they contain abundant proof of contemporaneous volcanic activity, and they show that, so far from there being any marked hiatus here, there is evidence of the persistence of eruptions even into the time of

FIG. 43.—Section of well-cleaved tuff, grit and breccia passing up into rudely-cleaved conglomerate and well-bedded cleaved fine conglomerate and grit. East side of Llyn Padarn.

the Llanberis Slates.[1] Considerable misapprehension has arisen from the attempt to make one of the conglomerates the base of the Cambrian series, and the real significance of the volcanic detrital strata in association with it was consequently missed. The conglomerate does not lie on one definite horizon. In truth, there are several zones of conglomerate, each with some difference of composition, thickness or extent.[2] These may be well studied both on the south and the north side of the porphyry ridge at the lower

[1] The sections in the Vale of Llanberis on either side of Llyn Padarn have been again and again described and fought over. Some of the papers are cited in the following pages, but it would be impossible in this volume to find room for a full discussion of the differences of opinion. What is stated in the text is the result of my own study of the rocks on the ground, coupled with a careful consideration of the work of other observers.

[2] I can find no evidence of unconformability beneath any of the conglomerates. The section described by Professor Green, *Quart. Journ. Geol. Soc.* vol. xli. (1885), p. 74, merely shows the difference between the effects of cleavage on the fine tuffs and the more massive resisting conglomerate which overlies them. This section is represented in Fig. 43. At first sight the conglomerate appears to be lying on the vertical edges of an older group of slates, but any one acquainted with cleavage can trace this structure from the tuffs into the conglomerate and resuming its course again in the finer sediments above. The whole series of deposits in the section is continuous and conformable. The section on the slate railway has deceived Mr. Blake as well as Professor Green (*Quart. Journ. Geol. Soc.* vol. xlix. (1893), p. 445). The correct interpretation is given by Professor Bonney and Miss Raisin (*op. cit.* vol. l. p. 592).

end of Llyn Padarn. They are intercalated among fine tuffs, grits, volcanic breccias and purple slates, sometimes full of fine ashy material. On the south-east side of the ridge, where the rocks have suffered intense cleavage, they assume a fissile unctuous character, and then resemble parts of the cleaved Cambrian tuffs at St. David's. But on the north-west side, where they have in large measure escaped the effects of the cleavage-movements, their original structures are well preserved.

One of the first features of these detrital deposits to arrest attention is the amount and variety of the fragments of igneous rocks in them. Some of the conglomerates, though enclosing pebbles of quartz, quartzite, granite and other rocks not found *in situ* in the immediate district, are mainly composed of the debris of the quartz-porphyry of the ridge. Indeed, this latter material appears to have contributed a large proportion of the detritus of which the general body of strata here is made up. But there are to be noticed among the contents of the conglomerates and breccias pieces of many volcanic rocks not to be found on the porphyry ridge. Among these, besides felsites showing sometimes beautiful flow-structure (rhyolites) and various quartz-porphyries, there occur abundant fragments of less acid lavas (andesites) and pieces of older tuffs. Some of the fragmental rocks are green in colour, probably from the abundance of fine basic volcanic dust in them. Certain bands are full of large angular pieces of shale, similar in character to the Cambrian slates, and doubtless due to the disruption of pre-existing Cambrian strata by volcanic explosions. It is clear that from vents in this neighbourhood there continued to be an abundant discharge of dust and various andesitic and other lapilli, which, falling on the sea-floor, mingled there with the ordinary mechanical sediment that was being deposited at the time.[1]

But we have evidence that, during the period when these showers of volcanic detritus were thrown out, streams of lava, though on a greatly

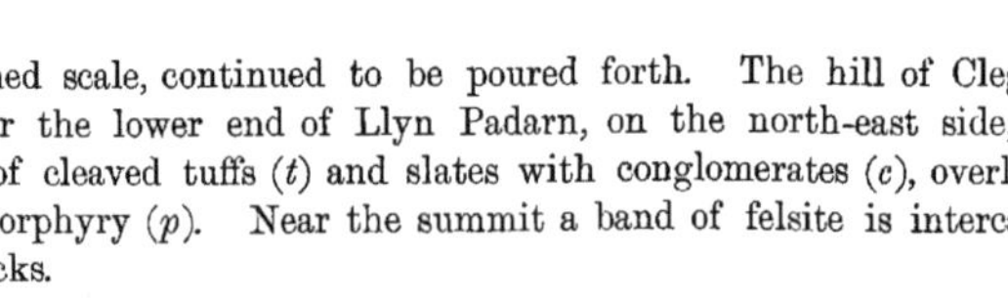

FIG. 44.—Section of Clegyr on the north-east side of Llyn Padarn, near the lower end.

diminished scale, continued to be poured forth. The hill of Clegyr (Fig. 44), near the lower end of Llyn Padarn, on the north-east side, consists mainly of cleaved tuffs (*t*) and slates with conglomerates (*c*), overlying the quartz-porphyry (*p*). Near the summit a band of felsite is intercalated in these rocks.

[1] On the composition of the conglomerates or breccias, see Professor Bonney and Miss Raisin, *Quart. Jour. Geol. Soc.* vol. l. (1894), p. 598.

Still more striking are the sections on the south-west side of the lake.[1] Starting from the porphyry of the ridge, we cross a zone of conglomerate and grit largely composed of the debris of the porphyry, until we reach a band of felsite or quartz-porphyry, which at its eastern end is about ten feet thick, while it seems to increase in dimensions westwards.[2] In the centre the rock is dark purplish-red, exceedingly compact or flinty, sprinkled with a variable proportion of quartz-blebs and felspar crystals. Towards its southern or upper edge (for the rocks, though nearly vertical, dip southwards) it has been cleaved into a variety of purple slate, and would there at once be classed among the ordinary slates of the neighbourhood. But the fissile character is merely a marginal structure which the rock shares with the highly-cleaved tuffs that follow it. Traced westwards, this bed is found to enclose a core of quartziferous porphyry, which, though it has escaped from the most complete results of crushing, is nevertheless cleaved along its margin as well as partially in its interior. It would not be possible to distinguish parts of this intercalated less crushed core from portions of the porphyry of the main ridge. The difference of colour does not count for much, for even in this core the purple tint gives place to greenish grey, and what in the centre at the east end is a solid dark purplish-red felsite passes westward into a greenish slate, like that already noticed on Mynydd y Cilgwyn.

The microscopical examination of this rock shows it to be a true felsite of the rhyolitic type, which in the central uncleaved part exhibits a wavy flow-structure like that found in the quartz-porphyry of the ridge. So intense has been the cleavage in its upper part that the original structure of the rock is there effaced. The immediately overlying tuffs, which are likewise so thoroughly cleaved that it is not easy to draw a sharp and continuous line between them and the intercalated lava, precisely resemble those found below the conglomerate on the opposite side of the lake. They include bands of coarse volcanic breccia as well as fine compact material, showing the varying intensity of the volcanic discharges. Their included stones consist of various felsites, andesites and slates.[3]

The thin sheet of interstratified quartz-porphyry here described is not the only one to be found in the section. Others thinner and more intensely cleaved lie among the tuffs higher up. They have been sheared into mere pale unctous slates, but the remains of their quartz-blebs may still be detected in them.

The tuffs, with their interstratified bands of porphyry, become more

[1] For various readings of these sections, compare Mr. Blake (*Quart. Jour. Geol. Soc.* vol. xlix. (1893), p. 450) with Professor Bonney and Miss Raisin (*op. cit.* vol. l. (1894), p. 581).

[2] See Professor Bonney and Miss Raisin, *op. cit.* p. 593 *et seq.*

[3] I see no reason to doubt that the less acid igneous fragments were ejected during the closing phases of volcanic action, even though no such rocks have been found at the surface *in situ*. We must remember how frequently mixtures of acid and basic materials are to be found in the same continuous series of volcanic ejections and even in the same vent, of which illustration will be given in subsequent pages. Nor should we forget what a mere fragment of a volcanic group is exposed at the surface in the Llanberis district. See Professor Bonney and Miss Raisin, *op. cit.* p. 596, *footnote.*

and more mingled with ordinary argillaceous and sandy sediment as they are followed in upward succession. Among them occur bands of grit and fine conglomerate containing pebbles of porphyry and pieces of slate. Some of these grits are mainly composed of white felspar, felsite and clear grains of quartz, evidently derived from the disintegration of a rock like the porphyry of the main ridge. As the ordinary sediment of the Llanberis group sets in, the tuffs are restricted to thinner and more widely-separated bands. Some thin layers of felspathic breccia, seen among the slates close to the Glyn Peris Hotel, probably mark the last discharges of the slowly-expiring vents of this region. Here, as at St. David's, from the most ancient of our volcanic records, striking evidence is furnished of the gradual extinction of volcanic action. Through many hundreds of feet of strata which now supervene, representing the closing ages of the Cambrian and the earlier ages of the Silurian period, no trace of volcanic material has been found in this district until we reach the Bala lavas and agglomerates of Snowdon and the Pass of Llanberis.

In the neighbourhood of Bangor another area of similar rocks wraps round the northern end of the western porphyry ridge. The Geological Survey map, in conformity with the ideas that governed its representation of the older rocks of Anglesey and Caernarvon, colours these as altered Cambrian. That this error should have been made, or, when made, should not have been speedily corrected, is all the more surprising when we consider the thorough mastery which the surveyors had acquired of the aspects and the interpretation of ancient volcanic rocks in Wales, and when, moreover, we remember that as far back as 1843, long before the Survey of Caernarvonshire was published, Sedgwick had pointed out the true volcanic nature of the rocks. That great pioneer recognized the presence of "trappean conglomerates" and "trappean shales (Schaalstein)" among these deposits at Bangor; but he could not separate them from the Cambrian series of the rest of Wales.[1] And in his section he represents them as undulating towards the east and passing under the great mass of the Caernarvonshire slates and porphyries.

This interpretation, which I believe to be essentially accurate, was modified by Professor Hughes, who, fixing on a conglomerate as the base of the Cambrian system, regarded all the rocks below it, or what he termed his "Bangor group," as pre-Cambrian.[2] He has been followed in this view by subsequent writers;[3] but Mr. Blake has argued that here, as in the Llanberis district, there is no evidence to separate the volcanic detrital deposits above the porphyry from the Cambrian system.[4]

A little southward from Bangor the quartz-porphyry is overlain by a most interesting group of fragmental rocks, the "Bangor group" of Professor Hughes. Largely of volcanic origin, they must be some hundreds of

[1] *Proc. Geol. Soc.* vol. iv. p. 212; *Quart. Journ. Geol. Soc.* vol. iii. (1847), p. 136.

[2] *Quart. Journ. Geol. Soc.* vol. xxxiv. (1878), p. 137.

[3] Prof. Bonney, *op. cit.* vol. xxxv. (1879), p. 316; Dr. Hicks, *ibid.* p. 296.

[4] *Op. cit.* vol. xliv. (1888), p. 278.

feet thick, and pass under the dark shales and grits of the Lower Silurian (Arenig) series. Some of the most persistent bands among them are conglomerates, which differ from each other in composition, but most of which consist largely of fragments of various igneous rocks. Some of the coarser masses might be termed agglomerates, for they show little or no trace of bedding, and are essentially made up of blocks of volcanic material. There are abundant beds of grit, sometimes pebbly or finely conglomeratic, alternating with tuffs and with bands of more ordinary sediment. Courses of purple shale and sandstone, green shale and dark grey sandy shale occasionally occur to mark pauses in the volcanic explosions. Perhaps the most striking feature in the pyroclastic materials is the great abundance of very fine compact pale tuffs (hälleflintas of some writers), sometimes thinly laminated, sometimes occurring in ribbon-like bands, each of which presents internally a close-grained, almost felsitic or flinty texture.[1]

A cursory examination of the contents of the conglomerates, breccias and grits shows them to consist largely of different felsites, with fragments of more basic lavas. Some of these might obviously have been derived from the rock of the porphyry ridge, but, as at Llyn Padarn, there is a far greater variety of material than can be found in that ridge. Some of the fragments show perfect flow-structure. Professor Bonney has described the microscopic characters of some of these fragments, and has especially remarked upon their glassy character. Among the slides prepared from specimens collected by myself, besides the abundant fragments of felsite (rhyolite), there are also numerous pieces of different andesitic lavas and fine tuffs, as well as grains of quartz and felspar, and sometimes a good deal of granular iron-ore.

That a large proportion of the material of the so-called "Bangor beds" was directly derived from volcanic explosions can hardly be doubted. There appears to have been a prolonged succession of eruptions, varying in intensity, and somewhat also in the nature as well as in the relative fineness of the material discharged. On the one hand, coarse massive agglomerates were probably accumulated not far from the active vents, as the result of more violent or transient explosions; on the other hand, exceedingly fine and well-stratified tuffs, which attain a great thickness, serve to indicate a phase of eruptivity marked by the long-continued discharge of fine volcanic dust. Ordinary sediment was doubtless drifted over the sea-bottom in this district during the volcanic episode, but the comparative infrequence of distinct interstratifications of shale or sandstone may be taken to imply that as a rule the pauses between the eruptions were not long enough to allow any considerable accumulation of sand or mud to take place.

No satisfactory proof has yet been obtained of any interstratified lavas among the tuffs of the Bangor district. Some rocks, indeed, can be seen

[1] The occurrence of flinty or cherty deposits, in association with volcanic rocks of Lower Silurian age, is well established in Britain, and will be more particularly referred to in the sequel.

on the road between the George Hotel and Hendrewen, which, if there were better exposures, might possibly furnish the required proof; but at present little can be made of them, for their relations to the surrounding rocks are everywhere concealed.

From what I have now adduced, it is obvious that while both felsitic and andesitic lavas existed within the volcanic foci, and were ejected in fragments to form the tuffs and breccias, the lavas poured out at the surface during the Cambrian period in Caernarvonshire were mainly, if not entirely, felsites (rhyolites) in which the chief porphyritic constituent was quartz. These lavas thus stand entirely by themselves in the volcanic history of Wales. Though felsites of various types were afterwards poured out, nothing of the same quartziferous kind, so far as we yet know, ever again appeared. Further south, in Merionethshire, as will be shown in Chapter xii., the Cambrian volcanic eruptions appear to have been on the whole less acid, and to have begun with the outpouring of andesitic lavas.

I have now to consider the relation of the volcanic group of Bangor to the strata which overlie it. The geological horizon of these strata is not, perhaps, very definitely fixed. It may be Arenig, possibly even older. But for my present purpose it will be sufficient to consider the strata in question as lying at the bottom of the Lower Silurian series. Professors Hughes and Bonney have taken as their base a marked but impersistent band of conglomerate. Mr. Blake, however, has more recently shown that, as this band is succeeded by tuffs like those below it, it cannot be claimed as marking the upper limit of the volcanic group. He therefore classes it in that group and traces what he thinks is an overlap or unconformability at the bottom of the Lower Silurian strata to the east. Mr. B. N. Peach, who accompanied me in an examination of this ground, agreed with me in confirming Mr. Blake's observation as to the position of the conglomerate, which is undoubtedly overlain by the same flinty felsitic tuffs as are found below it. But we were unable to trace any unconformability. According to the numerous observations which we made, there does not seem to be any discordance in strike or dip between the flinty tuffs and the overlying shales and grits. The two groups of rock appeared to us to be conformable and to pass into each other, as at Llyn Padarn.[1]

An unconformable junction here would, in some respects, have been welcome, for it would at once have accounted for the superposition of Lower Silurian strata directly upon the Cambrian volcanic series, and for the disappearance of the Llanberis slates and grits which form so conspicuous a feature above the tuffs and conglomerates at Llyn Padarn. In the absence of such a structure we must accept the order of succession as apparently unbroken, and rely on some such explanation as was proposed by Sir Andrew Ramsay to account for the overlap of the Arenig rocks on

[1] See Mr. Blake on this point, *Quart. Journ. Geol. Soc.* vol. xlviii. (1892), p. 252, *note*. I retain the opinion expressed above.

everything older than themselves as they are traced northwards.[1] But this explanation will not entirely remove the difficulties of the case. The inosculation of the volcanic group of Bangor with the base of the Lower Silurian series cannot be accounted for by any such overlap; it seems only explicable on the supposition that the volcanic activity, which ceased in the Llyn Padarn district about the time that the Llanberis Slates were deposited, was continued in the Bangor area until Arenig time, or was then renewed. The thick volcanic group of Bangor would thus be the stratigraphical equivalent not only of the thin volcanic group of Llyn Padarn, but of the overlying mass of strata up to the Arenig rocks. In confirmation of this view, I shall show in a later chapter that volcanic action seems to have been prolonged in Anglesey to a still later geological period, that it appeared during the deposition of the Arenig strata, and that it attained a great development throughout the time of the Bala group. That a series of volcanic rocks, with associated cherty strata, may be the stratigraphical equivalent of a great thickness of ordinary sediments in other districts will be dwelt upon in the description of the Lower Silurian volcanic geology of the Southern Uplands of Scotland.[2]

In the areas of North Wales which have now been described, volcanic action appears to have begun and ended within the limits of the Cambrian period. Southwards, in the district of Dolgelly, another distinct and, in some respects, very different development of Cambrian volcanic activity may be recognized. In that district there is evidence that the volcanoes which distinguished the earlier part of the Silurian period had already begun their eruptions during Cambrian time. As their records, however, are intimately linked with those of Silurian age, an account of them is deferred to the next chapter.

THE MALVERN HILLS

Although the chief surviving records of Cambrian volcanic action in Britain are found in Wales, there is no evidence that the volcanoes of the period lay chiefly in that region. It is certainly a suggestive fact that, in the few districts where Cambrian strata appear from under younger formations in England, they are generally accompanied with igneous rocks, though the age of the latter may be older or later than the Cambrian period. If the oldest Palæozoic rocks could be uncovered over the English counties, a more abundant development of volcanic materials might be laid bare than is now to be seen in Wales.

Taking, however, the extremely limited exposures of Cambrian strata, we find two tracts that specially deserve attention. Reference has already been made to the ancient eruptive rocks of the Malvern Hills, the

[1] *Mem. Geol. Surv.* vol. iii. 2nd edit. p. 252.

[2] A group of cherts and mudstones not more than 60 or 70 feet thick appear in that region to be stratigraphically equivalent to the great depth of sedimentary material which elsewhere constitutes the Upper Arenig and Lower and Middle Llandeilo formations. See *Annual Report of the Geological Survey for* 1895, p. 27 of reprint.

antiquity of which is proved by the position of the Cambrian fossiliferous strata that overlie them. But these strata themselves include certain igneous rocks which point to a recrudescence of eruptive energy in a far later geological period.

Nearly half a century has passed away since John Phillips mentioned the intercalation of igneous rocks in the series of strata which is now classed as Upper Cambrian in the Malvern Hills. Since that date hardly anything has been added to the information which he collected. The existence of a group of rocks of such high antiquity, asserted to be of truly volcanic origin, and the precise horizon of which could be fixed by the stratigraphical aid of organic remains, seems to have almost dropped out of sight. Phillips noted the occurrence of what he regarded as truly volcanic materials in the Hollybush Sandstone and the overlying dark (Lingula) shales, and he clearly recognized that a wide difference of age separated them from the far more ancient igneous rocks of the central core of the chain. The Hollybush Sandstones were observed by him to have "often a trappean aspect and to be traversed with felspathic dykes." He found

FIG. 45.—Section across the Cambrian formations of the Malvern Hills, showing the position of the intercalated igneous rocks (*p p*). After Phillips.

the overlying black shales to include "layers of trappean ashy sandstone." But it was at the top of these shales that he obtained what he regarded as the most conspicuous evidence of contemporaneous volcanic action. He there encountered a zone of "interposed trap rocks" varying up to 50 feet in thickness, consisting of "porphyritic and greenstone masses, which, erupted from below, have flowed in limited streams over the surface of the black shales." He recognized amygdaloidal and prismatic structures among them.[1] The position of these eruptive rocks is shown in Fig. 45.

These rocks were afterwards observed and described by Dr. Holl, who found what he considered to be four true lava sheets interstratified in the Hollybush Sandstones. He noted the intercalation of "numerous beds of volcanic ash, grit and lava" in the black shales.[2]

So far as I am aware, no more recent account of these rocks has been published. Their true stratigraphical and petrographical relations require to be more precisely determined. If they are really contemporaneous lavas, they point to volcanic eruptions at the time when the middle division of the Cambrian system was being deposited. If, on the other hand, they should prove to be intrusive, they would indicate probable volcanic activity in this part of England at some time later than the middle of the Cambrian period.

[1] *Mem. Geol. Survey*, vol. ii. part i. pp. 52, 55; also Horizontal Sections of the Geol. Survey, Sheet 13, No. 8, and Sheet 15. Reference to the igneous rocks of this area will be found in the remarkable essay by De la Beche in vol. i. of the *Mem. Geol. Surv.* pp. 34, 38.

[2] *Quart. Journ. Geol. Soc.* vol. xxi. (1865), pp. 87-91.

WARWICKSHIRE

Some fifty miles to the north-east of the Malvern Hills, in the heart of the rich Midlands, and among the coal-fields and the New Red Sandstone to which these Midlands owe so much of their manufacturing industry and their agricultural fertility, another little tract of Cambrian rocks rises to the surface on the east side of the Warwickshire coal-field between Nuneaton and Atherstone. So unobtrusively do these ancient strata take their place among their younger peers, that their venerable antiquity was for a long time undetected.[1] They were actually regarded as parts of the Carboniferous series, which at first sight they seem to underlie conformably. It was not until 1882 that the mistake was corrected by Professor Lapworth, who proved the rocks to be Cambrian by finding undoubted Upper Cambrian fossils in them.[2] Subsequent investigation enabled him to work out the detailed sequence of these strata. He found that the supposed "Millstone Grit" is a thick-bedded quartzite perhaps 1000 feet in thickness, and resembling the well-known quartzites of the Lickey and Caer Caradoc. The "Coal-shales" proved to be a series (possibly 2000 feet thick) of purple, green, grey and black shales, which from their fossils could be paralleled with the dark shales of the Upper Cambrian series of the Malvern Hills.[3] These shales are immediately overlain by the Coal-measures.

For our present inquiry, however, the chief feature of interest in these discoveries is the recognition of a group of volcanic rocks underneath the quartzite. This group was named the "Caldecote Volcanic Rocks" by Professor Lapworth, who first recognized its nature and relations. Its rocks have been studied by Mr. T. H. Waller[4] and Mr. F. Rutley,[5] and have been traced upon a revised edition of the Geological Survey map by Mr. A. Strahan.[6] They consist of a thin series of well-stratified tuffs apparently derived from andesitic lavas. Their base is not seen owing to the fault which brings down the New Red Sandstone against them. They are surmounted by the quartzite, which at its base is conglomeratic and contains blocks of the tuff. A mass of quartz-felsite is possibly intrusive in these strata, and is associated with a diabase-porphyrite. In these rocks, but still more in the shales which overlie them, numerous sills of diorite and diabase occur. The total thickness of rocks from the lowest

[1] Their antiquity was recognized by Yates as far back as 1825 (*Trans. Geol. Soc.* 2nd series, vol. ii. p. 261). They had been confounded with "Millstone Grit" and "Coal-shale" by Conybeare and Phillips, and this mistake was adopted on the maps and memoirs of the Geological Survey.

[2] *Geol. Mag.* (1882), p. 563.

[3] *Op. cit.* (1886), p. 319.

[4] *Op. cit.* p. 323.

[5] *Op. cit.* p. 557.

[6] *Geol. Mag.* (1886), p. 540. In this paper full references will be found to the previous papers on the geology of the district. Jukes had recognized that the rocks below the coal-bearing strata were "older than the Upper Silurian, perhaps older than any Silurian," *Mem. Geol. Survey*, "South Staffordshire Coal-field" (1859), p. 134.

visible part of the Caldecote volcanic series to the base of the Coal-measures is probably between 2000 and 3000 feet.

There can be no doubt as to the geological position of the dark fossiliferous shales and their underlying quartzite. The fact that the basement conglomerate of the quartzite is partly made up of the underlying volcanic series may possibly mark a wide difference of age between them, and may indicate that the eruption of the tuffs took place long before Upper Cambrian time. On the other hand, the tuffs have the same strike and angle of dip with the quartzite, and as Professor Lapworth admits, the break between them may not be of great moment. It is at least certain that the intrusive sills of the district are later than the tuffs, and later also than the sedimentary Cambrian groups.

BOOK IV

THE SILURIAN VOLCANOES

CHAPTER XII

CHARACTERS OF THE SILURIAN SYSTEM IN BRITAIN. THE ARENIG VOLCANOES

The Land and Sea of Silurian time—Classification of the Silurian System—General Petrography of the Silurian Volcanic Rocks—I. The Eruptions of Arenig Age.

THE next great geological period, to which Murchison gave the name of Silurian, has in Britain a fuller record than the period which preceded it. The rocks that tell its history are more varied in origin and structure. They are displayed at the surface over a far wider area, and, what gives them special interest and value, they contain a much larger assemblage of organic remains. For the immediate subject of the present volume, they have likewise the additional attraction that they include a singularly complete and widespread volcanic chronicle. They display in many admirable sections the piled-up lavas and tuffs of scores of volcanoes, scattered all over the three kingdoms, from the headlands of Kerry to the hills of Lammermuir. They thus enable us to form a truer conception of what the early Palæozoic volcanoes were than is possible from the more limited evidence furnished by the Cambrian system.

At the beginning of the Silurian period most of the area of the British Isles lay under the sea. But if we may judge from the sedimentary strata which represent the floor of that sea, the water, during most of the time, was of no great depth. There is evidence, indeed, that during a part of the period the sea was deep enough to admit of the accumulation of wide tracts of radiolarian ooze, with but little admixture of mechanical sediment. But, for the most part, sand and mud were drifted from neighbouring lands, the more important of which probably lay to the north, over what are now the Highlands of Scotland and the north and north-west districts of Ireland. No general change in topography or in physical conditions took place at the close of Cambrian time. The older era glided insensibly into the newer, unmarked by any such catastrophe as was once supposed to have intervened at the end of each great geological period. There are traces, indeed, of slight local disturbances, but these only make the general gradual transition more marked.

Of the vegetation which covered the Silurian lands hardly anything is known. Traces of lycopods and ferns have been detected, and these probably formed the chief constituents in what must have been rather a sombre and monotonous flora. The character of the terrestrial fauna is still hidden from us, though we do know that insects winged their way through those green flowerless forests, and that scorpions likewise harboured there. That these primeval arachnoids were air-breathers is shown by their breathing stigmata; and from the fact that they possessed a well-developed poison-gland and sting, we may believe that there were already living at the same time other land-animals, possibly of higher grade, on which they preyed. But of these ancestral types no actual relics have yet been discovered.

It is the life of the sea-floor that has mainly been chronicled among the sedimentary formations. Taking the Silurian system as a whole, we find it to be the repository of a remarkably varied assemblage of organisms. Among the simpler forms, Radiolaria deserve especial notice, from their wide range in space and time, and the comparative indestructibility of the highly-siliceous, fine-grained, flinty strata, which have preserved them in abundance and have a wide distribution over the British Isles. The Graptolites, so specially characteristic of the system, range entirely through it, and by their successive differences of specific and generic forms, furnish a basis for the division of the whole series of rocks into more or less definite stratigraphical zones. Hardly less important for purposes of correlation are the Trilobites which in the Silurian period reached the culmination of their development in regard to number of species and genera. These interesting extinct types of crustacean life must have swarmed over some parts of the sea-bottom, for their remains abound in its hardened silts. The Brachiopods are likewise numerously represented among Silurian strata; and since the vertical range of the species is generally not great, they serve as useful guides in fixing stratigraphical horizons. Lamellibranchs, Gasteropods, and Cephalopods become increasingly numerous and varied as we follow the succession of strata from the base to the summit of the Silurian system. That there were fishes also in the Silurian seas is proved by the occurrence of their remains, more particularly in the higher formations.

From the organic remains which have been preserved in the rocks, it may be inferred that the animal life of the globe became more varied in Silurian time; higher types made their appearance, until vertebrates took the place of pre-eminence which they have ever since maintained.

The volcanic activity that had marked the passage of Cambrian time in Britain was prolonged into the Silurian period. In North Wales, indeed, it is clear that though the eruptions began in the earlier era of geological history they continued to be comparatively feeble until they broke out into full activity in the succeeding epoch. There is no hiatus or essential difference between the volcanic phenomena, any more than there is between the sedimentary deposits, of the two periods.

Although it may be only owing to the fact that the Silurian formations come much more extensively to the surface of the land than the underlying Cambrian are permitted to do, yet it is at least noteworthy that the relics of Silurian volcanoes are spread over a far wider area of the British Isles than those of the earlier period. Throughout a large part of Wales they form some of the most prominent mountains, such as Cader Idris, the Arans, Arenig Fawr, Moel Wyn, Moel Siabod, and Snowdon. They rise into the picturesque hill-groups of the Lake District, they appear at many detached places throughout the south of Scotland, and form conspicuous eminences in Carrick. In Ireland they abound all down the east side of the island, and even reappear on the far western headlands of the Dingle coast-line.

To the same pioneers, by whom the foundations of our knowledge of the Cambrian volcanoes were laid, we are indebted for the first broad outlines of the history of volcanic action in Silurian time. The writings of Sedgwick and Murchison, but still more the detailed mapping of De la Beche, Ramsay, Selwyn, Jukes, and the other members of the Geological Survey, have given to the Silurian volcanic rocks of Wales a classic interest in the history of geology. To these labours further reference will be made in subsequent pages.[1]

The amount of material being so ample for the compilation of a record of volcanic action in Britain during Silurian time, it will be desirable to arrange it in stratigraphical order. For this purpose invaluable assistance is afforded by the evidence of organic remains, whereby the whole Silurian system has been subdivided into sections, each characterized throughout the whole region by certain distinctive fossils. The following tabular statement exhibits the chief stratigraphical divisions of the system, and the short black lines in it mark the positions of separate volcanic platforms in each of the three kingdoms:—

		England	Wales	Scotland	Ireland
Upper Silurian	Ludlow Group	...	...	...	...
	Wenlock Group	...	...	...	—
	Llandovery Group	?	...	...	...
Lower Silurian	Bala and Caradoc Group	—	—	—	—
	Llandeilo Group	—	—	—	—
	Arenig Group	—	—	—	—

It will be most convenient, following the combined stratigraphical and geographical arrangement of this table, to discuss first the volcanic history of the Lower Silurian period as recorded in each of the three kingdoms, and then that of the Upper Silurian.

[1] For references to the older literature see *ante*, p. 142.

I. THE ERUPTIONS OF ARENIG AGE

i. MERIONETHSHIRE

Placing the upper limit of the Cambrian system at the top of the Tremadoc group, we pass into the records of another series of volcanic eruptions which marked various epochs during the Silurian period over the area of the British Isles. The earliest of these volcanic episodes has left its memorials in some of the most impressive scenery of North Wales. To the picturesque forms sculptured out of the lavas and ashes of that early time, we owe the noble range of cliffs and peaks that sweeps in a vast semicircle through the heights of Cader Idris, Aran Mawddwy, Arenig, and Moel Wyn. To the east other volcanic masses, perhaps in part coeval with these, rise from amidst younger formations in the groups of the Berwyn and Breidden Hills, and the long ridges of the Shelve and Corndon country. Far to the south, traces of Silurian volcanoes are met with near Builth, while still more remote are the sheets of lava and tuff interstratified among the Lower Silurian rocks of Pembrokeshire, and those which extend into Skomer Island.

The most important of these districts is unquestionably that of Merionethshire. In this area, as was pointed out in the last chapter, the eruptions certainly began before the close of the Cambrian period, for traces of them occur in the Tremadoc and Lingula Flag groups. But below these strata, in the vast pile of grits and conglomerates of the Harlech anticline, there does not appear to be any trace of contemporaneous volcanic action.

At the time when the Geological Survey maps of this region were prepared, the Cambrian and Lower Silurian rocks had not been subdivided into the various palæontological groups which are now recognized. Nor had any attempt been made to separate the various kinds of contemporaneous igneous masses from each other and from the tuffs in so extensive and complicated a mountain-region. The task undertaken by the Survey was beset with difficulties, some of which geologists, furnished with the advantages of a later time, can hardly perhaps realize. The imperfections of the mapping were long ago recognized by the original surveyors, and various corrections of them were made from time to time. First of all, the volcanic rocks, which originally had been all massed under one colour, were traced out separately on the ground, according to their structure and mode of origin, and were distinguished from each other on the maps.[1] Subsequently divisional lines were followed out between some of the larger stratigraphical groups, the maps and sections were still further modified, and the results were summed up in the volume on the *Geology of North Wales.*[2]

[1] *Mem. Geol. Surv.* vol. iii. 2nd edit. p. 95, note.

[2] Some of the modifications introduced are, I think, to be regretted, for the earlier editions of the maps and sections are in certain respects more accurate than the later. On this point I concur with the criticism made by Messrs. Cole and Jennings, *Quart. Journ. Geol. Soc.* vol. xlv. (1889), p. 436.

But short of actually resurveying the whole of that rugged tract, it was impossible to bring the maps abreast of the onward march of science. They consequently remain, as a whole, very much as they were some thirty or forty years ago.

Sir Andrew Ramsay, in his great Monograph on the geology of North Wales, has described the Merionethshire volcanic district in considerable detail. He seems finally to have come to the conclusion that the eruptions of that area were included within the Arenig period.[1] He shows, indeed, that on Rhobell Fawr the ejected materials lie directly on disturbed Lingula Flags without the intervention of the Tremadoc group, which is nevertheless present in full development in the near neighbourhood.[2] And in trying to account for this remarkable fact he evidently had in his mind the possibility that volcanic eruptions had taken place long before as well as after the beginning of the deposition of the Arenig grit and slates.[3] He seems eventually, however, to have looked on the Rhobell Fawr sections as exceptional and possibly to be accounted for by some local disturbance and intrusion of eruptive rock.[4] He clearly recognized that there were two great epochs of volcanic activity during the Silurian period in Wales, one belonging to the time of the Arenig, the other to that of the Bala rocks, and he pointed out that the records of these two periods are separated by a thick accumulation of sedimentary strata which, being free from interstratifications of contemporaneous igneous rocks, may be taken to indicate a long interval of quiescence among the subterranean forces.[5]

The lower limit of the Arenig rocks has been fixed at a band or bands of grit or conglomerate (Garth grit) which can be followed with some slight interruptions all round the great dome of Cambrian strata from Llanegrin on the south to the shore at Criccieth on the north. The volcanic group doubtless lies, generally speaking, above that basement platform. But, besides the sections at Rhobell Fawr just referred to, where the volcanic materials lie on the Lingula Flags, the same relation may, I think, be observed on the north flank of Cader Idris. Messrs. Cole, Jennings, and Holland have come to the conclusion that the eruptions began at a rather earlier date than that assigned to them in the *Survey Memoirs*, and my own examination of the ground led me to accept their conclusion.[6] I inferred that the earliest discharges in the southern part of the region took place in Cambrian time, at or possibly before the close of the deposition of the Lingula Flags, and that intermittent outbursts occurred at many intervals during the time when the Tremadoc and Arenig rocks were deposited.

[1] *Mem. Geol. Survey*, vol. iii. 2nd ed., p. 96.

[2] The ashes and agglomerates of Rhobell Fawr can be seen in various places to rest on the highest members of the Lingula Flags. See Messrs. Cole and Holland, *Geol. Mag.* (1890), p. 451.

[3] *Op. cit.* p. 72.

[4] He was disposed to regard Rhobell Fawr as one of the great centres of eruption of the district. See *Memoir of A. C. Ramsay*, p. 81, and *Geology of North Wales*, 2nd edit. p. 98.

[5] *Op. cit.* pp. 71, 96, 105.

[6] *Quart. Journ. Geol. Soc.* vol. xlv. (1889), p. 436; *Geol. Mag.* (1890), p. 447. *Pres. Address Geol. Soc.* 1890, p. 107.

Important confirmation of this view of the Cambrian age of the earlier volcanic eruptions of the Cader Idris region has recently been obtained by Messrs. P. Lake and S. H. Reynolds who, in the ground intervening between the lower slopes of Cader Idris and Dolgelly, have ascertained the existence of a marked band of andesitic lava traceable for some distance in the upper Lingula Flags. They have also observed a higher volcanic group reposing upon the Tremadoc strata at the top of the Cambrian system, and consisting of rhyolite with rhyolite-tuffs.[1]

Some of the most stupendous memorials of the earlier eruptions are to be seen in the huge mountain mass of Rhobell Fawr (2403 feet). They consist mainly of agglomerates and tuffs, one of the most remarkable varieties of which is distinguished by its abundant scattered crystals of hornblende and of augite. The fragments of rock included in these rocks are scoriæ and lumps of various lavas, especially basaltic and trachytic andesites. The tuffs become finer towards the top of the mountain where they are interleaved with grits. Among the pyroclastic materials occasional lavas (basaltic andesites) occur which may be contemporaneous streams, but

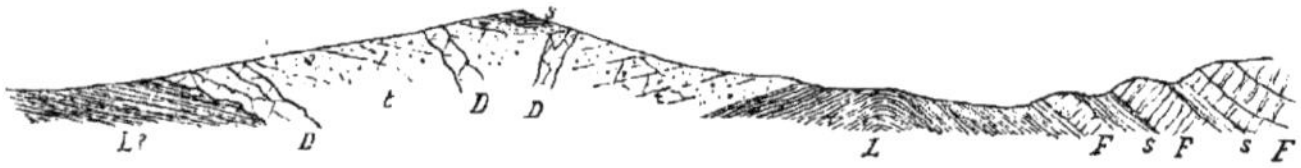

Fig. 46.—Section across Rhobell Fawr.[2]

L L, Lingula flags; *t*, tuffs and ashy slates; *s*, slates and grits; F F, Arenig volcanic series; D, dolerite.

most of the lava-form rocks appear to be intrusive. They include dolerites (augite-aphanites), basaltic andesites, and trachytic andesites.[3]

The materials from the Rhobell Fawr volcano are clearly distinguishable from those of the Arenig volcanoes in the neighbourhood. The latter begin to make their appearance among the black slates at the base of the northern declivities of Cader Idris, and extend upward through that mountain into the country beyond.

An upper limit to this volcanic group is not easily traceable; partly, no doubt, from the gradual cessation of the eruptions and partly from the want of any marked and persistent stratigraphical horizon near the top of the group. Sir Andrew Ramsay, indeed, refers to the well-known band of pisolitic iron-ore as lying at or near to the top of the Arenig rocks.[4] There can be no doubt, however, that the volcanic intercalations continue far above that horizon in the southern part of the district.

In spite of the extent to which the volcanic masses of the Arenig period have been covered by later Palæozoic formations, it is still possible to fix approximately the northern, western, and southern limits of the district over which the ashes and lavas were distributed. These materials die out

[1] *Quart. Journ. Geol. Soc.* vol. lii. (1896), p. 511.
[2] After Messrs. Cole and Holland, *Geol. Mag.* (1890), p. 450.
[3] Prof. Cole, *Geol. Mag.* (1893), p. 337.
[4] *Mem. Geol. Survey*, vol. iii. 2nd edit. pp. 249, 250.

as they are traced southwards from Cader Idris and north-westwards from Tremadoc.[1] The greatest diameter of ground across which they are now continuously traceable is about twenty-eight miles. They attain their greatest thickness, upwards of 5000 feet, in Aran Mawddwy, which rises from their most easterly escarpment. We may therefore infer that the main vent or vents lay somewhere in that direction. The noble range of precipices facing westwards shows how greatly the limits of the volcanic rocks have been reduced by denudation. There can be little doubt that at least the finer tuffs extended westwards as far as a line drawn from Tremadoc to Llanegrin—that is, some fifteen miles or more beyond the cliffs of Aran Mawddwy, thus stretching across much of the site of what is now the great Harlech anticline.

This compact, well-defined volcanic area, in spite of the faults which traverse it and the disturbed positions into which its rocks have been thrown, is, in many respects, one of the simplest and most easily studied among the Palæozoic formations of this country. Its main features have been delineated on the maps of the Geological Survey and have been described in Sir Andrew Ramsay's monograph. But these publications cannot be regarded as more than a first broad, though masterly, outline of the whole subject. There is an ample field for further and more minute research wherein, with the larger and better Ordnance maps now available, and with the advantage of the numerous modern petrographical aids, a more exhaustive account may be given of the district. The whole volcanic succession from base to summit is laid bare in innumerable magnificent natural sections along ranges of hills for a distance of some forty miles, and a careful study and re-mapping of it could not fail to add greatly to our knowledge of the early history of volcanic action.[2]

According to the observations of the Geological Survey, the Arenig volcanic rocks of Merionethshire naturally arrange themselves in three great bands, each of which is described as tolerably persistent throughout the whole district:—1st, a lower series of ashes and conglomerates, sometimes 3300 feet thick (Aran Mawddwy); 2nd, a middle group of "felstones" and "porphyries," consisting partly of true contemporaneous lava-streams and partly of intrusive sheets, and reaching a thickness of 1500 feet; 3rd, an upper series of fragmental deposits like that beneath, the extreme thickness of which is 800 feet (Arenig mountain). A re-mapping of the ground on the six-inch maps would, no doubt, show many local departures from this general scheme.

The pyroclastic members of this volcanic series present many features of interest both to the field-geologist and the petrographer; but they have as yet been only partially studied. At the southern end of the district it is remarkable to what a large extent the earliest eruptions must have

[1] *Op. cit.* p. 96.

[2] The excellent papers of Professor Cole, Mr. Jennings, Mr. Holland, Mr. G. J. Williams, Mr. P. Lake and Mr. S. H. Reynolds are illustrations of how the published work of the Geological Survey may be modified and elaborated.

been mere gaseous explosions, with the discharge of comparatively little volcanic material. Many of the tuffs that are interstratified with black slates (? Lingula Flags) at the foot of the long northern slope of Cader Idris, consist mainly of black-slate fragments like the slate underneath, with a variable proportion of grey volcanic dust.

The accompanying section (Fig. 47) represents the arrangement of the rocks exposed at the Slate Quarry of Penrhyn Gwyn. About 50 feet of black slate (*a*) are there seen, the bedding in which dips S. at 20°, while the cleavage is inclined towards S.W. at a slightly higher angle. The next 20 feet of slate (*b*) are distinguished by many intercalations of slate-tuff or breccia, varying from less than an inch to three feet in thickness. An intrusive sheet of andesite (*c*), which varies from two or three to ten feet in thickness, and is strongly cellular in the centre, interrupts the slates and hardens them. Above this sill the indurated slate and tuff (*d*), containing abundant felspar crystals, pass under a flinty porphyritic felsite (*e*) or exceedingly fine tuff, enclosing a band of granular tuff. Beyond this band the black slates with their seams of tuff continue up the hill and include a sheet of slaggy felsitic lava 8 or 10 feet thick.

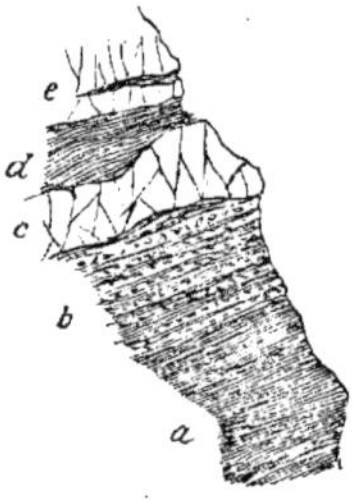

FIG. 47.—Section at the Slate Quarry, Penrhyn Gwyn, north slopes of Cader Idris.

This section, affording as it does the first glimpse of the volcanic history of Cader Idris, indicates a continued series of feeble gaseous discharges, probably from one or more small vents, whereby the black clay on the sea-floor was blown out, the fragments falling back again to be covered up under a gradual accumulation of similar dark mud. By degrees, as the vigour of eruption increased, lava-dust and detached felspar crystals were ejected, and eventually lava rose to the surface and flowed over the sea-bottom in thin sheets.

But elsewhere, and likewise at a later period in this same southern part of the district, the fragmental discharges consisted mainly of volcanic material. Sir Andrew Ramsay has described the coarse conglomerates composed of subangular and rounded blocks of different "porphyries," somtimes 20 inches in diameter, embedded in a fine matrix of similar materials. The true nature of the component fragments in these rocks has still to be worked out.

Messrs. Cole and Jennings have noticed that the grey volcanic dust of the older slate-tuff of Cader Idris is seen under the microscope "to abound in particles of scoriaceous andesite-glass, now converted into a green palagonite."[1] Their investigations show that while the same kinds of volcanic rocks continue to be met with from the bottom to the top, nevertheless there is an increase in the acid character of the lapilli as the section is traced upwards. Some of the fragments consist of colourless devitrified glass, with pieces of pumice, as if derived from the breaking up of previously-formed tuffs. Others resemble quartz-andesites, rhyolites, or

[1] *Quart. Journ. Geol. Soc.*, vol. xlv. (1889), p. 424 ; *Geol. Mag.* (1890), p. 447.

trachytes, while in at least one instance, somewhat low down in the section, quartz-grains with intruded material point to the existence of some fairly acid and vitreous lava.[1] On the south side of Llyn Cau, that is towards the top of the volcanic group, I found a coarse agglomerate with blocks of felsitic lavas, sometimes three feet across (see Fig. 48). This gradual increase of acidity in the lapilli of the tuffs finds an interesting confirmation in the contemporaneous lava-sheets to which I shall afterwards allude.

One of the most noticeable features in the tuffs of this volcanic group is the great abundance of entire and broken crystals dispersed through them. These crystals have certainly not been formed *in situ*, but were discharged from the vents as part of the volcanic dust. They usually consist of felspar which, at least in the southern portion of the district, appears generally to be plagioclase. Frequent reference to these crystals as evidence of volcanic explosions may be found in the publications of the Survey. Nowhere can they be better seen than in the black slate-tuffs of Cader Idris. They are there white, more or less kaolinized, and as they lie dispersed through the black base, they give the rock a deceptive resemblance to some dark porphyry. The large crystals of hornblende and augite abundantly scattered through much of the tuff of Rhobell Fawr have been already referred to.

In the central parts of the district thick bands of ashes were mapped by the Survey, and described as consisting almost wholly of volcanic materials, but containing occasional thin bands of slate which suffice to mark pauses in the eruptions, when ordinary sediment was strewn over the sea-bottom. In the Cader Idris ground, on the other hand, interstratifications of non-volcanic material are of such frequent recurrence as to show that there, instead of constant and vigorous discharges accumulating a vast pile of ashes, the eruptions followed each other after intervals of sufficient duration to allow of the usual dark sediment spreading for a depth of many feet over the sea-bottom.

One of the most interesting deposits of these interludes of quiescence is that of the pisolitic ironstone and its accompanying strata on the north front of Cader Idris (*i* in Fig. 48). A coarse pumiceous conglomerate with large slag-like blocks of andesite and other rocks, seen near Llyn-y-Gadr, passes upward into a fine bluish grit and shale, among which lies the bed of pisolitic (or rather oolitic) ironstone which is so widely diffused over North Wales. The finely-oolitic structure of this band is obviously original, but the substance was probably deposited as carbonate of lime under quiet conditions of precipitation. The presence of numerous small *Lingulæ* in the rock shows that molluscan life flourished on the spot at the time. The iron exists in the ore mainly as magnetite, the original calcite or aragonite having been first replaced by carbonate of iron, which was subsequently broken up so as to leave a residue of minute cubes of magnetite.[2]

Above the ironstone some more blue and black shale and grit pass under

[1] *Op. cit.* p. 429. A tuff lying below the ironstone near Cross Foxes, east of Dolgelly, likewise contains fragments of trachytic lavas.

[2] Messrs. Cole and Jennings, *op. cit.* p. 426.

a coarse volcanic conglomerate like that below, lying at the base of the high precipice of Cader Idris. Hence this intercalated group of sedimentary strata marks a pause in the discharge of ashes and lavas, during which the peculiar conditions of sedimentation indicated by the ironstone spread over at least the southern part of the volcanic area. Some few miles to the east, where the ironstone has been excavated near Cross Foxes, the band is again found lying among tuffs and grits full of volcanic lapilli.

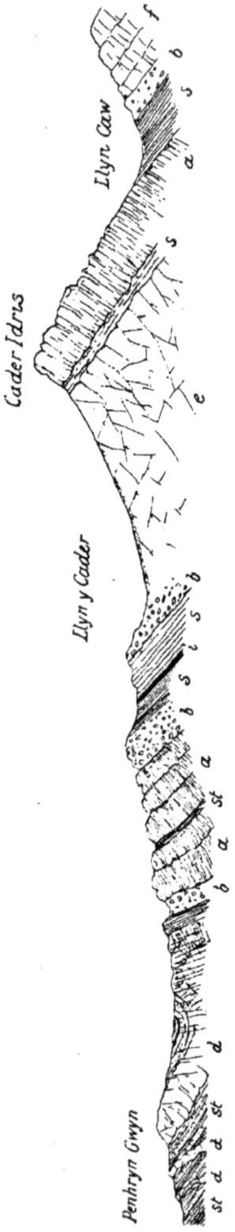

Fig. 48.—Sketch-section across Cader Idris.
st, slates and tuffs and ashy slates; s, slates and grits; i, ironstone; b, volcanic breccias; a, slaggy andesitic and more basic lavas; e, microgranite or eurite; f, felsites; d, "greenstone" (dolerites, diabases, etc.).

Between a lower and an upper band of tuff in the Arenig volcanic group the Maps and Memoirs of the Geological Survey distinguish a central zone of "felspathic porphyry," which attains a maximum thickness of 1500 feet (see Fig. 48). From Sir Andrew Ramsay's descriptions, it is clear that he recognized in this zone both intrusive and extrusive sheets, and that the latter, where thickest, were not to be regarded as one mighty lava-flow, but rather as the result of successive outpourings, with occasional intervals marked by the intercalation of bands of slate or of tuff. To a certain extent the intruded sheets are separated on the map from the contemporaneous lavas; but this has been done only in a broad and sketchy way. One of the most important, and at the same time most difficult, tasks yet to be accomplished in this district is the separation of the rocks which were probably poured out at the surface from those that were injected underneath it. My own traverses of the ground have convinced me that good evidence of superficial outflows may be found in tracts which have been mapped as entirely intrusive; while, on the other hand, some of the so-called "lavas" may more probably be of the nature of sills.

The petrography of the rocks, moreover, still requires much study. Among the so-called "felspathic porphyries" of the Survey maps a considerable variety of texture, structure and composition will doubtless be detected. In the *Descriptive Catalogue of Rock-Specimens in the Museum of Practical Geology* (3rd edit., 1862) the rocks that form the "lava-streams of Llandeilo age," in Merionethshire, are named "felstone," "felspar-porphyry," "felstone-porphyry," "felspathic-porphyry," and "calcareous amygdaloid."

The most interesting feature which my own slight personal acquaintance with the region has brought before me is the clear evidence of a succession from comparatively basic lavas in the lower part of the group to much more acid masses in the higher part. In the Survey map numerous sheets of intrusive "greenstone" are shown traversing the Lingula Flags, Tremadoc slates, and lower part of the volcanic group along the northern slopes of Cader Idris. The true intrusive nature of much of this material is clearly established by transgressive lines of junction and by contact-metamorphism, as well as by the distinctive crystalline texture of the rocks themselves. But the surveyors were evidently puzzled by some parts of the ground. Sir Andrew Ramsay speaks of "the great mass of problematical vesicular and sometimes calcareous rock which is in places almost ashy-looking." After several oscillations of opinion, he seems to have come finally to the conclusion that this vesicular material, which occurs also in the upper part of the mountain, passes into, and cannot be separated from, the undoubted intrusive "greenstones."[1]

The true solution of the difficulty will be found, I believe, in the recognition of a group of scoriaceous lavas among these greenstones. The presence of a cellular structure might not be sufficient to demonstrate that the rocks in which it appears are true lava-beds, for such a structure is far from unknown both among dykes and sills. But in the present case there is other corroborative testimony that some of these Cader Idris amygdaloids were really poured out at the surface. Below Llyn-y-Gadr—the dark tarn at the foot of the vast wall of Cader Idris—the beds of coarse volcanic conglomerate (*b* in Fig. 48), to which I have already alluded, are largely composed of blocks of the vesicular "greenstones" on which they lie. These "greenstones," moreover, have many of the most striking characteristics of true lavas (*a* in Fig. 48). They are extraordinarily cellular; their upper surfaces sometimes present a mass of bomb-like slags with flow-structure, and the vesicles are not infrequently arranged in rows and bands along the dip-planes.

A microscopic examination of two slides cut from these rocks shows them to be of a trachytic or andesitic type, with porphyritic crystals of a kaolinized felspar embedded in a microlitic groundmass. The rocks are much impregnated with calcite, which fills their vesicles and ramifies through their mass.

A few miles to the east some remarkable felsitic rocks take the place of these vesicular lavas immediately below the pisolitic iron ore. I have not determined satisfactorily their relations to the surrounding rocks, and in particular am uncertain whether they are interbedded lavas or intrusive sheets. Dr. F. H. Hatch found that their microscopic characters show a close resemblance to the soda-felsites described by him from the Bala series of the south-east of Ireland.

[1] *Mem. Geol. Surv.* vol. iii. 2nd edit. p. 36; see also pp. 31, 32.

The slopes of Cader Idris are partly obscured with debris, from above which rises the great precipitous face formed by the escarpment of "porphyry," here intrusively interposed among the Arenig volcanic rocks. This enormous sill will be referred to a little further on in connection with the other intrusive sheets of the region.

The remarkably cellular rock which forms the peak of Cader Idris is coloured on the Survey map as an intrusive sill of "greenstone," which in the Memoir is said to alter the contiguous slates and to appear to cut across them diagonally. I am disposed, however, to think that these appearances of intrusion are deceptive. On the southern declivity of the mountain this rock presents one of the most curious structures to be seen in the whole district. Its surface displays a mass of spheroidal or pillow-shaped blocks aggregated together, each having a tendency to divide internally into prisms which diverge from the outside towards the centre.[1] Some portions are extremely slaggy, and round these more solid portions finely crystalline parts are drawn, suggestive rather of free motion at the surface than of the conditions under which a subterranean sill must be formed. The idea occurred to me on the ground that while the band of rock marked as "greenstone" on the map is probably, in the main, an interstratified lava, there may nevertheless be basic intrusions along its course, as in the lower part of the mountain. The minute structure of this amygdaloid, as revealed by the microscope, shows it to be an epidiorite wherein the hornblende, paramorphic after augite, has been again partially altered along the margins into chlorite.

The highest lavas of Cader Idris, forming the ridge to the south of Llyn Cau, are separated from the amygdaloid just described by a thick zone of black slate with thin ashy intercalations, beyond which comes the coarse volcanic agglomerate already referred to as containing blocks of felsite a yard or more in diameter. These lavas are true felsites, sometimes beautifully spherulitic and exhibiting abundant flow-structure, like some of the felsites of the next or Bala volcanic period.[2] The petrography of these rocks still remains to be worked out.

The volcanic series of Cader Idris sweeps northward through the chain of Aran and Arenig, and then curves westward through the group of Manod and Moelwyn, beyond which it rapidly dies out. In its course of

[1] This peculiar structure of the more basic Arenig lavas, where the rock looks as if built up of irregularly-spheroidal, sack-like or pillow-shaped blocks, will be again referred to in connection with the Arenig (and Llandeilo) lavas of Scotland and Ireland. It appears to be widely distributed, and especially in connection with the occurrence of radiolarian cherts. The black slate above the Cader Idris amygdaloid would, in a similar position in Scotland, be associated with such cherts, but these have not yet been noticed at this locality. With the spheroidal internally-radiating prismatic structure of the Cader Idris rock, compare that of the lava at Acicastello already noticed on p. 26.

[2] Messrs. Cole and Jennings, *Quart. Journ. Geol. Soc.* vol. xlv. (1889), p. 430. From the examination of slices prepared from a few of the felsites of the Dolgelly district, Dr. Hatch observed a "striking difference between their characters and those of the Cambrian felsites of Caernarvonshire. The porphyritic constituent is now no longer quartz, but felspar (plagioclase), and the rocks belong, not to the rhyolitic, but rather to the less acid trachytes, perhaps even to the andesites."

about 45 miles it undergoes considerable variation, as may be seen by comparing a section through Moelwyn with that through Cader Idris already given. According to the researches of Mr. Jennings and Mr. Williams,[1] the main mass of volcanic material in the northern part of the region consists of fragmentary rocks varying in texture from agglomerates into fine tuffs, but showing some differences in the succession of beds in different localities.

The Tremadoc group of strata clearly underlies the volcanic series of these more northerly tracts. But it contains, so far as appears, no intercalation of volcanic material. The inference may thus be drawn that the eruptions began in the Cader Idris district, and did not extend into that of Manod and Moelwyn until after the beginning of the Arenig period. Above the Tremadoc group lies the well-marked and persistent band, about 13 feet thick, known as the Garth grit, which has been already referred to as a convenient base-line to the Arenig group.

In this northern district, among the sediments which overlie the Garth grit, layers of fine tuff begin to make their appearance, which north of Cwm Orthin thicken out into a considerable mass between the grit and the

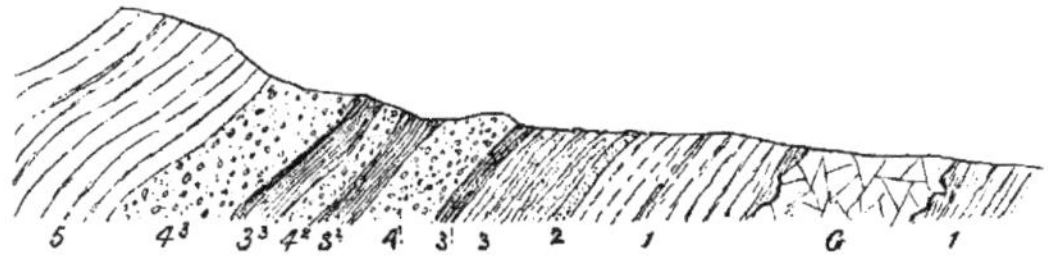

FIG. 49.—Section across the Moelwyn Range.[2]

1, Tremadoc Group; 2, Garth or Arenig grit (base of Arenig group); 3, Arenig slates, etc.; 3^1, Lower slate band; 3^2, Middle slate band; 3^3, Upper slate band; 4^1, Lower agglomerate; 4^2, Middle agglomerate; 4^3, Upper agglomerate; 5, Llandeilo group; G, Granite boss of Moel tan y Grisiau.

lowest of the great agglomerates. These tuffs, which mark the beginning of the volcanic eruptions of the district, are followed by a band of slate which in some places has yielded a *Lingula, Orthis Carausii*, and a *Tetragraptus*, and points to an interval of quiescence in the volcanic history. We now enter upon an enormous thickness of agglomerates and tuffs separated by several bands of slate. Taking advantage of the slaty intercalations, Messrs. Jennings and Williams have divided this great accumulation of fragmentary volcanic material into three beds (Fig. 49). The matrix of the agglomerates is compact and pale, so as to resemble and to have been called "felstone," but showing its fragmentary nature on weathered surfaces. The blocks imbedded in this paste range up to sometimes as much as 11 feet in length by 4 feet in width. Their minute petrographical characters have not been studied, but the blocks are stated to consist for the most part of "slaty and schistose fragments mixed with rounded pebbles of fine-grained 'felstone.'" They are heaped together as in true agglomerates. In the

[1] *Quart. Journ. Geol. Soc.* xlvii. (1891), p. 368.

[2] After Messrs. Jennings and Williams, *Quart. Journ. Geol. Soc.* vol. xlvii. (1891), p. 371, and Horizont. Sect. Geol. Surv. Sheet 28.

upper agglomerate, fragments of cleaved slate containing *Lingula* have been observed.

The name of "felstone" is restricted by Messrs. Jennings and Williams to certain fine-grained varieties of rock, of which a thin band lies at the base of the lower agglomerate, while another of considerably greater importance occurs in the middle of the upper agglomerate. These bands consist of a fine compact greenish base, and weather with a dull white crust; sometimes, as in the thicker sheet, a columnar structure shows itself. Whether these rocks are to be regarded as lavas or sills, or even as finer varieties of tuff, is a question that awaits further inquiry. But it is clear, from the investigation of the two observers just cited, that the pyroclastic constituents must vastly preponderate in the volcanic series over the northern part of the region. All these rocks, whether coarse or fine-grained, appear to be rather acid in composition, and no evidence has yet been obtained of a sequence among them from a more basic to a more acid series, as in Cader Idris.

The highest agglomerate bed of the Manod and Moelwyn area is covered by slates which contain Llandeilo graptolites. In this way, by means of palæontological evidence, the upward and downward limits of the Arenig volcanic series in this part of Wales are definitely fixed.

Hardly any information has yet been obtained as to the situation and character of the vents from which the lavas and ashes of Merionethshire were discharged. In the course of the mapping of the ground, the Geological Survey recognized that, as the greatest bulk of erupted material lies in the eastern and south-eastern parts of the region, the chief centres of emission were to be looked for in that quarter, and that possibly some of the intrusive masses which break through the rocks west of the great escarpment may mark the site of vents, such as Tyddyn-rhiw, Gelli-llwyd-fawr, Y-Foel-ddu, Rhobell Fawr, and certain bosses near Arenig.[1] The distribution of the volcanic materials indicates that there were certainly more than one active crater. While the southward thickening of the whole volcanic group points to some specially vigorous volcano in that quarter, the notable thinning away of the upper tuffs southward and their great depth about Arenig suggest their having come from some vent in this neighbourhood. On the other hand, the lower tuffs are absent at Arenig, while on Aran Mawddwy, only nine miles to the south, they reach a depth of 3000 feet. Still farther to the south these volcanic ejections become more and more divided by intercalated bands of ordinary sediment. One of the most important volcanoes of the region evidently rose somewhere in the neighbourhood of what is now Aran Mawddwy. There seems reason to surmise that the sites of the chief vents now lie to the east and south of the great escarpment, buried under the thick sedimentary formations which cover all that region.

If we are justified, on stratigraphical and petrographical grounds, in connecting the lowest volcanic rocks of the Berwyn range with those of

[1] *Mem. Geol. Surv.* vol. iii. 2nd edit. p. 98; see also pp. 44, 54, 58, 71.

Merionethshire, we may speculate on the existence of a group of submarine vents, coming into eruption at successive intervals, from some epoch during the period of the Lingula Flags up to that of the Bala rocks, and covering with lavas and ashes a space of sea-bottom at least forty miles from east to west by more than twenty miles from north to south, or roughly, an area of some 800 square miles.[1]

Besides the materials ejected to the surface, the ancient volcanic region of Merionethshire was marked by the intrusion of a vast amount of igneous rock between and across the bedding-planes of the strata deep underground. One of the most prominent features of the Geological Survey map is the great number of sills represented as running with the general strike of the strata, especially between the top of the Harlech grits and the base of the volcanic series. On the north side of the valley of the Mawddach, between Barmouth and Rhaiadr Mawddach, in a distance of twelve miles the Survey mapped "more than 150 intrusions varying from a few yards to nearly a mile in length."[2] This zone of sills is equally marked on the south side of the valley. It may be traced all round the Harlech anticline until it dies out, as the bedded masses also do, towards Towyn on the south and about Tremadoc on the north.

The presence of such a zone of intrusive sheets at the base of an ancient volcanic series is a characteristic feature in the geology of Britain. It is met with again and again among the Palæozoic systems, and appears on a striking scale in association with the Tertiary basaltic plateaux of Antrim and the Inner Hebrides. But nowhere, perhaps, is it more strongly developed than beneath the Arenig group of lavas and tuffs in North Wales. Abundant as are the protrusions marked on the Geological Survey map, they fall short of the actual number to be met with on the ground. Indeed, to represent them as they really are would require laborious surveying and the use of maps on a far larger scale than one inch to a mile.

The vast majority of these sills are basic rocks, or, in the old and convenient terminology, "greenstones." Those of the Cader Idris district have been examined by Messrs. Cole and Jennings, who found that, notwithstanding the considerable alteration everywhere shown by the abundant epidote and calcite, the coarser varieties may be recognized as having originally been dolerites approaching gabbro, with a well-developed ophitic character, the general range of structure being from dolerites without olivine and aphanites to andesitic rocks with an originally glassy matrix.[3] Dr. Hatch confirmed this diagnosis from slides prepared from my specimens. The ophitic structure is usually characteristic and well preserved, in spite of the alteration indicated by epidote, chlorite, uralite, and leucoxene.

That this zone of "greenstone" sills belongs to the period of the Merionethshire volcanoes may be reasonably concluded. The way in which they follow the line of the great escarpment, their almost entire absence from the Cambrian dome to the west, their cessation as the overlying lavas

[1] The Berwyn Hills, however, will be described in later pages as a distinct volcanic district.

[2] *Mem. Geol. Surv.* vol. iii. p. 26.

[3] *Quart. Journ. Geol. Soc.* vol. xlv. (1889), p. 432.

and tuffs die out laterally, and their scarcity above the lower part of the volcanic group, seem to indicate their close relationship to that group. Moreover, that they must have been as a whole later than the main part of the lavas and tuffs may be inferred from their position. The molten material of which they were formed could hardly have forced its way between and across the strata unless egress to the surface had been impeded by some thick overlying mass. The "greenstones" may therefore be regarded as lateral emanations from funnels of more basic lava towards the close of the volcanic period. Possibly some at least of the highly slaggy and vesicular bands to which I have referred may represent portions of this material, which actually flowed out as streams of lava at the surface.

But there is likewise evidence of extensive intrusion of more siliceous rocks. On the Geological Survey map, besides the numerous "greenstones," various sheets of "felspathic porphyry" are represented as running with the general strike of the region, but here and there breaking across it. One of the most remarkable of these acid sills is that which, in the noble precipice of Cader Idris, has a thickness of about 1500 feet and a length of three or four miles. It is shown on the map to be transgressive across other rocks, and, as seen on the ground, it maintains the uniformity of texture which is characteristic rather of sheets that have solidified underneath than of those which have congealed with comparative rapidity at the surface. On a fresh fracture the rock presents a pale bluish-grey tint, becoming yellowish or brownish as the result of weathering. Its texture is finely granular, with occasional disseminated felspars. Under the microscope a section of it was found by Dr. Hatch to exhibit the characteristic structure of a microgranite, a confused holocrystalline aggregate of quartz and felspar, with a few porphyritic felspars. Messrs. Cole and Jennings have proposed to revive for this rock Daubuisson's name "Eurite."[1]

A similar rock occurs at a lower horizon among the Lingula Flags at Gelli-llwyd-fawr, two miles south-west of Dolgelly,[2] and much microgranite has been injected along the slopes above Tyddyn-mawr.

The chronological relation of these acid sheets and bosses to the more basic intrusions has not yet been definitely determined. That some of them may have solidified in vents and may have been directly connected with the protrusion of the later or more highly siliceous lavas is not at all improbable. Others again would seem to belong to a much later geological period than the Arenig volcanoes. In this late series the well-known boss of Tan-y-grisiau near Festiniog should probably be included. This mass of eruptive material was mapped by the Geological Survey as "intrusive syenite." It has been more recently examined and described by Messrs. Jennings and Williams as a granitite.[3] These observers have noticed not only that it intrusively traverses and alters the Tremadoc group, but that its intrusion appears to have taken place subsequent to the cleavage which

[1] Mr. Harker speaks of the rock as a granophyre.

[2] Messrs. Cole and Jennings, *op. cit.* p. 435.

[3] *Quart. Journ. Geol. Soc.* vol. xlvii. (1891), p. 379.

affects the Llandeilo as well as older formations. This granitic boss has thus probably no connection with the Arenig volcanoes, but belongs to a later period in the volcanic history of the Principality.

The remarkable scarcity of dykes in the volcanic districts of Wales has been noticed by more than one observer. Among the intrusive "greenstones" of Merionethshire some occasionally assume the dyke form, and through the agglomerates and tuffs of Rhobell Fawr dykes of olivine-diabase have worked their way. In the Festiniog district various altered andesitic dykes have been noted. But there has been no widespread fissuring of the ground and uprise of lava in the rents, such as may be seen in the Archaean gneiss, and in the later Palæozoic, but still more in the Tertiary volcanic regions. This feature becomes all the more notable when it is viewed in connection with the great development of sills, and the evidence thereby afforded of widespread and extremely vigorous subterranean volcanic action.

In the Merionethshire region there certainly was a long period of quiescence between the close of the Arenig and the beginning of the Bala eruptions. Moreover, no evidence has yet been found that active vents ever again appeared in that district, the subterranean energy at its next outburst having broken out farther to the east and north. In Anglesey, however, where, as I shall point out, there is proof of contemporaneous tuffs among the Arenig rocks, it is possible that a continuous record of volcanic action may yet be traced from Arenig well onward into Bala time.

ii. SHROPSHIRE

About 35 miles to the south-east of the great volcanic range of Merionethshire a small tract of Arenig rocks rises from amidst younger formations, and forms the picturesque country between Church Stoke and Pontesbury. Murchison in his excellent account of this district clearly recognized the presence of both intrusive and interstratified igneous rocks.[1] The ground has in recent years been more carefully worked over by Mr. G. H. Morton[2] and Professor Lapworth.[3]

At the top of the Arenig group of this district lies a zone of well-stratified andesitic tuff and breccia (Stapeley Ash), with frequent intercalations of shales, and occasionally fossiliferous.[4] There is thus satisfactory proof of contemporaneous eruptions at intervals during the accumulation of the later Arenig sediments. That there were also outflows of lava is shown by the presence of sheets of augite- and hypersthene-andesite. These volcanic intercalations form marked ridges, having a general northerly trend. They are folded over the broad laccolitic ridge of Corndon, on the east side of which they are thrown into a synclinal trough, so that successive parallel outcrops of them are exposed. According to the mapping of

[1] *Silurian System* (1839), chap. xix.; *Siluria*, 4th edit. (1867), pp. 26, 49.
[2] *Proc. Liverpool Geol. Soc.* x. (1854), p. 62. [3] *Geol. Mag.* (1887), p. 78.
[4] Prof. Lapworth and Mr. W. W. Watts, *Proc. Geol. Assoc.* xiii. (1894), pp. 317, 337.

the Geological Survey they are thickest towards the west, and become more split up with intercalated sediments as they range eastward.

Volcanic eruptions in this Shropshire region continued from the Arenig into the Bala period. They are marked among the Llandeilo strata by occasional tuffs and by two massive beds of "volcanic grit," described by Murchison,[1] but they appear to have been rather less vigorous in the interval represented by this subdivision of the Silurian system. Those of Bala time gave forth abundant discharges of ash, of which the lowest accumulation, locally known as the Hagley Ash, consists of andesitic detritus. Occasional layers of tuff are intercalated in the overlying Hagley Shales, above which comes an important band called the Whittery Ash, "consisting of andesitic and rhyolitic breccias and conglomerates, fine ashes with curious spherulitic or pisolitic structures, and bands of shale often fossiliferous."[2] It is evident that the eruptions of the Shelve district came from independent vents in that neighbourhood, and never reached the importance of the great volcanoes of Arenig age in Montgomeryshire or of Bala age in Caernarvonshire.

Numerous dykes and sills traverse the rocks of this district. They consist chiefly of hypersthene-dolerite. They appear to belong to a much

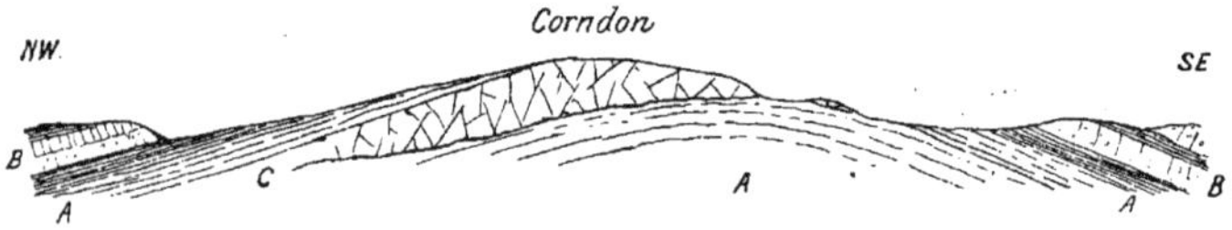

FIG. 50.—Section across the anticline of Corndon.[3]
A, Arenig flags and shales; B, andesites and tuffs; C, intrusive dolerite.

later period than the interstratified volcanic series; at least some of them are found altering the Pentamerus limestones, and these must be later than the Llandovery rocks.[4] The most important sill is that which forms Corndon, the central igneous mass of the district. This body of dolerite was ascertained by Mr. Watts not to be a boss but a laccolite, which wedges out both towards the north-west and south-east, as shown in Fig. 50.

Six miles to the north of the Shelve and Corndon district the Breidden Hills rise on the border of Shropshire and Montgomeryshire, and include a mass of volcanic material belonging to a distinct area of eruption. In the ridge that extends for about three and a half miles through Moel-y-golfa and Middletown Hill, a synclinal trough of volcanic rocks lies upon shales, which from their fossils have been placed in the Bala group. The volcanic series appears to exceed 1000 feet in thickness. The lowest part of it on Moel-y-golfa consists of andesitic lavas about 400 feet thick, followed by tuffs and volcanic conglomerates. The lavas resemble some of the "por-

[1] *Silurian System*, p. 229. [2] Messrs. Lapworth and Watts, *op. cit.* p. 318.
[3] After Prof. Lapworth and Mr. Watts, *op. cit.* p. 342. [4] *Op. cit.* p. 339.

phyrites" of the Old Red Sandstone, and contain two forms of pyroxene—one rhombic, probably enstatite, and the other monoclinic augite. There are likewise considerable masses of intrusive rock, which are varieties of diabase or dolerite.[1]

iii. SCOTLAND

From the centre of England we must in imagination transport ourselves into the Southern Uplands of Scotland, where a widely distributed series of Silurian volcanic rocks has been preserved. It was, until recently, supposed that the Silurian system north of the Tweed contains no contemporaneously erupted volcanic rocks. Yet, as far back as the year 1860, I pointed to the abundant existence of volcanic detritus in these strata throughout the southern counties as a probable indication of volcanic activity at the time and in the area within which the strata were deposited.[2] Some years later, when the microscope had been introduced as an aid to field-geology, I sliced some of the Silurian sediments of that region and found them, particularly certain shales and grits of Moffatdale, to contain a large admixture of perfectly fresh unworn felspar crystals, which I felt tolerably certain had been supplied by volcanic explosions. As no trace, however, had then been detected of an intercalated volcanic group in any part of the Silurian series of the south of Scotland, I used at that time to speculate on the possibility of the volcanic detritus having been wind-borne from the volcanoes of the Lake District. I had at that time no suspicion that its source was rather to be sought under my feet. The presence of volcanic rocks underneath the uplands of the south of Scotland would have been a welcome explanation of the frequent felspathic composition of many of the Silurian greywackes and shales of that region, and particularly the abundance of andesitic and felsitic fragments in them.

It had been long known that the Scottish Silurian formations, besides having undergone extensive plication, have also been injected by protrusions of igneous material of various kinds. The intrusive character of many of these is so obvious that a similar origin was attributed even to those bosses which could not be proved to be intrusive. Recent work of the Geological Survey, however, and more especially the numerous and careful traverses of my friend and colleague Mr. Peach, have revealed the unlooked-for and important fact that a large number of these supposed intrusions are really portions of a volcanic group brought up on the crests of anticlinal folds, and laid bare by denudation. This group can be traced for at least 100 miles from north-east to south-west over a belt of country sometimes 30 miles broad. Its original limits cannot be ascertained, but they obviously exceeded those within which the rocks can now be seen. Nevertheless the present boundaries embrace an area of nearly 2000 square miles. This Palæozoic volcanic region is thus one of the most extensive in the British Isles.

[1] See Mr. W. W. Watts on the Igneous and Associated Rocks of the Breidden Hills, *Quart. Journ. Geol. Soc.* vol. xli. (1885), p. 532.

[2] *Trans. Roy. Soc. Edin.* xxii. (1860), p. 636.

Owing, however, to the constant plication of the strata, and the wide space which the overlying sedimentary deposits are thus made to cover, the volcanic group only comes occasionally into view, and thus occupies but a mere fraction of the superficial extent of the region over which its scattered outcrops appear. These exposures, sometimes only a few square yards in extent, may always be looked for where the anticlinal folds bring up a sufficiently low portion of the Silurian system; they prove that a vast volcanic floor underlies the visible Lower Silurian grits and shales over the length and breadth of the Southern Uplands of Scotland.

Without anticipating details which will properly appear in the official *Memoirs* of the Geological Survey, I may briefly indicate the visible boundaries of the volcanic group, and refer to some of the localities where it may best be seen. The most easterly points where it has been recognized by Mr. Peach stand on the crests of some sharp anticlinal folds near St. Mary's Loch and near Leadburn and Winkstone in Peeblesshire. Farther westwards it appears at many places along the northern border of the Silurian territory, as at Romanno Bridge, Wrae, Kilbucho, Culter Water and Abington, the length and breadth of each exposure depending partly on the breadth of the anticline and partly on the depth to which it has been cut down by denudation. Near Sanquhar the volcanic series opens out for a breadth of more than a mile, and is seen at intervals across the wild moorlands of Carrick, until from the Stinchar valley it widens out seaward and occupies much of the coast-line of Ayrshire between Girvan and the mouth of Loch Ryan. It probably rises again along a fold near Portpatrick, and it is seen at various points along the southern borders of the Silurian uplands, as near Castle-Douglas, at Glenkiln, Bell Craig near Moffat, and the head of Ettrick-dale.

The best sections are those exposed along the coast to the north and south of Ballantrae. When that ground was first examined by the Geological Survey, the hypothetical views in regard to metamorphism already referred to were in full ascendant, and the rocks were mapped on the same general principles as those which had been followed in Wales. Professor Bonney, however, a few years later recognized the true igneous nature of many of the rocks. He found among them porphyrite lavas and agglomerates which he regarded as of Old Red Sandstone age, likewise intrusive serpentines and gabbros.[1]

The volcanic rocks of this wide district include both lavas and their pyroclastic accompaniments, as well as intrusive sills and bosses of various materials. They have recently been studied by Mr. J. J. H. Teall, and full descriptions of them by him will appear in a forthcoming volume of the *Memoirs* of the Geological Survey. He has ascertained that though generally more or less decomposed, the lavas would be classed by German petrographers as diabases and diabase-porphyrites. The former are compact dark-green non-porphyritic rocks, often containing numerous small spherical amygdales; while the latter are markedly porphyritic, enclosing

[1] *Quart. Journ. Geol. Soc.* vol. xxxiv. (1878), p. 769.

large phenocrysts of more or less altered plagioclase, often measuring half an inch across. These two groups of rock are connected by transitional varieties. They were probably, in the first instance, composed of plagioclase, augite, iron-ores, and a variable quantity of imperfectly crystallized interstitial matter.

Some of these rocks closely resemble in outward appearance the andesites ("porphyrites") of the Old Red Sandstone of the district not many miles to the north, that is, fine purplish-red rocks with a compact

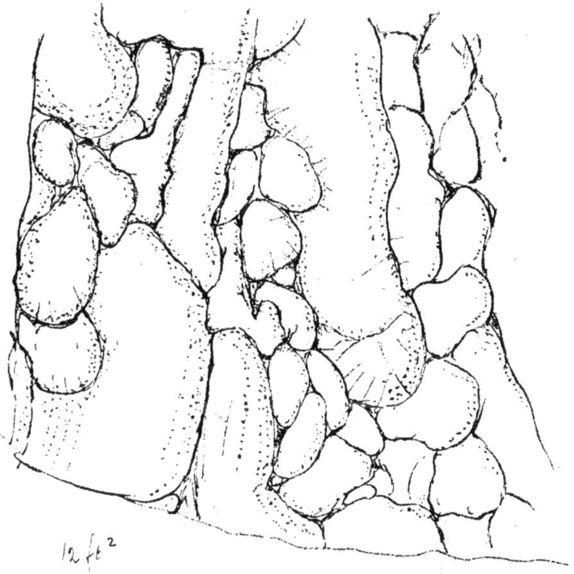

FIG. 51.—Structure in finely-amygdaloidal diabase lava, south of mouth of Stinchar River, Ayrshire. The fine dots and circles mark the lines of amygdales.

base through which porphyritic felspars are abundantly scattered. Occasionally they are markedly slaggy, and show even a ropy surface, while the breccias associated with them contain blocks of similar slag.

But the most characteristic external feature of these lavas is their tendency to assume irregularly-elliptical, sack-like or pillow-shaped forms On a weathered face they sometimes look like a pile of partially-filled sacks heaped on each other, the prominences of one projecting into corresponding hollows in the next. The general aspect of this structure is shown in Fig. 12, which represents a face of rock about eight feet high and six feet broad. The rocks exhibiting this peculiarity are usually finely amygdaloidal, and it may be observed that the vesicles are grouped in lines parallel to the outer surface of the pillow-like block in which they occur. The diagram in

Fig. 51 represents in ground-plan a surface about twelve feet square on the shore immediately to the south of the mouth of the River Stinchar. In the heart of the spheroids enclosed fragments of other lavas are sometimes observable.

This singular structure has already (p. 184) been referred to as strikingly displayed in a rock at the top of Cader Idris. It is found in dark basic lavas probably of Arenig age, which will be afterwards referred to as occurring along the southern flanks of the Scottish Highlands and also in the north of Ireland. It has been observed by Mr. Teall among the rocks of the Lizard, and has been described as occurring in Saxony and California.[1] In these different localities it is associated with jaspers and cherts, some of which contain abundant Radiolaria. The same structure has been found among the variolitic diabases of Mont Genèvre,[2] and likewise in some modern lavas, as in that of Acicastello already referred to (*ante*, p. 26).

The volcanic agglomerates and breccias, in the south-west of Ayrshire,

FIG. 52.—View of Knockdolian Hill from the east.

attain a great development in several centres probably at or near the original volcanic vents. They present several distinct petrographical types. The remarkable neck-like hill of Knockdolian in the Stinchar Valley is made of a coarse breccia composed mainly of angular pieces of dull greyish-green fine-grained diabase. The breccias and agglomerates of Bennane Head in some parts consist largely of broken-up shales, flinty mudstone, black radiolarian flint or chert, and abundant fragments of andesites and felsites. In other parts the volcanic material predominates, including angular and subangular fragments of various somewhat basic lavas, lumps of vesicular slag and pieces of pumice. Here and there much calcite is diffused through the matrix in strings, veins and patches, which enclose the lapilli. The agglomerate north of Lendalfoot possesses a greenish, somewhat serpentinous matrix, through which immense numbers of tabular felspar crystals are scattered. Similar crystals also occur abundantly in embedded blocks of

[1] Mr. J. J. H. Teall, *Roy. Geol. Soc. Cornwall*, 1894, p. 3. Mr. L. Ransome, *Bull. Depart. Geol. University of California*, vol. i. p. 106.

[2] Messrs. Cole and Gregory, *Quart. Journ. Geol. Soc.* vol. xlvi. (1890), p. 311.

one of the purplish diabase-porphyrites, which occurs in mass on the shore and inland, and closely resembles the rock of Carnethy in the Old Red Sandstone volcanic series of the Pentland Hills.

Yet another and very distinct type of agglomerate is to be seen on the Mains Hill south-east of Ballantrae. It is a coarse rock, enclosing blocks up to a yard or more in diameter, of a fine compact purplish porphyrite, with large crystals of plagioclase and smaller ones of augite. In some places immense numbers of the small lapilli in the matrix consist of an extremely fine vesicular pumice. Small perfect and larger broken crystals of augite are likewise abundant in some of the greenish, more basic parts of the mass. These greenish serpentinous parts and the numerous augite crystals point to the explosion of some tolerably basic pyroxenic lava. A similar dark green, almost black, rock, with augite crystals, which sometimes measure a quarter of an inch in diameter, occurs near Sanquhar in Nithsdale. It presents a close resemblance to the agglomerate of Rhobell Fawr, already alluded to. So far as these Scottish agglomerates have yet been microscopically examined, they have been found to be composed of crystals, crystal-fragments, and lapilli derived partly from lavas similar to those above described, and partly from felsitic and other rocks which have not yet been observed here in the form of lavas.

The finer tuffs show likewise a considerable range of composition. According to Mr. Peach's observations along the south-eastern parts of the volcanic area, the ejected materials have consisted largely of fine dust (probably in great measure felsitic), which towards the north-east is gradually interleaved with ordinary sediment till the ashy character disappears. As I have already remarked, there is reason to believe that the overlying greywackes and shales derived part of their material either directly from volcanic explosions or from the attrition of banks of lavas and tuffs exposed to denudation.

But besides the interstratified lavas and fragmental rocks there occur numerous intrusive masses which are so intimately associated with the volcanic series that they may with little hesitation be regarded as forming part of it. They consist of various gabbros and serpentines, which are especially developed where the volcanic series comes out in greatest force in the south-west of Ayrshire. They also include more acid intrusions which, as in the case of the rock of Byne Hill, near Girvan, even assume the characters of granite.

The dying out of the volcanic material towards the north-east probably indicates that the vents of the period lay rather in the central or south-western parts of the district. Unfortunately, the limited extent of the exposures of the rocks makes it a hopeless task to search for traces of these vents over by far the largest part of the area. There are two localities, however, where the search may be made with better prospect of success. One of these is a tract to the north of Sanquhar in Nithsdale, which still requires to be studied in detail with reference to the sequence and structure of its volcanic rocks. The other area is that south-western part of Ayrshire which has been already cited as displaying so large a development of the volcanic series. Here the coast-sections reveal the intercalation of fossiliferous bands which show the true stratigraphical horizons of the lavas and tuffs. Under Bennane Head, Professor Lapworth some years ago found, in certain hardened black shales, a group of graptolites which mark an undoubted Arenig platform.[1] Recently the ground has been carefully re-examined by Messrs Peach and Horne, who have detected a number of other fossiliferous zones which confirm and extend previous observations. They have also been able to unravel the complicated structure of the volcanic series, and to represent it on the 6-inch maps of the Geological Survey, of which a reduction on the scale of 1 inch to a mile is now in course of preparation. The following tabular summary, taken partly from notes made by myself during a series of traverses of the ground with Mr. Peach when the revision was begun, and partly from memoranda supplied by that geologist himself, may suffice as a general outline of the volcanic history of this exceedingly interesting and important region.

Llandovery.	Pentamerus grit. Conglomerate (Mulloch Hill).
Caradoc.	Shales, sandstones, grits, etc. (Ardmillan, Balcletchie). Thick conglomerate (Byne Hill, Bennane, etc.). Thick fossiliferous limestone (Stinchar, Girvan). (On this horizon come the perlitic felsites and soda-felsites of Winkstone and Wrae.) Sandstone (*Orthis confinis*) passing down into thick conglomerate.
	[Unconformability.]
Upper Llandeilo.	Green mudstones, grits and greywackes. Thin band of dark mudstone with Upper Llandeilo graptolites.
Arenig and Lower and Middle Llandeilo.	Group of Radiolarian cherts (about 70 feet) with alternating tuffs. Tuff or volcanic conglomerate, with occasional lava-flows. Black shale (10 feet) with Arenig graptolites. Volcanic breccias around local centres (Knockdolian, etc.). Thick group of porphyrite and diabase lavas. Red flinty mudstones with Arenig graptolites. Porphyrites, etc. Fine tuffs, etc., with Lower Arenig fossils. Diabase lavas, etc. (base not seen).

It will be noticed from this table that the bottom of the volcanic series is not reached, so that no estimate can be formed of its full thickness, nor on what geological platform it begins. Possibly its visible portions represent merely the closing scenes of a long volcanic history, which, over the area of the south of Scotland, extended into Cambrian time, like the contemporary series of Cader Idris.

Among the lowest lavas there are interstratified courses of fine tuffs,

[1] *Geol. Mag.* 1889, p. 22.

flinty shales and thin limestones, which sometimes fill in the hollows between the pillow-like blocks above referred to. Among the characteristic Lower Arenig graptolites of these intercalated layers are *Tetragraptus bryonoides*, *T. fruticosus*, *T. quadribrachiatus*, and *T. Headi* together with *Caryocaris Wrightii*. Considerable variation is to be seen in the development of the upper part of the volcanic series. In some places the lavas ascend almost to the top; in others, thick masses of breccia or agglomerate take their place. These fragmentary materials are locally developed round particular centres, which probably lie near the sites of active vents whence large quantities of pyroclastic material were discharged. One of the volcanic centres must have been situated close to the position of Knockdolian Hill already referred to. The exceedingly coarse breccia of that eminence is rudely stratified in alternations of coarser and finer material, which was probably to some extent assorted under water around the cinder-cone that discharged it. The date of the explosions of this hill has been ascertained by Mr. Peach from the intercalation of black shales containing Arenig graptolites among the breccias. Another vent lay somewhere in the immediate neighbourhood of

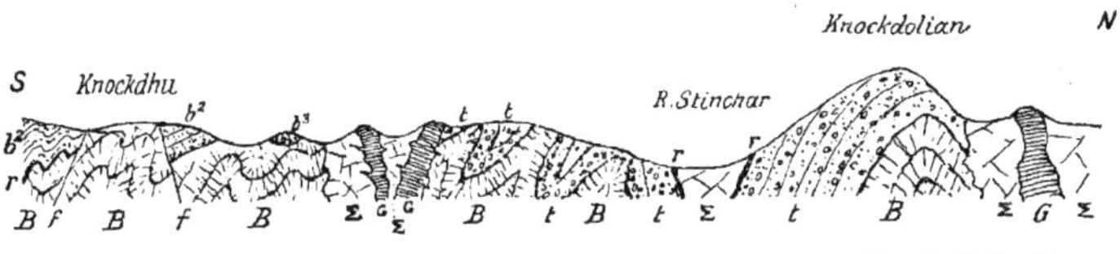

FIG. 53.—Section across the Lower Silurian volcanic series in the south of Ayrshire (B. N. Peach).
B, Interstratified lavas in Arenig group; *t*, tuffs; *r*, radolarian cherts; *b*2, Llandeilo group; *b*3, Caradoc group. Σ, Serpentine. G, Gabbro.

the Mains Hill agglomerate, if not actually on part of the site of that rock. Though probably not more than a mile from the Knockdolian volcano, and belonging to the same epoch of eruption, this vent, to judge from the peculiarities of its ejected material, must have been quite distinct in its source. A third vent lay somewhere in the immediate vicinity of Bennane Head, and threw out the extraordinary masses of agglomerate and the sheets of lava seen on the coast at that locality. A fourth may be traced by its separate group of fine tuffs on the coast three miles south of Ballantrae.

A feature of singular interest in the material erupted from these various centres of activity consists in the evidence that the explosions occurred at intervals during the deposition of the Lower Silurian formations, and that these formations were successively disrupted by submarine explosions. Mr. Peach has found, for example, abundant pieces of the peculiar and easily recognized radiolarian cherts imbedded in the volcanic series. That these cherts were deposited contemporaneously with the volcanic eruptions is proved by their intercalation among the breccias. Yet among these very breccias lie abundant fragments of chert which must have already solidified before disruption. It is thus evident that this siliceous ooze not only accumulated but set into solid stone on the sea-floor, between periods of volcanic outburst, and that such an occurrence took place several times in succession over the same area.

These facts derive further interest from the organic origin of the chert. It is now some years since Mr. Peach and his colleagues observed that between the Glenkiln Shale with its Upper Llandeilo graptolites and the top of the volcanic group in the central part of the Silurian uplands, alternations of green, grey or red shaly mudstones and flinty greywackes are interleaved with fine tuffs, and are specially marked by the occurrence in them of nodules and bands of black, grey and reddish chert. This latter substance, on being submitted to Dr. Hinde, was found by him to yield twenty-three new species of Radiolaria belonging to twelve genera, of which half are new. It thus appears that during the volcanic activity there must have been intervals of such quiescence, and such slow, tranquil sedimentation in clear, perhaps moderately deep water, that a true radiolarian ooze gathered over the sea-bottom.[1]

That the deposition of this ooze probably occupied a prolonged lapse of time seems clearly indicated by the evidence of the fossils that occur below and above the cherts. The graptolites underneath indicate a horizon in the Middle Arenig group, those overlying the cherts are unmistakably Upper Llandeilo. Thus the great depth of strata which elsewhere constitute the Upper Arenig and Lower and Middle Llandeilo subdivisions is here represented by only some 60 or 70 feet of radiolarian cherts. These fine siliceous, organic sediments probably accumulated with extreme slowness in a sea of some depth and over a part of the sea-floor which lay outside the area of the transport and deposit of the land-derived sediment of the time.[2]

As an illustration of some of the characteristic features in the succession of deposits in the volcanic series of the south-west of Ayrshire, the accompanying section (Fig. 54) is inserted. In descending order we come first upon a group of greywackes and grey shattery mudstones (*a*), followed by grey-green and dark banded cherts, containing Radiolaria and much plicated. Next comes a group of dark-grey, black and red cherts, with numerous partings and thin bands of tuff and volcanic conglomerate (*c*). The siliceous bands were certainly deposited during the volcanic eruptions, and they are moulded round the rugose, slaggy upper surface of the band of lavas (*d*) on which they directly lie. These lavas have the sack-like or pillow structure already described, and they enclose lumps of chert containing Radiolaria. A few yards to the west of the line of section bands of nodular tuff are interposed between the top of the lavas and the overlying cherts, with which also they are interstratified. These tuffs contain blocks

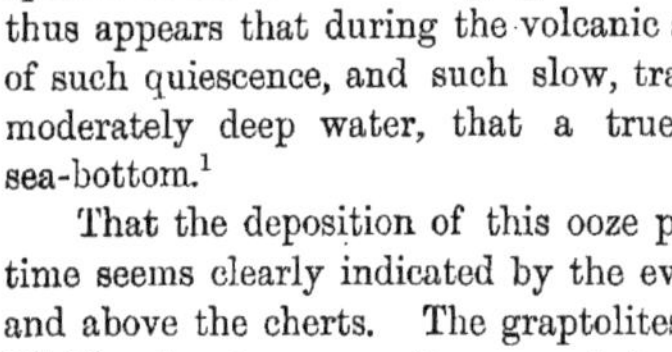

Fig. 54.—Section of part of the Arenig volcanic group, stream south of Bennane Head, Ayrshire.

[1] *Ann. Mag. Nat. Hist.* (1890), 6th ser. vi. p. 40.

[2] *Annual Report of the Geol. Surv. for* 1895, p. 27 of reprint.

of lava six inches or more in diameter. Below the belt of lavas come black cherts and shales (*e*) succeeded below by volcanic breccias and tuffs (*f*) alternating with shales in thin inconstant courses. These coarse detrital rocks are thoroughly volcanic in origin, and they contain fragments of the black cherts which lie still lower in the series. The whole depth of strata represented in this section does not amount to much more than 100 feet.

While in some parts of the Ayrshire district the coarse breccias that accumulated around their parent vents form most of the upper part of the volcanic series, in others the lavas are succeeded by fine tuffs which are intercalated among the ordinary sediments, and show a gradual decline and cessation of volcanic energy. South of Ballantrae, for example, the lavas occupy more than two miles of coast, in which space they display hardly any intercalations of sedimentary material, though they show more or less distinctly that they consist of many separate flows. Where they at last end, bands of nodular and fine tuff make their appearance, together with bands of ashy shale and the characteristic zone of the red radiolarian cherts or flints. Above these, in conformable sequence, come bands of black shale, containing abundant Upper Llandeilo graptolites, overlain by greenish or olive-coloured shaly mudstones, which pass upward into a thick overlying group of greywackes.

In this section the alternation of fine pyroclastic with ordinary sediment shows that the volcanic eruptions in the southern part of the Ballantrae district came to an end by a slowly-lessening series of explosions. The ashy material gradually dies out, and does not reappear all through the thick group of sandy and muddy sediments which here overlies the volcanic series.

We thus learn from the evidence of the Ayrshire sections that volcanic action was in full vigour in the south-west of Scotland during the Arenig period, but gradually died out before the end of the Llandeilo period. The rocks in which this volcanic history is chronicled have been very greatly disturbed and plicated, so that though from their frequently vertical position they might be thought to attain a vast depth, they very possibly do not exceed 500 feet in thickness.

As the volcanic series is followed north-eastwards it exhibits a gradual diminution in extent and variety, but this may be at least partly due to the much less depth of it exposed on the crests of the narrow anticlines that bring it to the surface. There is evidence in that region that the eruptions did not everywhere terminate in the Llandeilo period, but were in some districts prolonged into the age of the Bala rocks. Thus in the neighbourhood of Sanquhar volcanic breccias, tuffs and lavas have been found by Messrs. Peach and Horne intercalated in strata apparently belonging to the Bala group. Again, in the district of Hartfell, a moderately coarse volcanic agglomerate occurs in the heart of the so-called "barren mudstones" of the Hartfell black-shale group, which, from its graptolites, is placed on the horizon of the Bala rocks. At Winkstone, Hamilton Hill, and Wrae in Peeblesshire, perlitic felsites and soda-felsites have been detected by Messrs. Peach and Horne and determined by Mr. Teall. They are associated with the Bala limestone, which in some of its conglomeratic bands contains pebbles of felsite.

The intrusive rocks which accompany the Lower Silurian volcanic series of the south of Scotland are best displayed in the south-west of Ayrshire, between Girvan and Ballantrae, where they appear to be on the whole later than at least the great mass of the interstratified lavas and tuffs. The most abundant rocks and the earliest to be injected are complex basic masses which include serpentine, olivine-enstatite rock, troctolite, gabbro and other compounds, all which may be different modifications of the same original basic magma. They do not show a finer texture where they respectively meet, nor any other symptom of having been subsequently intruded into each other, though they do exhibit such structures along their lines of contact with the surrounding rocks, into which they are intrusive. These more basic masses have subsequently been invaded by irregular bosses and dyke-like protrusions, which, when small, are fine-grained dolerites, but when in larger bodies take the form of gabbro, sometimes exhibiting a mineral banding and foliated structure. These banded varieties much resemble the banded Tertiary gabbros of Skye and some parts of the Lewisian gneiss.

At the Byne Hill, near Girvan, a large intrusive boss or ridge displays on its outer margin a fine-grained texture, where it comes in contact with the serpentine. Further inwards it becomes a fine dolerite, passing into gabbro and increasing in coarseness of grain as well as in acidity of composition, through stages of what in the field would be called diorite and quartz-diorite, into a central granitic rock, whereof milky or blue quartz forms the prominent constituent. The intrusive rocks of this district have generally been injected parallel to the stratification-planes, and take on the whole the form of sills.

Some time after the close of the volcanic episode in the Silurian period of the south of Scotland, the rocks were locally subjected to considerable disturbance and elevation, whereby parts of the volcanic series were exposed to extensive denudation. Hence the overlying unconformable Caradoc conglomerates are in some places largely made up of the detritus of the volcanic rocks. It is interesting to find this evidence of waste during the very next stage of the Silurian period, for it affords good evidence that the extensive sheets of intrusive material could not have had any large amount of overlying strata resting upon them at the time of their injection. Pieces of these intrusive rocks, such as the serpentine, occur abundantly in the Caradoc conglomerates, some of which indeed are almost wholly composed of their detritus. Probably the total thickness of the overlying cover of rock under which the sills were injected did not amount to as much as 200 or 300 feet. Yet we see that among the sills were coarse gabbros and granitoid rocks. We may therefore infer that for the injection of such intrusive masses, great depth and enormous superincumbent pressure are possibly not always necessary.

During the progress of the Geological Survey along the southern borders of the Highlands, a remarkable group of rocks has been observed, intervening as a narrow interrupted strip between the schistose masses to the north and the great boundary-fault which brings the Old Red Sandstone in vertical strata against them. Between Cortachy in Forfarshire and Stonehaven on the east coast, these rocks have been mapped by Mr. G. Barrow, who has carefully worked out their relations. They appear again between Callander and Loch Lomond, where their extent and structure have been mapped by Mr. C. T. Clough. For the purpose of our present inquiry two chief features of interest are presented by these rocks. They include a group of sedimentary strata among which occur bands of jasper or chert containing radiolaria, and one of their most conspicuous members is a series of volcanic rocks consisting chiefly of dolerites and basalts, some of which have been much crushed and cleaved, but in which vesicular structures can still occasionally be recognized.

The striking resemblance of both the aqueous and igneous members of this marginal strip of rocks along the Highland border to the Arenig cherts and their accompanying lavas in the south of Scotland, the remarkable association of the same kinds of material in the same order of sequence, the occurrence of radiolaria in the siliceous bands in both regions, furnish strong presumptive evidence that a strip of Arenig rocks has been wedged in against the Highland schists.

In many respects, these dull green diabasic lavas of the Highland border resemble those of the Ayrshire coast. In particular, the same peculiar sack-like or pillow-shaped masses are conspicuous in the Forfarshire ravines. As in Ayrshire, igneous materials underlie the cherts which are doubled over and repeated by many successive folds. Unfortunately, it is only a narrow strip of these probably Arenig lavas that has been preserved, and no trace has been detected of tuffs, agglomerates or necks. If, however, we may regard the rocks as truly of Arenig age, they furnish interesting additional proof of the wide extent of the earliest Silurian volcanoes. The distance between the last Arenig volcanic outcrop in the Southern uplands and the band of similar lavas along the margin of the Highlands is about 50 miles. If the volcanic ejections were continuous across the intervening tract, the total area over which the lavas and tuffs of the Arenig volcanoes were distributed must be increased by at least 6000 square miles in Scotland.

But it is in the north of Ireland that this northern extension of what may probably be regarded as an Arenig series of volcanic rocks attains its greatest development. Of this Irish prolongation a brief account is given in Chapter xiv., where the whole of the Silurian volcanic rocks of the island are discussed.

CHAPTER XIII

THE ERUPTIONS OF LLANDEILO AND BALA AGE

i. The Builth Volcano—ii. The Volcanoes of Pembrokeshire—iii. The Caernarvonshire Volcanoes of the Bala Period—iv. The Volcanic District of the Berwyn Hills—v. The Volcanoes of Anglesey—vi. The Volcanoes of the Lake District; Arenig to close of Bala Period—vii. Upper Silurian (?) volcanoes of Gloucestershire.

THE stratigraphical subdivisions of geology are necessarily more or less arbitrary. The sequence in the sedimentary deposits of one region always differs in some degree from that of adjoining regions. In drawing up a table of stratigraphical equivalents for separate countries, we must be content to accept a general parallelism, without insisting on too close an identity in either the character of the strata or the grouping of their organic remains. We need especially to guard against the assumption that the limit assigned to a geological formation in any country marks a chronological epoch which will practically agree with that denoted by the limit fixed for the same formation in another country. The desirability of caution in this respect is well shown by the vagueness of the horizons between the several subdivisions of the Lower Silurian system. So long as the areas of comparison are near each other, no great error may perhaps be committed if their stratigraphical equivalents are taken to have been in a broad geological sense contemporary. But in proportion as the element of distance comes in, there enters with it the element of uncertainty.

Even within so limited a region as the British Isles, this difficulty makes itself strongly felt. Thus, in the typical regions of Wales, the several subdivisions of the Lower Silurian strata are tolerably well marked, both by lithological nature and by fossils. But as they are followed into other parts of the country, they assume new features, sometimes increasing sometimes diminishing in thickness, changing their sedimentary character, and altering the association or range of their organisms. The subdivisions into which the geologist groups them may thus be vaguely defined by limits which, in different parts of the region, may be far from representing the same periods of time.

Hence, in trying to ascertain how far the volcanic eruptions of one area during the Silurian period may have been contemporary with those of another area, we must be content to allow a wide margin for error. It is hardly possible to adhere strictly to the stratigraphical arrangement, for the geological record shows that in the volcanic districts the sedimentary formations by which the chronology might have been worked out are not infrequently absent or obscure. It will be more convenient to treat the rest of the Lower Silurian formations as the records of one long and tolerable definite section of geological time, without attempting in each case to distinguish between the eruptions of the successive included periods, so long as the actual volcanic sequence is distinctly kept in view. I will therefore take the history of each district in turn and follow its changes from the close of the Arenig period to the end of Upper Silurian time. The stages in the volcanic evolution of each tract will thus be clearly seen.

Above the Arenig group with its voluminous volcanic records comes the great group of sediments known as the Llandeilo formation, in which also there are proofs of contemporaneous volcanic activity over various parts of the sea-floor within the site of Britain. We have seen that in the south of Scotland the eruptions of Arenig time were probably continued into the period of the Llandeilo rocks, or even still later into that of the Bala group. But it is in Wales that the history of the Llandeilo volcanoes is most fully preserved. A series of detached areas of volcanic rocks, intercalated among the Llandeilo sediments, may be followed for nearly 100 miles, from the northern end of the Breidden Hills in Montgomeryshire, by Shelve, Builth, Llanwrtyd and Llangadock, to the mouth of the Taf river. But some 35 miles further west another group of lavas and tuffs appears on the coast of Pembrokeshire, from Abereiddy Bay to beyond Fishguard. The want of continuity in these scattered outcrops is no doubt partly due to concealment by geological structure. But from the comparative thinness of the volcanic accumulations and their apparent thinning out along the strike it may be inferred that no large Llandeilo volcano existed in Wales. There would rather seem to have been a long line of minor vents which in the south-east part of the area appear to have only discharged ashes. Certainly, if we may judge from their visible relics, these eruptions never rivalled the magnitude of the discharges from the Arenig volcanoes that preceded, or the Bala volcanoes that followed them.

i. THE VOLCANO OF BUILTH AND ITS NEIGHBOURS

So far as the available evidence goes, the most important volcanic centre down the eastern side of Wales during the Llandeilo period was one which lay not far from the centre of the long line of vents just referred to. Its visible remains form an isolated tract of hilly ground, some seven miles long, and four or five miles broad, immediately north from the town of Builth. This area is almost entirely surrounded by unconformable Upper Silurian strata, so that its total extent is not seen, and may be much more considerable than the area now laid bare by denudation.

The volcanic rocks of Builth were first described in the "Silurian System." Murchison clearly recognized that they included some which were "evolved from volcanic apertures during the submarine accumulation of the Lower Silurian rocks," and also "unbedded volcanic masses which had been intruded subsequently, dismembering and altering all the strata with which they came in contact."[1] These igneous rocks were mapped in some detail by the Geological Survey, and their general relations were expressed in lines of horizontal section.[2] They were likewise described by Ramsay in the *Catalogue of the Rock-specimens in the Jermyn Street Museum*, specimens of them being displayed in that collection.[3] The tuffs and lavas were distinguished, and likewise the intrusive "greenstones." But no attempt was made towards petrographical detail.

This interesting district has recently been studied by Mr. Henry Woods,[4] who has grouped the igneous rocks in probable order of appearance, as follows:—1st, Andesites; 2nd, Andesitic ash; 3rd, Rhyolites; 4th, Diabase-porphyrite; and 5th, Diabase.

Some of the andesites are described as intrusive in the Llandeilo strata. The ash in its lower part contains numerous well-rounded pebbles of andesite, usually five or six inches in diameter, but sometimes having a length of two feet. It contains fossils (*Orthis calligramma, Leptæna sericea, Serpulites dispar*, etc.), and as it is overlain with shales containing *Ogygia Buchii*, it may be regarded as probably of Lower Llandeilo age. The rhyolites are feebly represented, and some of them may possibly be intrusive. Among them a nodular variety has been noticed, the nodules being solid throughout, varying up to two inches in diameter, and formed of micro-crystalline quartz and felspar, with no trace of any radial or concentric internal arrangement. The diabase-porphyrite, the most conspicuous rock of the district, is intrusive in the andesites and ashes, and occurs in four separate masses or sills. The diabases are all intrusive and of later date than any of the other igneous rocks, and as they traverse also the Llandeilo shales, they are probably considerably later than the previous eruptions. But as they do not enter the surrounding Llandovery and Wenlock strata, they are regarded by Mr. Woods as of intermediate age between the time of the Llandeilo and that of the Upper Silurian formations.

About nine miles in a west-south-westerly direction from the southern extremity of the Builth volcanic area, another much smaller exposure of igneous rocks has been mapped by the Geological Survey at the village of Llanwrtyd. This tract is only about three miles long and half a mile broad. The volcanic rocks are represented as consisting of three or more bands of "felspathic trap" interstratified in the Lower Silurian strata, and folded into an anticline along the ridge of Caer Cwm. No published line of

[1] *Silurian System*, 1839, p. 330. The occurrence of "trappean ash" with fossils in the Builth district was noticed by De la Beche, *Mem. Geol. Surv.* vol. i. (1846), p. 31.

[2] See Sheet 56 of the one-inch map and Sheets 5 and 6 of the Horizontal Sections.

[3] Catalogue of Rock Specimens, 3rd edit. 1862, p. 36 *et seq.*

[4] *Quart. Journ. Geol. Soc.* vol. l. (1894), p. 566.

section runs across this ground, and the band of rock does not appear to have been described.[1]

Seventeen miles to the south-west a still feebler display of intercalated volcanic material occurs in the Llandeilo formation near the village of Llangadock. The Geological Survey map represents one or more bands of ash associated with limestone, and thrown into a succession of folds. In the *Horizontal Section* (Sheet III. Section 3) a band, 100 to 200 feet thick, of "trappean ash" with fossils is shown among the shales, limestones and grits, and in the *Catalogue of Rock-specimens* the same rock is referred to as brecciated ash in connection with specimens of it in the Museum, which are described as not purely ashy, but containing many slate-fragments and broken felspar-crystals together with organic remains.[2]

About twenty-four miles still further in the same south-westerly direction, two patches of "ash" are shown upon the Survey map, near the mouth of the river Taf. No description of these rocks is given.[3]

ii. THE VOLCANOES OF PEMBROKESHIRE

In north-western Pembrokeshire, the observations of Murchison, De la Beche and Ramsay showed the existence of an important volcanic district, where numerous igneous bands are interstratified among the Lower Silurian rocks, over an area extending from St. David's Head for thirty miles to the eastward.[4] On the maps of the Geological Survey, lavas, tuffs, sills and bosses were discriminated, but no description of these rocks was published. Since the publication of the Survey map very little has yet been added to our information on the subject.

There appear to have been at least three principal groups of vents. One may be indicated by the bands of "felspathic trap" which have been mapped as extending from near St. Lawrence for fourteen miles to the east. Another must have existed in the neighbourhood of Fishguard. A third is shown to have lain between Abereiddy Bay and Mathry, by the abundant bands of lava and tuff and intrusive sills there to be seen.

Of these areas the only one which has yet been examined and described in some detail is that of Fishguard, of which an account has recently been published by Mr. Cowper Reed.[5] This observer has shown that the eruptions began there during the deposition of the Lower Llandeilo rocks, and continued intermittently into the Bala period. The earliest consisted of felsites and tuffs intercalated between Lower Llandeilo black slates con-

[1] The locality is referred to by De la Beche, *Mem. Geol. Surv.* vol. i. p. 31, and by Ramsay in the *Descriptive Catalogue of Rock-specimens in the Museum of Practical Geology*, 3rd edit. p. 38, but no specimens from it are in the collection.

[2] *Op. cit.* p. 38.

[3] One of the patches was shown by J. Phillips in *Horizontal Section*, Sheet III. Section 6, as a "felspathic trap," near which the shales are bleached. The map, however, was subsequently altered, so as to make the igneous rocks pyroclastic.

[4] See *Silurian System*, p. 401; Sheet 40 of the Geological Survey Map; *Memoir of A. C. Ramsay*, p. 232 *et seq.*; De la Beche, *Trans. Geol. Soc.* 2nd series, vol. ii. part i. (1826), p. 3.

[5] *Quart. Journ. Geol. Soc.* vol. li. (1895), p. 149.

taining *Didymograptus Murchisoni*, the tuffs themselves being sometimes fossiliferous. A second great volcanic belt, composed of felsitic lavas, breccias and tuffs, lies at the base of the Upper Llandeilo strata and shows the maximum of volcanic energy. The breccias are partly coarse agglomerates, which probably represent, or lie not far from, some of the eruptive vents of the time. A higher band of lavas and breccias appears to be referable to the Bala formation. The whole volcanic series is stated to thin out towards the south-west, so that the chief focus of eruption probably lay somewhere in the neighbourhood of Fishguard.

The lavas may all be included under the general term felsite. Their specific gravity ranges from 2·60 to 2·76, and their silica percentage from 68 to 72. Mr. Cowper Reed observed among them three conspicuous types of structure. Some are characterized by a distinct arrangement in fine light and dark bands which rapidly alternate, and are sometimes thrown into folds and convolutions. A second structure, observed only at one locality, consists in the development of pale grey or whitish ovate nodules, about half an inch in length, with a clear quartz-grain in their centre, or else hollow. The third type is shown by the appearance of perlitic structure on the weathered surface.[1]

The tuffs and breccias are chiefly developed at the base and top of each volcanic group. Some of them contain highly vesicular fragments, as well as pieces of slate and broken crystals of quartz and felspar.

A characteristic feature of this volcanic district is the occurrence in it of sills and irregularly-intruded masses of "greenstone." Under that name are comprised basalts, dolerites, andesitic dolerites with tachylitic modifications, as well as diabases and gabbros.[2] Some of these rocks exhibit a variolitic structure. As regards age, some of the intrusions appear to have taken place before the tilting, cleavage and faulting of the strata. They have not been noticed in the surrounding Upper Silurian strata, and we may perhaps infer that here, as at Builth, they are of Lower Silurian date. Mr. Cowper Reed, however, is inclined to regard the large Strumble Head masses as later than the tilting and folding of the rocks.[3]

A few miles to the south-west of the Fishguard district, on the coast of Abereiddy Bay, good sections have been laid bare of the volcanic rocks of this region. Dr. Hicks has shown that the bands of tuff there displayed are intercalated among the black slates of the Lower Llandeilo group, and that there was probably a renewal of volcanic activity during the deposition of the upper group.[4] But the volcanic history of this area still remains to be properly investigated.

In southern Pembrokeshire two conspicuous bands of eruptive rocks have long been known and described. Their general characters and dis-

[1] Mr. Cowper Reed enters into a detailed account of the microscopic structures and chemical composition of these rocks. They have rather a high percentage of alumina, potash and soda, and are obviously akin to the keratophyres of other districts.

[2] Mr. Cowper Reed, *op. cit.* p. 180.

[3] *Op. cit.* p. 193.

[4] *Quart. Journ. Geol. Soc.* xxxi. (1875), p. 177.

tribution were sketched by De la Beche,[1] and further details were afterwards added by Murchison.[2] As traced by the officers of the Geological Survey, they were represented as consisting of "greenstone," "syenite" and "granite." The more northerly band was shown to run in a nearly east and west line from Lawrenny to the Stack Rock, west of Talbenny, a distance of about fourteen miles. The second band, placed a short way farther south, stretches in the same general line, from Milford Haven at Dall Road into Skomer Island, a distance of about seven miles.

The relations of these rocks to the surrounding formations and their geological age have been variously interpreted. De la Beche regarded the different masses as intrusive, and probably later than even the adjoining Coal-measures.[3] Murchison, on the other hand, considered the bedded eruptive rocks of Skomer Island to be undoubtedly lavas contemporaneous with the strata among which they are intercalated.[4]

The rocks have been studied petrographically by various observers. Mr. Rutley gave a full description of the remarkable nodular and banded felsites of Skomer Island.[5] Mr. Teall has also noticed these rocks, likewise "a magnificent series of basic lava-flows" in the same island, and a number of "porphyrites." The basic lavas seemed to him to contain too much felspar and too little olivine to be regarded as perfectly typical olivine-basalts, and he found them to lie sometimes in very thin and highly vesicular sheets. The "porphyrites" he placed "on the border-line between basic and intermediate rocks."[6]

More recently this southern district of Pembrokeshire has been examined by Messrs. F. T. Howard and E. W. Small, who have obtained further evidence of the interbedded character of the igneous series. Below an upper basalt they have noted the occurrence of bands of felsitic conglomerate, sandstone, shale and breccia lying upon and obviously derived from a banded spherulitic felsite, below which comes a lower group of basalts. The age of this interesting alteration of basic and acid eruptions has not been precisely determined, but is conjectured to be that of the Bala or Llandovery rocks.[7]

iii. THE CAERNARVONSHIRE VOLCANOES OF THE BALA PERIOD

Owing to the effects partly of plication and partly of denudation, the rocks of the next volcanic episode in Wales, that of the Bala period, occupy a less compact and defined area than those of the Arenig group in Merionethshire. From the latter they are separated, as we have seen, by a considerable depth of strata,[8] whence we may infer, with the Geological Survey, that the eruptions of Arenig, the Arans and Cader Idris were succeeded by a long period of repose, the Llandeilo outbreaks described in the foregoing

[1] *Trans. Geol. Soc.*, 2nd ser. vol. ii. (1823), p. 6 *et seq.* [2] *Silurian System*, p. 401 *et seq.*
[3] *Mem. Geol. Survey*, vol. i. p. 231. [4] *Silurian System*, p. 404.
[5] "The Felsitic Lavas of England and Wales," *Mem. Geol. Survey* (1885), pp. 16, 18.
[6] *British Petrography*, pp. 224, 284, 336. [7] *Rep. Brit. Assoc.* 1893, p. 766; *Geol. Mag.* 1896, p. 481.
[8] Estimated at from 6000 to 7000 feet, *Mem. Geol. Surv.* vol. iii. 2nd edit. p. 131.

pages not having extended apparently into North Wales. When the next outbreaks took place, the vents are found to have shifted northwards into Caernarvonshire, where they fixed themselves along a line not much to the east of where the Cambrian porphyries and tuffs now appear at the surface. The lavas and ashes that were thrown out from these vents form the highest and most picturesque mountains of North Wales, culminating in the noble cone of Snowdon. They stretch northwards to Diganwy, beyond Conway, and southwards, at least as far as the neighbourhood of Criccieth. They die out north-eastwards beyond Bala Lake, and there can be but little doubt that they thin out also eastwards under the Upper Bala rocks. The lavas and tuffs that rise up on a similar horizon among the Bala rocks of the Berwyn Hills evidently came not from the Snowdonian vents, but from another minor volcanic centre some miles to the east, while still more remote lay the vents of the Breidden Hills and the sheets of andesitic tuff that probably spread from them over the ground east of Chirbury (Map II.).

The Caernarvonshire volcanic group extends from north to south for fully thirty miles, with an extreme breadth of about fifteen miles; while, if we include the rocks of the Lleyn peninsula, the area will be prolonged some twenty miles farther to the south-west.

The general stratigraphical horizon of this volcanic group has been well determined by the careful mapping of Ramsay, Selwyn and Jukes on the maps of the Geological Survey. These observers brought forward ample evidence to show that the lavas and tuffs were erupted during the deposition of the Bala strata of the Lower Silurian series, that the Bala Limestone is in places full of ashy material, and that this well-marked fossiliferous band passes laterally into stratified volcanic tuffs containing the same species of fossils.[1] But the progress of stratigraphical geology, and the increasing value found to attach to organic remains as marking even minor stratigraphical horizons, give us reason to believe that a renewed and still more detailed study of the Bala rocks of North Wales would probably furnish data for more precisely defining the platforms of successive eruptions, and would thus fill in the details of the broad sketch which Sir Andrew Ramsay and his associates so admirably traced. Besides the Bala Limestone there may be other lithological horizons which, like the Garth grit and the pisolitic iron-ore of the Arenig group, might be capable of being followed among the cwms and crests as well as the opener valleys of Caernarvonshire. Until some such detailed mapping is accomplished, we cannot safely advance much beyond the point where the stratigraphy was left by the Survey.

From the Survey maps and sections it is not difficult to follow the general volcanic succession, and to perceive that the erupted materials must altogether be several thousand feet in thickness from the lowest lavas in the north to the highest on the crest of Snowdon. In that mountain the total mass of volcanic material is set down as 3100 feet. But this includes only the higher part of the whole volcanic group. Below it come the lavas of Y

[1] *Mem. Geol. Surv.* vol. iii. 2nd edit. pp. 126, 128, 131, 139, etc.

Glyder-Fach, which, according to the Survey measurements, are about 1500 feet thick, while still lower lie the ancient *coulées* of Carnedd Dafydd and those that run north from the vent of Y-foel-frâs, which must reach a united thickness of many hundred feet. We can thus hardly put the total depth of volcanic material at a maximum of less than 6000 to 8000 feet. The pile is, of course, thickest round the vents of discharge, so that no measurement, however carefully made at one locality, would be found to hold good for more than a short distance.

Though little is said in the Survey Memoir of the vents from which this vast amount of volcanic material was erupted, the probable positions of a number of these orifices may be inferred from the maps. From the shore west of Conway a series of remarkable eminences may be traced south-westwards for a distance of nearly forty miles into the peninsula of Lleyn. At the northern extremity of this line stands the prominent boss of Penmaenmawr, while southward beyond the large mass of Y-foel-frâs, with the smaller knobs west of Nant Francon, and the great dome of Mynydd-mawr, the eye ranges as far as the striking group of *puy*-like cones that rise from the sea around Yr Eifl and Nevin. Some of these hills, particularly Y-foel-frâs, were recognized by the Survey as vents.[1] But the first connected account of them and of their probable relation to the volcanic district in which they occur has been given by Mr. Harker in his exceedingly able essay on "The Bala Volcanic Series of Caernarvonshire,"[2]—the most important contribution to the volcanic history of Wales which has been made since the publications of the Geological Survey appeared. I shall refer to these vents more specially in the sequel. I allude to them here for the purpose of showing at the outset the marvellous completeness of the volcanic records of Caernarvonshire. So great has been the denudation of the region that the pile of lavas and tuffs which accumulated immediately around and above these orifices has been swept away. No trace of any portion of that pile has survived to the west of the line of bosses; while to the east, owing to curvature and subsequent denudation, the rocks have been dissected from top to bottom, until almost every phase of the volcanic activity is revealed.

The volcanic products discharged from these vents consist of a succession of lava-streams separated by bands of slate, tuff, conglomerate and breccia. These fragmental intercalations, which vary from a few yards to many hundred feet in thickness, are important not only as marking pauses in the emission of lava or in the activity of the volcanoes, but as affording a means of tracing the several lavas to their respective vents. Essentially, however, the volcanic materials consist of lava-flows, the intercalations of fragmentary materials, though numerous, being comparatively thin. The thickest accumulation of tuffs is that forming much of the upper part of Snowdon. It is set down by my predecessor at 1200 feet in thickness, but I should be inclined to reduce this estimate. I shall have occasion to show that the summit and upper shoulders of Snowdon are capped with andesites inter-

[1] *Op. cit.* pp. 137, 220.

[2] This was the Sedgwick Prize Essay for 1888, and was published in 1889.

stratified among the tuffs. Sir Andrew Ramsay has referred with justice to the difficulty of always discriminating in the field between the fine tuffs and some of the lavas.[1] Yet I am compelled to admit that, if the ground were to be re-mapped now, the area represented as covered by fragmental rocks would be considerably restricted. Mr. Harker is undoubtedly correct when he remarks that, taken "as a whole, the Bala volcanic series of Caernarvonshire is rather remarkable for the paucity of genuine ashes and agglomerates."[2]

The lavas of the Bala volcanic group, like those of the Arenig series, were mapped by the Survey as "porphyries," "felstones," or "felspathic traps." They were shown to be acid-lavas, having often a well-developed flow-structure comparable with that of obsidian and pitchstone, and to consist of successive sheets that were poured out over the sea-floor. Their petrography has subsequently been studied more in detail by many observers, among whom I need only cite Professor Bonney, Professor Cole, Mr. Rutley, Mr. Teall, and Miss Raisin; the most important recent additions to our knowledge of this subject having been made by Mr. Harker in the Essay to which I have just referred.

The great majority of these lavas are thoroughly acid rocks, and present close analogies of composition and structure to modern rhyolites, though I prefer to retain for them the old name of "felsites." Their silica-percentage ranges from 75 to more than 80. To the naked eye they are externally pale greyish, or even white, but when broken into below the thick decomposed and decoloured crust, they are bluish-grey to dark iron-grey, or even black. They break with a splintery or almost conchoidal fracture, and show on a fresh surface an exceedingly fine-grained, tolerably uniform texture, with minute scattered felspars.

One of their most striking features is the frequency and remarkable development of their flow-structure. Not merely as a microscopic character, but on such a scale as to be visible at a little distance on the face of a cliff or crag, this structure may be followed for some way along the crops of particular flows. The darker and lighter bands of devitrification, with their lenticular forms, rude parallelism and twisted curvature, have been compared to the structure of mica-schist and gneiss. One aspect of this structure, however, appears to have escaped observation, or, at least, has attracted less notice than it seems to me to deserve. In many cases it is not difficult to detect, from the manner in which the lenticles and strips of the flow-structure have been curled over and pushed onward, what was the direction in which the lava was moving while still a viscous mass. By making a sufficient number of observations of this direction, it might in some places be possible to ascertain the quarter from which the several flows proceeded. As an illustration, I would refer to one of the basement-felsites of Snowdon, which forms a line of picturesque crags on the slope facing Llanberis. The layers of variously-devitrified matter curl and fold over each other, and have been rolled into balls, or have been broken up

[1] *Op. cit.* p. 148. [2] *Bala Volcanic Rocks*, p. 25.

and enclosed one within the other (Fig. 55). The general push indicated by them points to a movement from the westward. Turning round from the crags, and looking towards the west, we see before us on the other side of the deep vale of ·Llyn Cwellyn, at a distance of little more than three miles, the great dome-shaped Mynydd-mawr, which, there is every reason to believe, marks one of the orifices of eruption. It might in this way be practicable to obtain information regarding even some of the vents that still lie deeply buried under volcanic or sedimentary rocks.

That these felsites were poured forth in a glassy condition may be inferred from the occurrence of the minute perlitic and spherulitic forms so characteristic of the devitrification of once vitreous rocks. Mr. Rutley was the first who called attention to this interesting proof of the close resemblance between Palæozoic felsites and modern obsidians, and other observers have since confirmed and extended his observations.[1]

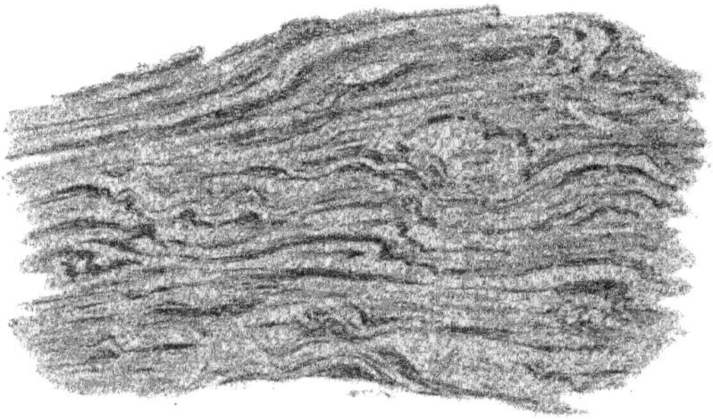

FIG. 55.—Flow-structure in the lowest felsite on the track from Llanberis to the top of Snowdon. Length about 4 feet, height 2½ feet.

Another remarkable aspect of the felsites is that nodular structure so often to be seen among them, and regarding the origin of which so much has already been written. I agree with Professor Cole and Mr. Harker in looking upon the "nodules" as derived from original spherulites by a process of alteration, of which almost every successive stage may be traced until the original substance of the rock has been converted into a flinty or agate-like material. If this be the true explanation of the structure, some of the original lavas must have exhibited perlitic and spherulitic forms on a gigantic scale. There can, I think, be little doubt that this peculiar structure was very generally misunderstood by the earlier observers, who naturally looked upon it as of clastic origin, and who therefore believed that large beds of rock consisted of volcanic conglomerate, which we should now map as nodular felsite (pyromeride).[2]

[1] *Quart. Journ. Geol. Soc.* vol. xxxv. (1879), p. 508.

[2] Another source of error may probably be traced in the occasional brecciated structure of the felsites, which has been mistaken for true volcanic breccia, but which can be traced disappearing into the solid rock. Sometimes this structure has resulted from the breaking up of the lenticles of flow, sometimes from later crushing.

While by far the larger porportion of the Caernarvonshire lavas consists of thoroughly acid rocks, the oldest outflows are much less acid than those erupted at the height of the volcanic activity, when the rocks of Snowdon were poured forth.[1] But towards the close of the period there was apparently a falling off in the acidity of the magma, for at the top of the group the andesitic lavas to which I have already alluded are encountered. Sir Andrew Ramsay has shown the existence of an upper "felstone" or "felspathic porphyry," almost entirely removed by denudation, but of which outliers occur on Crib-goch, Lliwedd, and other crests around Snowdon, and likewise on Moel Hebog.[2] Mr. Harker alludes to these remnants, and speaks of them as less acid than the older lavas, but he gives no details as to their structure and composition.[3] In an examination of Snowdon I was surprised to find that the summit of the mountain, instead of consisting of bedded ashes as hitherto represented, is formed of a group of lava-sheets having a total thickness of perhaps from 100 to 150 feet (6 in Fig. 56). The apex of Yr Wyddfa, the peak of Snowdon, consists of fossiliferous shale

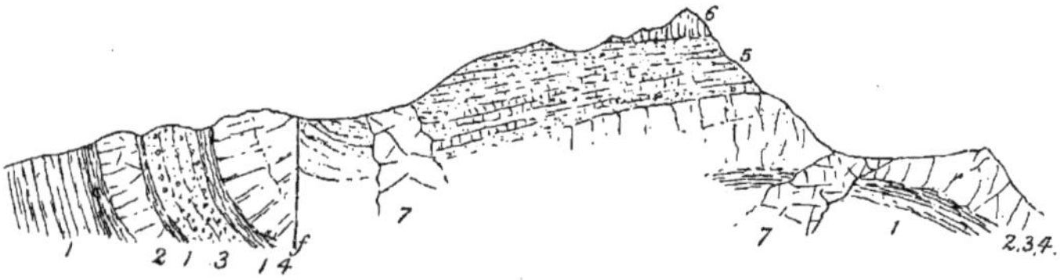

FIG. 56.—Section of Snowdon.[4]

1. Grits and slates; 2. Felsite with good flow-structure; 3. Volcanic tuffs; 4. Felsite; 5. Tuffs with sheets of felsite and andesite; 6. Group of andesitic lavas on summit of Snowdon; 7. Intrusive "greenstones."

lying on a dull grey rock that weathers with elongated vesicles, somewhat like a cleaved amygdaloid, but a good deal decomposed. A thin slice of this latter rock shows under the microscope irregular grains and microlites of felspar, with a few grains of quartz, the whole much sheared and calcified. Below this bed comes a felsite, or devitrified obsidian, showing in places good spherulitic structure, and followed by a grey amygdaloid. The latter is a markedly cellular rock, and, though rather decayed, shows under the microscope a microlitic felspathic groundmass, through which granules of magnetite are dispersed.

Underneath this upper group of lavas lie the tuffs for which Snowdon has been so long celebrated. But, as I have already stated, there does not appear to me to be such a continuous thickness of fragmental material as has been supposed. There cannot, I think, be any doubt that not only at the top, but at many horizons throughout this supposed thick

[1] Mr. Harker, *op. cit.* p. 127.

[2] *Mem. Geol. Surv.* vol. iii. 2nd edit. pp. 141, 144, 145, 147, 161.

[3] *Bala Volcanic Series*, pp. 10, 23, 125. He refers also to lavas occupying a similar position at Nant Gwynant and Moel Hebog; but he adds that he had not had an opportunity of studying them.

[4] After the Geological Survey Section (Horizont. Sect. Sheet 28), slightly modified.

accumulation of tuff, some of the beds of rock are really lava-flows. Some of these lavas have suffered considerably from the cleavage which has affected the whole of the rocks of the mountain, while the results of centuries of atmospheric disintegration, so active in that high exposed locality, have still further contributed to alter them. They consequently present on their weathered faces a resemblance to the pyroclastic rocks among which they lie. Where, however, the lavas are thicker and more massive, and have resisted cleavage better, some of them appear as cellular dull grey andesites or trachytes, while a few are felsites. Many instructive sections of such bands among the true tuffs may be seen on the eastern precipices of Snowdon above Glas-lyn.

It thus appears that the latest lavas which flowed from the Snowdonian vent were, on the whole, decidedly more basic than the main body of felsites that immediately preceded them. They occur also in thinner sheets, and are far more abundantly accompanied with ashes. At the same time it is deserving of special notice that among these less acid outflows there are intercalated sheets of felsite, and that some of these still retain the spherulitic structure formed by the devitrification of an original volcanic glass.

Far to the south-west, in the promontory of Lleyn, another group of volcanic rocks exists which may have been in a general sense contemporaneous with those of the Snowdon region, but which were certainly erupted from independent vents. Mr. Harker has described them as quartz-less pyroxene-andesites, sometimes markedly cellular, and though their geological relations are rather obscure, he regards them as lava-flows inter-bedded among strata of Bala age and occurring below the chief rhyolites of the district. If this be their true position, they indicate the outflow of much less highly siliceous lavas before the eruption of the acid felsites. In the Snowdon area any such intermediate rocks which may have been poured out before the time of the felsitic outflows have been buried under these.

The tuffs of the Bala series in Caernarvonshire have not received the same attention as the lavas. One of the first results of a more careful study of them will probably be a modification of the published maps by a reduction of the area over which these rocks have been represented. They range from coarse volcanic breccias to exceedingly fine compacted volcanic dust, which cannot easily be distinguished, either in the field or under the microscope, from the finer crushed forms of felsite. Among the oldest tuffs pieces of dark blue shale as well as of felsite may be recognized, pointing to the explosions by which the vents were drilled through the older Silurian sediments already deposited and consolidated. Sometimes, indeed, they recall the dark slate-tuffs of Cader Idris, like which they are plentifully sprinkled with kaolinized felspar crystals. The beds of volcanic breccia intercalated between the lower felsites of Snowdon include magnificent examples of the accumulation of coarse volcanic detritus. The blocks of various felsites in them are often a yard or more in diameter. Among the felsite fragments smaller scattered pieces of andesitic rocks may be found. This mixture of more basic materials appears to increase upwards, the highest ashes containing detritus of andesitic lavas like those which occur among them as flows.

The tuffs in the upper part of Snowdon are well-bedded deposits made up partly of volcanic detritus and partly of ordinary muddy sediment.[1] Layers of blue shale or slate interstratified among them indicate that the enfeebled volcanic activity marked by the fine tuffs passed occasionally into a state of quiescence. As is well known, numerous fossils characteristic of the Bala rocks occur in these tuffs. The volcanic discharges are thus proved to have been submarine and to have occurred during Bala time.

I have already alluded to some of the probable vents from which the lavas and tuffs were discharged, and to their position along a line drawn from Penmaen-mawr into the peninsula of Lleyn. It will be observed that they lie outside the area of the bedded volcanic rocks and rise through parts of the Silurian system older than these rocks. The largest and most important of them is unquestionably that formed by Y-foel-frâs and its neighbouring heights. As mapped by the Geological Survey, this mass of igneous rock is irregularly elliptical, measures about six square miles in area, and consists mainly of intrusive "felstone-porphyry" passing into "hornblendic greenstone."[2] Mr. Harker, however, has made an important correction of this petrography, by showing that a large part of the area consists of augitic granophyre, while the so-called "greenstone" is partly diabase and partly andesitic ashes and agglomerates. He suggests that an older vent has here been destroyed by a later and larger protrusion of igneous matter.[3] This high and somewhat inaccessible tract of ground is still in need of detailed mapping and closer study, for undoubtedly it is the most important volcanic vent now visible in North Wales. My former colleague in the Geological Survey, Mr. E. Greenly, spent a week upon it some years ago, and kindly supplied me with the following notes of his observations:—"The central and largest area of the neck is mainly occupied with diabases and andesites, while the ashes and agglomerates, which are intimately connected with them, seem to run as a belt or ring round them, and to occur in one or more patches in the midst of them. Portions of green amygdaloid run through the pyroclastic masses. Outside the ring of agglomerate and ashes an interrupted border of felsite can be traced, which may be presumed to be older than they, for a block of it was observed in them. The granophyre, on the other hand, which is interposed between the fragmental masses and the surrounding rocks on the western wall of the vent, seems to be of later date. Dykes or small bosses of diabase, like the material of the sills, pierce both the agglomerates and the rocks of the centre."[4]

[1] See the interesting account of these tuffs given by Sir A. Ramsay, *Mem. Geol. Survey*, vol. iii. 2nd edit. p. 142.

[2] *Mem. Geol. Survey*, vol. iii. 2nd edit. pp. 137, 139.

[3] *Bala Volcanic Series*, pp. 41, 71, 72, 123.

[4] Mr. Greenly has made a sketch map of this interesting locality. As he has now established his home in North Wales, I trust he may find an opportunity of returning to Y-foel-frâs and completing his investigations.

No agglomerate appears to have been noticed by any observer among the other supposed vents along the line that runs south-westwards from Penmaen-mawr, to the promontory of Lleyn. These bosses are rudely circular in ground-plan and rise vertically out of the Lower Silurian or Cambrian strata, or partake more of the nature of lenticular sheets or laccolites which have been thrust between the planes of bedding. There is usually an observable alteration of the surrounding rocks along the line of contact.

The material of these bosses is sometimes thoroughly acid, as is the granophyre of Y-foel-frâs, the microgranite of Mynydd-mawr with its riebeckite crystals, the augite-granite-porphyry of Clynog-fawr, and the granophyric and rhyolitic quartz-porphyries of the Rivals. In other cases the rock is of an intermediate grade, as in the enstatite-diorite of Penmaen-mawr, the pyroxene-andesite of Carn Boduan, and the quartz-augite-syenite of Llanfoglen.[1] A few bosses of still more basic material occur in the Sarn district, including hornblende-diabase and hornblende-picrite. Sometimes both the acid and the more basic rocks are found in the same boss, as in the large mass of Y-foel-frâs.

It must be confessed that there is no absolute proof that any of these masses mark the actual sites of eruptive vents, except probably the boss of Y-foel-frâs. Some of them may have been intruded without establishing any outlet to the surface.[2] But that a few of them really represent orifices from which the Bala volcanic group was erupted may be plausibly inferred from their neck-like form, from their positions with reference to the volcanic district, from the obvious thickening of the lavas and tuffs in the direction of these bosses, and from the petrographical relation that exists between their component materials and rocks that were discharged at the surface. This last-named feature has been well pointed out by Mr. Harker, who has established, by a study of microscopic slides, a gradation from the granophyric material of the bosses into structures greatly resembling those of the bedded felsites, and likewise a close similarity between the intermediate rocks of the other bosses and the andesites which have elsewhere been poured out at the surface.[3] But perhaps the most impressive evidence as to the sites of the chief centres of eruption is supplied by the lavas and tuffs themselves as they thicken in certain directions and thin away in others. This feature of their distribution has been well expressed in the maps and sections of the Survey, and has been clearly summarized by Mr. Harker.[4] The oldest lavas now visible lie at the northern end of the district, and the vents from which they proceeded may, with considerable probability, be placed somewhere in the tract which includes the chain of bosses of Penmaen-mawr, Y-foel-frâs, and Y Drosgl. The chief centre of eruption no doubt lay somewhere in the Snowdon tract, where the

[1] The geological relations and petrographical characters of these various rocks are given by Mr. Harker in the fourth and fifth sections of his Essay.

[2] Mr. Harker speaks of some of them as laccolites.

[3] *Op. cit.* pp. 57, 72.

[4] See especially pp. 9, 120 *et seq.*, and fig. 6 of his Essay.

lavas and tuffs attain their greatest thickness, and whence they thin away in all directions. The Mynydd-mawr boss may be presumed to have been one of the main vents. But there were not improbably others, now concealed under the deep cover of their own ejections.

More diligent search, with a special eye to the discovery of such vents, might indeed be rewarded, even in the midst of the volcanic district itself. To the north-east of Capel Curig, for example, there is a prominent knob of agglomerate,[1] which I visited with Mr. B. N. Peach, and which we regarded as probably marking one of the minor vents. The material of this eminence has a base which by itself would probably be regarded by the field-geologist as a felsite. But through this compact matrix are dispersed abundant stones of all sizes up to six inches or more in diameter. They are mostly subangular or somewhat rounded-off at the edges, and generally markedly cellular. Among them may be observed pieces of trachyte, felsite, and a rock that is probably a devitrified pitchstone or obsidian. The vesicles in these stones are sometimes lined with an acicular zeolite. Traces of rude bedding can be detected, dipping at high angles. On the north-east side of the hill finer agglomerate is seen to alternate with ashy grits and grey shales, which, dipping E.N.E. at 20°–30°, pass under a group of felsites, one at least of which retains a very fine perlitic structure and evidently flowed as a true glass. Some of these lavas are full of enclosed pieces of various flinty cellular and porphyritic felsites and andesites or trachytes, like the stones which occur abundantly in the agglomerate. The connection of these bedded lavas and tuffs with the agglomerate-neck seems obvious.

The Caernarvonshire volcanic area furnishes another admirable example of the intrusion of basic sills as a final phase of eruptivity. These masses have been carefully separated out on the maps of the Geological Survey, which present a striking picture of their distribution and their relation to the other igneous rocks. An examination of the maps shows at once that the basic sheets tend to lie parallel with the bedding along certain horizons. In the southern and western portions of the area they have forced themselves among the Lower Silurian sedimentary strata that underlie the Bala volcanic group—a position analogous to that taken by the corresponding sills of the Arenig series. But they likewise invade the volcanic group itself. Along the eastern borders of the district they abound, especially in the higher parts of the volcanic pile, where they have been injected between the flows, and have subsequently participated in the abundant plication of the rocks between the mountains and the line of the River Conway.

The curvatures into which the rocks of the region have been thrown, and the consequent breadth of country over which the volcanic sheets can now be examined, furnish a much better field than Merionethshire for the attempt to trace the probable centre or centres from which the basic magma of the sills was protruded. A study of the Survey maps soon leads to a conviction that the intrusions were not connected, except perhaps to a trifling

[1] This rock is referred to in the *Geological Survey Memoir* as "a short thick band of conglomeratic ash, which strikes northwards about half a mile and then disappears" (p. 134).

extent, with the great line of western vents. It is remarkable that the older strata which emerge from under the volcanic group on its western outcrop are, on the whole, singularly free from sills, though some conspicuous examples are shown opposite to Mynydd-mawr, while a few more occur further north along the same line. Their lenticular forms, their short outcrops, and their appearance on different horizons at widely separated points seem to indicate that the sills probably proceeded from many distinct subterranean pipes. Their greater abundance along the eastern part of the district may be taken to indicate that the ducts lay for the most part considerably to the eastward of the line of western vents. They may have risen in minor funnels, like that of Capel Curig.

It is noteworthy that so abundant an extravasation of basic material should have taken place without the formation of numerous dykes. We have here a repetition of the phenomena that distinguished the preceding Arenig volcanic period in Merionethshire, and it will be remembered that the Llandeilo eruptions of Builth were likewise followed by the injection of large bodies of basic rock. As an enormous amount of igneous magma may thus be impelled into the Earth's crust without the formation of dykes, it is evident that the conditions for the production of sills must be in some important respects different from those required for dykes.

No evidence has yet been obtained that any one of these sills established a connection with the surface. Not a trace can be found of the outpouring of any such basic lava-streams, nor have fragments of such materials been met with in any of the tuffs. On the other hand, there is abundant proof of the usual contact-metamorphism. Though the sills conform on the whole to the bedding, they frequently break across it. They swell into thick irregular masses, and thin out rapidly. In short, they behave as true intrusive sheets, and not as bedded lavas.

In regard to their internal character, they show the customary uniformity of texture throughout each mass. They are mapped under the general name of "greenstones" by the Geological Survey, and are described in the *Memoir* as hornblendic.[1] The more precise modern methods of examination, however, prove them to be true diabases, in which the felspar has, as a rule, consolidated before the augite, giving as a result the various types of diabasic structure.[2]

The date of the intrusion of these basic sills can be fixed by the same process of reasoning as was applied to those of the Arenig volcanic group. Their connection with the other igneous rocks of Caernarvonshire is so obvious that they must be included as part of the volcanic history of the Bala period. But they clearly belong to a late stage, perhaps the very latest stage, of that history. They probably could not have been injected into their present positions, unless a considerable mass of rocky material had overlain them. Some of them are certainly younger than the tuffs of Snowdon and Moel Hebog, which belong to a late part of the volcanic period. On the other hand, they had been intruded before the curvature and compression of the

[1] *Op. cit.* p. 156. [2] Mr. Harker, *Bala Volcanic Series*, p. 83.

region, for they share in the foldings and cleavage of the rocks among which they lie. The terrestrial movements that produced this disturbance have been proved to have occurred after the time when the uppermost Bala rocks were deposited, and before that of the accumulation of the Upper Silurian formations.[1] The epoch of intrusion is thus narrowed down to some part of the Upper Bala period. With this subterranean manifestation, volcanic action in this part of the country finally died out.

iv. THE VOLCANIC CENTRE OF THE BERWYN HILLS

Among the thick group of sedimentary formations which overlies the great volcanic ridge of the Arans and Arenig, and undulates eastwards across the Bala Valley, occasional thin intercalations of tuff point to the existence of another centre of volcanic activity which lay somewhere in the region of the Berwyn Hills. The structure of this ground, first indicated by Sedgwick, was investigated in detail by J. B. Jukes and his colleagues, whose work was embodied in the Maps, Sections and Memoirs of the Geological Survey.[2] The distinguishing characteristics of the volcanic rocks of this district are the occurrence of both lavas and tuffs as comparatively thin solitary bands in the midst of the ordinary sediments, and the persistence of these bands for a distance of sometimes more than 24 miles. The position of the vent or vents from which this extensive outpouring of volcanic material took place has not been revealed. As the bands tend to thin away eastwards, it may be surmised that the chief focus of eruption lay rather towards the west, perhaps under the trough of Upper Silurian strata somewhere in the neighbourhood of Llandderfel. There was probably another in the Hirnant district.

The mapping of the officers of the Survey showed that in the Berwyn Hills there are representatives of both the great volcanic periods of North Wales. A lower series of "felstones and greenstones" probably belongs to the older period, which began towards the end of Cambrian time and lasted in some districts even into the time of the Llandeilo formation. An upper group of tuffs, lying among the Bala rocks, is evidently equivalent, on the whole, to the much thicker volcanic series of the Snowdon region.

The lowest visible volcanic rocks occur among the hills to the north-west of Llanrhaiadr yn Mochnant. They are described as consisting of felstone of a pale greenish-grey colour and compact texture, like those of Arenig, and ashes distinctly interstratified with the slates. No exact petrographical examination of these rocks has yet been made. From the account given in the Survey *Memoir* there appears to be here a group of lavas and tuffs intercalated in Llandeilo perhaps partly in Upper Arenig, strata which form the broken dome of the Berwyn anticline. The lavas are represented as lying on four or five platforms, a single band reaching a

[1] *Mem. Geol. Sur.* vol. iii. 2nd edit. p. 326. See also Mr. Harker's *Bala Volcanic Series*, p. 76.

[2] See Sheet 74 of the one-inch map; Sheets 32, 35, 37 and 38 of the Horizontal Sections; and chapter xxxi. of the *Memoir* on the Geology of North Wales.

thickness of 300 feet and separated from the next band by sometimes 1000 or 1500 feet of non-volcanic sediment.

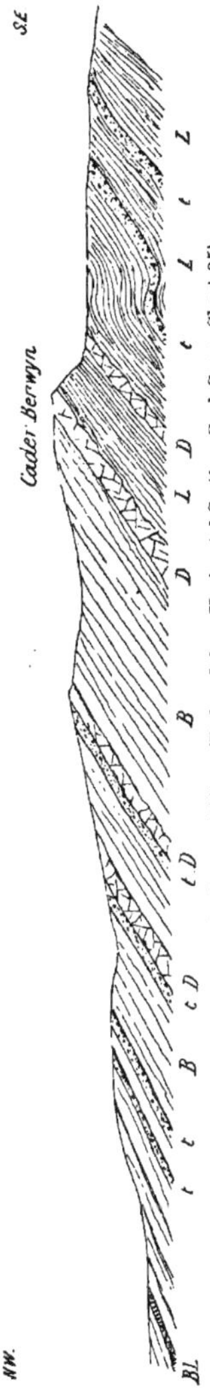

FIG. 57.—Section across the Berwyn Hills. (Reduced from Horizontal Section, Geol. Surv., Sheet 35). *L*, Llandeilo Flags; *B*, Bala group; *B L*, Bala Limestone; *t t*, volcanic tuffs; *D*, intrusive "greenstones."

These lower lavas, according to the measurements of Jukes, are overlain by more than 4000 feet of sedimentary strata before the upper or Bala volcanic series is reached. Three successive "ash-beds" constitute this upper series. Of these the lowest band, about 50 or 60 feet thick, was named a "greenstone ash" in contradistinction to a felstone ash, and was not traceable for more than a short distance. Above it, after an intervening thickness of several hundred feet of sedimentary strata, comes a second and much more continuous band of tuff, known as the "Lower ash-bed," about 100 feet thick on the west front of the Berwyn range. Still higher, after an interval of about 1500 feet of slates, lies the "Upper ash-bed," which on the same line of section has a thickness of about 200 feet. This is the most persistent of all the volcanic horizons, for it can be followed continuously round the whole range of the Berwyns until it is overlain by the Carboniferous Limestone near Selattyn, a distance of not less than twenty-four miles. The same band, but much more feebly developed, has been traced through the faulted country on both sides of Bala Lake, where it formed a useful platform in the investigation of the complicated geological structure of that area. Along the north side of the Berwyn Hills another thin band of tuff lies from 150 to 200 feet still higher up in the series, and has been traced for a distance of about twelve miles. The Bala limestone comes in about 800 or 1000 feet above the "Upper ash-bed."

Besides the rocks now enumerated, the Survey maps show the intercalation of four or five sheets of "greenstone," which are represented as following with marked regularity the strike of the strata. Until these sheets have been more precisely examined it is impossible to decide regarding their true petrographical character, or to determine whether they are sills, or interstratified lavas, or include rocks of both these types.

V. THE VOLCANOES OF ANGLESEY

We now turn to another part of the country, about which much has been written and keen controversy has arisen. In the centre of Anglesey, among the rocks grouped together by the Geological Survey as "altered Cambrian," there occur masses of breccia, the probable volcanic origin of which was, so far as I know, first suggested by Professor Hughes.[1] Dr. Callaway regards them as pre-Cambrian,[2] while Professor Blake places them in his "Monian system."[3] When I went over them some years ago, I accepted the view that they are volcanic agglomerates.[4] Subsequent examination, however, has convinced me that notwithstanding their remarkable resemblance to true agglomerates they are not really of volcanic origin, but are essentially "crush-conglomerates," like those in the Isle of Man, so well described by Mr. Lamplugh.[5]

But though their present coarse, agglomerate-like structure is, I think, entirely due to the mechanical crushing of the rocks *in situ* and not to volcanic explosions, it does not follow that the rocks which have been broken up do not contain evidence of volcanic action contemporaneous with their original formation. Obviously, pyroclastic materials may be subjected to deformation and disruption as well as any other components of the earth's crust, and may be equally converted into crush-conglomerates. And in Anglesey it can, I think, be shown that some of the rocks which have been broken up were originally tuffs and volcanic breccias.

Throughout Anglesey the stratified rocks present evidence of having undergone very great compression, deformation and rupture. Thus at Llanerchymedd thick-bedded Lower Silurian grits, with their intercalations of shale, have been broken up by numerous small faults, and have been pushed over each other in large irregular blocks, the shales being now pinched out, and now pressed up into the interstices between the dislocated harder and more resisting grits. This condition of rupture may be regarded as one of the stages towards the formation of a conglomerate by the crushing together of rocks *in situ*. A few miles further south at the beginning of the railway cuttings of Llangefni, green, red and purple slates and grits appear in a rather more crushed state, and immediately beyond these strata come the coarse breccias. Neither in their composition nor in their structural condition do these Llangefni strata appear to be marked off from the undoubted Lower Silurian rocks as parts of a different system.

The railway cuttings at Llangefni reveal a series of rocks which appear to have been originally shales, with thin bands of siliceous grit. The argillaceous portions of this series are now green and phyllitic, and remind one of the finer parts of some basic tuffs among the older Palæozoic systems. They include, however, pale flinty bands, such as might have been derived from fine felsitic dust. The grits are for the most part fine-grained and highly siliceous, but they include also coarser varieties with clear quartz-grains. The enormous deformation which these strata have undergone is

1 *Proc. Camb. Phil. Soc.* vol. iii. (1880), p. 347.
2 *Quart. Journ. Geol. Soc.* 3 *Op. cit.*
4 *Presidential Address Geol. Soc.* vol. xlvii. (1891), p. 130.
5 *Quart. Journ. Geol. Soc.* vol. li. (1895), p. 563. See *Geol. Mag.* 1896, p. 481.

at once apparent. They seem to have been plicated, ruptured and thrust over each other, the harder parts surviving longest, but being eventually broken into small fragments. Every stage may be traced from a recognizable band of grit down to the rounded or elliptical pebbles of the same material entirely isolated in this phyllitic matrix of crushed shale.

But while the volcanic origin of these coarsely-fragmental masses cannot be maintained, there is elsewhere evidence that the older Palæozoic rocks of Anglesey include relics of contemporaneous volcanic eruptions. Seven miles to the south-east of Holyhead, in the basal Lower Silurian conglomerates which, as before referred to, Mr. Selwyn found lying unconformably on the green schists, there occur abundant fragments of volcanic rocks, besides the prevalent detritus of the schists of the neighbourhood. Some of the bands have somewhat the character of volcanic breccias or tuffs, and they show an evident resemblance to portions of the Bangor group and the rocks of Llyn Padarn, though they are doubtless of much later age. That these volcanic fragments were not derived from the waste of rocks of a much earlier period is made tolerably certain by the intercalation of true tuffs among the black shales higher up in the order of succession. Here, then, we have evidence of contemporaneous volcanic action in the very basement Lower Silurian strata of Anglesey, which by their fossil contents are shown to be on the horizon of the lowest Arenig or even Tremadoc group.

But still further and fuller evidence of Silurian volcanism is to be obtained by an examination of the northern coast-line. I have already referred to the elliptical fault which is marked on the Geological Survey map as running from the north-western headland to the eastern coast beyond Amlwch. The necessity for inserting this fault, apart from any actual visible trace of its occurrence, arose when the conclusion was arrived at that the rocks of the extreme north of Anglesey were essentially altered Cambrian strata.[1] For immediately to the south of these rocks black shales, obviously Silurian, were seen to dip to the north—a structure which could only be accounted for by a dislocation letting them down into that position. The same necessity for a fault has of course been felt by all writers who have subsequently treated the northern area as pre-Cambrian. But it is deserving of notice that in the original mapping of the Survey no continuous abrupt hiatus is shown by the line which was afterwards marked as a continuous line of fault. On the contrary, on one of the field-maps in, I believe, Mr. Selwyn's handwriting the remark occurs :—"The gradual passage from the black shale to the upper green gritty slates of Llanfechell is best seen at Bothedd, on road from Llanfaethlu to Llyn-llygeirian."[2]

It is no part of my aim to disprove the existence of faults along the line

[1] I have fully considered the evidence adduced by Dr. Callaway and Professor Blake, and have examined the ground, and can come to no other conclusion than that stated in the text. But see Mr. Blake's remarks, *Geol. Mag.* 1891, p. 483.

[2] There is no continuous section now visible at this place, but the two groups of rock can be traced to within a few feet of each other, both inclined as usual in the same direction, and the black shales appearing to pass under the others.

referred to. These may quite well exist; but there is assuredly no one gigantic displacement, such as the theory I am combating would require; while any faults which do occur cannot be greatly different from the others of the district, and do not prevent the true relations of the rocks from being discoverable.

Where the supposed elliptical fault reaches the shore at Carmel Point, the two groups of rock seem to me to follow each other in unbroken sequence.[1] The black slates, which are admittedly Lower Silurian, dip underneath a breccia and greenish (Amlwch) slates. Not only so, but bands of similar black slates occur higher up, interstratified with and shading-off into tuffs and greenish slates. Further, bands of coarse volcanic breccia occur among the black slates south of the supposed break. These, in accordance with the exigencies of theory, are represented as separated by a network of faults from the black slates amid which they lie. But good evidence may be found that they are truly interbedded in these slates. In short, the whole of the rocks in that part of Anglesey form one great series, consisting partly of black slates, partly of greenish slates, with abundant intercalations of volcanic detritus. The age of the base of this series is moreover determined by the occurrence of Bala fossils in a band of limestone near Carmel Point.

The rocks which extend eastward along the coast from the north-western headland of Anglesey are marked on the Survey map as "green, grey and purple slates with conglomeratic and siliceous beds." The truly volcanic nature of a considerable proportion of these strata has been clearly stated by Mr. Blake.[2] As they dip in a general northerly direction, higher portions of the series present themselves as far as the most northern projection of the island near Porth Wen (Fig. 58). They have been greatly crumpled and crushed, so that the slates pass into phyllites. They include some thick seams of blue limestone and white quartzite, also courses of black shale containing Lower Silurian graptolites. Among their uppermost strata several (probably Bala) fossils, including *Orthis Bailyana*, have been obtained by Professor Hughes. It has been supposed that the higher bands of black shale may also have been brought into their present positions by faults, and that they do not really belong to the series of strata among which they lie. But this suggestion is completely disproved by the coast-sections, which exhibit many thin interstratified leaves of black shale, sometimes less than an inch thick. These and the ashy layers containing the *Orthis* and other fossils form an integral part of the so-called "Amlwch slates."[3]

As evidence of the regular intercalation of the black shales and tuffs in this sedimentary series, a portion of the coast section at Porth Wen is here

[1] I cannot admit that there is any evidence of a thrust-plane here. To insert one is merely to modify field-evidence to suit theory. See *Geol. Mag.* 1891, p. 483.

[2] *Quart. Journ. Geol. Soc.* vol. xliv. (1888), p. 517. See his further remarks in *Geol. Mag.* 1891, p. 483.

[3] The Amlwch slates exhibit on a great scale the puckering that points to intense compression. This "gnarled" structure, as Prof. Hughes called it, has been illustrated by Mr. Harker, *British Assoc. Report* (1885), pp. 839, 840.

given (Fig. 58). The lowest member (1) of the series is a white quartzite much jumbled in its bedding, but yet distinctly interstratified with the other sediments, and containing intercalated courses of green tuff and highly carbonaceous shale. Markings like worm-pipes are here and there to be seen. The next group of strata (2) consists of black shale followed by yellow conglomeratic sandstone and pebbly tuffs. The shales enclose rounded and angular fragments of quartzite. The sandstone passes upward into pinkish and yellowish conglomerate (3), with an abundant lustrous phyllitic matrix, which when free from pebbles closely resembles some of the tuffs of Llyn Padarn. The next band (4) is one of yellow, sandy, felspathic grit, quartz-conglomerate and fine tuffs, with leaves of dark shale towards the base. It was in the lower part of this band that the *Orthis* above mentioned was found. The black shales contain markings which are probably graptolites. Reddish quartzite and quartz-conglomerate (5) next succeed. These strata have the same phyllitic base just noticed. The highest group here shown is one of black, yellow and green shales mixed with patches and bands of volcanic breccia and tuff, the whole being greatly disturbed, cleavage and bedding seeming as it were to be struggling for the mastery.

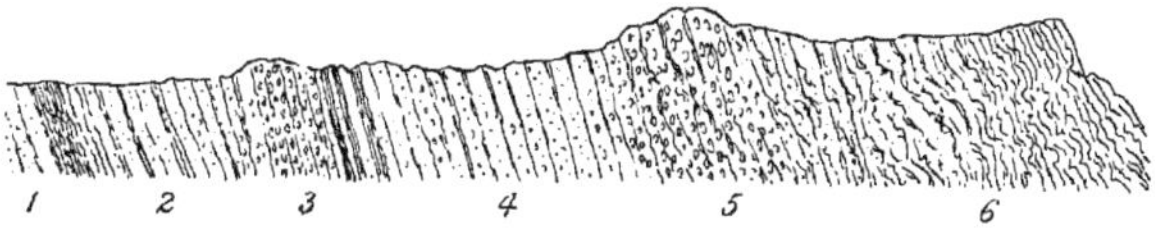

FIG. 58.—Section of the strata on the shore at Porth Wen, west of Amlwch.

These last strata look as if they were about to pass up vertically into the ordinary dark Lower Silurian shales or slates.

There can be no doubt regarding the serious amount of crushing which the rocks of this coast-line have undergone. Some of the bands might even be described as "crush-conglomerates." Yet the intercalation of seams of black shale and limestone, and the occurrence of the exactly similar but thicker group of black shales at Porth Prydd, which are admitted to be Lower Silurian, unite the whole series of strata as parts of one formation.

It thus appears that the area coloured "altered Cambrian" on the Survey map, and regarded as pre-Cambrian by some later observers, is proved by the evidence of fossils at its base, towards its centre and at its top, to belong to the Lower Silurian series, probably to the Bala division. That this was the geological horizon of part at least of the area was recognized by Sir A. Ramsay, though he confessed himself unable "precisely to determine on the north coast of Anglesey how much of the strata are of Silurian and how much of Cambrian age."[1] Professor Hughes was the first to suggest that the whole of these rocks should be referred to the Bala group.[2]

I have dwelt on the determination of the true geological age of the rocks

[1] *Mem. Geol. Surv.* vol. iii. 2nd edit. p. 242.

[2] *Proc. Camb. Phil. Soc.* vol. iii. (1880), pp. 341-348.

of the north of Anglesey because of the diversity of opinion respecting them, and because of their great interest in regard to the history of volcanic action in Wales. These rocks contain a record of volcanic eruptions, probably contemporaneous on the whole with those of the Bala period in Caernarvonshire, yet independent of them and belonging to a different type of volcanic energy. Some of the vents probably lay in the north-western part of Anglesey. The materials ejected from them were, so far as we know, entirely of a fragmentary kind. Vast quantities of detritus, largely in the form of fine dust, were thrown out; but no trace has yet been found of the outflow of any lava. The lower part of this volcanic series consists of bedded breccias which are sometimes remarkably coarse. Their included stones, ranging up to six inches or more in diameter, are usually more or less angular, and consist mainly of various felsites. Layers of more rounded pebbles occasionally occur, while the bedding is still further indicated by finer and coarser bands, and even by intercalations of fine tuffs and ashy shales. Towards their upper limits some of these volcanic bands shade off into pale grey or greenish ashy shale, followed by black sandy shale of the usual kind. The relation of the peculiar greenish shale of the Amlwch type to these tuffs and breccias is well shown east of Carmel Point. This shale is interleaved with tuff and contains frequent repetitions of finer or coarser volcanic breccia, as well as occasional seams of black shale. An illustration of this structure is given in Fig. 59, where some yellow decomposing breccias (1), cut by a fault (*f*), are overlain by about 40 or 50 feet of black shale (2), above which lies a flinty felsitic rock (3) that appears to run in bands or dykes through the agglomerate. At Carmel Point (Fig. 60) a similar structure may be observed to that at Llyn Padarn already referred to (p. 163). The cleavage, which is well developed in the green slates (*a*), is much more faintly marked in the overlying breccia (*b*), but the bedding can still be detected in both rocks running parallel to their mutual boundary-line. Beyond Porth Padrig, which lies east from

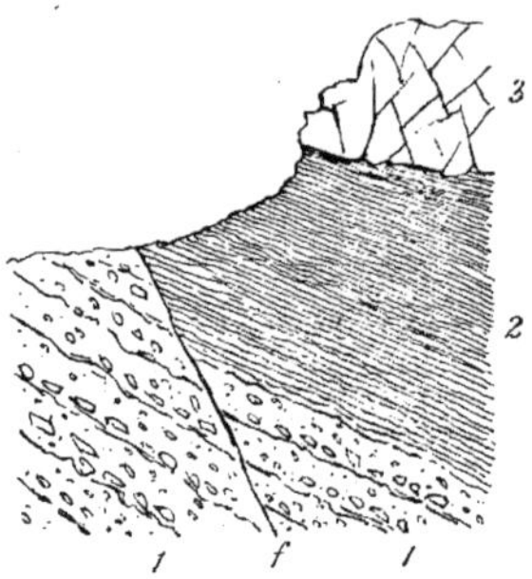

FIG. 59.—Section of intercalated black shale in the volcanic series at Porth yr hwch, south of Carmel Point.

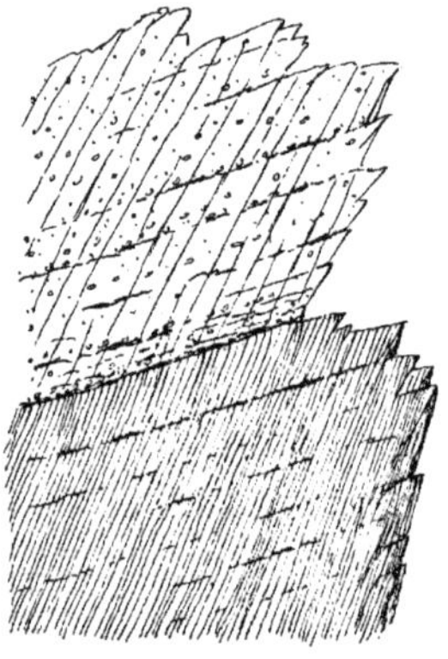

FIG. 60.—Green slates overlain with volcanic breccia, Carmel Point, Anglesey.

Carmel Point, the section may be seen which is shown in Fig. 61. Here the blue or lead-coloured shale or slate (*a*) marked as Silurian on the Geological Survey map passes up into a mass of fine yellowish felsitic tuff and breccia (*b*). The shale at the junction intercalates in thin leaves with the tuff.

The breccias south of Carmel Point, though they are chiefly made up of felsitic detritus, sometimes show a preponderance of fragments of shale. They vary also rapidly in texture and composition. These variations may indicate that the vent or vents from which their materials were derived stood somewhere in the near neighbourhood, if indeed they are not to be recognized in some of the boss-like eminences that rise above the shore. At the same time, the enormous amount of crushing and shearing which

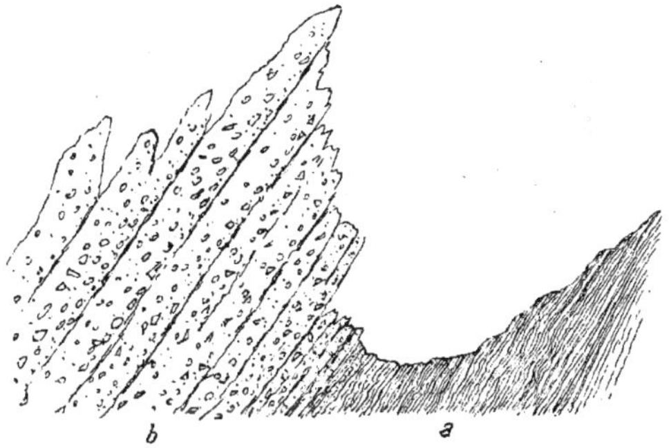

FIG. 61.—Blue shale or slate passing into volcanic breccia east of Porth Padrig, near Carmel Point, Anglesey.

the rocks of this region have undergone has doubtless introduced crush-conglomerates into the structure of the ground. And some patient labour may be required before the nature and origin of the different fragmental masses are determined.

Certain remarkably coarse, tumultuous breccias, exposed on the coast at Mynyddwylfa and Cemmaes, were formerly regarded by me as volcanic agglomerates. But more recent examination has satisfied me that these, like the breccias at Llangefni, are not of volcanic origin but are crush-conglomerates.[1]

While the lower breccias are sometimes tolerably coarse, the volcanic detritus becomes much finer in the higher parts of the Amlwch slates. Above the limestones and black shales of Cemmaes volcanic breccias and ashes, with limestone, quartzite, conglomerate and thin seams of black shale, continue to the extreme northern headlands. The amount of fine volcanic detritus distributed through these strata is very great. We can clearly make

[1] Presidential Address, *Quart. Journ. Geol. Soc.* vol. xlvii. p. 134; *Rep. Brit. Assoc.* 1896, Section C; *Geol. Mag.* 1896, p. 481.

out that while ordinary sedimentation was in progress, an almost constant but variable discharge of fragmental materials took place from the vents in the neighbourhood. Sometimes a special paroxysm of explosion would give rise to a distinct band of breccia or of tuff, but even where, during a time of comparative quiescence, the ordinary sand or mud predominated, it was generally mingled with more or less volcanic dust.

Some bands of conglomerate in this group of strata deserve particular notice. The most conspicuous of these, already referred to as seen at Porth Wen, is made up of quartz and quartzite blocks, embedded in a reddish matrix largely composed of ashy material, and recalling the red spotted tuffs of Llyn Padarn. The occurrence of strong conglomerates near the top of a volcanic series has been noted at St. David's, Llyn Padarn and Bangor. In none of these localities, as I have tried to show, do the conglomerates mark an unconformability or serious break between two widely-separated groups of rock. The Anglesey section entirely supports this view, for the conglomerates are there merely intercalations in a continuous sequence of deposits; they are succeeded by tuffs and shales like those which underlie them. The interposition of such coarse materials, however, may undoubtedly indicate local disturbance, connected, perhaps, in this and the other localities, with terrestrial readjustments consequent upon the waning of volcanic energy.

The detailed geological structure of Anglesey is still far from being completely understood. Besides the serious crushing here referred to, there is reason to suspect that considerable plication, perhaps even inversion, of the strata has taken place, and that, by denudation, detached portions of some of the higher groups have been left in different parts of the island. The occurrence of Upper Silurian fossils in several localities adds to the perplexity of the problem by indicating that, among the folds and hardly distinguishable from the older slates, portions of Upper Silurian formations may have been caught and preserved. These difficulties, moreover, involve in some obscurity the closing phases of volcanic activity in Wales; for until they are, to some extent at least, removed, we shall be left in doubt whether the vents in the north of Anglesey, which were in eruption probably during Bala time, were the last of the long succession of Welsh volcanoes. If the black shales of Parys Mountain are really referable to the horizon of the Mayhill Sandstone, the two great igneous bands between which they lie would seem to mark an outbreak of volcanic energy during Upper Silurian time. No other indications, however, of eruptions of that age having been met with in Great Britain (though they occur in the south-west of Ireland and possibly in Gloucestershire), more careful investigation is required before such a position can be safely assigned to any rocks in Anglesey.

Putting these doubtful rocks aside for the present, we may, in conclusion, contrast the type of eruption in Anglesey with that of the great Snowdonian region. While the Caernarvonshire volcanoes were pouring forth their volumes of felsitic lava, and piling them up for thousands of feet on the sea-floor, the northern Anglesey vents, not more than some five-and-twenty miles away, threw out only stones and dust, but continued their intermittent explosions until they had strewn the sea-bottom with detritus to a depth of many hundred feet.

There is yet another feature of interest in this independent group of submarine vents in Anglesey. Their operations appear to have begun before the earliest eruptions of the Bala volcanoes in Caernarvonshire. Their first beginnings may, indeed, have been coeval with the explosions that produced the older Arenig tuffs of Merionethshire; their latest discharges were possibly the last manifestations of volcanic energy in Wales. They seem thus to bridge over the vast interval from Tremadoc to Upper Bala, possibly even to Upper Silurian time. But we may, perhaps, connect them with the still earlier period of Cambrian volcanism. I have referred to the evidence which appears to show that the vents whence the lavas and tuffs of Moel Trefan and Llyn Padarn were erupted gradually moved northwards, and continued in eruption until after the beginning of the deposition of the black slates that are generally regarded as Arenig. The Anglesey tuffs and breccias may thus be looked upon as evidence of a still further shifting of the active orifices northward. In this view, while the Aran and Cader Idris volcanoes broke out in Upper Cambrian and continued through Arenig time, and the Snowdonian group was confined to Bala time, a line of vents opened to the north-west in the Cambrian period before the epoch of the Llanberis slates, and, dying out in the south, continued to manifest a minor degree of energy, frequently discharging fragmental materials, but no lava, over the sea-bottom, until, towards the close of the Bala period, possibly even in Upper Silurian time, they finally became extinct.

vi. THE VOLCANOES OF THE LAKE DISTRICT (ARENIG TO CLOSE OF BALA PERIOD)

From the time of the appearance of Sedgwick's classic letters to Wordsworth, no volcanic area of Britain has probably been so well known in a general sense to the ordinary travelling public as the district of the English Lakes. Many geologists have since then visited the ground, and not a few of them have published additions to our knowledge respecting what is now known as the Borrowdale Volcanic Series. The most elaborate and detailed account of any part of it is that given by the late Mr. J. C. Ward in the *Geological Survey Memoirs*, wherein he embodied the results of his minute investigation and mapping of the northern portion of the district.[1] Notices of the petrography of some of the more interesting rocks have subsequently been published by Mr. Rutley, Professor Bonney, Mr. Harker, Mr. Marr, Mr. Hutchings and others. But up to the present time no complete memoir on the volcanic geology of the Lake District as a whole has

[1] Sheet 101 S.E. of the Geological Survey of England and Wales and Explanation illustrating the same; and papers by him in *Quart. Journ. Geol. Soc.* vols. xxxi. xxxii. (1875-76). See also Messrs. Aveline and Hughes, *Mem. Geol. Survey*, Sheet 98 N.E. (Kendal, Sedbergh, etc.).

appeared. The sheets of the Geological Survey map present a graphic view of the general distribution of the rocks, but so rapid has the progress of certain branches of geology been since these sheets were published, that the map is even now susceptible of considerable improvement.

In estimating the area over which the volcanic rocks of the Lake District are spread, geologists are apt to consider only the tract which lies to the south of Keswick and stretches southward to a line drawn from the Duddon Sands to Shap. But it can easily be shown that this area falls far short of the extent of that wherein the rocks can still be traced, and yet further short of that over which the lavas and ashes originally spread. For, in the first place, the volcanic group can be followed round the eastern end of the mountain-group which culminates in Skiddaw, and along the northern base of these heights to Cockermouth, though only a narrow fringe of it emerges from underneath the Carboniferous series. It is thus manifest that the volcanic rocks once stretched completely across Skiddaw and its neighbours, and that they extend northwards below the Whitehaven Coal-field. But, in the next place, far beyond these limits, volcanic rocks, which there can be little doubt were originally continuous with those of the Lakes, emerge from beneath the base of the Cross Fell escarpment,[1] and still further to the east a prolongation of the same group rises for a brief space to the surface from under the great limestone sheets of Upper Teesdale. Between the north-western and south-eastern limits within which the rocks can now be seen there intervenes a distance of some 11 miles, while the extreme length of the tract from south-west to north-east is about 50 miles. Even if we take these figures as marking the approximate boundaries of the region covered by the volcanic ejections, it cannot be less than 550 square miles. But this is probably much less than the original area.

The thickness of the accumulated volcanic materials is proportionate to the large tract of country over which they have been spread. From various causes, it is difficult to arrive satisfactorily at any precise statement on this question. In a volcanic series bedding is apt to be obscure where, as in the present case, there are no interstratified bands of ordinary sedimentary strata to mark it off. It tends, moreover, to vary considerably and rapidly within short distances, not only from subsequent unequal movements of subsidence or elevation, but from the very conditions of original accumulation. Mr. Ward considered that the maximum thickness of the volcanic group of the Lake District might be taken to range from 12,000 to 15,000 feet.[2] Professors Harkness and Nicholson, on the other hand, gave the average thickness as not more than 5000 feet.[3] My own impression is that the truth is to be found somewhere between these two estimates, and that the maximum thickness probably does not exceed 8000 or 9000 feet. In any case there cannot, I think, be much doubt that we have here the

[1] For an account of the Cross Fell inlier of Silurian rocks see the paper by Professor Nicholson and Mr. Marr, with the petrographical appendix by Mr. Harker. *Quart. Journ. Geol. Soc.* vol. xlvii. (1891), pp. 500, 512.

[2] Ward, *op. cit.* p. 46.

[3] *Brit. Assoc.* (1870) *Sectional Reports*, p. 74.

thickest accumulation of volcanic material, belonging to a single geological period, anywhere known to exist in Britain.

The geological age of this remarkable volcanic episode is fortunately fixed by definite palæontological horizons both below and above. The base of the volcanic group rests upon and is interstratified with the upper part of the Skiddaw Slate,[1] which from the evidence of its fossils is paralleled with the Arenig rocks of Wales. The highest members of the group are interstratified with the Coniston Limestone, which, from its abundant fauna, can without hesitation be placed on the same platform as the Bala Limestone of Wales, and is immediately followed by the Upper Silurian series. Thus the volcanic history comprises the geological interval that elapsed between the later part of the Arenig period and the close of the Bala period. It begins probably not so far back as that of the Arenig group of Merionethshire, and its termination was perhaps coincident with the dying out of the Snowdonian volcanoes. But it contains no record of a great break or interval of quiescence like that which separated the Arenig from the Bala eruptions in Wales.

The materials that form this enormous volcanic pile consist entirely of lavas and ashes. No intercalations of ordinary sedimentary material have been met with in it, save at the bottom and at the top. The lower lavas, well seen among the hills to the south of Keswick, were shown by Mr. Ward to be intermediate between felsites and dolerites in regard to their silica percentage, and he proposed for them the name of felsi-dolerites. They are comprised in the group of the andesites or "porphyrites." From the analyses published by Mr. Ward, the amount of silica appears to range up to about 60 per cent.[2] They are usually close-grained, dull dark-grey to black rocks, breaking, where fresh, with a splintery or conchoidal fracture, showing a few minute striated felspars, apt to weather with a pale-brown or yellowish-grey crust, and sometimes strongly vesicular or amygdaloidal. They present many external resemblances to some of the "porphyrites" or altered andesites of the Lower Old Red Sandstone of Scotland. A microscopic examination of specimens collected by Dr. Hatch and myself from the hills to the south of Keswick showed the rocks to be true andesites, composed of a multitude of slender laths (sometimes large porphyritic crystals) of

[1] Mr. Dakyns has expressed his belief that the volcanic group lies unconformably on the Skiddaw Slate (*Geol. Mag.* 1869, pp. 56, 116), and Professor Nicholson has formed the same opinion (*op. cit.* pp. 105, 167; *Proc. Geol. Assoc.* vol. iii. p. 106). Mr. Goodchild, however, has shown that in the Cross Fell inlier the oldest tuffs are interstratified with the Skiddaw Slates (*Proc. Geol. Assoc.* vol. xi. (1889), p. 261). Mr. Ward in mapping the district inserted a complex series of faults along the junction-line between the volcanic series and the Skiddaw Slates. When I went over the ground with him some years before his death I discussed this boundary-line with him and could not adopt his view that it was so dislocated. More recent re-examination has confirmed me in my dissent. A large number of the faults inserted on the Geological Survey map to separate the Skiddaw Slates from the Borrowdale volcanic series cannot be proved, and probably do not exist. Others may be of the nature of "thrust-planes." But see Mr. Ward's explanation of his views, *op. cit.* p. 48.

[2] *Quart. Journ. Geol. Soc.* vol. xxxi. (1875) p. 408, vol. xxxii. (1876) p. 24. Geology of Northern Part of Lake District (*Mem. Geol. Survey*), p. 22. In a subsequent paper the more basic lavas of Eycott Hill are compared with dolerites (*Monthly Microscopical Journ.* 1877, p. 246).

felspar with a brownish glassy groundmass, and with some chloritic material probably representing augite, but with no trace of quartz.[1]

Another type of andesite has been found by Mr. Hutchings to occur abundantly at Harter Fell, Mardale, between the Nan Bield Pass and High Street, and in the cliffs on the right side of the Kentmere Valley. It consists of rocks mostly of a grey-green or grey-blue colour with resinous lustre and extremely splintery fracture. They are augite-andesites of a much more vitreous nature than the dominant type of lavas of the Lake District. Their groundmass under the microscope is seen to have originally varied from a wholly glassy base to an intimate mixture of glass and exceedingly minute felspar-microlites. This groundmass is permeated with chlorite in minute flakelets, and encloses numerous porphyritic sharply-defined felspar-crystals, together with chlorite-pseudomorphs after augite.[2] Gradations from these rocks to the ordinary more coarse-grained andesites may be observed.

Some of the andesites appear to have a trachytic facies, where the felspars of the groundmass consist largely of untwinned laths and appear to be mainly orthoclase.[3]

Among the lavas of the Lake District there occur many which are decidedly more basic than the andesites, and which should rather be classed among the dolerites and basalts, though they do not appear to contain olivine. These rocks occur at Eycott Hill, above Easedale Tarn, Scarf Gap Pass, Dale Head, High Scawdell, Seatoller Fell and other places. Analyses of those from Eycott Hill were published by Mr. Ward, and their silica percentage was shown to range from 51 to 53·3.[4] The microscopic characters of the group have been more recently determined by Mr. Hutchings[5] and Messrs. Harker and Marr.[6]

The andesitic and more basic lavas are particularly developed in the lower and central part of the volcanic group. They rise into ranges of craggy hills above the Skiddaw Slates, and form, with their accompanying tuffs, the most rugged and lofty ground in the Lake District. They extend even to the southern margin of the volcanic area at one locality to the south-west of Coniston, where they may be seen with their characteristic vesicular structure forming a succession of distinct flows or beds, striking at the Coniston Limestone which lies upon them with a decided, though probably very local, unconformability.[7] One of the flows from this locality was found by Dr. Hatch, under the microscope, to belong to the more basic series. It

[1] These rocks were mapped as tuffs by Mr. Ward. Their microscopic characters have been described by Messrs. Harker and Marr, *Quart. Journ. Geol. Soc.* xlvii. (1891), p. 292 ; by Mr. Harker, *op. cit.* p. 517 ; and by Mr. W. M. Hutchings, *Geol. Mag.* 1891, p. 537 ; 1892, pp. 227, 540.

[2] Mr. Hutchings, *Geol. Mag.* 1891, p. 539. This observer describes a quartz-andesite or dacite from near Dunmail Raise.

[3] *Op. cit.* p. 543.

[4] *Monthly Microscopical Journal*, 1877, p. 246.

[5] *Geol. Mag.* 1891, p. 538.

[6] *Quart. Journ. Geol. Soc.* vol. xlix. (1893), p. 389. Mr. Harker, *op. cit.* vol. xlvii. (1891).

[7] This unconformability has been described and discussed by various observers. The general impression has been, I think, that the break is only of local importance. Mr. Aveline, however, believed it to be much more serious, and he regarded the volcanic rocks which were ejected during the deposition of the Coniston Limestone series as much later in date than those of the Borrowdale group. See *Mem. Geol. Survey*, Explanation to Sheet 98 N.E. 2nd edit. p. 8 (1888).

approaches a basalt, containing porphyritic crystals of fresh augite instead of the usual felspars, and showing a groundmass of felspar microlites with some granules of augite and dispersed magnetite. This local increase of basic composition is interesting as occurring towards the top of the volcanic group. A porphyritic and somewhat vesicular andesite, with large crystals of striated felspar in a dark almost isotropic groundmass, occurs under the Coniston Limestone near Stockdale.

Mr. Ward was much impressed with the widespread metamorphism which he believed all the volcanic rocks of this region had undergone, and as a consequence of which arose the difficulty he found in discriminating between close-grained lavas and fine tuffs. There is, of course, a general induration of the rocks, while cleavage has widely, and sometimes very seriously, affected them. There is also local metamorphism round such bosses as the Shap granite, but the evidence of any general and serious metamorphism of the whole area does not seem to me to be convincing.[1]

With regard to the original structure and subsequent alteration of some of the andesitic lavas, an interesting section has recently been cut along the road up Borrowdale a little south of the Bowder Stone. Several bands of coarse amygdaloidal lava may there be seen interstratified among tuffs. The calcite amygdales in these rocks are arranged parallel to the bedding and therefore in the planes of flow, while those lined with chlorite are more usually deformed parallel to the direction of the cleavage. This difference suggests that before the cleavage took place, not improbably during the volcanic period, the rocks had been traversed by heated water producing internal alteration and rearrangements, in virtue of which the vesicles along certain paths of permeation were filled up with calcite, so as then to offer some resistance to the cleavage, while those which remained empty, or which had been merely lined with infiltrated substance, were flattened and pulled out of shape. Messrs. Harker and Marr have shown that the amygdaloidal kernels had already been introduced into the cellular lavas before the intrusion of the Shap granite. In the account to be given of the Tertiary plateau-basalts (Chapter xxxvi.) evidence will be adduced that this filling up of the steam-cavities of lava may take place during a volcanic period, and that it is probably connected with the passage of heated vapours or water through the rocks.

Though acid lavas are not wholly absent from the central and lower parts of the volcanic group, it is at the top that their chief development appears to occur. These rocks may be grouped together as felsites or rhyolites. They probably play a much larger part in the structure of the southern part of the volcanic area than the published maps would suggest, and a detailed survey and petrographical study of them would well reward the needful labour.[2] A fine series of felsites is interbedded in the lower

[1] The metamorphism of all the rocks, aqueous and igneous, around the Shap granite has been well worked out by Messrs. Harker and Marr, *Quart. Journ. Geol. Soc.* vol. xlvii. (1891) p. 266, xlix. (1893) p. 359.

[2] See Mr. F. Rutley, "The Felsitic Lavas of England and Wales," *Mem. Geol. Surv.* 1885, pp. 12-15; also the description of Messrs. Harker and Marr, *Quart. Journ. Geol. Soc.* xlvii. (1891), p. 301.

part of the Coniston Limestone, and spreads out underneath it along the southern margin of the volcanic district from the Shap granite south-westward for some miles[1] (Fig. 62). Between the valleys of the Sprint and Kent these felsites (which farther east are said to be 700 feet thick) may be seen interposed between the limestone and the fossiliferous calcareous shales below it, while from underneath the latter other sheets rise up into the range of hills behind.

These acid lavas are generally grey, cream-coloured, or pink, with a white weathered crust. Their texture when fresh is flinty or horny, or at least extremely fine-grained and compact. They are seldom markedly porphyritic. They frequently display good flow-structure, and sometimes split up readily along the planes of flow. Occasionally the flow-lines on the outer crust have broken up in the movement of the rock, giving rise to irregular fragments which have been carried forward. Short, extremely irregular, branching veins of a fine cherty felsitic substance, which occasionally shows a well-marked flow-structure parallel to the walls, traverse certain parts of a dark-grey felsite, near Brockstones, between the valleys of

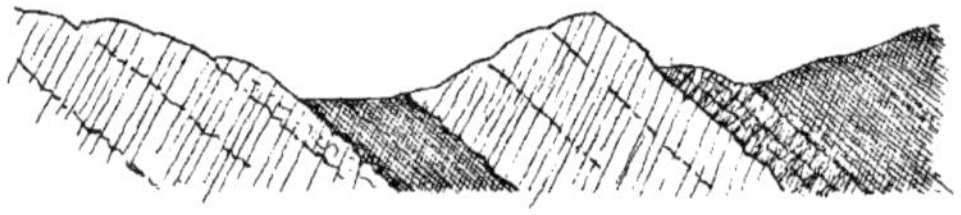

FIG. 62.—Section of felsites on the Coniston Limestone group, west of Stockdale.
a, Felsites more or less cleaved ; *b*, Calcareous shales with fossils, much cleaved ; *c*, Cleaved felsite ; *d*, Coniston Limestone ; *e*, Stockdale Shales (with graptolites).

the Kent and Sprint.[2] Occasionally a distinct nodular structure may be observed in these acid lavas, sometimes minute, like an oolite, in other parts, as on Great Yarlside, presenting large rounded balls. This nodular structure is not confined to the lava-flows, but has been detected by Messrs. Harker and Marr in what appears to be an intrusive rock near Shap Wells. The microscopic characters of some of the Lake District rhyolites were described by Mr. Rutley, who found them to exhibit beautiful perlitic and spherulitic structures.[3] That such rocks as these were poured out in a vitreous condition, like obsidian or pitchstone, cannot be doubted. Chemical analysis shows that the Lake District rhyolites agree exactly with those of North Wales in their composition. They contain about 76 per cent of silica.[4]

The rhyolitic lavas have been seriously affected by the general cleavage of the region. In some places they have been so intensely cleaved as to become a kind of fissile slate, and there seems good reason to believe that in

[1] Unfortunately these acid lavas are not distinguished from the others in the Geological Survey maps.

[2] Compare the structure described by Mr. Harker from the Cross Fell inlier, *Quart. Journ. Geol. Soc.* xlvii. (1891), p. 518.

[3] "Geology of Kendal," etc., *Mem. Geol. Survey*, Sheet 98 N.E. 2nd edit. p. 9.

[4] Messrs. Harker and Marr, *op. cit.* p. 302.

this altered condition they have often been mistaken for tuffs. Where they assume a nodular structure, the nodules have sometimes been flattened and elongated in the direction of the prevalent cleavage.

The abundance and persistence of thoroughly acid lavas along the southern edge of the volcanic area where the youngest outflows are found, is a fact of much interest and importance in the history of the eruptions of this region. It harmonizes with the observations made in Wales, where in the Arenig, and less distinctly in the Bala group, a marked increase in acidity is noticeable in the later volcanic products. At the same time, as above mentioned, there is evidence also of the discharge of more basic materials towards the close of the eruptions, and even of the outflow of a lava approaching in character to basalt.

According to the Geological Survey maps, by far the largest part of the volcanic district consists of pyroclastic materials. When my lamented friend, the late Mr. Ward, was engaged in mapping the northern part of the district, which he did with so much enthusiasm, I had an opportunity of going over some of the ground with him, and of learning from him his ideas as to the nature and distribution of the rocks and the general structure of the region. I remember the difficulty I had in recognizing as tuff much of what he had mapped as such, and I felt that had I been myself required, without his experience of the ground, to map the rocks, I should probably have greatly enlarged the area coloured as lava, with a corresponding reduction of that coloured as tuff. A recent visit to the district has revived these doubts. It is quite true, as Mr. Ward maintains, that where the finer-grained tuffs have undergone some degree of induration or metamorphism, they can hardly, by any test in the field, be distinguished from compact lavas. He was himself quite aware of the objections that might be made to his mapping,[1] but the conclusions he reached had been deduced only after years of unremitting study in the field and with the microscope, and in the light of experience gained in other volcanic regions. Nevertheless I think that he has somewhat exaggerated the amount of fragmental material in the northern part of the Lake District, and that the mapping, so consistently and ably carried out by him, and followed by those members of the Survey who mapped the rest of the ground, led to similar over-representation there. Some portions of the so-called tuffs of the Keswick region are undoubtedly andesites; other parts in the southern tracts include intercalated bands of felsite as well as andesite.

But even with this limitation, the pyroclastic material in the Lake District is undoubtedly very great in amount. It varies in texture from coarse breccia or agglomerate, with blocks measuring several yards across, to the most impalpable compacted volcanic dust. In the lower parts of the

[1] He says: "I shall be very much surprised if my mapping of many parts of the district be not severely criticized and found fault with by those who examine only one small area and do not take into consideration all the facts gathered together, during the course of several years, from every mountain flank and summit" (*op. cit.* p. 25). Mr. Hutchings has expressed his agreement with the opinions stated in the text. He likewise coincides in the belief that there are many of these Lake District volcanic rocks, regarding which it is impossible to decide whether they are lavas or ashes (*Geol. Mag.* 1891, p. 544).

group some of the tuffs abound in blocks and chips of Skiddaw Slate. Some good examples of this kind may be seen in Borrowdale, below Falcon Crag and at the Quayfoot quarries. Where the tuff is largely made up of fragments of dark blue slate, it much resembles the slate-tuffs of Cader Idris. Some of the pieces of slate are six or eight inches long and are now placed parallel to the cleavage of the rock. Among the slate debris, however, felspar crystals and felsitic fragments may be observed. Bands of coarser and finer green tuff show very clearly the bedding in spite of the marked cleavage (Fig. 63).

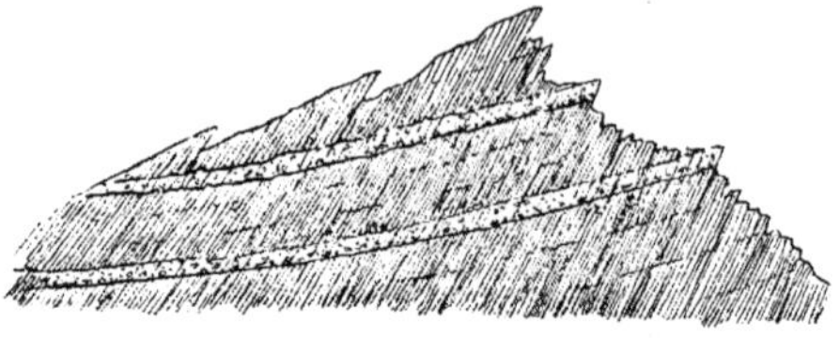

FIG. 63.—Fine tuff with coarser bands near Quayfoot quarries, Borrowdale.

The highly-inclined fine lines show the cleavage. The more gently dipping bands and lines mark the bedding.

But throughout the whole volcanic group the material of the tuff is chiefly of thoroughly volcanic origin, and its distribution appears to agree on the whole with that of the bedded lavas. In the older portions of the group it is probably mainly derived from andesitic rocks, though with an occasional intermingling of felsitic or rhyolitic detritus, while in the higher parts many of the tuffs are markedly rhyolitic. Among the lapilli minute crystals of felspar, broken or entire, may be detected with the microscope. Some of the ejected ash must have been an exceedingly fine dust. Compacted layers of such material form bands of green slates, which may occasionally be seen to consist of alternations of coarser and finer detritus, now and then false-bedded. Such tuffs bring vividly before the mind the intermittent explosions, varying a little in intensity, by which so much of the fabric of the Lake mountains was built up.

Breccias of varying coarseness are likewise abundant, composed of fragments of andesite and older tuffs in the central and lower parts of the volcanic group, and mainly of felsitic or rhyolitic detritus in the upper parts. Some of these rocks, wherein the blocks measure several yards across, are probably not far from the eruptive vents, as at Sourmilk Gill and below Honister Pass. Generally the stones are angular, but occasionally more or less rounded. Stratification can generally be detected among these fragmental rocks, but it is apt to be concealed or effaced by the cleavage, while it is further obscured by that widespread induration on which Mr. Ward has laid so much stress. The extreme state of comminution of the volcanic dust that went to form the tuffs has probably caused them to be more liable to metamorphism than the lavas.[1]

Little has yet been done in identifying any of the vents from which the vast mass of volcanic material in the Lake District was ejected. Mr.

[1] The microscopic and chemical characters of the Ash-Slates of the Lake District have been investigated by Mr. Hutchings, *Geol. Mag.* 1892, pp. 155, 218.

Ward believed that the diabase boss forming the Castle Head of Keswick marks the site of "one of the main volcanic centres of this particular district,"[1] whence the great lava sheets to the southward flowed out. There are obviously two groups of bosses on the northern side of the district, some of which may possibly mark the position of vents. A few of them are occupied by more basic, others by more acid rocks. It is not necessary to suppose that the andesitic lavas ascended only from the former and the felsites from the latter. While the felsites on the whole are younger than the more basic lavas, they may have been erupted from vents which had previously emitted andesites, so that the present plug may represent only the later and more acid protrusions.

Besides the boss of Castle Head there are numerous smaller basic intrusions farther down the Derwent Valley on either side of Bassenthwaite Lake. Among these are the highly basic rocks forming the picrite on the east side of the Dash Beck and the dykes on Bassenthwaite Common. All these bosses, sills, and dykes rise through the Skiddaw Slates, but there is no positive proof that they belong to the Lower Silurian volcanic series; they may possibly be much later.

The most important and most interesting of all the intrusive masses of basic material is that which constitutes a large part of the eminence that culminates in Carrock Fell. The remarkable variations in the composition of this mass have been already referred to. Mr. Harker has shown that while the centre of the mass is a quartz-gabbro, it becomes progressively more basic towards the margin. Through the gabbro a mass of granophyre has subsequently made its way, and along the line of junction has incorporated into its own substance so much of the basic rock as to undergo a marked modification in its structure and composition. Whether these intruded bodies of basic and acid material have ascended in one of the old volcanic funnels and have been injected laterally in laccolitic fashion has not been ascertained. Mr. Harker, indeed, is rather inclined to refer the intrusions to a time not only later than the Borrowdale volcanoes, but later even than the terrestrial movements that subsequently affected the district and gave the rocks their present cleaved and faulted structures. Besides the gabbro and granophyre of this locality, igneous activity has manifested itself in the uprise of numerous later dykes and veins, intermediate to basic in composition. Some of these are glassy (tachylyte) and spherulitic or variolitic.[2]

Throughout the Lake District a considerable number of bosses of more acid rocks rise through the Skiddaw Slates, and likewise through the volcanic group even up to its highest members. Some of these bosses may possibly indicate the site of volcanic vents. Two of them, which form conspicuous features on either side of the Vale of St. John, consist of microgranite, and rise like great plugs through the Skiddaw Slates, as well as through the

[1] *Op. cit.* p. 70.

[2] Mr. Harker, *Quart. Journ. Geol. Soc.* vol. l. (1894) p. 312, li. (1895) p. 125. *Geol. Mag.* 1894, p. 551.

base of the volcanic group. The view of the more eastern hill, as seen from the west, is at once suggestive of a "neck." These masses measure roughly about a square mile each.

With the acid intrusions may possibly be associated some of the other masses of granophyre, microgranite and granite (felsite, felstone, quartz-felsite, syenitic granite, quartz-syenite, elvanite), which have long attracted attention in this region. The largest of these intrusions is the tract of granite which stretches from Eskdale down to near the sea-coast as a belt about eleven miles long and from one to three miles broad. Another large mass is the granophyre or "syenite" of Ennerdale. Numerous other intrusions of smaller dimensions have been mapped.

To what extent any of these eruptive masses were associated with the volcanic phenomena remains still to be worked out. There seems to be little doubt that a number of them must belong to a much later period. Mr. Harker has expressed his belief that the intrusion of some of these igneous rocks was intimately associated with the post-Silurian terrestrial movements of which cleavage is one of the memorials.[1] The Skiddaw granite, though it does not touch any part of the volcanic group, but is confined to the underlying Skiddaw Slates, was erupted after the cleavage of the district, which affects the volcanic as well as the sedimentary series. In other instances also, as in that of Carrock Fell, the intrusion seems to have been later than the disturbances of the crust.[2] The amount of metamorphism around some of the bosses of granite is considerable. That of the Skiddaw region has been well described by J. C. Ward,[3] while that of the volcanic group by the Shap granite has been carefully worked out by Mr. Harker and Mr. Marr.[4]

The Shap granite comes through the very highest member of the volcanic series, and even alters the Upper Silurian strata. It must thus be of much younger date than the volcanic history of the Lake District. It presents some features in common with the granite bosses of the south of Scotland. Like these, it is later than Upper Silurian and older than Lower Carboniferous or Upper Old Red Sandstone time. Its protrusion may thus have been coeval with the great volcanic eruptions of the period of the Lower Old Red Sandstone. It will accordingly be again referred to in a later chapter.

It must be confessed that none of the large bosses of massive rocks, whether diabases, gabbros, felsites, granophyres, or granites, appear to afford any satisfactory proof of the position of the vents which supplied the lavas and tuffs of the Lake District. Nor can such a decided accumulation of the volcanic materials in certain directions be established as to indicate the quarters where the centres of eruption should be sought. On the contrary, the confused commingling of materials, and the comparative shortness of the

[1] *Quart. Journ. Geol. Soc.* vol. li. (1895), p. 144. [2] *Op. cit.* p. 126.

[3] "Geology of Northern Part of the English Lake District," *Mem. Geol. Surv.* 1876, chap. iii. The metamorphism around the diorites and dolerites, and the granophyres and felsites, is described in the same chapter.

[4] *Quart. Journ. Geol. Soc.* xlvii. (1891) p. 266, xlix. (1893) p. 359.

outcrop of the several sheets which have been traced, rather suggest that if any one great central volcano existed, its site must lie outside of the present volcanic district, or more probably, that many scattered vents threw out their lavas and ashes over no very wide area, but near enough to each other to allow their ejected materials to meet and mingle. The scene may have been rather of the type of the Phlegræan fields than of Etna and Vesuvius. If this surmise be true, we may expect yet to recognize little necks scattered over the volcanic district and marking the positions of some of these vanished cones.

What appears to have been one of these small vents stands near Grange at the mouth of Borrowdale, where I came upon it in 1890. In the little Comb Beck, the Skiddaw Slates are pierced by a mass of extremely coarse agglomerate, forming a rudely-circular boss. The slates are greatly disturbed along the edges of the boss, so much so, indeed, that it is in some places difficult to draw a line between them and the material of the agglomerate. That material is made up of angular blocks, varying in size up to three feet long, stuck in every position and angle in an intensely-indurated matrix formed apparently of comminuted debris like the stones. The blocks consist of a finely-stratified shale, which is now hardened into a kind of hornstone, with some felsitic fragments. I could see no slags or bombs of any kind. There is no trace of cleavage among the blocks, nor is the matrix itself sensibly cleaved. I believe this to be a small volcanic neck and not a "crush-conglomerate." It has been blown through the Skiddaw Slates, and is now filled up with the debris of these slates. Its formation seems to have taken place before the cleavage of the strata, and its firm position and great induration enabled it to resist the cleavage which has so powerfully affected the slates and many members of the volcanic group.

It was the opinion of my predecessor, Sir Andrew Ramsay, and likewise of Mr. Ward, that the Cumbrian volcanic action was mainly subaerial. This opinion was founded chiefly on the fact that, save at the bottom and top of the series, there is no evidence of any interstratified sediment of non-volcanic kind. The absence of such interstratification may undoubtedly furnish a presumption in favour of this view, but, of course, it is by no means a proof. Better evidence is furnished by the unconformability already mentioned between the Coniston Limestone and the lavas on which it lies. Besides angular pieces of lava, probably derived from direct volcanic explosion, this limestone contains fragments of amygdaloidal andesite, and also rolled crystals of striated felspar.[1] These ingredients seem to indicate that some part of the volcanic group was above water when the Coniston Limestone was deposited.

The absence of interstratifications of ordinary non-volcanic sediment in the Borrowdale group might conceivably arise from the eruptions following each other so continuously on the sea-floor, and at so great a distance from land that no deposition of sand or mud from the outside could sensibly affect the accumulation of volcanic material. Certainly some miles to the

[1] Messrs. Harker and Marr, *Quart. Journ. Geol. Soc.* vol. xlvii. (1891), p. 310.

east at the Cross Fell inlier, as already mentioned, there is evidence of the alternation of tuffs with the sandy and muddy sediment of the sea-bottom. Here, at the outer confines of the volcanic district, the ejected materials evidently fell on the sea-floor, mingled there with ordinary sediment, and enclosed the same organic remains. The well-defined stratification of many of the fine tuffs is rather suggestive to my mind of subaqueous than of subaerial accumulation. At the same time, there seems no reason why, here and there at least, the volcanic cones should not have risen above the water, though their materials would be washed down and spread out by the waves.

One of the most marked points of contrast between the Cumbrian and the Welsh volcanic districts is to be found in the great paucity of sills in the former region. A few sheets of diorite and diabase have been mapped, especially in the lower parts of the volcanic group and in the underlying Skiddaw Slates. On the other hand, dykes are in some parts of the district not unfrequent, and certainly play a much more prominent part here than they do in the Welsh volcanic districts. The majority of them consist of felsites, quartz-porphyries, diorites, and mica-traps. But there is reason to suspect that where they are crowded together near the granite, as around Shap Fells, they ought to be connected with the uprise of the post-Silurian granitic magma rather than with the history of the volcanic group.[1] If this series of dykes be eliminated, there remain comparatively few that can with any confidence be associated with the eruption of the Borrowdale rocks.

vii. UPPER SILURIAN (?) VOLCANOES OF GLOUCESTERSHIRE

A remarkable group of igneous materials has long been known to rise among the Silurian rocks of the Tortworth district at the north end of the Bristol coal-field. They were believed to be aqueous deposits in the Wernerian sense by Weaver.[2] Murchison regarded them as intrusive sheets;[3] Phillips looked on them as partly intrusive and partly interstratified.[4] They consist largely of coarsely-amygdaloidal basalts, some of which have been microscopically examined.[5] But their field-relations as well as their petrography have not yet been adequately determined. They are represented on the Geological Survey Map as forming a number of parallel bands in strata classed as Upper Llandovery. If, as seems probable, some of them are really interstratified, they form the youngest group of Silurian volcanic rocks in England, Scotland, or Wales.

[1] For a description of the dykes around the Shap granite see the paper by Messrs. Harker and Marr, *Quart. Journ. Geol. Soc.* vol. xlvii. (1891), p. 285.

[2] *Trans. Geol. Soc.* 2nd ser. vol. i. (1819), pp. 324-334.

[3] *Silurian System* (1839), p. 457.

[4] *Mem. Geol. Surv.* vol. ii. part i. (1848), p. 194.

[5] "Geology of East Somerset," etc., in *Mem. Geol. Surv.* (1876), p. 210; descriptions by Mr. F. Rutley.

CHAPTER XIV

THE SILURIAN VOLCANOES OF IRELAND

ABUNDANT as are the volcanic records of the Silurian period in England, Wales and Scotland, the description of them would be incomplete without an account of those of Ireland. The eruptions of Arenig, Llandeilo and Bala time, which we have followed from the south of Caermarthenshire to the borders of the Scottish Highlands, had their counterparts all down the east of Ireland. The Irish register of them, however, supplies some details which are less clearly preserved in the sister island. But the most distinctive feature of the Silurian volcanic history in Ireland is the preservation of memorials of eruptions during the Upper Silurian period. In no part of Great Britain has any unquestionable trace been found of volcanic activity during that part of the geological record, the last eruptions of which the age is known being those of the Bala rocks. But in the south-west of Ireland there is evidence that for a time active vents appeared over the sea-floor on which the earlier deposits of Upper Silurian time were laid down.

I. THE LOWER SILURIAN SERIES

i. *Eruptions probably of Arenig Age*

It is in that part of Ireland which lies east of a line drawn from Strabane to Dungarvan Harbour that the records of Lower Silurian volcanic activity are to be found. In the north the development of volcanic rocks resembles that in Scotland, in the south it corresponds rather with the volcanic districts of Wales.

The Irish Silurian volcanic rocks have been traced with more or less detail on the maps of the Geological Survey. Since these maps were published, however, great advances have been made in the study of the petrography of volcanic rocks, as well as in the art of tracing their structure upon maps. Much, therefore, now remains to be done to bring our knowledge of the older volcanic history of Ireland abreast of that of the rest of the British Isles. In the following summary I have had to rely mainly on my own traverses of the ground, guided by the maps and memoirs of the Survey, and with the personal assistance of some of my colleagues.

The remarkable zone of crushed cherts, igneous rocks and sandstones, probably of Lower Silurian age, which I have referred to (p. 201) as wedged in between the schists and the Old Red Sandstone along the southern margin of the Highlands of Scotland, reappears in Ireland. It occupies an area in the County Tyrone, about 24 miles long and about 9 miles broad at the broadest part, but disappearing towards the north-east and south-west.[1] Lying between the Palæozoic formations on the south and the schists on the north, it occupies a similar position to the Scottish belt, but presents a much broader area, and thus affords greater facilities for examining the rocks. It presents the same indefinite or faulted boundaries as in Scotland, so that its relations to the rocks along its flanks have not been satisfactorily determined. That the rocks of this area are older than the Silurian strata to the south of them seems to be established by the occurrence of fragments of them in these strata, and that they are younger than the schists may be inferred from their non-foliated character. But they have undoubtedly undergone considerable crushing by powerful terrestrial movements which have placed them in their present position.

The special feature of interest in this Irish area is the remarkable development of volcanic materials which is there to be seen, spreading over a far wider area than in Scotland. The rocks include lavas associated with tuffs and agglomerates, likewise a varied series of intrusive masses.

The lavas are chiefly dull greenish, fine-grained rocks, having the general character of diabases and "porphyrites." They are sometimes quite slaggy, and where the amygdaloidal kernels remain, these are usually of calcite. Under the microscope, the diabases show in some parts that their lath-shaped felspars, and the augite which these penetrate, are tolerably fresh, while in other parts fibrous chlorite, granular epidote and veins of calcite bear witness to the metamorphism which they have undergone.

One of the most conspicuous features in some of these lavas is the occurrence of the same sack-like or pillow-shaped structure which has been already referred to as so marked among the Arenig lavas of Scotland. Though the vesicles of these rocks are often quite uncrushed, showing that there has been no general subsequent deformation of the whole mass, there occur local tracts where evidence of considerable movement may be noticed. Thus close to a mass of gneiss, and elsewhere along their margin, the lavas are apt to be much jointed and broken with numerous lines of shear, along which the crushed material assumes more or less of a schistose structure. Yet in the solid cores between these bands of crushing the original forms of the vesicles are retained.

These greenish lavas are occasionally interleaved with grey flinty mudstones, cherts and red jaspers, which are more particularly developed immediately above. In lithological character, and in their relation to the

[1] This area was mapped by Mr. J. Nolan for the Geological Survey, and was described by him in the *Geol. Mag.* for 1879. I visited it in company with my colleagues, Mr. B. N. Peach and Mr. A. M'Henry, in 1890 and again in 1894. My first conclusion was that the volcanic rocks should be regarded as part of the schistose series lying to the north of them (*Pres. Address Geol. Soc.* 1891, p. 77). But on the second visit, after having studied the rocks of the border of the Scottish Highlands, I formed the opinion stated in the text.

diabases, these siliceous bands bear the closest resemblance to those of Arenig age in Scotland. But no recognizable Radiolaria have yet been detected in them.

Besides the more basic lavas, there occur also, but less abundantly, platy felsitic rocks which have suffered much from shearing, and consequently have acquired a fissile slaty structure.

The agglomerates are made up of angular, subangular and rounded fragments imbedded in a matrix of similar composition. This matrix has in places become quite schistose, and then closely resembles some parts of the "green schists" of the Scottish Highlands. Of the inclosed stones the great majority consist of various felsites, which, weathering with a thick white opaque crust, are internally close-grained, dull-grey or even black, sometimes showing flow-structure, and of all sizes up to eight inches in diameter or more. There are also fragments of the basic lavas, and likewise pieces of chert and jasper. On many of the rocky hummocks no distinct bedding can be made out in the agglomerate, but in others the rock is tolerably well stratified.

The tuffs are fine silky schistose rocks, and seem to have been largely derived from basic lavas. They have suffered more than any of the other rocks from mechanical deformation, for they pass into green chloritic schists. Some portions of them are not unlike the slaty tuffs of Llyn Padarn in Caernarvonshire.

Accompanying the fragmental volcanic rocks, some ordinary sedimentary intercalations are to be found—red shales and pebbly quartzites, that seem to have escaped much crushing. The true order of succession in the volcanic series has not yet been determined. But apparently above this series come some dark shales, such as might yield graptolites, pale grits and occasional limestones.

Later than the lavas and the pyroclastic material are various intrusive masses, which in bands and bosses form numerous craggy hills throughout the area. So far as I have been able to observe, these rocks include two groups. Of these the older consists of basic injections, such as gabbros and allied rocks, some of which remind me of the so-called "hypersthene-rock" of Lendalfoot, in Ayrshire. The coarser varieties, as at Carrickmore or Termon rock, are sometimes traversed by fine-grained veins from an inch to several feet in breadth. Portions of the slaggy diabases may be observed inclosed in these intrusive masses. The younger group is of more acid composition (granite, quartz-porphyry, etc.), and sends veins into the older.

ii. *Eruptions of Llandeilo and Bala Age*

Into the east of Ireland the Lower Silurian rocks are prolonged from Scotland, from the Lake District and from Wales. Though greatly concealed under younger formations across the breadth of the island, and occasionally interrupted by what are regarded as older strata of Cambrian age, they nevertheless occupy by much the larger part of the maritime counties from Belfast Lough to the southern coast-line of Waterford, even as far as Dungarvan Harbour. With the same lithological types of sedimentary deposits as in other parts of the United Kingdom, they carry with them here also their characteristic records of contemporaneous volcanic action. Though nowhere piled into such magnificent mountain-masses as in Westmoreland and North Wales, these records become increasingly abundant and interesting as they are traced southwards, until they are abruptly terminated by the coast-line along the south of the counties of Wexford and Waterford.

While much remains to be done, both in the field and in the laboratory and microscope-room, before our acquaintance with the Irish Silurian volcanic rocks is as complete as our knowledge of their equivalents in other portions of the United Kingdom, a serious preliminary difficulty must be recognized in the fact that the several geological horizons of these rocks have only been approximately fixed. Great difficulty was experienced by the Geological Survey in drawing any satisfactory line between the Llandeilo and Bala formations. This arose not so much from deficiency of fossil evidence as from the way in which the fossils of each group seemed to occur in alternating bands in what were regarded as a continuous series of strata. Indeed, in some localities it almost appeared as if the occurrence of one or other *facies* of fossils depended mainly on lithological characters indicative of original conditions of deposit, for the Llandeilo forms recurred where black shales set in, while Bala forms made their reappearance where calcareous and gritty strata predominated.[1] More recent work among the Silurian formations in England and Scotland, however, indicates that the parallel repetition of the two types of fossils is due to rapid and constant plication of the rocks, whereby the two formations, neither of them, perhaps, of great thickness, have been folded with each other in such a way that without the evidence of an established sequence of fossils, or the aid of continuous sections, it becomes extremely difficult to make out the stratigraphical order in any district. When the ground is attacked anew in detail, with the assistance of such palæontological and lithological horizons as have permitted the complicated structure of the southern uplands of Scotland to be unravelled, we may be enabled to tabulate the successive phases of the volcanic history of the region in a way which is for the present impossible. We have as yet no palæontological evidence that in the Silurian region of the east of Ireland, which extends from Belfast Lough to the south coast of County Waterford, any of the anticlinal folds bring up to the surface a portion of the Lower Arenig formation, though possibly some of the lowest visible strata may be of Upper Arenig age. A considerable part of the region must be referred to the Llandovery and other Upper Silurian formations, but the precise limits of the two divisions of the Silurian system have not yet been determined, except for the region north of Dublin, which has recently been re-examined for the Geological Survey by Mr. F. W. Egan and Mr. A. M'Henry.

[1] Jukes was disposed to regard the two faunas as essentially coeval, but inhabiting different kinds of sea-bottom. See his note, Explanation of Sheets 167, 168, 178, 179, p. 30.

These observers have ascertained that, as in Southern Scotland, by far the larger part of the Silurian region of the north-east of Ireland is occupied by strata belonging to the upper division of the system. The Lower Silurian formations, including the Llandeilo and Bala groups, form a belt varying up to six miles in breadth, which stretches from the coast of Down, between the mouth of Belfast Lough and Copeland Island, in a south-westerly direction to near the valley of the Shannon in County Longford. South of this belt the Lower Silurian rocks rise to the surface only here and there on the crests of anticlinal folds, and it is in these scattered "inliers" that the volcanic and intrusive rocks are found. So far as the available evidence goes, the volcanic history of this part of Ireland is entirely to be assigned to Lower Silurian time, and more especially to the interval between the beginning of the Llandeilo and the close of the Bala period. I must for the present content myself with this general limit of geological chronology, and make no attempt to trace the relative antiquity of the igneous rocks in the several districts in which they are distributed.[1]

Viewing the volcanic region of Eastern Ireland as a whole, we are first struck by the feebleness of the manifestations of eruptivity in the north, and their increasing development as we advance southwards. At the northern end of the Silurian area in County Down, thin bands of "felstone" and "ash" have been mapped by the Geological Survey as interstratified with the highly inclined and plicated Silurian rocks.[2] As the latter are plainly a continuation of the strata which have been mapped out zone by zone in the south of Scotland, their igneous intercalations may be looked upon as probably equivalents of some of those in the Silurian districts of Wigtonshire and Kirkcudbrightshire. But in County Down no representative has yet been detected of the Arenig and Llandeilo volcanic series of the southern uplands of Scotland. Nor has more precise petrographical examination confirmed the reference of any of the igneous rocks in the Silurian area of that district to truly contemporaneously intercalated volcanic rocks. All the eruptive material appears to be of an intrusive character. It occurs in the form of dykes of lamprophyre or mica-trap belonging to the groups of minettes and kersantites. Nothing definite is known of the age of these intrusions: they are possibly referable to the time of the Lower Old Red Sandstone.[3]

Far in the interior several bands of "felspathic ash" and "massive agglomerate" are shown on the Survey map as running through the counties of Monaghan and Cavan.[4] In one locality south of the Drumcalpin Loughs a large exposure of this ash is visible: "brown crumbly beds, with small

[1] The task of revising the Irish maps and tracing out the respective areas of Upper and Lower Silurian rocks over the whole island is now in progress by the Geological Survey, Mr. Egan and Mr. M'Henry being entirely engaged on it.

[2] See Sheet 49 Geol. Survey, Ireland, and Explanation thereto (1871), pp. 16, 37, 39. The so-called "ashes" of the Explanation are probably parts of dykes which have been more or less crushed.

[3] *Guide to the Collection of Rocks and Fossils belonging to the Geological Survey of Ireland*, by Messrs. M'Henry and Watts, Dublin, 1895, p. 74.

[4] Sheet 69 Geol. Survey, Ireland, and Explanation of Sheets 68 and 69, pp. 9, 13, 15.

rounded pebbles, give place to a massive bed of agglomerate, the enclosed blocks of which are always of one species of felstone, sometimes measuring 10 × 12 × 18 inches, and not always rounded." South of Carrickatee Lough, and a few miles farther to the south-west, near Lackan Bridge, considerable exposures of these rocks occur. One crag in particular displays a thickness of more than 70 feet of "tough flaky breccias," "thick agglomerates with small and large blocks of felstone," and "thin beds of fine pale green compact grit without pebbles, and a few flags." "One of the flaky beds contains numerous white worn crystals of felspar"; "the imbedded blocks of felstone are of the usual kind—pale compact matrix showing dark oblong patches, vesicular and amygdaloidal, the cavities being filled with chlorite."

Further south a more extensive area of igneous rocks has been mapped on the borders of Louth and Meath, where, according to the Geological Survey map, a group of lavas and tuffs extends for about twelve miles near Slane.[1] Other bands of "ash" and "felstone" have been mapped in the Silurian area south of Drogheda. Thus at Hilltown, west from the race-course, a "bluish crystalline felstone, showing in places lines of viscous flow," is stated to be overlain by "indurated felspathic ash and tuff, felstone, and indurated shale" in alternating beds.[2] On a recent visit to this locality I found that the "porcellanite or indurated shale" is a greenish-grey chert, full of Radiolaria and finely-diffused volcanic dust. This association of radiolarian chert with contemporaneous volcanic activity is of much interest, as showing the extension of the same physical conditions of the Lower Silurian sea from Scotland into Ireland. The Lower Llandeilo age of the volcanic intercalations in County Meath is further indicated by the occurrence of *Didymograptus Murchisoni* in grey shales in the same neighbourhood with the radiolarian cherts. In the Lower Silurian district of Balbriggan numerous intrusive bosses and sills have been mapped by the Geological Survey. I have found, however, that among these rocks there occur bands of volcanic breccia, containing abundant angular fragments of a minutely-vesicular pumice, and also that some of the diabase-masses display the pillow-structure and amygdaloidal texture. Hence, though most of the igneous rocks are no doubt intrusive, they appear to include lavas and tuffs of Bala age.

When the numerous Silurian cores of the mountain-groups in the interior of Ireland shall have been searched for traces of contemporaneous volcanic action, it is not improbable that these will be found. One of the smaller Silurian inliers which diversify the great Carboniferous plain, that

[1] *Ibid.* Sheets 81 and 91. These rocks are chiefly augitic andesites, a few are basalts, and some seem related to felstones. Probably many of them are intrusive sills of uncertain age. The "ashes" contain fragments of felsite and porphyrite often of considerable size (*Guide to Irish Rock-Collection*, p. 36).

[2] *Ibid.* Sheets 91 and 92 and Explanation to these Sheets (1871), p. 10; *Guide to Irish Rock-Collection*, p. 36. Some of these lavas are andesites, others are felsites. Mr. M'Henry has contended that certain "ashes" and "agglomerates," particularly those exposed on the coast at Portraine, opposite Lambay Island, are "crush-conglomerates" due to terrestrial disturbances, which have affected both intrusive igneous rocks and the sedimentary series into which these have been injected.

of the Chair of Kildare, has long been known to have igneous rocks associated with its abundantly fossiliferous Bala limestone.[1] On recently visiting this locality I found that, besides the amygdaloidal and porphyritic andesites and basalts described by Jukes and Du Noyer, the fossiliferous conglomerates contain pebbles of rocks like those of the Chair, together with worn crystals of felspar, while intercalated with them are thin courses of volcanic tuff. There is thus evidence here of contemporaneous volcanic activity during the accumulation of the Bala group of strata. The limited area over which the rocks are exposed, however, affords merely a glimpse of this volcanic centre.

Crossing over the broad belt of Carboniferous Limestone through which the Liffey flows into Dublin Bay, we come to the great continuous tract of older Palæozoic rocks which stretches southward to the cliffs of Waterford. Through this tract runs the huge ridge of the Wicklow and Carlow granite. On the west side of this intrusive mass, bands of "greenstone-ash," as well as "felspathic ashes," have been traced among the Silurian rocks by the Geological Survey. But it is on the south-east side of the granite that the volcanic intercalations are best displayed. Indeed, from Wicklow Head to Dungarvan Harbour there is an almost continuous development of igneous rocks, rising into rocky eminences, trenched into ravines by the numerous streams, and laid bare by the waves in fine coast-cliffs. It is in this south-eastern region, comprising the counties of Wicklow, Wexford and Waterford, that the Irish Lower Silurian igneous rocks can best be studied.

There are obviously various distinct centres of eruption in this long belt of country. The Rathdrum and Castletimon tract forms one of these. Another of less size culminates in Kilpatrick Hill, a few miles to the southward. Arklow Head marks the position of a third. The lavas and tuffs which set in a few miles to the south of that promontory, and may be said to extend without interruption to the south coast, were probably thrown out by a series of vents which, placed along a north-east and south-west line, united their ejections into one long submarine volcanic bank. There can be no doubt that the most active vents lay at the southern end of the belt, for there the volcanic materials are piled up in thickest mass, and succeed each other with comparatively trifling intercalations of ordinary sedimentary material. Some of these vents, as I shall relate in the sequel, have been cut open by the sea along a range of precipitous cliffs.

The comparatively feeble character of the volcanic energy during Lower Silurian time over the greater part of the south-east of Ireland is shown by the great contrast between the thickness of the volcanic intercalations there and in Wales and the Lake country, but still more strikingly by innumerable sections where thin interstratifications of fine tuff or volcanic breccia occur among the ordinary sedimentary strata, and are sometimes crowded with Bala fossils. Some interesting illustrations of this feature are to be seen in the Enniscorthy district, where layers of fine felsitic tuff, sometimes less than

[1] See Explanation to Quarter Sheet 35 N.E. (Sheet 119 of newer numeration) of Geol. Survey Ireland (1858), p. 16. (See note, p. 256.)

an inch in thickness, lie among the shales. In some of the tuffs the lapilli are fragments of trachytic or andesitic rocks.

A striking example of rapid alternations of pyroclastic material with ordinary sediment lies far to the south in County Waterford, close to Dunhill Bridge, where a group of fine volcanic breccias and grits has been laid bare by quarrying.[1] These strata consist of coarser and finer detritus, enclosing angular fragments of felsites and grey and black shale. The felsite-lapilli vary in texture, some of them presenting beautiful flow-structure. The stones are stuck at random through each bed, the largest being often at the bottom. The beds of breccia vary from a few inches to a foot or more in thickness. There can, I think, be little doubt that each of these breccia-bands points to a single volcanic explosion, whereby felsitic fragments were thrown out, mingled with pieces of the Silurian strata through which the vents were drilled. In a vertical thickness of some fifty feet of rock there must thus be a record of ten or twelve such explosions.

Nearer the active vents the fragmental deposits become, as usual, coarser and thicker. But I have not observed any thick masses of tuff like those of North Wales. So far as my examination has gone, the tuffs are mainly felsitic. The so-called "greenstone-ash" of the Survey maps is certainly in many cases not a true tuff. This term was proposed by Jukes for certain apple-green to olive-brown flaky fissile rocks only found "in association with masses of greenstone."[2] Some years ago I had occasion to make a series of traverses in Wicklow and Wexford, and then convinced myself that in that part of the country the "greenstone-ashes" were probably crushed bands of basic sills. Dr. Hatch has proved this to be their origin from a series of microscopic slides prepared from specimens collected by himself on the ground.[3] In other cases the "greenstone-ashes" seem to be excessively-cleaved or sheared felsites, which have acquired a soapy feel and a dull green colour; but they also do include true tuffs. Thus, in one instance, at Ballyvoyle cross-roads, in the south of County Waterford, a "greenstone-ash" is a dull green tuff full of fragments of felspar (chiefly plagioclase) and pieces of dark andesitic lavas. Another example may be found to the west of the Metal Man, near Tramore, where the tuff is full of fragments of felspar and shale cemented in a greenish-yellow material which may be palagonite.

The felsites of the south-east of Ireland form by much the largest proportion of the whole volcanic series. They occur as lenticular sheets from a few feet to several hundred feet in thickness, and occasionally traceable for some miles. On the whole, they are compact dull grey rocks, weathering with a white crust. A geologist familiar with the contemporary lavas of North Wales cannot fail to be struck with the absence of the coarse flow-structure so often characteristic of the felsites in that region. This structure, indeed, is not entirely absent from the Irish rocks, but it occurs, so far, at least, as I have seen, rather as a fine streakiness than in the bold

[1] See Explanation of Sheets 167, 168, 178 and 179, Geol. Surv. Ireland, p. 56.

[2] Explanation of Sheets 129, 130, p. 13 (1869).

[3] Explanation of Sheets 138, 139.

lenticular bands so common in Caernarvonshire. In like manner the nodular structure, though not entirely absent, is rare.[1]

Until these felsites have been subjected to more detailed investigation, little can be said as to their petrography, and as to the points of resemblance or difference between them and those of other Lower Silurian districts in the United Kingdom. An important step, however, in this direction was taken by Dr. Hatch, who studied them on the ground, in the laboratory, and with the microscope. He found that some of them were soda-felsites or keratophyres (with albite as their felspar), that others were potash-felsites (with orthoclase as their felspar), while a third group contained both soda and potash, the last-named greatly preponderating.[2] The existence of soda-felsites had not been previously detected among British volcanic rocks, and it remains to be seen how far they may occur in the large and somewhat varied group of rocks combined under the general term "felsites." Dr. Hatch believed that these rocks probably graduate into the normal or orthoclase felsites; but it has not yet been possible to test this view on the ground, nor to ascertain whether there is any essential difference between the mode of occurrence of the two types.

Besides the more abundant felsites, occasional bands of andesite have been detected. Various other eruptive rocks occur, probably in most or all cases intrusive. Such are quartz-mica-diorites, quartz-diorites, augite-diorites or proterobases, dolerites, gabbros, diabases and epidiorites.[3]

I have said that the chief theatre of eruption lay towards the south-west end of the volcanic belt of the south-east of Ireland. The coast-line of County Waterford, from Tramore westward to Ballyvoyle Head—a distance of nearly fifteen miles—presents, perhaps, the most wonderful series of sections of volcanic vents within the British Islands. No one coming from the inland is prepared for either the striking character of the cliff scenery or the extraordinary geological structure there presented, for the country is, on the whole, rather featureless, and much of it is smoothed over and obscured by a covering of drift, through which occasional knobs of the harder felsites protrude. The cliffs for mile after mile range from 100 to 150 or 200 feet in height, and present naked vertical walls of rock, trenched by occasional gullies, through which a descent may be made to the beach. Throughout the whole distance agglomerates and felsites succeed each other in bewildering confusion, varied here and there by the intercalation of Lower Silurian shales and limestones involved and pierced by the igneous rocks. Hardly any bedded volcanic material is to be recognized from one end to the other. The sea has laid bare a succession of volcanic vents placed so close to each other that it will be difficult or impossible to separate them out. A careful study and detailed mapping of this marvellous coast-section, however, is a task well worthy of the labour of any one desirous

[1] In Waterford nodular felsites occur with concretions varying from the size of a pea to several inches in diameter. Explanation to Sheets 167, 168, 178 and 179, p. 11.

[2] Explanation of Sheets 138, 139, p. 49; and *Geol. Mag.* 1889, p. 545.

[3] *Guide to Irish Rock-Collections*, pp. 34, 35.

of making himself acquainted with some of the conditions of volcanism during older Palæozoic time.

At the east end of the section, black shales containing Llandeilo graptolites, and calcareous bands full of Bala fossils, dip westward below a group of soda-felsites and felsitic tuffs, which seem to lie quite conformably on these strata. Here, then, we start with proof that the volcanic eruptions of this locality began during some part of the Bala period. But immediately to the west, these bedded igneous rocks are broken through by a neck of coarse agglomerate stuck full of chips and blocks of shale, some of them a foot long, with abundant fragments of scoriform and flinty felsites. Some columnar dykes of dolerite cut through the neck, and a larger intrusion seems to have risen up the same funnel. The bedded tuffs appear again for a short distance, but they are soon replaced by a tumultuous mass of agglomerate. And from this part of the coast onwards for some distance all is disorder.

The agglomerates are crowded with blocks of various felsites and micro-granites sometimes 18 inches in diameter, many of them presenting the most exquisite streaky flow-structure. The angularity of these stones and the abrupt truncation of their lines of flow prove that they were derived from the shattering of already consolidated rocks. In other places the ejected materials consist almost wholly of black shale fragments, but with an intermixture of felsite-lapilli.

It is difficult to convey an adequate idea of the way in which the agglomerates are traversed by dykes, veins and bosses of various felsites, and of how these break in endless confusion through each other. Some of the intrusive rocks are compact and amorphous, others are vesicular, others close-grained and columnar. Again and again they present the most perfect flow-structure, and it is noticeable that the lines of flow follow the inequalities of the walls of the fissure up which the rock has ascended, and not only so, but even of the surfaces of detached blocks of shale or felsite which have been caught up and enclosed in the still moving mass.

A few of these intrusive rocks were examined in thin slices by Dr. Hatch. Most of them appear to be soda-felsites, but they include also rather decomposed rocks, some of which are probably diorites and quartz-diorites. Occasionally, thoroughly basic dykes (dolerite) may be observed.

In the midst of this tumultuous assemblage of volcanic masses, representing the roots of a group of ancient vents, there occur occasional interspaces occupied by ordinary stratified rocks. In the eastern part of the section these consist mainly of black shale, sometimes with calcareous bands, from which a series of Bala fossils has been obtained.[1] A very cursory examination suffices to show that these intercalations do not mark pauses in the volcanic eruptions. They are, in fact, portions of the marine accumulations under the sea-floor through which the vents were blown;

[1] But see the *Geol. Survey Memoir* on Sheets 167, 168, 178 and 179, Ireland (1865), p. 28, for a description of the association of Bala and Llandeilo fossils on that coast-line.

they have been tossed about, crushed and invaded by dykes and veins of felsite.

But certain other intercalated strips of stratified rocks present a special interest, for they bring before us examples of volcanic ashes that gathered on the sea-floor, but which were disrupted by later explosions. Thus, at the Knockmahon headland, well-bedded felspathic grits and ashy shales occur, thrown in among the general mass of eruptive material. As I have already remarked, it is difficult or impossible to fix the horizons of the stratified patches that are involved among the igneous ejections of this coast-section, save where they contain recognizable fossils, but the intercalation of true bedded tuffs among them is a proof that volcanic action had been in operation there long before the outbreak of the vents which are now laid bare along the cliffs.

In the south-east of Ireland there is the usual association of acid and basic sills with the evidence of a superficial outpouring of lavas and ashes. But these intrusive masses play a much less imposing part than in Wales. They may be regarded, indeed, as bearing somewhat the same proportion to the comparatively feeble display of extrusive rocks in this region that the abundant and massive sheets of Merionethshire and Caernarvonshire do to the enormous piles of lavas and tuffs which overlie them.

Among the acid intrusive sheets the most conspicuous are those mapped by the Survey as "elvans." These rocks, as they occur in Wicklow and Wexford, have been examined by Dr. Hatch, who finds them to be micro-granitic in structure, occasionally exhibiting micropegmatitic or granophyric modifications.[1] The true stratigraphical relations of these rocks have not yet been adequately investigated. Those of them which occur on the flanks of the great granite ridge are not improbably connected with that mass, and if so are much younger than the Lower Silurian volcanoes.[2]

The basic sills, or "greenstones," consist largely of diabase, frequently altered into epidiorite; they include also varieties of diorite.[3] That they were intruded before the plication and cleavage of the rocks among which they lie is well shown by their crushed and sheared margins where they are in thick mass, and by their cleaved and almost schistose condition where they are thinner. The intense compression and crushing to which they have been subjected are well shown by the state of their component minerals, and notably by the paramorphism of the original augite into hornblende.

The scarcity of dykes associated with Silurian volcanic action is as noticeable in the south-east of Ireland as it is in Wales. I have observed a considerable number, indeed, but they are confined to the line of old vents on the Waterford coast, and, but for the clear cliff-sections cut by the sea, they would certainly have escaped observation, for they make no

[1] Explanation of Sheets 138 and 139, p. 53.

[2] The Leinster granite is certainly later than the Lower Silurian rocks and older than the Carboniferous rocks of the south-east of Ireland. It may belong to the great epoch of granite protrusion during the Old Red Sandstone period.

[3] Dr. Hatch, *op. cit.* p. 49.

feature on the ground in the interior. They are sometimes distinctly columnar, and vary from less than a foot to many yards in width. They traverse both the agglomerates and the intrusive felsites. Most of them are of felsite, sometimes cellular; but in some cases they are dolerites. There is obviously no clue to the dates of these dykes.

That some at least of the vents along the south coast of County Waterford may be vastly younger than the Lower Silurian rocks through which they have forced their way is suggested, if not proved, by a section which is in some respects the most extraordinary of the whole of this remarkable series. The occurrence of a group of red strata was carefully noted by the late Mr. Du Noyer at Ballydouane Bay, when he was engaged in carrying on the Geological Survey of that part of the country. At first he regarded them as belonging to the Old Red Sandstone, which comes on in great force only a few miles to the west; but he subsequently arrived at the belief that they are really an integral part of the Lower Silurian rocks of the district. Professor Jukes had previously expressed himself in favour of this latter idea, which was thought to receive support from the occurrence of some reddish strata in the Lower Silurian rocks of Tagoat, County Wexford.[1]

The occurrence of red rocks among Silurian strata, which are not usually red, might quite reasonably be looked for in the neighbourhood of Old Red Sandstone, Permian or Triassic deposits. If these deposits once spread over the Silurian formations, a more or less decided "raddling" of the latter may have taken place. But in the present instance, though the Old Red Sandstone begins not many miles to the west, no such explanation of the colour of the strata is possible. The cliffs of Ballydouane Bay consist of red sandstone, red sandy shale and conglomerate. The red tint is of that dull chocolate tone so characteristic of the Lower Old Red Sandstone. The conglomerates are immense accumulations of ancient shingle, consisting largely of pieces of white vein-quartz and quartzite, sometimes a foot long and often well water-worn. Some of the sandy beds are full of large scales of white mica, as if derived from some granitic or schistose region at no great distance. Taken as a whole, the strata are much less indurated and broken than the Silurian grits and shales of the district; some of them, indeed, weather into mere incoherent sand that crumbles under the fingers. There does not appear to be any positive proof that the red rocks are truly bedded with the ordinary Silurian strata, the junctions being faulted or obscured by intrusive igneous masses.

Nowhere in the British Islands, so far as I am aware, is there a similar group of strata among the Lower Silurian rocks. If they belong to so ancient a series, they show that in the south of Ireland, during Lower Silurian time, there arose a set of peculiar physical conditions precisely like those that determined the accumulation of the Old Red Sandstone in the same region at a later geological period. And in that case it is hardly

[1] Explanation of Sheets 167, 168, 178 and 179 of the Geological Survey of Ireland (1865), pp. 10, 59.

possible to conceive that these conditions could have been confined to the extreme south of Ireland. We should certainly expect to meet with evidence of them elsewhere, at least in the same Silurian region.[1]

While I hesitate to express a decided opinion in opposition to the conclusions of such experienced observers as Jukes and Du Noyer, I incline to believe that the rocks in question really belong to the Old Red Sandstone. If such shall finally be determined to be their geological position, they will supply evidence that some at least of the volcanic vents of the coast-line cannot be older than the Old Red Sandstone. They are pierced by masses of soda-felsite and by a coarse red agglomerate containing abundant pieces of felsite. These volcanic rocks belong to the same type as those which break through the undoubted Silurian rocks on either side. They may thus come to prove a recrudescence of volcanic energy in this same district at a much later geological period; and a new problem will arise to task the skill of the most accomplished field-geologist and petrographer—to unravel the structure and history of this chain of volcanic vents, and, in so doing, to detect and separate the eruptions of Lower Silurian time from those of the Lower Old Red Sandstone.

In the far west of Ireland, another group of Lower Silurian volcanoes has left its remains in the mountainous tract of country between the western shores of Lough Mask and Killary Harbour.[2] There appear to have been at least three separate centres of eruption along a line stretching in a north-easterly direction for about 16 miles from the western end of Lough Nafooey to the hamlet of Derrindaffdery beyond Tourmakeady, where the older rocks are unconformably overlain by the lower Carboniferous strata. As shown by the mapping of the Geological Survey, the most northerly area, which may be called the Tourmakeady centre, has a breadth of about a mile, and dies out southward after a course of nearly six miles. About a mile to the south-west of the last visible prolongation of its rocks, we encounter a second volcanic centre which occupies an area of about a square mile in the valley of Glensaul. The third centre stretches from the western shores of Lough Mask across Lough Nafooey, where it forms a mass of high rugged ground, and reaches a length of some six or seven miles before it finally dies out.[3]

[1] The nearest approach of any Silurian group of strata to the character of these conglomerates is furnished by the remarkably coarse conglomerates, boulder-beds and pebbly grits of the Bala and Llandovery series in the region between Killary Harbour and Lough Mask, to which further reference is made in a later part of this chapter.

[2] This group was placed in the Upper Silurian series by the officers of the Geological Survey who mapped the region (see Sheets 84, 85, 94 and 95 of the Geological Map of Ireland and accompanying Explanation), and on their testimony I formerly referred to the volcanic rocks as of Upper Silurian age. Mr. Baily, however, had pointed out that the limestone associated with the lavas and agglomerates contains Bala fossils. Yet, in spite of this palæontological testimony, the fossils were considered to be "derivative," and the rocks were removed from the series of formations to which they would naturally be assigned. A recent examination of the ground, in company with Mr. J. R. Kilroe of the Geological Survey, has satisfied me that the volcanic rocks are interstratified with sedimentary deposits of Bala age, and must consequently be grouped with the rest of the Lower Silurian series of Ireland. The results of this examination are given in the text.

[3] These areas were carefully mapped for the Survey by Mr. Nolan, and the lines of division marked by him fairly represent the general distribution of the rocks.

The rocks in each of these three areas are similar. One of their distinguishing features is the intercalation among them of a fossiliferous limestone and calcareous fossiliferous tuffs, which contain well-preserved species of organisms characteristic of the Bala division of the Lower Silurian rocks.[1] There cannot be any question that these organisms were living at the time the strata in which their remains occur are found. The most delicate parts of the sculpture on *Illænus Bowmanni* and *Orthis elegantula* are well preserved. Nor have the limestones been pushed into their present places by volcanic agency, or by faults in the terrestrial crust. They are not only regularly intercalated among the volcanic rocks, but the limestone in some places abounds in volcanic dust, while above it come calcareous tuffs, also containing the same fossils. It is thus clearly established that the volcanic series now to be described has its geological age definitely fixed as that of the Bala period.

The lavas of the Lough Mask region consist of felsites and andesites with rocks of probably more basic composition. The felsites are generally quartziferous porphyries, which occupy a considerable space in each of the three districts. To what extent they are intrusive rather than interstratified remains for investigation. Some of them have undoubtedly invaded other members of the volcanic series. But, on the other hand, fragments of similar quartz-porphyries and felsites abound in the intercalated bands of volcanic breccia.

The andesites and more basic lavas are finely-crystalline or compact, dull-green to chocolate-purple rocks, often resembling the "porphyrites" of the Old Red Sandstone. Some of them are strongly vesicular, the cavities being filled with calcite on fresh fracture, though empty on weathered surfaces. The sack-like or pillow structure, already referred to as characteristic of many Lower Silurian lavas, appears conspicuously among some of these rocks. At Bohaun, nine miles south from Westport, where a prolongation of the volcanic series rises to the surface from under the overlying coarse conglomerates, I observed that, owing to the compression which the rocks have there undergone, the pillow-shaped blocks have been squeezed together into rudely polygonal forms, while their vesicles have been greatly drawn out in the direction of tension. Where the rocks have been still more sheared, the distinct pillow-shaped blocks with their vesicular structure disappear, while the more fine-grained crusts that surround them have been broken up and appear as fragments involved in a matrix of green schist.

Intercalated with the lavas are numerous bands of volcanic breccia and fine tuff. The stones in these breccias consist chiefly of various felsites with andesites and more basic lavas. But pieces of jasper, chert, shale and grit are not infrequent. In some places abundant blocks of black shale are to be noticed, probably derived from the Llandeilo group which exists below, and which has here and there been ridged up to the surface in the midst of the

[1] See the list of fossils as determined by Mr. Baily in *Explanatory Memoir* to accompany Sheets 73, 74, 83 and 84 of the Geological Survey of Ireland, p. 68 (1876).

volcanic rocks.[1] Near Shangort I noticed in one of these breccias one block measuring 12 feet, another 20 feet in length and 3 or 4 feet thick, composed of alternating bands of grit and slate. It is interesting to note that these strata had already undergone cleavage before disruption, the bands of slate being strongly cleaved obliquely to the bedding. None of the Llandeilo or other rocks in the neighbourhood display this structure. The blocks seem to have been derived from some deeper group of strata. They are laid down parallel with the rude bedding of the breccia in which they lie.

The fine tuffs and thin ashy limestones associated with the thicker band of limestone show the renewal of volcanic explosions after the interval marked by the calcareous deposit which is sometimes 30 or 40 feet thick. In many places this limestone is brecciated and much mingled with volcanic dust and lapilli. At Shangort, for example, the thick tolerably pure limestone is truncated on the west and north sides by a coarse agglomerate probably filling a volcanic vent. A few hundred yards further north, beyond the interrupting agglomerate, the limestone reappears on the same line of strike, but is then found to be nodular and brecciated and much mingled with volcanic detritus. It lies among ashy grits and tuffs.

The general structure of the ground occupied by the Lough Mask volcanic rocks is diagrammatically represented in Fig. 64. The thickness of the

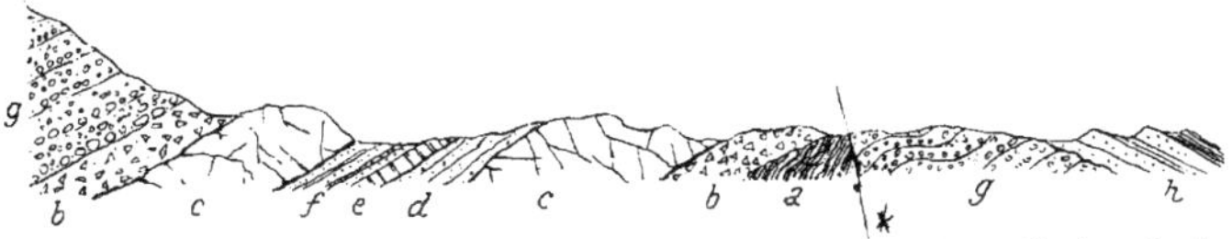

FIG. 64.—Diagram of the general relations of the different groups of rock in the Lower Silurian volcanic district along the western shore of Lough Mask.

a, Llandeilo shales, cherts and grits; *b*, Volcanic breccias; *c*, Felsites and andesites; *d*, Tuffs and ashy grits and shales; *e*, Limestone with Bala fossils; *f*, Calcareous tuffs and thin bands of ashy limestone with fossils; *g*, Coarse conglomerate and grits; *h*, Wenlock strata resting unconformably on the Bala rocks and passing southwards from these to overlie an older series of schists; *, Fault.

volcanic series must amount to many hundred feet, but it has not been precisely determined. The uppermost parts of the series pass under a great thickness of coarse conglomerates and pebbly grits which form the ridge of Formnamore, and stretch thence westwards along Killary Harbour and through the Mweelrea mountains. These strata are classed as the Upper Silurian on the Geological Survey map. Since, however, they conformably overlie rocks containing Bala fossils, and in the Killary district include green shales which have yielded fossils of the same age, they doubtless belong in large part to the Lower Silurian division. The remarkable coarseness of these conglomerates towards the south, and their rapid passage into much finer grits and shales towards the north, probably

[1] In re-examining this region, Mr. Kilroe has found in the stream west of the monastery, Tourmakeady, an uprise of graptolitic black shale containing forms belonging to the very lowest Llandeilo or Upper Arenig strata, and a similar band above Leenane, Killary Harbour.

indicate that they were formed close to the shores of a land composed of schistose rocks, quartzite and granite, of which the mountainous tracts of Connemara are the last relics.

A base to the volcanic series is found in the occasional uprise of a short axis of Llandeilo, or perhaps even upper Arenig strata, containing bands of dark chert and black graptolitic shales. Unfortunately the relations of these underlying rocks to the volcanic masses are not very clear, being obscured by superficial accumulations and also by faulting. It is thus hardly possible to be certain whether they pass up conformably into the base of the volcanic series, or are covered by it unconformably.

The position of this isolated volcanic district in the far west of Ireland, the abundance, variety and thickness of the erupted materials, and the definite intercalation of these materials in the Bala or highest division of the Lower Silurian series, acquire a special interest from the history of the nearest Silurian volcanic area which has now to be described—that of the western shores of the Dingle promontory.

II. The Upper Silurian Series

The latest volcanic eruptions of Silurian time yet definitely known took place during the accumulation of the Wenlock and Ludlow rocks in the far west of Ireland. No satisfactory record of any contemporaneous phenomena of a like kind has yet been met with in any other Upper Silurian district in the British Isles, unless at Tortworth in Gloucestershire, as above described. So far as at present known, only one centre of activity has been preserved. It lies among the headlands of Kerry, where the land projects furthest west into the stormy Atlantic. The occurrence of volcanic rocks in this remote area and their geological horizon have been clearly indicated on the maps of the Geological Survey. More than thirty years, however, have elapsed since some of the mapping was done, and we must therefore be prepared to find it, more especially in its petrography, capable of modification and improvement now.

In the country known as the Dingle promontory, these traces of contemporaneous volcanic rocks are to be observed at various localities and on several horizons. To the east, near Anascaul, on the northern shore of Dingle Bay, some tuffs occur in what are believed to be Llandovery strata. But it is on the western coast, among the headlands and coves that lie to the north and south of Clogher Head, that the best sections are to be seen. The succession of the rocks in this locality was well worked out by Du Noyer, and the Memoir prepared by him, with the general introduction by Jukes, is an invaluable guide to the geologist who would explore this somewhat inaccessible region.[1] The most important correction that will require to be made in the work arises from a mistake as to the true nature of certain rocks which were described as pisolitic tuffs, but which are nodular felsites.

[1] Sheets 160 and 171 of the one-inch map, and Memoir on Sheets 160, 161, 171 and 172.

By far the most striking geological feature of this singularly interesting and impressive coast-line is to be found in the interstratification of lavas with bands of tuff among abundantly fossiliferous strata which, from their organic contents, are unmistakably of the age of the Wenlock group. These lavas occur in a number of sheets, separated from each other by tuffs and other fragmental deposits. They thus point to a series of eruptions over a sea-bottom that teemed with Upper Silurian life. They consist for the most part of remarkably fine typical nodular felsites. The nodules vary in dimensions from less than a pea to the size of a hen's egg. They are sometimes hollow and lined with quartz-crystals. They vary greatly in number, some parts being almost free from them and others entirely made up of them. The matrix, where a fresh fracture can be obtained, is horny in texture, and often exhibits an exceedingly beautiful and fine flow-structure. On weathered faces there may be seen thick parallel strips and lenticles of flow-structure like those of the Snowdon lavas. The upper portions of some of the sheets enclose fragments of foreign rocks. The microscopic examination of a few slices cut from these lavas shows them to be true felsites (rhyolites) composed of a microcrystalline aggregate of quartz and felspar, with layers and patches of cryptocrystalline matter, and only occasional porphyritic crystals of orthoclase and plagioclase.

The pyroclastic rocks associated with these lavas vary from exceedingly fine tuff to coarse agglomerate. Some of the finer tuffs contain pumiceous fragments and pieces of grey and red shale; they pass into fine ashy sandstones and shales, crowded with fossils, and into gravelly breccias made up of fragments of different volcanic rocks.

But the most extraordinary of these intercalated fragmental strata is a breccia or agglomerate, about 15 feet thick, which lies in a thick group of fossiliferous dull-yellow, ashy and ochreous sandstones. The stones of this bed consist chiefly of blocks of different felsites, varying up to three feet in length. Some of them show most perfect flow-structure; others are spongy and cellular, like lumps of pumice. The calcareous sandstone on the top of the breccia is crowded with fossils chiefly in the form of empty casts, and the same material, still full of brachiopods, crinoids, corals, etc., fills up the interstices among the blocks down to the bottom of the breccia, where similar fossiliferous strata underlie it.

Nowhere has the volcanic history of a portion of Palæozoic time been more clearly and eloquently recorded than in this remote line of cliffs swept by the gales of the Atlantic. We see that the ordinary sedimentation of Upper Silurian time was quietly proceeding, fine mud and sand being deposited, and enclosing the remains of the marine organisms that swarmed over the sea-bottom when volcanic eruptions began. First came discharges of fine dust and small stones, which sometimes fell so lightly as not seriously to disturb the fauna on the sea-floor, but at other times followed so rapidly and continuously as to mask the usual sediment and form sheets of tuff and volcanic gravel. Occasionally there would come more paroxysmal explosions, whereby large blocks of lava were hurled forth until they gathered in a thick layer over the bottom. But the life that teemed in the sea, though temporarily destroyed or driven out, soon returned. Corals, crinoids and shells found their way back again, and fine sediment carried their remains with it and filled up the crevices. The ejected volcanic blocks are thus enclosed in a highly fossiliferous matrix.

A succession of lava-streams, of which the strongly-nodular sheet of Clogher Head is the thickest and most conspicuous, mark the culmination of the volcanic energy, and show how at this late part of the Silurian period felsites that reproduce some of the most striking peculiarities of earlier time were once more poured out at the surface. A few more discharges of tuff and the outflow of a greenish flinty felsite brought this series of eruptions to an end, and closed in Britain the long and varied record of older Palæozoic volcanic activity.[1]

[1] As this sheet is passing through the press, the interesting paper by Messrs. S. H. Reynolds and C. J. Gardiner, "On the Kildare Inlier" has appeared (*Quart. Journ. Geol. Soc.* vol. lii. p. 587). These authors give petrographical details regarding the lavas, which they show to be andesites and basalts of Bala age, associated with highly fossiliferous tuffs.

The Edinburgh Geographical Institute

Copyright

J. G. Bartholomew.

BOOK V

THE VOLCANOES OF DEVONIAN AND OLD RED SANDSTONE TIME

CHAPTER XV

THE DEVONIAN VOLCANOES

THROUGHOUT the whole region of the British Isles, wherever the uppermost strata of the Silurian system can be seen to graduate into any later series of sedimentary deposits, they are found to pass up conformably into an enormous accumulation of red sandstones, marls, cornstones, and conglomerates, which have long been grouped together under the name of "Old Red Sandstone." In England and Wales, in Scotland and in Ireland, this upward succession is so well shown that at first British geologists were naturally disposed to believe it to represent the normal order of the geological record. When, however, Sedgwick and Murchison demonstrated that in the counties of Devon and Cornwall a very different group of strata contained an abundant assemblage of organic remains, including types which Lonsdale showed to be intermediate between those of the Silurian and the Carboniferous systems; when, moreover, this palæontological facies of the south-west of England, termed by its discoverers "Devonian," was found to be abundantly developed on the Continent, and to be there indeed the prevalent stratigraphical type of the formations intervening between Silurian and Carboniferous, geologists began to perceive that the Old Red Sandstone must be regarded as the record of peculiar local conditions of sedimentation, while the Devonian type was evidently the more usual development of the same geological period.

From the remote Shetland Isles, across the whole of Scotland and England, down to the northern shores of the Bristol Channel, the Old Red Sandstone maintains its general characters. Nowhere, indeed, are these characters more typically developed than in South Wales, where many thousands of feet of red sediments, almost entirely devoid of organic remains, emerge from under the escarpments of Carboniferous Limestone, and stretch into broad uplands until they are lost at the top of the Silurian system.

But when the geologist crosses the Bristol Channel to the opposite shores of North Devon, he encounters a remarkably different assemblage of rocks. It is true that he has not yet been able to detect there any equivalents of the uppermost Silurian strata of Glamorganshire, nor does he find any conspicuous band of Carboniferous Limestone, such as that which encircles the Welsh Coal-field. He is thus unable to start from a known definite horizon in the attempt to work out the order of succession, either in an upward or downward direction. Lithological characters likewise afford him no means of establishing any satisfactory parallelism. As he follows the Devonian strata, however, he finds them to disappear conformably under the Culm-measures, which, though strangely unlike the Carboniferous strata on the opposite coast, are yet proved by their fossils to belong to the Carboniferous system. Hence the Devonian type, like the Old Red Sandstone, is proved to be immediately anterior to, and to graduate into, the Carboniferous rocks.

There is no stratigraphical change in Britain so rapid and complete as that from the Old Red Sandstone on the one side of the Bristol Channel to the Devonian series on the other. No satisfactory explanation has yet been found for this sudden transformation, which still remains one of the unsolved problems in British geology.

As the observer follows the Devonian assemblage across the land to the southern coast-line, he is conscious that its general characters, both lithological and palæontological, depart more and more from the type of the Old Red Sandstone, and approach more closely to the common Devonian facies of the Continent. He is forced to admit that the Old Red Sandstone, notwithstanding its extensive development in Britain, must be regarded as an exceptional type of sedimentation, while the Devonian facies represents that which is most widely prevalent, not only in Europe, but generally over the world.

The broad estuary of the Bristol Channel unfortunately conceals from view the tract which lies between the typical Old Red Sandstone of Glamorganshire and the typical Devonian formations of Devonshire. Whether this intervening space of some fifteen miles was occupied by a physical barrier, which separated the respective areas of deposit of these two types, or the circumstances of sedimentation in the one region merged insensibly into those of the other, must remain matter for speculation.

The geographical conditions betokened by the Old Red Sandstone will be considered in the next chapter. There can be no doubt that those indicated by the Devonian system were marine. The organic remains so plentifully distributed through the argillaceous and arenaceous sediments of that system, and so crowded together in its limestones, were obviously denizens of the open sea. Yet the only tract of Britain over which this sea can be shown to have spread was the south of England. To the north of that belt, the site of Britain during Devonian time appears to have been partly land and partly wide water-basins in which the Old Red Sandstone was deposited.

In that half terrestrial half lacustrine territory that stretched northwards to beyond the Shetland Isles, many volcanoes were active, of which the chronicles will be described in later pages. The most southerly of these centres of eruption yet known was the district of the Cheviot Hills. Throughout the rest of England and Wales no trace of any contemporary volcanic action has been detected in the Old Red Sandstone. It is true that over most of that region rocks of this age have been concealed under younger formations. Yet throughout Wales, where the Old Red Sandstone attains so vast a thickness, and covers so wide an area, it has not yet yielded a vestige of any contemporaneous volcanic eruptions.

But over the sea-floor that covered the south of England, and stretched thence into the heart of Europe, abundant volcanoes have left behind them proofs of their activity. The first geologist who recognized these proofs and traced their extent on the ground appears to have been De la Beche, who, by his detailed maps and careful description of the igneous rocks of Devonshire, did so much to advance the study of ancient volcanic action. This great pioneer not only determined the former existence of Devonian volcanoes, but he was likewise the first to detect and map the volcanic rocks associated with the Carboniferous and "New Red Sandstone" formations of the same region. The broad outlines traced by him among the volcanic products of these three geological periods in the south-west of England still remain but little changed. Nor are they likely to be much improved until the ground is resurveyed on a larger and more accurate map, and with better petrographical equipment than were available in his day.

Not long after the observations of De la Beche came those of A. C. Godwin-Austen, who devoted much time to a sedulous exploration of the rocks of South Devon, and satisfied himself that contemporaneous volcanic sheets were intercalated among the limestones of that district. "The coral limestones," he says, "are in many places superincumbent on great sheets of volcanic materials, with which, in some instances, as at North Whilborough, they alternate." He pointed out that the interstratified volcanic rocks are of two periods, one Devonian and the other Carboniferous.[1]

In his Geological Maps of Devon and Cornwall, which are to the present time those issued by the Geological Survey, De la Beche made no attempt to discriminate between the varieties of igneous rocks, save that the basic "greenstones" were distinguished from the acid bosses of granite and the elvans. But in his classic "Report" much more detail was inserted, showing that he clearly recognized the existence both of volcanic ashes and of lavas, as well as of intrusive sheets. At the outset of his account of the "Grauwacke," he remarks that the sedimentary deposits are accompanied with igneous products, "a portion of which may also be termed sedimentary;

[1] *Trans. Geol. Soc.* 2nd ser. vol. vi. (1842), pp. 465, 470, 473.

inasmuch as it would seem to have been deposited in beds among contemporaneous rocks of the former description by the agency of water, after having been ejected from fissures or craters in the shape of ashes and cinders, precisely as we may now expect would happen with the ashes and cinders ejected from volcanoes, particularly insular and littoral volcanoes, into the sea.[1] Again he speaks of "two kinds of trappean rocks having probably been erupted, one in the state of igneous fusion, and the other in that of ash, during the time that the mud, now forming slates, was deposited, the mixtures of volcanic and sedimentary materials being irregular from the irregular action of the respective causes which produced them; so that though the one may have been derived from igneous action, and the other from the ordinary abrasion of pre-existing solid rocks, they were geologically contemporaneous."[2] He recognized the origin of the amygdaloidal varieties of rock, and by dissolving out the calcite from their cells showed how close was their resemblance to modern pumice.[3]

Since these early researches many geologists have studied the igneous rocks of Devonshire. I would especially refer to the labours of Mr. Allport,[4] the late J. A. Phillips,[5] Mr. Rutley,[6] the late Mr. Champernowne,[7] Mr. W. A. E. Ussher,[8] Mr. Hobson,[9] and General M'Mahon.[10] Mr. Champernowne in particular has shown the abundance of volcanic material among the rocks of Devonshire, and the resemblance which in this respect they offer to the Devonian system of North Germany.

Unfortunately the geological structure of the Palæozoic rocks of the South-west of England has been complicated to an amazing extent by plication and fracture, with concomitant cleavage and metamorphism. Hence it is a task of extreme difficulty to trace out with any certainty definite stratigraphical horizons, and to determine the range of contemporaneous volcanic action. Mr. Ussher has shown with what success this task may be accomplished when it is pursued on a basis of minute mapping, combined with a sedulous collection and determination of fossils.[11] But years must necessarily elapse before such detailed work is carried over the whole Devonian region, and probably not till then will the story of the volcanic history of the rocks be adequately made out.

In the meantime, it has been established that while there is a singular

[1] "Report on the Geology of Cornwall, Devon and West Somerset," *Mem. Geol. Survey*, 1839, p. 37. [2] *Op. cit.* p. 57. [3] *Op. cit.* pp. 57, 61.

[4] *Quart. Journ. Geol. Soc.* xxxii. (1876), p. 418.

[5] *Op. cit.* xxxi. (1875) p. 325, xxxii. (1876) p. 155, xxxiv. (1878) p. 471.

[6] "Brent Tor," *Mem. Geol. Surv.* p. 18; *Quart. Journ. Geol. Soc.* lii. (1896), p. 66.

[7] See in particular his last paper "On the Ashprington Volcanic Series of South Devon," *Quart. Journ. Geol. Soc.* vol. xlv. (1889), p. 369.

[8] This geologist has spent many laborious years in the investigation of the geology of Devonshire, and has published numerous papers on the subject, in the *Transactions of the Devonshire Association* and of the *Royal Cornwall Geological Society*, in the *Proceedings of the Somersetshire Archæological and Natural History Society*, and of the *Geologists' Association*, in the *Geological Magazine*, and the *Quarterly Journal of the Geological Society*. Reference may especially be made to his Memoir in the last named journal, vol. xlvi. (1890), p. 487.

[9] *Quart. Journ. Geol. Soc.* xlviii. (1892), p. 496.

[10] *Op. cit.* xlix. (1893), p. 385. [11] See Memoir cited in a previous note.

absence of igneous rocks in North Devon, a strip of country extending from Newton Abbot and Torquay westwards by Plymouth across Cornwall to Penzance contains abundant records of volcanic action. It has not yet been possible to map out, among what were formerly all grouped together as "greenstones," the respective limits of the bedded lavas and the tuffs, to distinguish the true sills, and to fix on the position of the chief vents of eruption. So intense have been the compression and shearing of the rocks that solid sheets of diabase have been crushed into fissile schists, which can hardly be distinguished from tuffs. Moreover, owing perhaps in large measure to the mantle of red Permian (or Triassic) strata, which has been stripped off by denudation from large tracts of this region once overspread by it, the Devonian rocks have been deeply "raddled," or stained red. But probably one of the main sources of difficulty in studying the petrography of the area is to be found in the results of atmospheric weathering. Devonshire lies in that southern non-glaciated strip of England, where the rocks have been undergoing continuous decay since long before the Ice Age. No ploughshare of ice has there swept off the deep weathered crust, so as to leave hard surfaces of rock, fresh and bare, under a protecting sheet of boulder-clay. It is seldom that a really fresh piece of any igneous rock can be procured among the lanes and shallow pits of Devon, where alone, for the most part, the materials are exposed.

Much, therefore, remains to be done, both in the stratigraphy and petrography of the Devonian volcanic rocks of this country. To the late J. A. Phillips geology is indebted for the first detailed chemical and microscopical investigation of these rocks. He clearly showed the truly volcanic origin of many of the so-called "greenstones." He believed that certain "slaty blue elvans," which he found to have a composition identical with that of altered dolerites, might be highly metamorphosed tuffs, and that others might have been originally sheets of volcanic mud. After studying the chemical composition and minute structure of a large collection of "greenstones," he demonstrated that in all essential particulars they were dolerites, though somewhat altered from their original character.[1] Subsequently they were studied by Dr. Hatch, who found the fresher specimens generally to possess an ophitic structure, while some are granular, others porphyritic.[2]

Although the rocks have undergone so much crushing, solid cores of them, showing the original structure, may be obtained, also examples of the amygdaloidal, vesicular or slaggy character. They occur in sheets either singly or in groups, and appear generally to be regularly interstratified in the slates and grits. While some of these intercalations, especially the amygdaloidal sheets, may be true superficial lavas, it can hardly be doubted that others are sills, especially those which assume the crystalline structure

[1] See especially *Quart. Journ. Geol. Soc.* vols. xxxii. and xxxiv.

[2] A few of the eruptive rocks of Devonshire have recently been studied by K. Busz. He finds most of his specimens (chiefly from the Torquay district) to be varieties of diabase, but describes a palæopicrite from Highweek near Newton Bushel, and a kersantite from South Brent on the south-east edge of Dartmoor (*Neues Jahrb.* 1896, p. 57).

and composition of gabbros, and show an entire absence of the vesicular structure. But no one has yet attempted to separate the two types from each other.

With these rocks are associated abundant diabase-tuffs (schalstein), frequently mingled with ordinary non-volcanic detrital matter, and shading off into the surrounding grits and slates. There is thus clear evidence of the outpouring of basic lavas and showers of ashes during the Devonian period in the south-west of England, under conditions analogous to those which characterized the deposition of the Devonian system in Nassau and the Harz.

The exact range of these eruptions in geological time has still to be ascertained. So far as at present determined, volcanic activity was not awakened during the accumulation of the Lower Devonian formations. It was not until the sporadic coral-reefs and shell-banks had grown up, which form the basement limestones of the Middle Devonian group, that the first eruptions took place. As Godwin Austen, Champernowne and Mr. Ussher have shown, some of these reefs were overwhelmed with streams of lava or buried under showers of ashes. The volcanic discharges, however, were peculiarly local, probably from many scattered vents, and never on any great scale. Some districts remained little or not at all affected by them, so that the growth of limestone went on without interruption, or at least with no serious break. In other areas again the place of the limestone is taken by volcanic materials.

The chief epoch of this volcanic action, marked by the "Ashprington Volcanic Series," appears to have occurred about midway in the Middle Devonian period. But in certain districts it extended into Upper Devonian time. Intrusive sills of diabase may mark the later phases of the volcanic history. But the occurrence of such sills even in the Upper Devonian rocks, and the alteration of the strata in contact with them (spilosite), point to the continuance or renewal of subterranean disturbance even in the later Devonian ages, if not in subsequent geological time. That volcanic activity accompanied the deposition of the Carboniferous rocks of Devonshire has long been well known (see Chapter xxix.).

CHAPTER XVI

THE VOLCANOES OF THE OLD RED SANDSTONE

Geological Revolutions at the close of the Silurian Period—Physical Geography of the Old Red Sandstone—Old Lake-basins, their Flora and Fauna—Abundance of Volcanoes—History of Investigation in the Subject.

We now enter upon the consideration of the records of a notable era in the geological evolution of north-western Europe. Up to the close of the Silurian period the long history embodied in the rocks presents a constant succession of slowly sinking sea-floors. Wide tracts of ocean stretched over most of Europe, and across the shifting bottom, sand and mud, washed from lands that have long vanished, spread in an ever-accumulating pile. Now and then, some terrestrial movement of more than usual potency upraised this monotonous sea-bed, but the old conditions of ceaseless waste continued, and fresh sheets of detritus were thrown down upon the broken-up heaps of older sediment. All through the vast cycles of time denoted by these accumulations of strata, generations of sea-creatures came and went in long procession, leaving their relics amidst the ooze of the bottom. Genera and families, once abundant, gradually died out, and gave place to others, the onward march of life being slow but uninterrupted. Of the land of the time or of the plants and animals that lived on its surface, hardly anything is known. The chronicles that have come down to us are almost wholly records of the vicissitudes of the ocean-bed.

Over the centre and south of Europe, the marine conditions of Silurian time were prolonged, as we have seen, into the next period, when the Devonian formations were deposited. In that wide region, no marked break has been traced between either the sedimentation or the animal life of the Silurian and Devonian periods. But in the north-west of Europe a striking departure took place from the protracted monotony of marine conditions. By a series of terrestrial movements that affected the area lying to the north of the line of the Bristol Channel, and extended not only to the furthest limit of the British Isles, but probably as far as Norway, and perhaps even into northern Russia, the previous widespread conditions of marine sedimentation were entirely altered. Instead of the fine oceanic silts and sands with their abundant organic remains, and the thick limestones with their masses of coral and crowds of crinoids, there were now laid down, over these northern regions, vast piles of deep red sediment, from which traces of animal life are almost wholly absent. The shelving land against which these ferruginous sands and gravels gathered can still in part be recognized. As the observer follows its margin, notes the varying local peculiarities of its sediment, and detects, sometimes in great abundance, remains of the vegetation which clothed it, the conviction grows in his mind that the remarkable contrast between these deposits, known as the Old Red Sandstone, and those of the Silurian and Devonian systems is not to be accounted for by any mere rearrangement of the sea-bottom, or re-distribution of the land that supplied that sea-bottom with sediment. It has long been the general belief among geologists that the subterranean movements which, over the greater part of Britain, brought the deposition of the Upper Silurian formations to a close, led to a total alteration of the geography of the region affected, that the sea-floor was elevated, and that, over the upraised tract, large lakes or inland seas were eventually formed, in which the peculiar sediments of the Old Red Sandstone were accumulated.

The records of this series of geographical changes are too fragmentary to enable us to follow, except in a very general way, the sequence of events in the transformation of the Silurian sea into the peculiar topographical conditions in which the Old Red Sandstone was laid down. While there was a wide-spread elevation of the sea-floor, and of such ridges of insular land as may have risen above sea-level, the upheaval appears to have been of a somewhat complicated kind, and to have been combined with many local subsidences. The area of disturbance was probably thrown into a series of parallel ridges and troughs, the former continuing to be pushed upward, while the latter tended to subside. The ridges thus became land surfaces, and their prolonged elevation may have more or less compensated for the denudation to which, on their emergence, they were necessarily exposed. The troughs, on the other hand, which sank down, may in many cases have subsided below the sea. But where the general upheaval of the crust was most pronounced, some of the depressions would be isolated above sea-level and become lake-basins in the terrestrial areas.

Of some of these water-basins the outlines can still in some measure be defined. The rocks that rose into hills around them, and from which their enormous accumulations of detritus were derived, still partially survive. We can explore these piles of sediment, and from them can form some idea of the condition of the water in the lakes, and the nature of the vegetation on the surrounding land. The frequent occurrence and exceeding coarseness of the conglomerates, which appear on many successive horizons throughout the deposits of these basins, probably indicate contemporaneous terrestrial disturbances. The same causes that led to the wrinkling of the crust into parallel ridges and troughs no doubt still continued in operation. From time to time the ridges, much worn down by prolonged denudation, were pushed upward, either by gradual uprise or by more rapid jerks. The troughs may in like manner have been still affected by their old tendency to subsidence. Hence, in spite of the effects of degradation and deposition, it is possible that the ridges may not, on the whole, have varied much in height, nor the basins in depth, during the time when thousands of feet were stripped off the land and strewn in detritus over the bottoms of the lakes.

Let us try to realize a little more definitely the general aspect of the region in which the Old Red Sandstone water-basins lay. As the axes of the folds into which the crust of the earth was thrown ran in a north-east and south-west direction, they gave this trend to the lakes and to the tracts of land that separated them. These intervening ridges must in some instances have been hilly or even mountainous. Thus, the Scottish Highlands rose between two of the lakes, and poured into them an abundant tribute of gravel, sand and silt. The terrestrial vegetation of the time has been partially preserved. The hills seem to have been clothed with conifers, while the lower slopes and swamps were green with sigillariæ, lepidodendra and calamites. One of the most characteristic plants was *Psilophyton*, of which large matted sheets were drifted across the lakes and entombed in the silt of the bottom. A grass-like vegetation, with long linear leaves, seems to have grown thickly in some of the shallows of the lakes.

Of the land animals we have still less knowledge than of the vegetation. Doubtless various forms of insect life flitted through the woodlands, though no relics of their forms have yet been recovered. But the remains of myriapods have been found in Forfarshire.[1] These early relics of the animal life of the land inhabited the woodlands, like our modern gally-worms, and were swept down into the lakes, together with large quantities of vegetation.

Some of the lakes, especially in the earlier part of their history, abounded in eurypterid crustacea. These animals inhabited the seas in Upper Silurian time, and appear to have been isolated in the water-basins of the Old Red Sandstone. Certain species of Pterygotus, a Silurian genus found also in the Lower Old Red Sandstone, reached a length of six feet. But the most abundant forms of animal life were fishes. These furnish additional evidence in favour of the lacustrine nature of the waters in which they lived. Such characteristically marine forms as the sharks and rays of the Silurian seas were replaced by genera of Acanthodians, Ostracoderms, Dipnoids, Teleostomes, Placoderms, and Palæoniscids, which abounded in the more northerly waters. The distinctive outward characters of many of these early vertebrates were their bony scales and plates. Some of them had their heads encased in an armature of bone, of large size and massive thickness. In several genera the bone was coated with a layer of glittering enamel. Even now, after the vast lapse of time since their day, the cuirasses and scale-armour of these fishes keep their bright sheen in the hardened sand and mud from which they are disinterred.

A difference is observable between the faunas of the different water-basins. Even where the same genus occurs in two adjacent areas, the

[1] Mr. B. N. Peach, *Proceedings of Royal Physical Society of Edinburgh*, vol. vii. (1882).

species are often distinct. Two large lakes, separated by the tract of the Scottish Highlands, had each its own assemblage of fishes, not a single genus being common to the two basins. Such contrasts, whether the two lakes were geologically contemporaneous, or the northern arose later than the southern, undoubtedly indicate long-continued isolation and the gradual evolution of new forms under different conditions of environment.[1]

Such, in brief, were the aspects of the physical geography of the time on the further consideration of which we are now to enter. The subterranean disturbances, so characteristic of the period, were accompanied by a display of volcanic activity more widespread, perhaps, than any which had yet taken place in the geological history of Britain. Nevertheless, it is worthy of remark that this manifestation of underground energy did not begin with the commencement of these displacements of the crust. The earliest eruptions only took place after the geography of the region had been completely changed; at least no trace of them is to be found in the earliest portions of the Old Red Sandstone. After the last lingering Silurian volcanoes in the west of Ireland had died out, a protracted quiescence of the subterranean fires ensued. In the latest ages of Silurian time there was not in Britain, so far as at present known, a single volcanic eruption. Not until after the inauguration of the Old Red Sandstone topography, when the lakes had taken shape and had begun to be filled with sediment from the surrounding hills, did a series of new volcanoes burst into activity over the northern half of Britain. Rising in the midst of the lakes in groups of separate cones, these vents poured out floods of lava, together with clouds of ashes and stones. Their sites, the history of their eruptions, and the piles of material ejected by them, can still be ascertained, and I shall now proceed to give some account of them.

The thick mass of sedimentary material known as the Old Red Sandstone, lying between the top of the Silurian and the base of the Carboniferous system, has been divided into two sections, which, however, are of unequal dimensions, and doubtless represent very unequal periods of time. The older series, or Lower Old Red Sandstone, is by far the more important and interesting in its extent, thickness, palæontological riches, and, what

[1] In my memoir "On the Old Red Sandstone of Western Europe" (*Trans. Roy. Soc. Edin.* vol. xxviii. 1878), I argued for the probable geological contemporaneity of the conglomerates, sandstones and flagstones on either side of the Grampian chain, even although their organic contents were so unlike. The stratigraphical evidence favours this view. In each case a thick series of strata is covered unconformably by Upper Old Red Sandstone, containing *Holoptychius nobilissimus* and other fishes. The question cannot perhaps be definitely settled by the data available in Scotland. It is quite possible that the basin on the northern side of the Grampians, which I have termed "Lake Orcadie," came into existence after that on the southern side. But I do not think the differences in their respective faunas are to be accounted for simply by lapse of time and the gradual organic evolution in progress over one continuous region. The more the Old Red Sandstone is studied, the more local do its various fish-faunas appear to have been. These strongly-marked diversities appear to me rather to point to prolonged isolation of the basins from each other, as stated above. Dr. Traquair has drawn attention to the remarkable fact that, even in what appears to be one continuous series of strata of no great thickness forming the Upper Old Red Sandstone of the Moray Firth basin, the fishes found about Nairn are entirely different from those met with in the rest of the region.

specially concerns us in the present inquiry, in its volcanic records. Wherever its true base can be seen, this series passes down conformably into Upper Silurian strata. It sometimes reaches a thickness of 15,000 and even 20,000 feet. There is generally a marked break between its highest visible strata and all younger formations. Even the upper division of the Old Red Sandstone rests unconformably upon the lower.[1] Such a hiatus undoubtedly points to a considerable lapse of geological time, and to the advent of important geographical changes that considerably modified the remarkable topography of the older part of the period.

The younger division or Upper Old Red Sandstone passes upward conformably into the base of the Carboniferous system. Its red and yellow sandstones, conglomerates and breccias, covering much more restricted areas, and attaining a much less thickness than those of the lower division, indicate the diminution and gradual effacement of the lakes of the older time, and the eventual return of the sea to the tracts from which it had been so long excluded. So vast an interval elapsed between the time recorded in the deposits respectively of the two sections of the Old Red Sandstone that the characteristic forms of animal life in the earlier ages had entirely passed away, and their places had been taken by other types when the diminished lake-basins of the second period began to be filled up. Volcanic action also dwindled to such a degree that in contrast to the abundant vents of the older period, only one or two widely scattered groups of vents are known to have existed in the area of the British Isles during the later period, and these, after a feeble activity, gave way to a prolonged volcanic quiescence, which lasted until the earlier ages of the succeeding or Carboniferous period.

Although geologists are in the habit of grouping the Old Red Sandstone and the Devonian rocks as equivalent or homotaxial formations, deposited in distinct areas under considerably different conditions of sedimentation, the attempt to follow out the sequence of strata in Devonshire, and to trace some analogy between the Devonian succession and that of the Old Red Sandstone, presents many difficulties for which no obvious solution suggests itself. Into these problems it is not needful to enter further than was done in the last chapter. We may assume that not improbably some of the eruptions now to be described were coeval with those of Devonian time in the south-west of England, though we may hesitate to decide which of them should be brought into parallelism.

As we trace the shore-lines of the ancient basins of the Lower Old Red Sandstone, and walk over the shingle of their beaches, or as we examine the silt of their deeper gulfs, and exhume the remains of the plants that shaded their borders, and of the fishes that swarmed in their waters, we gradually learn that although the sediments which accumulated in some of these basins

[1] *Quart. Journ. Geol. Soc.* vol. xvi. (1860), p. 312. In Wales no break has actually been discovered between the two divisions of the Old Red Sandstone, though it is suspected to exist there also.

amount to many thousand feet in thickness; yet from bottom to top they abound in evidence of shallow-water conditions of deposit. The terrestrial disturbances above referred to continued for a vast interval, and while, as already suggested, the floors of the basins sank, and the intervening tracts were ridged up, as the results of one great movement of the earth's crust, the denudation of the surface of the land contributed to the basins such a constant influx of sediment as, on the whole, compensated for the gradual depression of their bottoms.

We need not suppose that these movements of subsidence and upheaval were uninterrupted and uniform. Indeed, the abundant coarse conglomerates, which play so prominent a part in the materials thrown into the basins, suggest that at various intervals during the prolonged sedimentation subterranean disturbances were specially vigorous. But the occurrence of strong unconformabilities among the deposits of the basins sets this question at rest, by proving that the terrestrial movements were so great as sometimes to break up the floor of a lake, and to place its older sediments on end, in which position they were covered up and deeply buried by the succeeding deposits.[1]

It is not surprising to discover, among these evidences of great terrestrial disturbance, that eventually groups of volcanoes rose in long lines from the waters of most of the lakes, and threw out enormous quantities of lava and ashes over tracts hundreds of square miles in extent. So vast, indeed, were these discharges, across what is now the Midland Valley of Scotland, that the portions of sheets of lava and tuff visible at the surface form some of the most conspicuous ranges of hills in that district, stretching continuously for 40 or 50 miles and reaching heights of more than 2000 feet above the sea. Exposed in hundreds of ravines and escarpments, and dissected by the waves along both the eastern and western coasts of the country, these volcanic records may be studied with a fulness of detail which cannot be found among earlier Palæozoic formations.

It might have been supposed that a series of rocks so well displayed and so full of interest, would long ere this have been fully examined and described. But they can hardly be said to have yet received, as a whole, the attention they deserve. Without enumerating all the writers who, each in his own measure, have added to the sum of our knowledge of the subject, I may refer to the labours of Jameson,[2] Macknight[3] and Fleming,[4] among the observers who began the investigation. But of the early pioneers, by far the most important in regard to the igneous rocks of the Old Red Sandstone was Ami Boué. While attending the University of Edinburgh, where he took the degree of M.D. in the year 1816, he imbibed from Jameson a love of mineralogy and geognosy, and for several years spent his leisure time in personally visiting many parts of Scotland, in order to study the geological

[1] An unconformability of this kind occurs between the south end of the Pentland Hills and Tinto in Lanarkshire, and another in Ayrshire.

[2] *Memoirs of the Wernerian Society*, vol. ii. (1811), pp. 178, 217, 618; vol. iii. (1820), p. 220, 225.

[3] *Op. cit.* vol. ii. pp. 123, 461.

[4] *Op. cit.* vol. i. (1808), p. 162; vol. ii. (1811), pp. 138, 339.

structure of the country. Probably in 1820 he published in French his now classic *Essai*.[1] The value of this work as an original contribution to the geology of the British Isles has probably never been adequately acknowledged. For this want of due recognition the author himself was no doubt in some measure to blame. He refers distinctly enough to various previous writers, notably to Jameson and Macculloch, but he mingles the results of his own personal examinations with theirs in such a way that it is hardly possible to ascertain what portions are the outcome of his own original observations. Less credit has accordingly been given to him than he could fairly have claimed for solid additions to the subjects of which he treated. In the later years of his life I had opportunities of learning personally from him how extensive had been his early peregrinations in Scotland, and how vivid were the recollections which, after the lapse of half a century, he still retained of them. Judged simply as a well-ordered summary of all the known facts regarding the geology of Scotland, his *Essai* must be regarded as a work of very great value. Especially important is his arrangement of the volcanic phenomena of the country, which stands far in advance of anything of the kind previously attempted. Under the head of the "Terrain Volcanique," he treats of the basaltic formations, distinguishing them as sheets (*nappes, coulées*) and dykes; and of the felspathic or trachytic formations, which he subdivides into phonolites, trachytes, porphyries (forming mountains and also sheets) and felspathic or trachytic dykes. In the details supplied under each of these sections he gives facts and deductions which were obviously the result of his own independent examination of the ground, and he likewise marshals the data accumulated by Jameson, Macculloch and others, in such a way as to present a more comprehensive and definite picture of the volcanic phenomena of Scotland than any previous writer had ventured to give.

The account which Boué wrote of the Old Red Sandstone and its associated igneous rocks marked the first great forward step in the investigation of this section of the geological record. He was the earliest observer to divide what he calls the "roches feldspathiques et trappéennes" into groups according to their geological position and mineralogical character, and to regard them as of igneous origin and of the age, or nearly of the age, of the red sandstone of Central Scotland.

Of later writers who have treated of the volcanic rocks of the Old Red Sandstone, my old friend Charles Maclaren deserves special recognition. His survey and description of the Pentland Hills embodied the first detailed and accurate investigation of any portion of these rocks, and his *Geology*

[1] *Essai géologique sur l'Écosse* (Paris; no date, but probably about 1820). He acknowledges his indebtedness to Jameson, whose demonstrations of the geology of the Edinburgh district he partly reproduced in his book. Jameson's early writings in the *Wernerian Memoirs* and in separate works were mere mineralogical or "geognostical" descriptions. His later lectures became more valuable but were never published, save indirectly in so far as they influenced the opinions of his pupils who published writings on the same subjects. See, for instance, Hay Cunningham's *Geology of the Lothians*, p. 59, footnote. Compare an article on Boué, *Edinburgh Review* for May 1823 (vol. xxxviii. p. 413).

of Fife and the Lothians may still be read with pleasure and instruction.[1] Boué had indicated roughly on the little sketch-map accompanying his *Essai* the chief bands of his felspathic and trappean rocks of the Old Red Sandstone, but their position and limits were more precisely defined in Macculloch's "Geological Map of Scotland," which was published in 1840, five years after the sudden and tragic death of its author. The observers who have more recently studied these rocks have been chiefly members of the Geological Survey, and to some of the more important results obtained by them I shall refer in the sequel.

For many years I have devoted much time to the investigation of the Old Red Sandstone and its volcanic rocks. In the year 1859 I ascertained the existence of the great hiatus between the Lower and Upper divisions of the system.[2] A first sketch of the volcanic history of the Old Red Sandstone was given by me in 1861,[3] which was subsequently enlarged and filled in with more detail in 1879.[4] But it was not until 1892 that I published a somewhat detailed outline of the whole subject, tracing the history of volcanic action during the period of the Old Red Sandstone, the distribution of the volcanoes, and the character of the materials erupted by them.[5] This outline I now proceed to amplify, filling in details that were necessarily omitted before, though there are still several districts regarding which information is scanty.

In arranging the treatment of the subject I shall divide the record into two main sections, the first and much the more important being devoted to the Lower and the second to the Upper Old Red Sandstone. In the first of these divisions it will be convenient to begin by taking note of the distribution of the various districts over which the geological evidence is spread. We may then proceed to consider the general character of the volcanic rocks and the manner in which they are arranged in the stratigraphy of the country, taking in consecutive order (1) the superficial lavas and tuffs; (2) the vents; (3) the dykes and sills. From these general considerations we may pass to the detailed history of events in each of the separate volcanic areas, and thus obtain, as far as the evidence at present permits, a broad view of the progress of volcanic action during the time of the Lower Old Red Sandstone in Britain.

[1] *Geology of Fife and the Lothians*, 1839. More detailed reference will be made in later pages to this classic.

[2] "On the Old Red Sandstone of the South of Scotland," *Quart. Journ. Geol. Soc.* xvi. (1860), p. 312.

[3] "On the Chronology of the Trap-Rocks of Scotland," *Trans. Roy. Soc. Edin.* vol. xxii. (1861), p. 63.

[4] Article "Geology," in Ninth Edition of the *Encyclopædia Britannica*, vol. x. (1879), p. 343. Reprinted in my *Text-Book of Geology*, of which the first edition appeared in 1882.

[5] "Presidential Address to the Geological Society," *Quart. Journ. Geol. Soc.* vol. xlviii. (1892).

CHAPTER XVII

DISTRIBUTION OF THE VOLCANIC CENTRES IN THE LOWER OLD RED SANDSTONE—CHARACTERS OF THE MATERIALS ERUPTED BY THE VOLCANOES

i. DISTRIBUTION OF VOLCANIC CENTRES

THE area within which volcanic rocks belonging to the Lower Old Red Sandstone appear is one of the most extensive regions over which the volcanic eruptions of any geological period can be traced in the British Isles (Map I.). Its northern limit reaches as far as the islet of Uya in Shetland, and its southern appears in England in the Cheviot Hills—a distance of about 250 miles. But volcanic rocks of probably corresponding age occur even as far to the south as the hills near Killarney. The most easterly margin of this area is defined by the North Sea on the coast of Berwickshire, and its extreme western boundary extends to near Lough Erne in the north of Ireland—a distance of some 230 miles. If we include the post-Silurian bosses and dykes, like those of Shap, and likewise the Devonian volcanic rocks of Devon and Cornwall, as contemporaneous with those of the Old Red Sandstone, the area of eruption will be greatly enlarged. But leaving these out of account for the present, and confining our attention to the Lower Old Red Sandstone series, we find that, within the wide limits over which the volcanic rocks are distributed, a number of distinct and often widely separated centres of eruption may be traced. Taking these as they lie from north to south, we may specially enumerate the following:—

1. The Shetland and Orkney Islands, together with the basin of the Moray Firth. This region includes several distinct volcanic groups, of which the most northerly extends through the centre to the north-western headlands of the mainland of Shetland, another lies in the island of Shapinshay, one of the Orkneys, while at least two can be recognized on the south side of the Moray Firth. To this wide region of Old Red Sandstone I have given the general designation of "Lake Orcadie."[1]

2. The basin of Lorne, on the west of the mainland of Argyllshire, extending from Loch Creran to Loch Melfort and the hills on the west side of Loch Awe.

[1] *Trans. Roy. Soc. Edin.* vol. xxviii. (1878), p. 354.

3. The great central basin of Scotland, which, for the sake of distinctness, I have called "Lake Caledonia,"[1] stretching between the Highlands and the Southern Uplands, from the east coast south-westwards across Arran and the south end of Cantire into Ireland as far as Lough Erne. Numerous distinct volcanic groups occur in this great basin, and their volcanic history will be discussed in detail in later chapters (see Map III.).

4. The basin of the Cheviot Hills and Berwickshire, with these hills as the chief area, but including also other tracts, probably independent, which are cut off by the sea along the eastern coast of Berwickshire between St. Abb's Head and Eyemouth.

5. The Killarney tract, including the hills lying around Lough Guitane in the east of County Kerry.

At the outset we may take note of a feature in the volcanic history of Britain, first prominently noticeable in the records of the Old Red Sandstone, and becoming increasingly distinct during the rest of the long sequence of Palæozoic eruptions, namely, the persistence with which the vents have been opened in the valleys and have avoided the high grounds. I formerly dwelt on this relation, with reference to the Carboniferous volcanic phenomena,[2] but the observation may be greatly extended. With regard to the Old Red Sandstone of Central Scotland, though the lavas and tuffs that were discharged over the floor of the sheet of water which occupied that region gradually rose along the flanks of the northern and southern hills, yet it was on the lake-bottom and not among the hills that the orifices of eruption broke forth.

So far as I am aware, no undoubted vents of the age of the Lower Old Red Sandstone have been detected among the high grounds of the Highlands on the one hand, or among the Silurian uplands on the other, although a fringe of the lavas may be traced here and there along the base of the hills.[3] In some cases, doubtless, the position of the valleys may have been determined by lines of fault that might well serve as lines of relief along which volcanic vents would be opened. But in many instances it can be proved that, though the vents have risen in valleys and low grounds, they have not selected lines of fault visible at the surface, even when these existed in their neighbourhood. Any fissures up which the volcanic ejections made their way must have lain at great depths beneath the formations that now form the surface rocks.

[1] *Op. cit.*

[2] *Trans. Roy. Soc. Edin.* vol. xxix. (1879), p. 454.

[3] Certain remarkable necks of breccia have been detected by Mr. J. R. Dakyns rising through the schists at the upper end of Loch Lomond; but there is not sufficient evidence to connect them with the volcanic series of the Lower Old Red Sandstone. Some of the younger granite bosses are not improbably to be referred to this volcanic series. The latest granites of the eastern Grampians, as already stated, have lately been found by Mr. Barrow cutting the band of probably Lower Silurian strata along the southern border of the Highlands. Those of Galloway are younger than the Upper Silurian formations, which they invade, and older than the conglomerates of the Upper Old Red Sandstone, which contain pebbles of them. These eruptive bosses will be further discussed in the sequel.

ii. CHARACTERS OF THE MATERIALS ERUPTED BY THE VOLCANOES

A general summary of the petrographical characters of the igneous rocks of the Lower Old Red Sandstone may here find a place. Further details will be given in the account of "Lake Caledonia," which is the typical area for them; but, on the whole, the prevailing types in one region are found to be repeated in the others.

1. *Bedded Lavas.*—Beginning with the lavas which were poured out at the surface, we have to notice a considerable range of chemical composition among them, although, as a rule, they are characterized by general similarity of external appearance. At the one end, come diabases and other ancient forms of basalt or dolerite, wherein the silica percentage is below or little above 50. By far the largest proportion of the lavas, however, are porphyrites or altered andesites, having about 60 per cent of silica. With these are associated lavas containing more or less unstriped felspar and a somewhat higher proportion of silica, which may be grouped as trachytes, though no very sharp line can be drawn between them and the andesites. In the Pentland Hills, and some other areas, orthophyres flowed out alternately with the more basic lavas, and were associated with felsitic tuffs and breccias.

It is noteworthy that the lava-sheets of the Lower Old Red Sandstone, if we consider the character of the prevalent type, hold an intermediate grade between the average chemical composition of those of Silurian and of those of later Carboniferous time. On the one hand, they rarely assume the character of thoroughly acid rocks, like the nodular rhyolites of the Bala and Upper Silurian series;[1] on the other hand, they seldom include such basic lavas as the basalts, so common among the puy-eruptions of the Carboniferous system, and never, so far as I know, contain varieties comparable to the "ultra-basic" compounds which I shall have occasion to allude to as characteristic of a particular volcanic zone in that system.

(*a*) The DIABASE-LAVAS are typically developed in the chain of the Pentland Hills, where they form long bands intercalated between felsitic tuffs — a remarkable association, to which I shall make more detailed reference in a later chapter. They range in texture from a compact dark greenish base to a dull earthy amygdaloid. One of their most remarkable varieties is a fine-grained green porphyry, with large flat tabular crystals of plagioclase arranged parallel to the direction of flow (Carnethy Hill). Most of them, however, are more or less amygdaloidal, and some of them (Warklaw Hill) strongly so. The following analyses, made in the laboratory of the Royal School of Mines under the direction of Prof. E. Frankland, show the chemical composition of some of the diabases of the Pentland Hills:[2]—

[1] The only examples known to me are those of Benaun More and other hills in County Kerry.

[2] For analyses of some Shetland diabases of Old Red Sandstone age, see Mr. R. R. Tatlock, *Trans. Roy. Soc. Edin.* vol. xxxii. (1887), p. 387.

		SiO_2	Al_2O_3	Fe_2O_3	FeO	CaO	MgO	K_2O	Na_2O	H_2O	P_2O_5	CO_2
Carnethy Hill*	.	51·16	22·27	2·94	4·02	5·61	3·46	2·42	2·58	3·42	0·48	1·28
Buiselaw. Sp. grav. 2·80.	Soluble in HCl .	...	1·30	1·53	1·14	2·43	0·98	...	...	...	0·32	...
	Insoluble in ditto .	52·00	17·46	7·85	...	6·80	1·06	1·66	4·17	2·68	...	...
Warklaw Hill. Sp. grav. 2·77.	Soluble in HCl .	...	5·23	7·32	...	7·88	3·65	...	...	...	0·12	5·01
	Insoluble in ditto .	47·77	13·08	0·84	...	4·07	0·30	1·17	2·30	2·48	...	...

* There was a trace of manganous oxide in this specimen.

(*b*) The ANDESITES, or, as they were formerly called, PORPHYRITES, which constitute by far the largest proportion of the lavas, have a characteristic but limited range of lithological varieties. The prevailing type presents a close-grained, rather dull texture, and a colour varying from pinkish grey, through many shades of green and brown, to purplish red, which last is, on the whole, the predominant hue. Minute lath-shaped felspars may frequently be detected with the naked eye on fresh surfaces, while scattered crystals, which are generally hæmatitic pseudomorphs after some pyroxene, occasionally after hornblende or mica, may often be observed. The usual porphyritic constituents are plagioclase felspars, occasionally in abundant tabular crystals measuring half an inch or more across, also one or more pyroxenes (augite, enstatite), and sometimes brown or black mica. Where large felspar-crystals occur in a compact green matrix, the rock assumes a resemblance to the *verde antique* of the ancients.[1] One of the Cheviot andesites lying at the bottom of the series is distinguished by its large and abundant platès of black mica.[2]

The texture of the andesites occasionally becomes faintly resinous, where a considerable proportion of glass still remains undevitrified, as in the well-known varieties from the Cheviot Hills, and in another pitchstone-like rock from above Airthrey Castle in the Ochil Hills, near Bridge of Allan. It sometimes presents a nodular or coarsely perlitic character, weathering out in nut-like balls, like the rock of Buckham's Wall Burn in the Cheviot Hills.[3] Much more frequent is a well-developed amygdaloidal structure, which indeed may be said to be the most obvious characteristic of these rocks as a whole. The steam-vesicles, now filled with agate, quartz, calcite or zeolite, vary in size from mere granules up to large irregular cavities a foot or more in diameter. Where the kernels are coated with pale-green earth and lie in a dark brown matrix, they give rise to some of the most beautiful varieties of rock in any volcanic series in this country, as may be seen on

[1] An instance of this rock occurs in Kincardineshire, from which the large flat twins of labradorite have been analyzed by Dr. Heddle (*Trans. Roy. Soc. Edin.* vol. xxviii. (1879), p. 257).

[2] C. T. Clough, "The Cheviot Hills," *Mem. Geol. Survey* (1888), p. 12.

[3] *Ibid.* p. 11.

the Ayrshire coast at Culzean and Turnberry. Some rocks contain the vesicles only as rare individuals, others have them so crowded together as to form the greater part of the cubic contents of the mass. When the infiltration-products have weathered out, some of the amygdaloids present a striking resemblance to recent slaggy brown lavas; lumps of them must have been originally light enough to float in water.

My colleague in the Geological Survey, Mr. J. S. Grant Wilson, some years ago made for me a large series of determinations of the specific gravity of the volcanic rocks of the Lower Old Red Sandstone of Scotland. He found that the andesites collected from various districts to illustrate the more typical varieties of rock averaged about 2·66. He also made a series of chemical analyses of a number of the same rocks from the Cheviot Hills, where they are well preserved. The results are shown in the following table:—

	SiO_2	Al_2O_3	Fe_2O_3	FeO	MnO	CaO	MgO	K_2O	Na_2O	H_2O	H_2SO_4	Loss.
Scawd Law .	59·29	16·30	1·77	3·70	·41	4·81	3·15	4·19	3·44	3·84	...	...
Rennieston .	62·81	16·40	·55	3·27	·81	4·46	1·64	3·60	3·02	4·04	...	...
Cunrieston .	63·38	15·77	·73	2·65	·08	4·44	1·88	1·88	4·54	4·69	...	...
Duncan's Dubs	59·44	16·15	1·05	2·83	·37	6·70	2·46	3·18	3·70	3·35	...	...
Whitton Hill .	60·70	17·98	·66	2·58	·20	7·07	2·20	3·57	2·95	3·45	...	...
Cuddies' Tops	60·58	12·25	1·01	4·13	·15	4·40	2·86	2·19	3·61	...	·55	2·15
Cocklawfoot .	62·29	17·03	·93	2·44	·21	3·92	2·71	1·14	3·20	·29*	·37	4·81
Morebattle .	59·82	16·96	·20	6·57	·15	4·73	2·84	2·63	3·04	...	trace	1·98

* This is CO_2.

The microscopic structure of the andesites of the Lower Old Red Sandstone has been partially investigated, especially those of the Cheviot Hills, by Mr. Teall[1] and by Dr. Petersen,[2] who both give chemical analyses of the rocks. Much, however, still remains to be done before our knowledge of this branch of British petrography can be regarded as adequate. The groundmass in some of the rocks consists mainly of a brown glass with a streaky structure (as in the well-known variety of Kirk Yetholm, and in the rock, still more like pitchstone, from near Airthrey Castle in the Ochil chain); more usually it has been devitrified more or less completely by the appearance of felspathic microlites, until it presents a confused felspar aggregate. The porphyritic felspars are often large, generally striped, but sometimes including crystals that show no striping. They are frequently found to be full of inclusions of the base, and these sometimes consist of glass. The ferro-magnesian constituents are usually rather decomposed, being now represented by chloritic pseudomorphs; but augite, and perhaps still more frequently enstatite, may be recognized, or its presence may be

[1] *Geol. Mag. for* 1883, pp. 100, 145, 252.

[2] *Mikroskopische und chemische Untersuchungen am Enstatit-porphyrit aus den Cheviot Hills*, Inaug. Dissert. Kiel, 1884. Descriptions have also been published of detached rocks from other districts, such as those by Prof. Judd and Mr. Durham of specimens from the Eastern Ochils, *Quart. Journ. Geol. Soc.* vol. xlii. (1886), p. 418.

inferred among them. The beautiful resinous or pitchstone-like rock from near Airthrey Castle has been found by Mr. Watts to be a glassy hypersthene-augite-andesite, since among its phenocrysts of plagioclase, augite and hypersthene both occur. Magnetite is commonly traceable, and apatite may be occasionally detected. As the result of decomposition, calcite, chlorite and limonite are very generally diffused through the rocks.[1]

(*c*) The lavas which may be separated as TRACHYTES offer no distinctive features externally by which they may be distinguished from the andesites. Indeed, both groups of rocks appear to be connected by intermediate varieties. In the Cheviot Hills some of the lavas are found, on microscopic examination, to contain a large admixture of unstriped porphyritic felspars, which can occasionally be recognized as sanidine in Carlsbad twins. The groundmass is sometimes a brown glass, but is usually more or less completely devitrified, portions of it being inclosed in the large felspars. Chlorite, pseudomorphic after augite or enstatite, may be detected, and sometimes a brown mica. A specimen of one of these rocks, from a locality to the north-west of Whitton, near Jedburgh, was found by Mr. J. S. Grant Wilson to have the following composition :—

	SiO_2	Al_2O_3	Fe_2O_3	FeO	MnO	CaO	MgO	K_2O	Na_2O	H_2O	Total.
N.W. of Whitton Hill, Jedburgh (No. 1938) Sp. gr. 2·55.	62·44	18·99	3·35	1·8	·25	1·84	1·37	5·02	2·65	2·48	100·19

(*d*) Acid rocks such as FELSITES and RHYOLITES are rare among the lavas poured out at the surface during the time of the Lower Old Red Sandstone. They occur in the Pentland Hills, also near Dolphinton in the Biggar district, and in the Ochil Hills near Auchterarder, associated with extensive accumulations of felsitic tuffs and breccias. They are usually so much decomposed that it is hardly possible to procure fresh specimens of them. Some of them display beautiful flow-structure. They appear to be generally orthoclase-felsites or orthophyres. Dull, fine-grained to flinty in texture, they hardly ever display free quartz, so that they can seldom be placed among the typical rhyolites, though in their banded flow-structure they often strongly resemble some lithoid varieties of these rocks, especially such varieties as that represented in Fig. 9.

Mr. Watts, to whom I submitted, for microscopic examination, a number of specimens from the Pentland and Ochil Hills, has found them to "consist of a brown felsitic groundmass in which are embedded a generation of small stumpy prisms of orthoclase and a set of larger phenocrysts, generally consisting of orthoclase and plagioclase in equal proportions. Brown mica is usually present and zircons are not uncommon." The rocks,

[1] Dr. F. H. Hatch supplied notes on microscopic structure which are incorporated in the text, together with particulars afterwards furnished by Mr. Watts.

when they undergo weathering, pass into the varieties formerly comprised under the name claystone.

The only nodular felsite of this age which I have met with is that of Lough Guitane among the "Dingle Beds," near Killarney, to which reference will be made in later pages.

2. *Intrusive Bosses, Sills and Dykes.*—While the interbedded lava-sheets are mainly andesites, the intrusive rocks are generally more acid, and most of them may be grouped under the convenient head of felsites. Some intrusive andesites, and even more basic rocks, do indeed occur in dykes and sills, as well as also filling vents. But the rule remains of general application over the whole country that the materials which have consolidated in the volcanic orifices of the Old Red Sandstone, or have been thrust among the rocks in dykes, bosses or sills, are decidedly acid. In this series of rocks a greater range of types may be traced than among the extrusive lavas. At the one end we find true granites or granitites, as in the intrusive bosses of Spango Water and of Galloway, which, for reasons which I will afterwards adduce, may with some probability be assigned to the volcanic history of the Lower Old Red Sandstone period. Among the bosses, many of which probably mark the positions of eruptive vents, orthophyres are especially prominent. These rocks frequently contain no mica, but, on the other hand, they sometimes show abundant quartz in their groundmass. The augite-granitite of the Cheviot Hills, so well described by Mr. Teall, has invaded the bedded andesites of that region.[1] A similar rock has been noticed by my brother, Prof. James Geikie, associated with the Lower Old Red Sandstone volcanic rocks of the east of Ayrshire. A remarkable petrographical variety has been mapped by Mr. B. N. Peach, rising as a small boss through the lower part of the great lava-sheets of the Ochil Hills, above Tillicoultry. It is a granophyric quartz-diorite, which, under the microscope, is seen to be composed of short, thick-set prisms of plagioclase, with abundant granophyric quartz, a pleochroic hypersthene,. and needles of apatite. Sometimes the pyroxene is replaced by green chloritic pseudomorphs.[2]

At the other end of the series come the felsites, quartz-porphyries, mica-porphyrites, minettes, vogesites, "hornstones" and "claystones" (or decayed felsites), which have a close-grained texture, often with porphyritic felspars, quartz or black mica, generally a whitish, pale buff, orange, pink or purplish-grey colour, and a specific gravity of about 2·55.[3]

Though I class these rocks as intrusive, I am not prepared to assert that in none of the instances where they occur as sheets may they possibly have been erupted at the surface as lavas. In one or two cases the evidence either way is doubtful, but as the great majority of the acid rocks can be

[1] *Geol. Mag. for* 1883, pp. 100, 145, 252; and *British Petrography*, pp. 272, 278.

[2] Notes by Dr. Hatch.

[3] The intrusive "porphyry" of Lintrathen in Forfarshire (which may be younger than the Old Red Sandstone) is a bright red rock with porphyritic felspar, quartz, white mica and a very singular black mica (Mr. Teall's *British Petrography*, p. 286).

shown to be intrusive in their behaviour, I have preferred to keep them all in the same category. I am prepared to find, however, that, as so vast an amount of felsitic debris was ejected to form the tuffs, more of this material may have flowed out in streams of lava than is at present recognized.

The following table shows the chemical composition of some acid sills and dykes from the Lower Old Red Sandstone, as determined in the laboratory of Prof. E. Frankland:[1]—

	SiO_2	Al_2O_3	Fe_2O_3	MnO	CaO	MgO	K_2O	Na_2O	P_2O_5	H_2O
"Hornstone." Torgeith Knowe, Pentlands . .	73·91	14·41	·76	·07	1·21	4·90	3·36	1·57	...	·90
"Hornstone." Braid Hills*. .	64·73	17·01	2·35	·24	4·19	·66	3·27	3·75	·26	2·78
Tinto, Lanarkshire: Soluble in hydrochloric acid .	·04	1·01	1·24	...	·92	·52	...	...	·16	...
Insoluble in ditto .	70·28	12·54	·43	...	·91	...	3·92	5·84	...	1·99

* This specimen also yielded 0·13 of ferrous oxide, and 2·42 of carbon dioxide.

The rock of Tinto, which may be considered typical of the prevailing acid intrusive rocks of the series, presents several slightly different varieties. Dr. Hatch, as the result of his examination of a number of microscopic slides prepared from specimens taken by me from various parts of the hill, found some to be minettes, showing small isolated crystals of orthoclase and rare flakes of biotite, sometimes granules of quartz, imbedded in a brown, finely microlitic groundmass of felspar powdered over with calcite; while other specimens had a granular instead of a microlitic groundmass, and contained a considerable amount of quartz in addition to the constituents just mentioned. A conspicuous knob on the south side of Tinto, called the Pap Craig, is a mass of augite-diorite, which has risen through the other rocks[2] (see Fig. 93). The sills in the same region show still further differences. Some are true "felspar-porphyries," and "quartz-porphyries" varying in the relative abundance and size of their porphyritic orthoclase and quartz, while others, by the introduction of hornblende or pseudomorphs after that mineral, pass into vogesites.

Basic sills and bosses are chiefly developed among the Ochil and Sidlaw Hills. They may generally be classed as diabases. But sometimes their pyroxenic constituent is partly hypersthene, as in a coarsely crystalline boss about a mile south of Dunning, which has been determined by Mr. Watts

[1] Two analyses of rhyolites from Shetland by Mr. Tatlock will be found in *Trans. Roy. Soc. Edin.* vol. xxxii. (1887), p. 387. Their silica percentage is 72·32 and 73·70. An analysis of a quartz-felsite from the Cheviot Hills by Mr. T. Waller is given in the Geological Survey Memoir on the Cheviot Hills, p. 25. The proportion of silica in this rock is 67·9.

[2] This rock differs considerably from the other intrusive masses in its neighbourhood. Dr. Hatch found it to be composed chiefly of lath-shaped striped felspar, with some granular augite, magnetite and interstitial quartz.

to "consist of augite and hypersthene imbedded in and occurring amongst large plagioclase prisms. Some iron-ore is also present; the rock is a hyperite."

3. *Tuffs and Agglomerates.*—The fragmental materials, ejected from or filling up the vents, vary from the finest compacted dust up to some of the coarsest agglomerates in this country. In general they consist mainly of detritus of andesite, and have been derived from the blowing up of already consolidated masses of that rock. The fragments are usually angular, and range from minute grains up to blocks as large as a cottage. The tuffs are often more or less mixed with ordinary non-volcanic sediment, and as they are traced away from the centres of eruption they pass insensibly into sandstones and conglomerates.

But while, as might be expected, the tuffs are most commonly made up of debris of the same kind of lavas as those that usually form the sheets which were poured out at the surface, they include also bands of material derived from the destruction of much more acid rocks. Throughout the chain of the Ochil Hills, for example, in the midst of the bedded andesite-lavas, many of the thin courses of fine tuff consist largely of felsitic fragments, with scattered felspar crystals. The most remarkable examples of this nature, however, are to be met with at the great vent of the Braid Hills, in the chain of the Pentland Hills which runs south-westward from it, and in the Biggar volcanic district still further south. These acid tuffs are generally pale flesh-coloured or lilac in tint, and compact in texture, but, like the felsitic lavas from which they were derived, they are apt to weather into yellow or buff "claystones." The finer varieties are so compact as to present to the naked eye no distinguishable grains; they might be mistaken for felsites, and indeed, except where they contain recognizable fragments of rock or broken crystals of felspar, can hardly be discriminated from them. They consist of an exceedingly fine compacted felsitic dust. Here and there, however, the scattered crystals of felspar and small angular fragments of felsite, which may be detected in them, increase in number until they form the whole of the rock, which is then a brecciated tuff or fine volcanic breccia, made up of different felsites, among which, even with the naked eye, delicate flow-structures may be detected. In these pale acid tuffs, fragments of different andesites may often be observed, which increase in number as the rocks are traced away from the main vents of eruption.

At my request my colleague, Mr. George Barrow, determined the silica percentages in a few specimens which I selected as showing some of the more characteristic varieties of these tuffs from the Braid and Pentland Hills. His results are exhibited in the following table:—

	Silica percentage.
1. Quarry above Woodhouselee	63·3
2. South-west side of Castlelaw Hill	73·15
3. Quarry on road, ½ mile N.E. of Swanston (Braid Hill vent)	74·1
4. South-west side of Castlelaw Hill	75·0
5. Castlelaw Hill	76·00
6. South side of White Hill Plantation . . .	90·00

From these analyses it may be inferred that the average amount of silica in the more typical varieties is between 70 and 75 per cent. The last specimen in the table, with its abnormally high percentage of acid, must be regarded as an exceptional variety, where there has either been an excessive removal of some of the bases, or where silica has been added by infiltration.

The microscopic examination of these rocks has not added much to the information derivable from a study of them in the field. In their most close-grained varieties, as above remarked, they are hardly to be distinguished from felsites. But they generally show traces of the minute detrital particles of felsite of which they are essentially composed. The brecciated varieties exhibit finely-streaked flow-structure in some of the fragments. Pieces of andesite, grains of quartz, and other extraneous ingredients appear in these rocks towards the southern limits of the volcanic area of the Pentland Hills, where the acid tuffs are associated with and pass laterally and vertically into ordinary non-volcanic sedimentary strata. Further details as to the part which these tuffs play in the volcanic history of the regions wherein they occur will be given in later pages.

CHAPTER XVIII

STRUCTURE AND ARRANGEMENT OF THE LOWER OLD RED SANDSTONE VOLCANIC ROCKS IN THE FIELD

We have now to consider the manner in which the various volcanic products have been grouped around and within the orifices of discharge. The first feature to arrest the eye of a trained geologist who approaches them as they are displayed in one of the ranges of hills in Central Scotland is the bedded aspect of the rocks. If, for example, he looks eastward from the head of the Firth of Tay, he marks on the right hand, running for many miles through the county of Fife, a succession of parallel escarpments, of which the steep fronts face northwards, while their long dip-slopes descend towards the south. On his left hand a similar but higher series of escarpments, stretching far eastwards into Forfarshire, through the chain of the Sidlaw Hills, repeats the same features, but in opposite directions. If he stands on the alluvial plain of the Forth, near Stirling, and looks towards the north, he can trace bar after bar of brown rock and grassy slope rising from base to summit of the western end of the Ochil Hills. If, again, from any height on the southern outskirts of the city of Edinburgh, he lets his eye range along the north-western front of the chain of the Pentland Hills, especially towards evening, he can follow the same parallel banding as a conspicuous feature on each successive hill that mounts above the plain. Or if, as he traverses the west of Argyllshire, he comes in sight of the uplands of Lorne, he at once recognizes the terraced contours of the hills between Loch Awe and the western sea, presenting so strange a contrast to the rugged and irregular outlines of the more ancient schist and granite mountains all around (see Fig. 99).

i. BEDDED LAVAS AND TUFFS

On a nearer inspection, the dominant topographical features are found to correspond with a well-marked stratification of the whole volcanic series. Where two sheets of andesite are separated by layers of tuff, sandstone or conglomerate, a well-marked hollow will often be found to indicate the junction-line; but even where the lavas follow each other without such interstratifications, their differences of texture and consequent variations in mode and amount of weathering usually suffice to mark them off from each other, and to indicate their trend along the surface in successive terraces. Even where the angles of inclination are high, the bedded arrangement can generally be detected.

It is in the picturesque and instructive coast-sections, however, that the details of this bedded structure are most clearly displayed. On both sides of the country, along the shores of Ayrshire on the west, and those of Kincardineshire and Forfarshire on the east, the volcanic group has been admirably dissected by the waves. The lava-beds have been cut in vertical section, so that their structure and their mode of superposition, one over another, can be conveniently studied, while at the same time, the upper surfaces of many of the flows have been once more laid bare as they existed before they were buried under the sedimentary accumulations of the waters in which they were erupted.

Though distinctly bedded, the Lavas show little of the regularity and persistence so characteristic of those of Carboniferous and of Tertiary time. Some of them are not more than from four to ten feet thick, and generally, on the coast-cliffs, they appear to be less than fifty feet. A continuous group of sheets can sometimes be traced for ten miles or more from the probable vent of discharge.

That many of these lavas were erupted in a markedly pasty condition may be inferred from certain of their more prominent characteristics. Sometimes, indeed, they appear as tolerably dense homogeneous masses, breaking with a kind of prismatic jointing; but more frequently they are strongly amygdaloidal, and sometimes so much so that, as already stated, the amygdales form the larger proportion of their bulk. Where the secondary infiltration-products have weathered out, the rough scoriform rock looks as if it might only recently have been erupted. In a few instances I have observed an undulating rope-like surface, which reminded me of well-known Vesuvian lavas. Usually the top and bottom of each sheet assume a strikingly slaggy aspect, which here and there is exaggerated to such an extent that between the more solid and homogeneous parts of two consecutive flows an intermediate band occurs, ten or twelve feet thick, made up of clinker-like lumps of slag, the interspaces being filled in with hardened sand. In some cases these agglomeratic layers may actually consist in part of ejected blocks; but the way in which many of the lavas have cooled in rugged scoriaceous surfaces is as conspicuous as on any modern *coulée*. The loosened slags, or the broken-up cakes and blocks of lava, have sometimes been caught up in the still moving, pasty current, which has congealed with its vesicles drawn out round the enclosed fragments, giving rise to a mass that might be taken for a breccia or agglomerate. Now and then we may observe that the upper slaggy portion of a sheet has assumed a bright red colour from the oxidation of its ferruginous minerals; and from the contrast it thus presents to the rest of the rock we may perhaps legitimately infer that the disintegration took place before the outflow of the next succeeding lava. If this inference be well founded, and it is confirmed by other evidence which will be subsequently adduced, it points to the probable lapse of considerable intervals of time between some of the outflows of lava.

FIG. 65.—Veins and nests of sandstone due to the washing of sand into fissures and cavities of an Old Red Sandstone lava. Turnberry Point, Ayrshire.

But perhaps the most singular structure displayed by these lavas is to be seen in the manner in which they are traversed by and enclose portions of sandstone. Since I originally observed this feature on the Ayrshire coast, near Turnberry Point, many years ago,[1] I have repeatedly met with it in the various volcanic districts of the Lower Old Red Sandstone across the whole of the Midland Valley of Scotland. The first and natural inference which a cursory examination of it suggests is that the molten rock has caught up and carried along pieces of already consolidated sandstone. But a little further observation will show that the lines of stratification in the sandstone, even in what appear to be detached fragments, are marked by a general parallelism, and lie in the same general plane with the surface of the bed of lava in which the sandy material is enclosed. In a vertical section the sandstone is seen to occur sometimes in narrow dykes with even, parallel walls, but more usually in irregular twisting and branching veins, or even in lumps which, though probably once connected with some of these veins, now appear as if entirely detached from them (Fig. 65). Frequently, indeed, the nodular slaggy andesite and the sandstone are so mixed up that the observer may hesitate whether to describe the mass as a sandstone enclosing balls and blocks of lava, or as a scoriaceous lava permeated with hardened sand. A close connection may be traced between these sandstone-inclosures and the beds of sandstone interstratified between the successive lavas. We can follow the sandy material downwards from these intercalated beds into the andesites below them. On exposed upper surfaces of the lava, an intricate reticulation of sandstone veins may be noticed, in each of which the stratification of the material runs across the veins, showing sometimes distinct current-bedding, but maintaining a general parallelism with the bedding of the volcanic sheets and their fragmentary accompaniments (Fig. 66). If we could remove the sandstone-veinings and aggregates, we should find the upper surfaces of these igneous masses to present a singularly fissured

[1] See Jukes' *Manual of Geology*, 3rd edit. (1872), Fig. 111, p. 276.

and slaggy appearance, reminding us of the rugged, rent and clinker-loaded slopes of a modern viscous lava, like some of those in the Atrio del Cavallo on Vesuvius. There cannot, therefore, be any doubt that the sandstone, so irregularly dispersed through these lavas, was introduced originally as loose sand washed in from above so as to fill the numerous rents and cavernous interspaces of the volcanic rock. A more striking proof of the subaqueous character of the eruptions could hardly be conceived. This interesting feature in lavas erupted under water is not confined to the volcanic series of the Old Red Sandstone. We shall find that it is hardly less distinct among the basic lavas of the Permian series both in Scotland and in Devonshire.

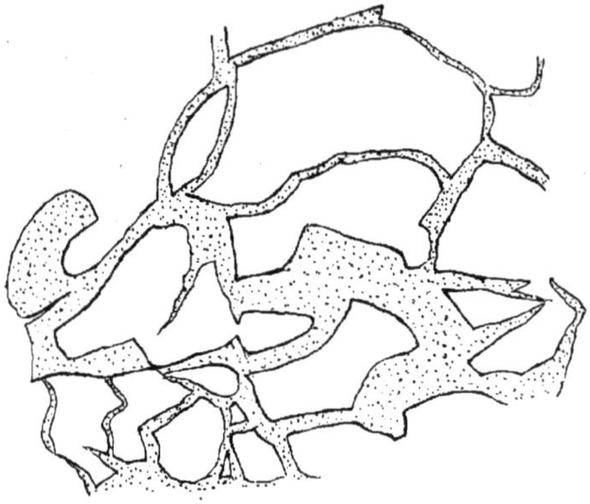

FIG. 66.—Ground-plan of reticulated cracks in the upper surface of an Old Red Sandstone lava filled in with sandstone. Red Head, Forfarshire.

A remarkable exception to the general type of dark basic and intermediate lavas is furnished by the pale, decomposing felsites of the Pentland and Dolphinton Hills. Those which issue from the great eruptive centre of the Braid Hills, alternate with the andesites and the diabases, gradually diminishing like these in a southward direction and dying out in some six or seven miles. Beyond the limits of these lavas, another similar thick group was erupted from a separate vent at the northern end of the Biggar district near Dolphinton. The same occurrence has been ascertained also in the area of the Ochil chain. Fuller reference will be made to these interesting rocks in the descriptions to be afterwards given of the structure and history of the volcanic areas of the Pentland Hills, the Biggar centre and the Ochil Hills.

It is certainly a notable feature in the volcanism of Old Red Sandstone time that from the same, or from closely adjoining vents, lavas should be alternately poured forth, differing so much from each other, alike in chemical composition and petrographical characters, as andesites and diabases on the one hand, and felsites on the other. Additional examples, from widely different geological systems, will be cited in subsequent pages. It will be shown that even in the very latest volcanic period in Britain, that of older Tertiary time, highly basic and markedly acid materials were ejected from the same centres of eruption.

The part taken by the Tuffs in the structure of the ground agrees with what might have been expected in the accompaniments of extremely slaggy and viscid lavas. These pyroclastic intercalations are, in most of the volcanic districts, comparatively insignificant in amount, by far the largest proportion of solid material ejected from the various vents having consisted of streams of lava. Round or within some of the vents the fragmentary materials attain a remarkable coarseness, as may be seen in the great agglomerates of Dumyat, near Stirling, the largest of which is more than 700 feet thick. These massive accumulations doubtless represent a long series of explosive discharges from the summit of the lava column in one or more adjacent vents. Traced away from the orifices of emission, the tuffs rapidly grow finer in grain, less in thickness, and more mixed with ordinary detritus, until they pass into ordinary non-volcanic sediment or die out between the lava-sheets.

Good sections, showing the nature and arrangement of the thin intercalations of andesite-tuff between the successive outpourings of lava, may be examined on the coast. Thus, near Turnberry Point, in Ayrshire, upwards of a dozen successive flows of lava, with their sandy and ashy intervening layers, are exposed in plan upon the beach, and partly also in section along the cliffs on which the ruins of the historic castle of Turnberry stand. (Figs. 95, 96, 97). Again, along the coast of Forfarshire, from the Red Head to Montrose, the numerous sheets of andesite are separated by layers of dull purplish tuff passing into conglomerate, with blocks of porphyrite a yard or more in diameter.

The most remarkable interstratified tuffs in the Lower Old Red Sandstone are the felsitic varieties. Those which proceed from the great vent of the Braid Hills, extend south-westwards for eight or nine miles, and their peculiar materials, mixed with ordinary sediment, may be traced several miles further. They occur in successive sheets, which, from a maximum thickness and number at the north end, gradually thin away southwards, like the felsitic lavas which they accompany, and from the explosion of which they no doubt were derived. They consist to a large extent of extremely fine volcanic dust, and since they are generally much decomposed, it is often, as already remarked, hardly possible to distinguish between them and the equally decayed felsites. In some parts of the hills they present a distinct fissile bedding; but still more satisfactory is the occasional fine brecciated structure which they assume, when they are seen to consist of angular lapilli of different felsites.

The amount of volcanic material ejected from the more important vents was much greater than the height of the present hills would lead us to suppose. The rocks have generally been tilted into positions much more inclined than those which they originally occupied, so that to measure their actual thickness we must take a line approximately perpendicular to the dip. In this way we ascertain that the accumulated mass of lavas and tuffs immediately outside the vent at the north end of the Pentland Hills must be at least 7000 feet thick, for the base of the series is concealed under the unconformable overlap of the Lower Carboniferous Sandstones, while the top is cut off by a fault which brings down the Carboniferous formations against the eastern flank of the hills. Probably not less voluminous is the pile of ejected material in the Ochil Hills, where, though the base of the whole is concealed by the fault which throws down the coalfield, some 6500 feet of lavas, tuffs and conglomerates can be seen. There were thus, during the time of the Lower Old Red Sandstone, more than one volcano in Central Scotland which might be compared in bulk of ejected material to Vesuvius.

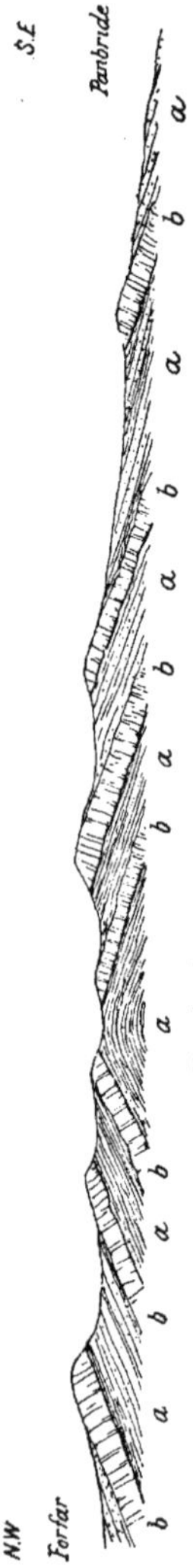

FIG. 67.—Section across the volcanic series of Forfarshire.
a, conglomerates, sandstones and flagstones; *b*, sheets of andesitic lava.

That the eruptions were mainly subaqueous is indicated, as I have shown, by the intercalated bands of sandstone and conglomerate between the successive lavas, as these are traced away from the centres of discharge, and likewise, even more impressively, by the hardened sand which has been washed into former fissures and crevices in the lava. But that, in some cases, the volcanic cones were built up above the surface of the lake may be legitimately inferred from the remarkable volcanic conglomerates which occur, more particularly in the great chain of the Ochil and Sidlaw Hills. These thick accumulations of well-rounded and water-worn blocks are interspersed between sheets of andesite, and are mainly made up of andesite fragments. Impressive sections of them may be seen along the Kincardineshire coast. The conglomerates are sometimes so remarkably coarse, many of their blocks exceeding two feet in diameter, and so rudely bedded, that it is only by noting the position of oblong boulders that one can make out the general direction of the stratification. In their smooth rounded forms, these blocks resemble the materials of storm-beaches on an exposed coast. The trituration of the andesite fragments has given rise to a certain amount of green paste, which firmly wraps round the stones, and retains casts of them after they have dropped out. It is further deserving of remark that while in some districts, as in the central Ochils, the materials were entirely derived from the destruction of volcanic rocks, in others a large proportion of non-volcanic materials is mingled with the debris of the lavas. South of Stonehaven, for example, large boulders of quartzite form a conspicuous feature in the conglomerates, of which in places they make up quite half of the total constituents. There can be little doubt, I think, that the materials of these coarse detrital accumulations were gathered together as shingle-beaches, and were derived in part from volcanic cones which had risen above the level of the lake. They seem to suggest considerable degradation of these cones by breaker-action, whereby blocks of rock a yard or more in diameter could be rounded and smoothed.

Another inference deducible from such conglomerates, and to which I have already alluded, is that considerable intervals of time took place between some of the eruptions. Round the vents, indeed, where the successive sheets of volcanic material follow each other continuously, it is perhaps impossible to form any definite opinion as to the relative chronological value of the lines of separation between different ejections. But where some hundreds of feet of coarse conglomerate, chiefly composed of well-rounded andesite blocks, intervene between two streams of lava, we may conclude that the interval between the outpouring of these streams must have been of considerable duration. Other evidence of a similar tendency may be recognized in the intercalation of groups of varied sedimentary accumulations, such as those which were deposited over the site of Eastern Forfarshire and Kincardineshire during the time that elapsed between two successive floods of lava. In the Den of Canterland, for example, in the midst of the volcanic sheets we find interesting evidence of one of these intervals of quiescence, during which layers of fine olive shales were laid quietly down, while macerated vegetation, drifting over the lake-bottom, was buried with remains of fishes, and abundant gally-worms (*Kampecaris, Archidesmus*), washed from the neighbouring land.[1] So undisturbed were the conditions of deposition that calcareous sediment gathered round some of the organisms and encased them in limestone nodules.

In some of the districts the discharges of volcanic material were so abundant or so continuous that no recognizable deposition of ordinary sediment has taken place between them. Thus, at the north end of the Pentland Hills the rocks are entirely of volcanic origin, and though, as we trace them southwards away from the centre of eruption, they diminish in thickness, they include hardly any interstratified sandstones and conglomerates until they finally begin to die out.

The distances to which the lavas and tuffs have been erupted from the chief vents of a district vary up to 15 or 20 miles. Those of the Pentland Hills extend from the Braid Hill vent for 10 miles to the south-west. Those of the Biggar centre stretch for about 16 miles to the north-east. Those of the Ochil Hills, which probably came from a number of distinct vents, can be traced for nearly 50 miles.

ii. VENTS

On the whole the actual vents of the volcanoes of Lower Old Red Sandstone time are less clearly distinguishable than those of subsequent volcanic

[1] An abundant organism in some of these deposits, named *Parka*, was first regarded as a plant, was afterwards believed to be the egg-packets of crustacea, and is now pronounced by competent authorities to belong to an aquatic plant with creeping stems, linear leaves and sessile sporocarps.

periods. This deficiency doubtless arises from the geological structure of the districts in which the formation is chiefly developed. Thus, in the great Midland Valley of Scotland, where the Old Red Sandstone covers a large part of the surface, the vents seem to have been placed along the central parts of the long trough rather than among the older rocks on either margin. Hence they are in large measure buried either under the volcanic and sedimentary accumulations of their own period or under Carboniferous strata.

Certain bosses of massive rocks lying well within the volcanic area may with some confidence be regarded as the sites of eruptive centres. They occur either singly or in groups, and may be specially noticed along the chain of the Ochil and Sidlaw Hills. Yet it seems to me probable that these visible bosses, even if we are correct in regarding them as marking the positions of true vents, do not indicate the chief orifices of discharge. If we consider their size and their distribution with reference to the areas of lava and tuff discharged at the surface, we are rather led to

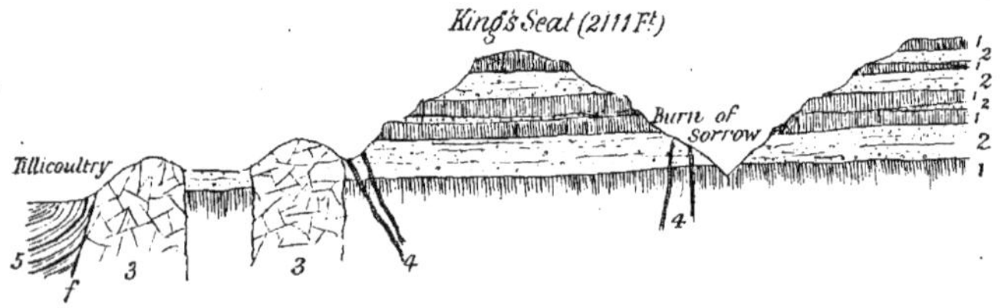

FIG. 68.—Section across two necks above Tillicoultry, Ochil Hills.

1 1, Andesite lavas; 2 2, Tuffs and volcanic conglomerates; 3 3, The two necks; 4 4, Dykes of felsite, etc.; 5, Coal-measures; *f*, Fault.

look upon them as subsidiary vents, the more important orifices, from which the main bulk of the eruptions took place, being still concealed under the Carboniferous rocks of the Midland Valley. The bosses which rise through different portions of the volcanic series are obviously not the oldest or original vents. The great felsitic mass of Tinto in Lanarkshire (Fig. 93), indeed, pierces strata which lie near the base of the Lower Old Red Sandstone, but the smaller cone of Quothquan in its neighbourhood appears in the midst of the lavas (Fig. 92). In the south-western part of the Ochil chain the bosses or necks are chiefly small in size, seldom exceeding half a mile in diameter. They have been filled sometimes with crystalline, sometimes with fragmental materials. Two of them, containing the remarkable granophyric quartz-diorite already referred to, emerge from among the tuffs in a low part of the series, immediately above the village of Tillicoultry in Clackmannan (Fig. 68). Two or three more, which are occupied by orthophyres and quartz-felsites, pierce the volcanic group a few miles to the west of Loch Leven. The whole of the visible bosses of the Ochil Hills may be regarded as one connected group, subsidiary to the main orifices which lay rather further to the south and west. More particular reference to this district will be made in the following chapter (p. 303).

Vents which have been filled up with agglomerate, and which thus furnish the most obvious proofs of their connection with the eruptions of the volcanic series, though not frequent, may be observed in a number of the volcanic districts. Their fragmentary materials generally consist mainly of the detritus of andesites or diabases like those which form the bedded lavas. But where more acid lavas have risen to the surface, fragments of felsite may occur more or less abundantly. In the great vent of the Braid Hills the tuffs and breccias are almost wholly acid. Non-volcanic materials may often be found in the agglomerates, and occasionally even to the exclusion of volcanic detritus. Thus, in the far north of Scotland several examples occur among the Shetland Isles of necks filled entirely with blocks of the surrounding flagstones and sandstones. Such cases, as has been already pointed out, probably represent incompleted volcanoes, when the explosive vapours were powerful enough to drill orifices in the crust of the earth and eject the shattered debris from them, but were not sufficiently vigorous or lasting to bring up any solid or liquid volcanic material to the surface. These Shetland examples are further noticed on p. 345.

Necks of agglomerate in the Lower Old Red Sandstone vary in size from a great orifice measuring two miles across to little plugs only a few yards in diameter. They may be found in limited numbers in most of the volcanic districts. No examples have been observed rising through older rocks than the Old Red Sandstone, all the known instances being eruptive through some part of the volcanic series or of the sandstones, and therefore not belonging to the earliest eruptions.

The largest, and in some respects the most interesting, vent in the Lower Old Red Sandstone, that of the Braid Hills near Edinburgh, described in Chapter xx., covers an area of more than two square miles, and is filled with felsitic breccias and tuffs, through which bosses and veins of acid and basic rocks have been injected. It completely truncates the bedded lavas and tuffs of the Pentland Hills, and not improbably marks the chief centre from which these rocks were erupted. Several smaller necks rise a little beyond its southern margin, marking, perhaps, lateral cones on the main volcano.

In the small area of Lower Old Red Sandstone lying between Campbeltown and the Mull of Cantyre, several necks of agglomerate occur, which have been partly dissected by the waves along the shore, thus revealing their internal structure and their relation to the surrounding conglomerates. An account of them will be found at p. 311. One of the series, which lies back from the coast-line, forms a prominent rounded hill measuring about 400 yards in its longest diameter. Its general contour is represented in Fig. 82.

Of the eruptive bosses of massive rock outside the limits of the Old Red Sandstone which may be plausibly referred to the volcanic phenomena of the period, though they cannot be proved to be actually part of them, the most notable are the bosses of granite and other acid material which rise through the Silurian strata of the Southern Uplands of Scotland.[1] The largest are the well-known masses of Galloway (Fig. 69), with which must be grouped the bosses near New Cumnock, that of the Spango Water (Fig. 94), and those of Cockburn Law and Priestlaw in Lammermuir, together with a number of masses of felsitic material scattered over the same region, such as the Dirrington Laws of Berwickshire (Fig. 70). These bosses present some points of structure in common with true vents. They come like great vertical columns through highly-folded and puckered strata, and, as they truncate the Llandovery and Wenlock formations, they are certainly younger than the greater part of the Upper Silurian series. They must be later, too, than the chief plication and cleavage of these strata; but they are older than the Upper Old Red Sandstone or basement Carboniferous rocks which contain pebbles of them. Their date of eruption is thus narrowed to the interval between the later part of the Upper Silurian period and the beginning of the Upper Old Red Sandstone. I have myself little doubt that they are to be asso-

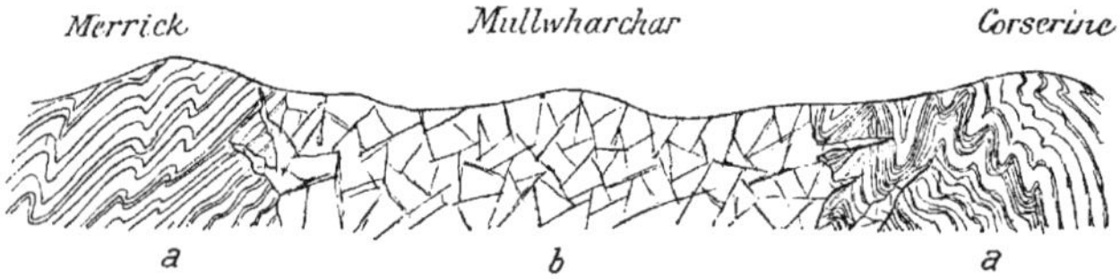

FIG. 69.—Section of the granite core between Merrick and Corscrine.
a, Silurian greywackes, grits and shales; *b*, granite.

ciated with the volcanic epoch we are now considering, as it was the only known great episode of igneous activity in this region during the interval within which the protrusion of these granites must have taken place. In the Cheviot Hills, indeed, we have evidence of the eruption of a large mass of augite-granitite through the porphyrite-lavas of the Lower Old Red Sandstone, with abundant veins projecting from it into them, as will be narrated in later pages.[2]

Not improbably many other granite protrusions throughout the British Isles are to be referred to the volcanic operations of the Lower Old Red Sandstone. Such are those of the Lake District, notably that of Shap,[3] the granites of Newry and Leinster in the east of Ireland, which are later than the Silurian rocks and older than the Carboniferous Limestone, and the younger Grampian granites, which pierce the presumably Arenig belt along the Highland border. Whether or not these granitic protrusions were

[1] I suggested this possible connection many years ago in *Trans. Geol. Soc. Edin.* vol. ii. (1874) p. 21.

[2] The volcanic geology of the Cheviot Hills is described by Mr. Teall, *Geol. Mag. for* 1883, p. 106; and by Mr. Clough, *Mem. Geol. Survey*, "Geology of the Cheviot Hills," Sheet 108 N.E., 1888, p. 24.

[3] See the descriptions of the Shap granite by Messrs. Marr and Harker, *Quart. Journ. Geol. Soc.* xlvii. (1891) p. 266, and xlix. (1893) p. 359.

connected with superficial volcanic discharges of which no remains have survived, they seem to indicate the wide extent and remarkable vigour of the subterranean igneous action of this geological period.

Viewed as a whole, the materials which now occupy the vents of the volcanic chains in the Lower Old Red Sandstone of the British Isles are more acid than the lavas erupted at the surface. In the Pentland district, indeed, and in some other areas this acid material was ejected at intervals in abundant discharges of dust and lapilli and in outflows of felsitic lavas, while between these successive discharges copious streams of diabasic and

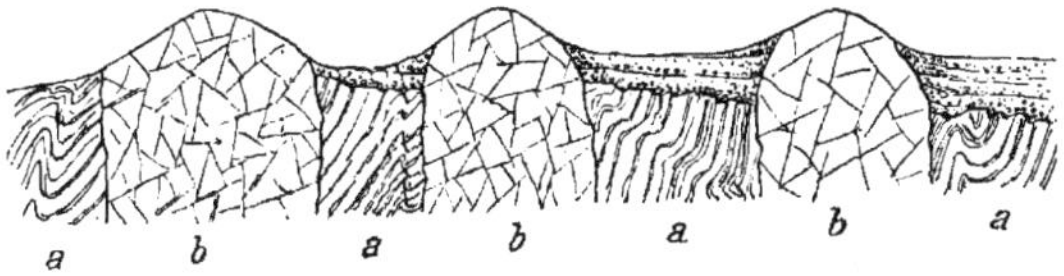

FIG. 70.—Section across the three Dirrington Laws, Berwickshire.
a, Upper Silurian strata; *b*, Necks probably of Lower Old Red Sandstone age; *c*, Upper Old Red Sandstone lying unconformably both on *a* and *b*.

andesitic lavas, either from the same or from some closely-adjoining vent, were poured out. Throughout the whole region, however, as a closing phase of the volcanic history, the acid magma rose after the outpouring of the more basic lavas and filled such chimneys of the volcanoes as were not already blocked with agglomerate. It was probably after these pipes were plugged that the final efforts of volcanic energy were expended in the protrusion of the acid material as sills between the bedding-planes of the surrounding rocks, and as dykes and veins in and around the vents.

iii. SILLS AND DYKES

Nowhere throughout the volcanic tracts of the Lower Old Red Sandstone is there any such development of sills as may be seen beneath the Silurian volcanic sheets of North Wales. Those which occur are most abundant in the Lanarkshire district, to the north-west and south-west of Tinto, and in the south of Ayrshire. From the village of Muirkirk to the gorge of the Clyde, below the Falls, the Upper Silurian and Lower Old Red Sandstone strata are traversed by numerous intrusive sheets of pink and yellow felsite, quartz-porphyry, minette, lamprophyre and allied rocks, which are no doubt to be regarded as part of the volcanic phenomena with which we are here concerned. In the south of Ayrshire, between the villages of Dalmellington and Barr, there is a copious development of similar sills, especially along one or more horizons near the base of the Old Red Sandstone. Garleffin Fell, Glenalla Fell, Turgeny and other heights are conspicuous prominences formed of these rocks; above the sills lie thick conglomerates and sandstones on which the great andesite-sheets rest.

In the Pentland Hills, as will be described in Chapter xx., a massive felsitic sill forms a conspicuous feature along the north side of the chain, and there are probably others which have not yet been separated from the felsitic tuffs and orthophyres which they so much resemble.

Fig. 71.—Section of Papa Stour, Shetlands, showing sill of spherulitic felsite traversing Old Red Sandstone and bedded porphyrites (Messrs. Peach and Horne).

1. Red sandstones and flagstones; 2. Purple diabase-porphyrites; 3. Great sheet of pink spherulitic felsite.

Fig. 72.—Section across Northmavine, from Okrea Head to Skea Ness, Shetland, showing dykes and connected sill of granite and felsite (Messrs. Peach and Horne).

1. Schists, etc.; 2. Serpentine; 3. Granite and quartz-felsite; 4. Breccia of serpentine fragments; 5. Bedded andesites and tuffs. *f*, Fault.

Perhaps the most remarkable acid sills in the Old Red Sandstone of Britain are those which occur at the extreme northern end of the region among the volcanic phenomena of the Shetland Isles (Figs. 71, 72). The largest of them, consisting mainly of granite and felsite, is believed to reach a length of 20 and a breadth of from three to four miles.[1]

A group of sills composed of a bright red quartz-porphyry has been traced along the southern flanks of the Highlands for upwards of 18 miles.[2] This rock, already referred to as the "Lintrathen porphyry," lies chiefly among the conglomerates and sandstones, but also intersects the lavas, and may be later than the Old Red Sandstone (p. 277). An extension of it is found even on the north side of the boundary fault, cutting the andesites which there lie unconformably on the schists.

Examples, however, occur of sills much less acid in composition. In the Dundee district, for instance, the intrusive sheets are andesites and diabases. They send veins into and bake the sandstones among which they have been intruded, and are sometimes full of fragments of such indurated sandstone, as may be well seen on the northern shore of the Firth of Tay, west of Dundee.

A conspicuous characteristic of most of the volcanic tracts of the Lower Old Red Sandstone is the comparative scarcity of contemporaneous dykes. In the band of acid sills between Muirkirk and the Clyde, a considerable number of dykes

[1] Messrs. B. N. Peach and J. Horne, *Trans. Roy. Soc. Edin.* xxxii. (1884), p. 359.

[2] See Sheet 56 of the Geological Survey of Scotland.

have been mapped, which must be regarded as due to the same series of movements and protrusions of the magma that produced the adjacent sills. Throughout the length of the Southern Uplands dykes of felsite, minette, lamprophyre, vogesite and other varieties, which may also be connected with the volcanic phenomena of the Lower Old Red Sandstone, not infrequently occur among the Silurian rocks. On the Kincardineshire coast, south of Bervie, a number of dykes of pink quartz-porphyry traverse the conglomerates and sandstones. The coast south of Montrose displays some singularly picturesque sections, where a porphyry dyke running through andesitic lavas and agglomerates stands up in wall-like and tower-like projections. On the shore at Gourdon, as well as inland, intrusive dykes of serpentine occur. A line of these, possibly along the same fissure, has been traced for more than a dozen of miles from above Cortachy Castle to near Bamff. But there is no evidence to connect them with the volcanic phenomena of the Old Red Sandstone. Not improbably they belong to a later geological period.

One would expect to meet with a network of dykes in and around the volcanic vents; but even there they are usually not conspicuous either for number or size. In the great vent of the Braid Hills only a few have been noticed. In the Ochil Hills groups of dykes of felsite and andesite may be observed, especially near the necks. They are fairly numerous in the neighbourhood of Dollar (see Fig. 68). One of the most abundant series yet observed traverses the tract around the granite boss of the Cheviot Hills, from which many dykes of granite, felsite, quartz-porphyry and andesite radiate. This district will be more fully referred to in Chapter xxi. Another remarkable development of dykes occurs in Shetland (Fig. 72), where they consist of granite, felsite and rhyolite, and are associated with the acid sills above referred to.

CHAPTER XIX

VOLCANOES OF THE LOWER OLD RED SANDSTONE OF "LAKE CALEDONIA"

Description of the several Volcanic Districts: "Lake Caledonia," its Chains of Volcanoes —The Northern Chain: Montrose Group, Ochil and Sidlaw Hills, the Arran and Cantyre Centre, the Ulster Centre.

I NOW propose to give some account of each of the districts which have been separate areas of volcanic action during the time of the Lower Old Red Sandstone, tracing its general structure, the arrangement and sequence of its volcanic rocks and the history of its eruptions. As by far the most varied development of the Old Red Sandstone is to be found in the great Midland Valley of Scotland, and as it is there that the remarkable volcanic phenomena of the system have been most abundantly displayed and are most clearly recorded, I shall begin my description of the volcanic eruptions of the Lower Old Red Sandstone with a detailed account of the different centres of volcanic activity in that region. The phenomena are so fully displayed there that a more summary treatment of the subject will suffice for the other regions.

Under the designation of "Lake Caledonia," as already remarked, I include the whole of the Midland Valley of Scotland between the Highlands and the Southern Uplands, likewise the continuation of the same ancient hollow by Arran and the south of Cantyre across the north of Ireland to Lough Erne.[1] Throughout most of the area thus defined, the present limits of the Lower Old Red Sandstone are sharply marked off by large parallel faults. On the north-west side one, or rather a parallel series, of such dislocations runs from Stonehaven along the flank of the Highland mountains to the Clyde, thus traversing the whole breadth of the island. On the

[1] My own investigations of this region have been continued over an interval of forty years. Besides personally traversing every portion of it, I have mapped in detail, for the Geological Survey, many hundreds of square miles of its area from the outskirts of Edinburgh south-westwards into Lanarkshire, in Ayrshire, and in the counties of Fife, Perth and Kinross. The Geological Survey maps of the volcanic tracts of the Sidlaw Hills have been prepared by my brother, Prof. James Geikie, and Messrs. H. M. Skae and D. R. Irvine. The Western Ochils were mapped chiefly by Mr. B. N. Peach, partly by Prof. J. Young, Mr. R. L. Jack and myself; the Eastern Ochils were surveyed mainly by Mr. H. H. Howell; while the volcanic belt between the tracts mapped by me in Lanarkshire and in Ayrshire was chiefly traced out by Mr. Peach. As a rule, each of these geologists has described in the Survey Memoirs the portions of country surveyed by him.

south-east side another similar series of faults, which there skirts the edge of the Silurian table-land, has nearly the same effect in precisely defining the margin of the Old Red Sandstone. As thus limited, the tract has a breadth of about 50 miles in Scotland, while the portion of it now visible in the British Isles has an extreme length of about 280 miles (Map III.).

But though the boundary-faults determine, on the whole, the present limits of the tract of Old Red Sandstone, they do not necessarily indicate the shore-lines of the sheet of water in which that great series of deposits was laid down. They point to an enormous subsidence of the tract between them—a prolonged and extensive sagging of the strip of country that stretches across the Midland Valley of Scotland into the north of Ireland.[1] This downward movement began as far back as the close of the Silurian period, but the marginal fractures and the disruption and plication of the thick masses of sandstone and conglomerate which were accumulated in the lake chiefly took place after the close of the period of the Lower Old Red Sandstone. I think we may reasonably connect these movements with the general sinking of the area consequent upon the enormous outpouring of volcanic materials during that period.

Along both the northern and southern margins of the basin there occur, on the farther side of the boundary faults, outlying patches of Lower Old Red Sandstone that rest unconformably on the rocks forming the flanks of the hills. These areas possess a peculiar interest, inasmuch as they reveal some parts of the shore-line of the lake, and show the relation between the earlier rocks and the sediments of the Old Red Sandstone. We learn from them that the shore-line was indented with wide bays, but nevertheless ran in a general north-easterly direction. It thus corresponded in trend with the present Midland Valley, with the axes of plication among the schists of the Highlands as well as among the Silurian rocks of the Southern Uplands, and with the subsequent faulting and folding of the Old Red Sandstone.

I may remark in passing that the conglomerates and other associated

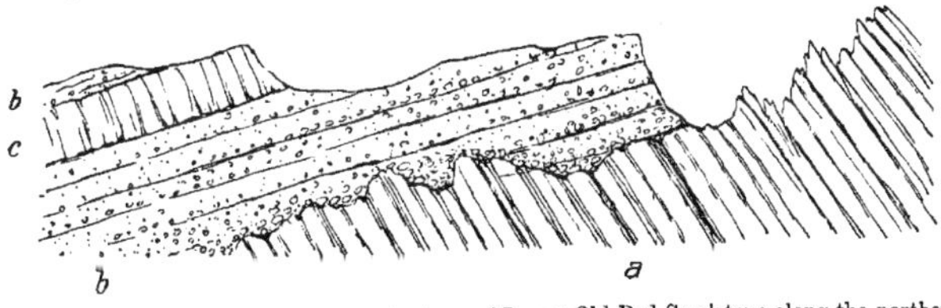

FIG. 73.—Section at the edge of one of the bays of Lower Old Red Sandstone along the northern margin of Lake Caledonia, near Ochtertyre.

a, slates and phyllites; *b*, volcanic conglomerates; *c*, andesite-lava.

materials which have been preserved in these bays and hollows beyond the

[1] In some of the dislocations along the Highland border, the Old Red Sandstone is bent back upon itself, and the older schists are thus made to recline upon it, as if there had been a push over from the Highland area.

lines of the great faults, though they lie unconformably on the rocks beneath, are not the basement portions of the Old Red Sandstone. On the contrary, where their probable stratigraphical horizons can be recognized or inferred, they are found to belong to parts of the series considerably above the base of the whole. They point to the gradual sinking of the basin and the creeping of the waters with their littoral shingles further and further up the slopes of the hills on either side (Fig. 73).

But this is not all the evidence that can be adduced to show that the limits of the lake extended considerably beyond the lines of dislocation between which the present area of Old Red Sandstone mainly lies. No one can look at the noble escarpments of the Braes of Doune on the one

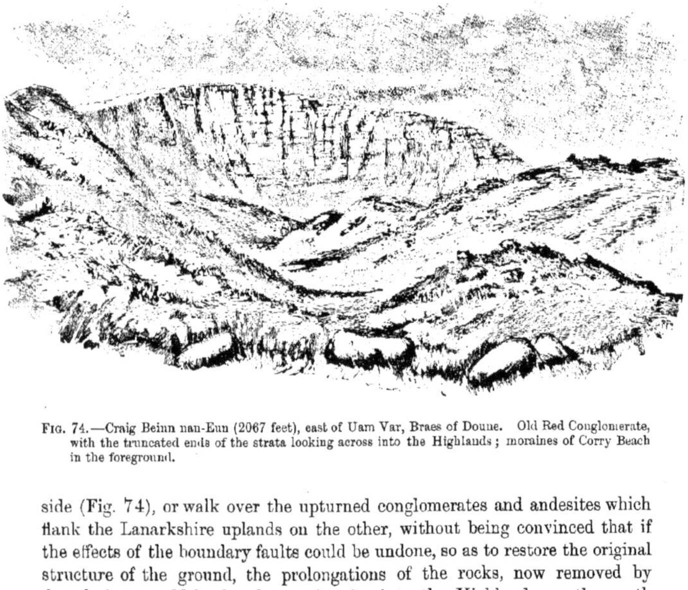

FIG. 74.—Craig Beinn nan-Eun (2067 feet), east of Uam Var, Braes of Doune. Old Red Conglomerate, with the truncated ends of the strata looking across into the Highlands; moraines of Corry Beach in the foreground.

side (Fig. 74), or walk over the upturned conglomerates and andesites which flank the Lanarkshire uplands on the other, without being convinced that if the effects of the boundary faults could be undone, so as to restore the original structure of the ground, the prolongations of the rocks, now removed by denudation, would be found sweeping far into the Highlands on the north and into the Silurian Uplands on the south.

If the area of "Lake Caledonia" were taken to be defined by the boundary faults, it covered a space of about 10,000 square miles. But, as we know that it certainly stretched beyond the limits marked by these faults, it must have been of still greater extent. We shall probably not exaggerate if we regard it as somewhat larger than the present Lake Erie, the superficies of which is about 9900 square miles. In this long narrow basin the remarkable volcanic history was enacted of which I now proceed to give some account.

The Lower Old Red Sandstone of Central Scotland may be conveniently divided into three great groups, each of which marks a distinct epoch in the history of the basin wherein they were successively accumulated. The lowest of these groups indicates a time of quiet sedimentation during which the basin was defined by plication of the terrestrial crust, and when, by the same subterranean movements, some parts of the floor of the lake were pushed upward above water, and were then denuded and buried. The middle group consists largely of volcanic rocks. It points to the existence of lines of active volcanic cones situated along the length of the lake. The uppermost group records the extinction of volcanic action and the gradual obliteration of the lake, partly by the pouring of sediment into it, and partly no doubt by the continued terrestrial movements which had originally produced the basin.

It is evident from these records that though volcanic activity continued vigorous for a vast period of time, it had entirely ceased in "Lake Caledonia" long before the last sediments of the Lower Old Red Sandstone were laid down. The great cones of the Ochil Hills, for example, sank below the waters of the lake in which they had long been a conspicuous feature, and so protracted was the subsidence of the lake-bottom that the site of these volcanoes was buried under 8000 or 9000 feet of sandstones and conglomerates, among which no trace of any volcanic eruptions has yet been found. The sagging of the terrestrial crust over an area from which such an enormous amount of volcanic products had been discharged would doubtless be a protracted process. Long after the subsidence of the lake-bottom and the accumulation of its thick mass of sediments, after even the entire effacement of the topography and the deposition of the thick Carboniferous formations over its site, the downward movement showed itself in the production of gigantic north-east faults, and the sinking of the Carboniferous rocks for several thousand feet. These dislocations, as was natural, have run through the heart of some of the volcanic groups, carrying much of the evidence of the ancient volcanoes out of sight, and leaving us only fragments from which to piece together the records of a volcanic period which is by no means the least interesting in the geological history of this country.

Confining our attention for the present to the records of the middle or volcanic group, we find evidence of a number of distinct clusters of volcanoes ranged along the whole length of the basin. The independence of these volcanic districts may be inferred from the following facts:—1st, The actual vents of discharge may in some cases be recognized; 2nd, Even where these vents have been buried, we may often observe, as we approach their probable sites, a marked increase in the thickness of the volcanic accumulations, as well as a great development of agglomerates and tuffs; 3rd, Traced in opposite directions, the volcanic materials are found to thin away or even to disappear. Those from one centre of discharge may be observed now and then to overlap those from another, but the two series remain distinct.

Reasoning from these data and studying the distribution of the various volcanic areas, we are led to recognize the former existence of two parallel chains of vents, running along the length of the lake at a distance from each other of somewhere about twenty miles. They may be conveniently distinguished as the northern and the southern chain.

The northern band runs from the coast-line near Stonehaven south-westward through the Sidlaw and Ochil Hills. It is then abruptly truncated by a large fault and by the unconformable superposition of the Carboniferous formations. But 60 miles further to the south-west, where the Old Red Sandstone comes out on the west side of the Firth of Clyde, a continuation of the volcanic band has recently been detected by Mr. W. Gunn of the Geological Survey in the Island of Arran. Twenty-five miles still further in the same direction a much ampler development of the volcanic rocks occurs to the south of Campbeltown in Cantyre. If we cross the 22 miles of sea that separate the Argyllshire coast-sections from those of Red Bay in Ireland, we find near Cushendall a repetition of the Scottish volcanic conglomerates, while still further along the same persistent line, some 50 miles into the interior, the hills of Tyrone include sheets of lava precisely like those of Central Scotland. The total length of this northern chain of volcanoes is thus not much less than 250 miles, and as its north-eastern end is now cut off by the North Sea it must have been still longer. It ran parallel to the north-western coast-line of the lake, at a distance which, over the site of the Midland Valley of Scotland, seems to have varied from 10 to 20 miles, but which greatly lessened further to the south-west.

At a distance of some twenty miles to the south of the northern belt, the second parallel chain of volcanoes ran in a nearly straight line, which is now traceable from the southern suburbs of Edinburgh to the coast of Ayrshire, a distance of about 75 miles, but as its north-eastern end is concealed by Carboniferous formations, and its south-western passes under the sea, its true length is probably considerably more.

If the areas which present evidence of distinct and independent vents are grouped according to their positions on these two lines, they naturally arrange themselves as in the following list:—

I. NORTHERN CHAIN OF VOLCANOES

1. The Montrose Centre.
2. The Sidlaw and Ochil Group.
3. The Arran and Cantyre Centre.
4. The Ulster Centres.

II. SOUTHERN CHAIN OF VOLCANOES

5. The Pentland Volcano.
6. The Biggar Centre.
7. The Duneaton Centre.
8. The Ayrshire Group.

The distribution of these various volcanic areas will be most easily understood from an examination of Map III. accompanying this volume.

I. THE NORTHERN CHAIN OF VOLCANOES IN "LAKE CALEDONIA"

1. *The Montrose Centre*

Beginning at the north-eastern end of the area, we first encounter a series of volcanic rocks which attain their maximum thickness in Forfarshire around the town of Montrose. The main vents probably lay somewhere to the east of the present coast, under the floor of the North Sea; at least no clear indication of their existence either on the coast or inland has been detected. From Montrose, both to the north-east and south-west, the lavas thin away, becoming intercalated among the sandstones, flagstones and conglomerates, and gradually dying out. The total length of the volcanic belt is about 18 miles, that is nine miles from the central thick mass in a north-easterly and the same distance in a south-westerly direction.[1] The volcanic pile must be several thousand feet thick, but owing to the prolongation of the great Ochil anticline, the lavas roll over and do not allow their base to be seen. The axis of the fold must pass out to sea, through the hollow on which the town of Montrose stands. The volcanic series consists of andesite-sheets with volcanic conglomerates. It contains little ordinary tuff, but the conglomerates no doubt partly represent ejected fragmental material, as well as the waste of exposed lavas. A section across the anticlinal fold from Forfar to Panbride, a little to the south-west of Montrose, would reveal the structure shown in Fig. 67.

In the north-eastern prolongation of the volcanic series from the Montrose centre, successively lower members are exposed along the coast-line. But the lavas are dying out in that direction, and sometimes many hundreds of feet of ordinary sediment intervene between two successive flows. It was in one of these long pauses near the top of the whole pile of lavas that the strata of Canterland were deposited, to which reference has already been made. South-west from Montrose the thick volcanic mass rapidly diminishes, and is prolonged to the end only by three or four bands separated by sandstones and flagstones. It is in these intercalated groups of sedimentary material that the "Forfarshire flags" occur.

Nowhere can the details of the Old Red Sandstone volcanic rocks be more conveniently studied than along the coast-section in this district from the Red Head to Stonehaven. The rocks have not only been cut into vertical cliffs, but along many parts of the shore they have been also laid bare in ground-plan, so that a complete dissection of them is presented to the geologist. At the south end, the top of the volcanic series appears at the

[1] The south-western part of this area from Arbroath to Johnshaven was mapped for the Geological Survey by the late Mr. H. M. Skae, the north-eastern part by Mr. D. R. Irvine. My account of it is mainly taken from notes made by myself on the ground preliminary to the commencement of the mapping of the Survey.

bold promontory of the Red Head. There, at the base of the cliffs of red sandstone, the accompanying section may be seen. Beneath the red false-bedded and sometimes pebbly sandstones (*e*), which form nearly the whole precipice, lies a band of dull purplish ashy conglomerate (*d*), composed almost wholly of fragments of different andesites, imbedded in a paste of the same comminuted material. Towards the south, this rock rapidly becomes coarser, until it passes into a kind of agglomerate, in which the andesite blocks are sometimes a yard or more in diameter. It includes bands of sandstone, which increase in number and thickness towards the north, and sometimes intervene underneath the conglomerate.

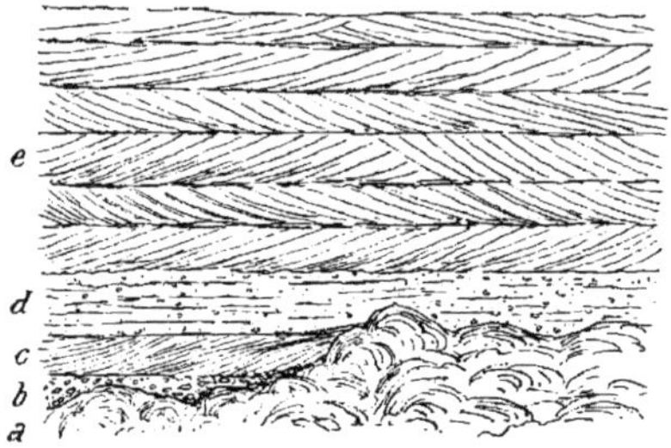

FIG. 75.—Section showing the top of the volcanic series at the foot of the precipice of the Red Head, Forfarshire.

a, Top of slaggy andesite; *b*, coarse volcanic conglomerate; *c*, Red sandstone; *d*, Tuff and volcanic conglomerate; *e*, Red sandstones.

The lowest rocks here visible are sheets of andesite or "porphyrite" (*a*), separated from each other by irregular bright red layers of tufaceous sand and agglomerate. These lavas are dull purplish-grey to green, some of them being tolerably compact, others highly amygdaloidal, with large steam-cavities often drawn out in the direction of flow.

One of the most striking features in the andesites of this coast is the remarkable manner in which they include the veinings of pale green and red sandstone already described (see Figs. 65, 66). Some of the sheets have in cooling cracked into rude polygons. They are likewise traversed by large cavernous spaces and intricate fissures or steam-cavities. Into all these openings the sand has been washed, filling them up and solidifying into well-stratified sandstone, the bedding of which is generally parallel with that of the rocks that enclose it, the dip of the whole series of strata being gently seawards. But a still more intimate mixture of the sand with the lava-sheets is to be remarked where these rocks assume their most slaggy character. In some of them the upper part, to a depth of ten or twelve feet, consists of mere rugged lumps of slag which, while the mass was in motion, were probably in large measure loose, and rolled over each other as they were borne onward. The sand has found its way into all the interstices of these clinker-beds, and now binds the whole mass firmly together. At first sight, these bands might be taken for agglomerates of ejected blocks, and as already suggested, some of the slags may have been thrown out as loose pieces, but a little examination will show that in the main the rough scoriaceous lumps are pieces of the lava underneath. In these instances, also, it is clear that the blocks were in position before the fine sand was sifted into their interspaces, for the pale green sandstone is horizontally stratified through its intricate ramifications among the pile of dark clinkers.

The seaward inclination of the rocks allows the succession of lavas to be seen as the coast is followed westward into Lunan Bay. On the further side of that inlet, after passing over a group of sandstones that underlie the volcanic series of the Red Head, the observer meets with a second and lower succession of lavas which in the five miles northward to Montrose Harbour are admirably exposed both along coast-cliffs and on the beach. They resemble those of the Red Head, being made up of alternations of highly vesicular andesite with more compact varieties, and showing similar sandstone veinings. Here and there, as at Fishtown of Usar, the sea has cut them down into a platform from which the harder parts rise as fantastic half-tide stacks. In some cases, the more durable rock consists of the slaggy upper portions of the flows, and in one case this material stands up as a rude pillar twelve feet high, composed of clinkers firmly cemented with veinings of sandstone. The geologist who wanders over this coast-line is arrested at every turn by the marvellously fresh volcanic aspect of many of the lavas. Their upper parts are so cellular that if the calcite, chalcedony and other infiltrated minerals were removed from their vesicles, they would be transformed into surfaces of mere slag. In one respect would their antiquity still be evident. These slaggy bands are generally a good deal reddened, as if they had been long exposed to oxidation before being covered by the overlying sheets of lava—a feature already cited, as probably indicating the lapse of some considerable interval of time between successive outflows.

Along this coast-section the absence of intercalated tuffs is soon remarked. The volcanic ejections seem to have consisted almost entirely of andesitic lavas, though it is possible that here and there the very slaggy bands between the more solid parts of the sheets may include a little pyroclastic material. The lowest portion of the volcanic group here visible is reached at Montrose Harbour, where, in the flagstones and shales of Ferryden, the late Rev. Hugh Mitchell obtained some of the fossil-fishes of the formation.

A space of more than three miles now intervenes where the rocks are concealed by blown sand and other superficial accumulations. It is through this hollow, as already stated, that the great Ochil anticline runs out to sea. On the north side of the North Esk River, we again come upon the same band of lavas as to the south of Montrose, but with a dip to the north-west. This inclination, however, soon bends round more westerly, and the result of the change is to expose a slowly descending section all the way to the Highland fault at Stonehaven.

A picturesque line of high inland cliff, running northwards beyond St. Cyrus, reveals with great clearness the bedded structure of the andesites. But as one moves northward, owing to the change in the direction of dip, one finally passes out of this volcanic belt and begins gradually to descend into the thick Kincardineshire Old Red Sandstone. The amount of conglomerate exposed along this part of the coast-line probably considerably surpasses in thickness any other conglomerate series in the Lower Old Red Sandstone of Britain. Throughout the enormous depth of sedimentary material, the conglomerates are well-bedded, consisting of a dull green paste, composed in large degree of comminuted andesitic debris, and interstratified with green felspathic sandstones. They are often remarkably coarse, the pebbles sometimes measuring three feet in length. Interposed among them are some ten or twelve bands, probably often single outflows of andesite, sometimes compact and porphyritic, at other times highly amygdaloidal. Such is the succession of rocks for many miles along the shore; and as the inclination varies from a little north of west to west, or even west by south, the observer gradually passes over a thickness of rather more than 2000 feet from the base of the St. Cyrus andesites to Gourdon. In this accumulation of coarse, well water-worn material, with abundant intercalations of finer sandstone and occasional sheets of lava, there is the record of prolonged and powerful denudation with intermittent volcanic activity. Dykes of a quartziferous porphyry cut the conglomerates, and at Gourdon they are pierced by the intrusion of serpentine above referred to.

The proportion of andesite fragments in the conglomerates of this part of the coast varies, but is generally much lower than that of the rocks from the Highlands. Thus at Johnshaven, out of 100 blocks, broken promiscuously from the conglomerate, I found that only 8 per cent were of andesite, while 44 per cent were of quartzite, and the remainder consisted of various quartz-porphyries, granites and schists. It is evident, therefore, that some area of crystalline rocks was subjected to enormous waste, and that its detritus was strewn over the floor of Lake Caledonia, at the same time that from the Montrose volcanic vents many streams of andesitic lava were poured forth.

A vast mass of coarse conglomerate intervenes between Gourdon and Dunnottar, and forms a nearly continuous line of precipices which in some places rise 200 feet above the waves. The bedding is everywhere distinctly marked, so that there is no difficulty in following the succession of the strata, and estimating their thickness. From the last of the lavas at Gourdon to the base of the conglomerates near Stonehaven, there lies an accumulation of conglomerate at least 8000 feet thick. The boulders and pebbles in these deposits are generally well-rounded, and vary up to four feet or more in length. I observed one of quartz-porphyry at Kinneff which measured seven feet long and six feet broad. The proportion of andesite fragments in these conglomerates continues to be small. I ascertained that in the coarsest mass at Kinneff they numbered only 14 per cent; at Todhead Point, a mile and a half to the north, 20 per cent, and at Caterline, three quarters of a mile further in the same direction, 21 per cent.

In the midst of this gigantic accumulation of the very coarsest water-worn detritus, there are still records of contemporaneous volcanic action. Near Kinneff the beautiful andesite, with large tabular crystals of plagioclase, alluded to on p. 274, occurs in the conglomerate.[1] South of Caterline

[1] For an analysis of the felspar in this rock, see Prof. Heddle's paper, *Trans. Roy. Soc. Edin.* xxviii. (1879), p. 257.

two flows, lying still lower in the system, project into the sea. One of these presents a section of much interest. It shows a central solid portion, jointed into rudely prismatic blocks, with an indefinite platy structure, which gives it a roughly-bedded aspect. Its upper ten or twelve feet are sharply marked off by their slaggy structure, ending upwards in a wavy surface like that of the Vesuvian lava of 1858. Into its fissures, steam-cavities and irregular hollows, fine sand has been washed from above, as at Red Head, while immediately above it comes a coarse conglomerate of the usual character (Fig. 76). Still lower down, beneath some 900 feet of remarkably coarse conglomerate, another group of sheets of andesite abuts at Crawton upon the coast, with which, at a short distance inland, it runs parallel for more than two miles, coming back to the sea at Thornyhive Bay and at Maidenkaim. We have then to pass over about 5000 feet of similar conglomerates, until, after having crossed several intercalated sheets of andesite, we meet with the last and lowest of the whole volcanic series of this region in the form of some bands of porphyrite at the Bellman's Head, Stonehaven. The peculiar geographical conditions that led to the formation of the coarse conglomerates appear to have been established at the same time that the volcanic eruptions began, for as we descend in the long coast section, we find that the coarse sediment and the intercalated lavas cease on the same general horizon. Below that platform lie some 5000 feet of red sandstones and red shales, yet the base of the series is not seen, for the lowest visible strata have been faulted against the schists of the Highlands. It is thus obvious that more than 5000 feet of sediment had been laid down over this part of the floor of Lake Caledonia before the first lavas were here erupted.

FIG. 76.—Andesite with sandstone veinings and overlying conglomerate. Todhead, south of Caterline, coast of Kincardineshire.

2. *The Sidlaw and Ochil Group*

The volcanoes which poured out the masses of material that now form the chain of the Ochil and Sidlaw Hills appear to have been among the most vigorous in the whole region of Lake Caledonia. Their chief vents probably lay towards the south-west in the neighbourhood of Stirling, where the lavas, agglomerates and tuffs discharged from them reach a thickness of not less than 6500 feet, without revealing their bottom. From that centre the lavas range continuously for nearly fifty miles to the north-east, until they reach the sea at Tayport; but they are prolonged on the north side of the Firth of Tay from Broughty Ferry to near Arbroath, so as to overlap those of the Montrose group. They thus attain a total length of nearly sixty miles in a north-easterly line. How far they stretched south-west cannot now be ascertained, for they have been dislocated and buried in that direction under the Carboniferous formations of the Midland Valley.

It will be observed from the map (No. III.) that the great volcanic ridge of the Ochil Hills continues unbroken for twenty-two miles, from Stirling to Bridge of Earn. Thereafter it branches into two divergent portions, one of which runs on through the north of Fife to the southern promontory of the estuary of the Tay, while the other, after sinking below the alluvial plains of the Earn and the Tay, mounts once more into a high ridge near Perth, and thence stretches eastward into Forfarshire as the chain of the Sidlaw Hills. This bifurcation is due to the opening out and denudation of the great anticlinal fold above mentioned. The rocks in the northern limb dip north-westward, those in the southern limb dip south-eastward. The lower members of the Old Red Sandstone, underlying the volcanic series, ought to be seen beneath them along the crest of the anticline. Unfortunately, however, partly by the action of faults along the boundaries of the volcanic bands, but chiefly from the unconformable overspread of Upper Old Red Sandstone and Lower Carboniferous rocks across the plains of the Carse of Gowrie and of the Earn, the lower parts of the system are there concealed (see Fig. 78). As already remarked, this important anticlinal fold runs to the north-east across Forfarshire, and passes out to sea north of Montrose.

Through the Ochil chain the fold runs obliquely in a south-westerly direction, until it is truncated by the great fault which lets down the Clackmannan coalfield. The total traceable length of this anticline is thus about sixty miles. It flattens down towards the south-west; consequently the rocks in the western part of the Ochil Hills are so gently inclined that the same bands may be followed winding round the sides of the valleys, and giving to the steep declivities the terraced contours to which allusion has already been made (see Fig. 68). Another result of this structure is that the base of the volcanic series is entirely concealed by its higher portions.

From an examination of the map it will be further obvious that the whole wide plain of Strathmore—that is the great hollow, more than 80 miles long and about ten or twelve miles broad, which stretches between the base of the Highland mountains and the north-western slopes of the Ochil and Sidlaw chain—is underlain with volcanic rocks of Lower Old Red Sandstone age. This plain lies on a broad synclinal fold, along the south-east side of which the lavas, tuffs and conglomerates of the Ochil and Sidlaw Hills dip under a thick accumulation of red sandstone and flagstone. On the north-west side similar lavas and tuffs rise again to the surface, both on the southern side of the great boundary faults, and also in the little bays which here and there survive on the northern side of the dislocations (Fig. 77). I have already alluded to these interesting relics of the shore-line of Lake Caledonia, and to the fact that though they lie unconformably on the Highland schists, they do not belong to the actual basement members of the Old Red Sandstone (*ante*, p. 295, and Fig. 73). We have seen that below the bottom of the volcanic series a thickness of 5000 feet of sandstones and shales emerges on the Stonehaven coast, and yet that even there the base of the whole system is not visible, owing to the effect of the Highland boundary fault.

It is thus evident that over the bottom of Lake Caledonia a very thick deposit of tolerably fine sedimentary material was spread before the commencement of the Ochil and Sidlaw eruptions,—that when the lavas were poured out and the coarse conglomerates began to be formed, these materials overlapped the older deposits and gradually encroached upon the subsiding area

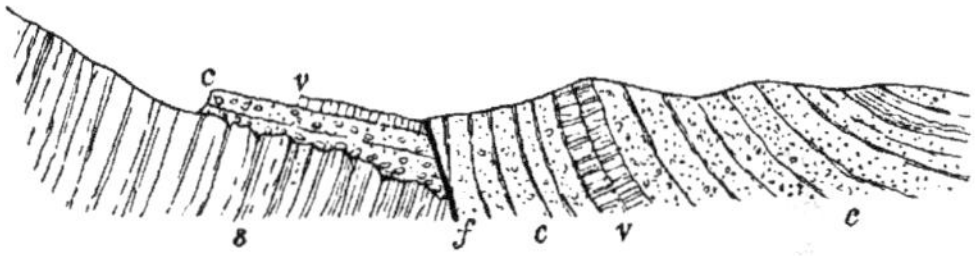

FIG. 77.—Section across the Boundary-fault of the Highlands at Glen Turrit, Perthshire.
s, Crystalline schists of the Highlands; c c, conglomerates and sandstones (Lower Old Red Sandstone) with interstratified volcanic rocks (v v); f, fault.

of the Highlands. The lavas rolled across the floor of the lake and entered the successive bays of the northern coast-line, where their outlying patches may still be seen.

From these facts it is clear that to the actually visible area of volcanic material in the Ochil and Sidlaw region, and to the anticlinal tract whence the andesites have been removed by denudation, we have to add the area that lies under the plain of Strathmore, which may be computed to be at least 800 square miles, making a total of probably not less than 1300 square miles. But it will be remembered that practically only one side of the anticlinal fold is accessible to observation. We cannot tell how far in a southerly direction the lavas of the Ochil Hills may extend. It is quite possible that not a half of the total area covered by the eruptions of this volcanic group is now within reach, either of observation or of well-founded inference.

One further general characteristic of this volcanic district will be obvious from an inspection of the map. While the thickest mass of lavas and tuffs, lying towards the south-west, points to the existence of the most active vents in that part of the area, the actual positions of these vents have not been detected. Probably they lie somewhere to the south of the edge of the Ochil chain, under the tract which is overspread with the coalfield. But other and possibly minor orifices of eruption appear to have risen at irregular intervals towards the north-east along the length of the lake. Thus there are numerous bosses of felsitic and andesitic rocks among the central Ochils, some of which may mark the positions of active vents. For some miles to the east of that area an interval occurs, marked by the presence of only a few small intrusive masses. But as the broad anticline of the Firth of Tay opens out and allows the lower or pre-volcanic members of the Old Red Sandstone to approach the surface, another group of bosses emerges from the lower sandstones and flagstones. Some of these cover a considerable space at the surface, though a portion of their visible area may be due to lateral extravasation from adjacent pipes, the true dimensions of which are thereby obscured. Some of the masses are undoubtedly sills. In the case of Dundee Law we probably see both the pipe and the sill which proceeded from it; the prominent, well-defined hill marking the former, while the band of rock which stretches from it south-westwards to the shore belongs to the latter. The material that forms the bosses and sills in this neighbourhood is generally a dark compact andesite. The rock of Dundee Law was found by Dr. Hatch to show under the microscope "striped lath-shaped felspars abundantly imbedded in a finely granular groundmass, speckled with granules of magnetite, but showing no unaltered ferro-magnesian constituents." Here and there in the same district a solitary neck may be observed filled with agglomerate (Fig. 78).

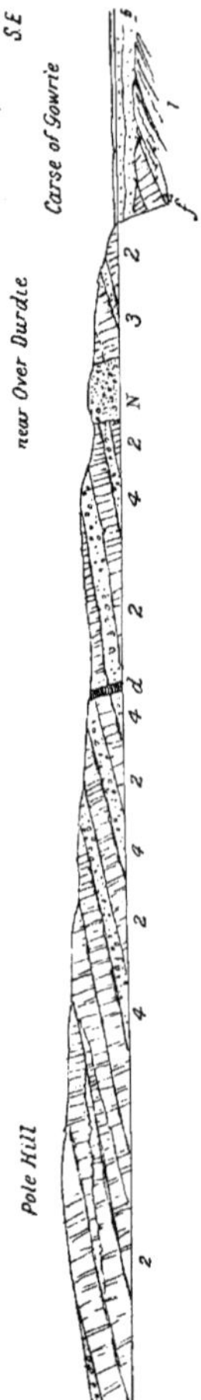

FIG. 78.—Section across the chain of the Sidlaw Hills, near Kilspindie.

1. Lower Old Red Flagstones and Sandstones; 2. Andesite lavas; 3. Volcanic tuff; 4. Volcanic conglomerates and sandstones; N, Volcanic neck; 5. Upper Old Red Sandstone under Carse of Gowrie, lying unconformably on the lower division; *f*, Fault; *d*, Basic dyke.

The variations in the structure of the Ochil and Sidlaw volcanic group will be most easily understood from a series of parallel sections. Beginning on the north-eastern or Sidlaw branch of the volcanic band, we find the arrangement of the rocks to be as is shown in the accompanying figure[1] (Fig. 78). As is usually the case in this region, the base of the volcanic series is here concealed by the fault which brings down the Upper Old Red Sandstone under the alluvial deposits of the Carse of Gowrie. The total thickness of the series in this section is about 2500 feet. The rocks consist of successive sheets of andesite of the familiar types, varying in colour through shades of blue, purple and red, and in texture from a dull compact almost felsitic character to more coarsely crystalline varieties. They are often amygdaloidal, especially in the upper and lower portions of the individual flows. They are not infrequently separated from each other by courses of conglomerate or ashy

[1] This section and the notes accompanying it have been supplied by Prof. James Geikie, who mapped the western half of the Sidlaw range for the Geological Survey. The eastern half was mapped by the late Mr. H. M. Skae.

sandstone and grit. Of these intercalations four are of sufficient thickness and persistence to be mapped, and are shown on the Geological Survey Sheet 48. The stones in the conglomerates vary up to blocks two feet in diameter, and consist chiefly of andesites, but include also some pink felsites and pieces of greenish hardened sandstone. Generally they are more or less well-rounded; but occasionally they become angular like those of volcanic agglomerates.

One of the most interesting features in this section is the neck which at Over Durdie rises through the volcanic series. Oval in form, it measures 630 yards in one diameter and 350 in another, and is filled with pinkish granular tuff, full of andesitic lapilli and blocks. A much smaller neck of similar material lies about 100 yards further to the south-west. There seems no reason to doubt that these necks mark two of the volcanic vents belonging to a late part of the volcanic history of the district.

The structure of the Sidlaw range is repeated among the hills of east Fife on the southern side of the great anticlinal fold.[1] Thus a section from near Newburgh on the Firth of Tay southward to near Auchtermuchty in Strathedeu gives the arrangement of rocks shown in Fig. 79. In this

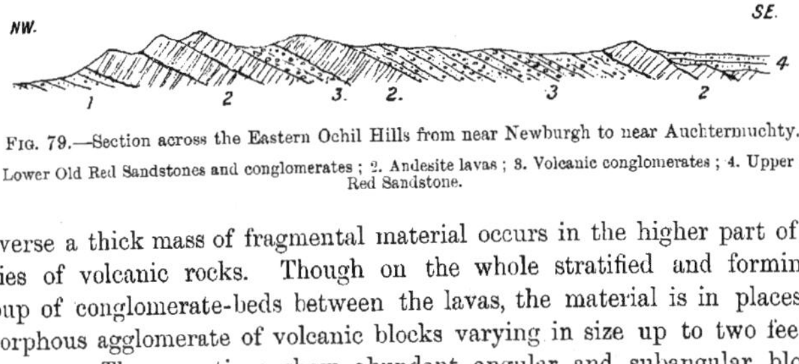

FIG. 79.—Section across the Eastern Ochil Hills from near Newburgh to near Auchtermuchty.

1. Lower Old Red Sandstones and conglomerates; 2. Andesite lavas; 3. Volcanic conglomerates; 4. Upper Old Red Sandstone.

traverse a thick mass of fragmental material occurs in the higher part of the series of volcanic rocks. Though on the whole stratified and forming a group of conglomerate-beds between the lavas, the material is in places an amorphous agglomerate of volcanic blocks varying in size up to two feet in diameter. These portions show abundant angular and subangular blocks, many of which, after having undergone some attrition, have been finally broken across before reaching their present resting-places. Sharply fractured surfaces can be picked out of the felspathic ashy matrix. The stones are chiefly varieties of andesite, but they include also pink felsites and pieces of some older fine-grained tuff.

These fragmental materials form a local deposit about nine miles long, and probably not less than 1700 feet thick. They are partly interstratified with flows of andesite. Though, from the rounded forms of some of the pebbles, wave-action may be inferred to have been concerned in their accumulation, they seem to be mainly due to volcanic explosions. No trace, however, has been found of the vent from which the eruptions took place. Not improbably its site lies somewhere to the south in the area now concealed under the Upper Old Red Sandstone and Carboniferous formations. The large size of many of the blocks suggests that they do not lie far from their

[1] The eastern part of the Ochils was mapped for the Geological Survey by Mr. H. H. Howell and Mr. B. N. Peach.

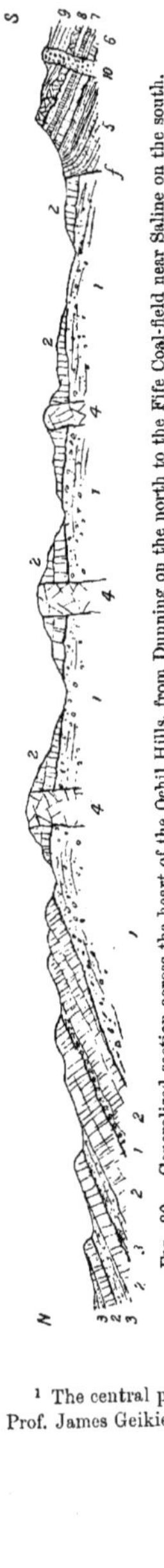

FIG. 80.—Generalized section across the heart of the Ochil Hills, from Dunning on the north to the Fife Coal-field near Saline on the south.

1. Volcanic tuffs and agglomerates; 2. Andesite lavas; 3. Lower Old Red Sandstone and conglomerate; 4. Necks of felsitic rocks; 5. Upper Old Red Sandstone and Calciferous Sandstones; 6. Representative of the Plateau lavas and tuffs of the Lower Carboniferous series; 7. Hurlet (Carboniferous) Limestone; 8. Dolerite sill; 9. Sandstones, shales and coals of the Carboniferous Limestone series; 10. Neck of the Puy series (Carboniferous); *f*, Fault.

parent focus of discharge. It is impossible to tell how much of the volcanic series is here concealed by the unconformable overlap of the younger formations.

A section across the centre of the Ochil chain,[1] from Dunning in Strathearn to the Crook of Devon and the Fife Coal-field, gives the structure which is generalized in Fig. 80. At the north end the volcanic series is found to be gradually split up into separate lava-sheets until it dips under the red sandstones of Strathearn. Traced southwards the rocks become entirely volcanic. Some of their most conspicuous and interesting members are pale felsitic tuffs, which occupy a considerable tract of ground about Craig Rossie, south-east of Auchterarder. As the dip gradually lessens the harder lavas are able to spread over wider tracts of ground, capping the hills and ridges, while underneath them thick masses of tuff and conglomerate are laid bare in the valleys. A number of bosses of orthophyre rise through these rocks and are accompanied by many veins and dykes of similar material. It is not improbable that some of these bosses, as already suggested, may represent vents. They are especially prominent among the hills due south of Auchterarder. One of these eminences, known as the Black Maller, is composed of a typical orthoclase-felsite without mica. Another, about four and a half miles further south, forms the conspicuous summit of Ben Shee overlooking Glen Devon, and consists of a similar rock with a characteristic platy structure.

No necks of agglomerate have been observed in this part of the chain. It will be seen from the section that the lowest visible parts of the Ochil volcanic series are here truncated by a fault which brings in the lower part of the Carboniferous system. By a curious conjuncture, immediately on the south side of this fault, a band of tuff appears, lying on the platform of the Carboniferous "plateau-lavas," to be hereafter considered, and passing below the well-known Hurlet seam of the Carboniferous Limestone, while through these strata rises one of the puys belonging to the second phase of volcanic activity in Carboniferous time in Scotland.

The best sections to show the nature and sequence

[1] The central portion of the Ochils was mapped for the Geological Survey by Mr. B. N. Peach, Prof. James Geikie, Prof. J. Young, Mr. R. L. Jack and myself.

of the volcanic series of the Ochil Hills are to be observed at the west end of the chain. But as the whole succession of rocks cannot conveniently be obtained along one line, it is better to make several traverses, starting in each case from a known horizon. In this way, by means of three parallel sections, we may obtain the whole series of lavas and tuffs in continuous order. The first line of section starts in the lowest part of the tuffs represented at the bottom of the group in Fig. 80, and runs up to the first thick ashy intercalation among the lavas. Following this bed south-westward to the Burn of Sorrow, we make from that horizon a second traverse across the strike to the summit of King's Seat Hill (2111 feet above the sea), where we meet with a well-marked lava which can be traced south-westwards, gradually descending the southern escarpment of the hills until it reaches the boundary fault near the village of Menstrie. Starting again from this definite horizon, we take a third line across the top of Dumyat (1373 feet) to the plain of Sheriffmuir, and there pass beyond the volcanic series into the overlying red sandstones. Arranged thus in continuous vertical sequence the succession is found to be as represented in Fig. 81. The total thickness of volcanic material amounts to more than 6500 feet.

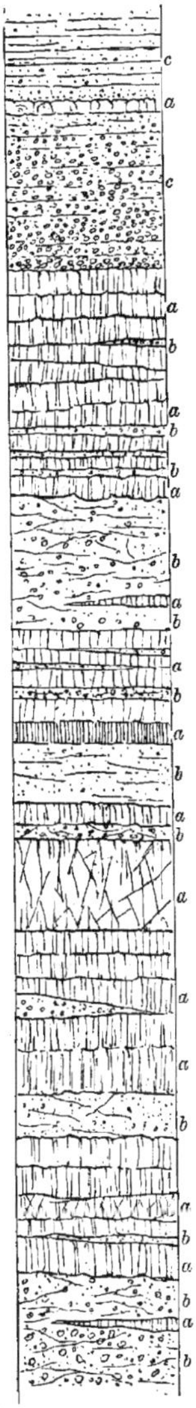

FIG. 81.—Diagram of the volcanic series of the Western Ochil Hills.

The bands with vertical lines are various lavas (*a*); the tuffs and volcanic breccias are shown by the dotted bands (*b*); the uppermost portion of the section above the last thick group of lavas consists of conglomerates and sandstones (*c*) with a sheet of lava.

In this vast pile of volcanic ejections the lavas are almost entirely andesites of the usual characters. They include many slaggy and amygdaloidal varieties, some beautiful porphyries with large tabular felspars, likewise the resinous or glassy variety already referred to as occurring above Airthrey Castle. Their upper and under surfaces show the same structure as already described in those of the coast-sections in the Montrose tract. They include also more acid lavas, like the pale pink decomposing felsites of the Pentland Hills.

The tuffs and conglomerates occur on many platforms throughout the succession of lava-sheets. They form the lowest visible part of the whole volcanic series, but they are most abundant towards the top, and are best displayed at the western end of the hills. In Dumyat they form a conspicuous feature. The whole of that hill consists of a constant alternation of lavas (chiefly slaggy andesites, but including also one felsitic flow) with bands of coarse and finer tuff and volcanic conglomerate. The greatest continuous mass of this fragmental material is 600 or 700 feet thick. From the extraordinary size of its included blocks it obviously must have been formed of ashes, stones and huge pieces of lava ejected from some vent in the near neighbourhood. Some of the individual blocks in this mass are as large as a Highland crofter's cottage.

The uppermost lavas of Dumyat dip under a still higher series of coarse volcanic conglomerates entirely made up of andesitic debris and reaching a thickness of about 1000 feet. This enormous accumulation was probably due partly to the abrasion of exposed cones and lava-ridges, and partly to volcanic discharges of fragmentary materials. Yet it is worthy of note that even amidst these evidences of the most vigorous volcanic activity we have also proofs of quiet sedimentation and traces of the fishes that lived in the waters of the lake. This particular zone of coarse conglomerate as it extends in a south-westerly direction becomes finer, and its upper part passes into a chocolate-coloured sandstone which has been quarried at Wolfe's Hole, Westerton, Bridge of Allan, at a distance of about three miles from where the line of section runs, which is embodied in the diagram, Fig. 81. It was from this locality that the specimens of *Encephalaspis*, *Pteraspis* and *Scaphaspis* were obtained which were described by Professor Ray Lankester.[1]

Above the last-named thick group of coarse volcanic conglomerates a solitary sheet of dark slaggy andesite may be observed. This lava is then overlain by the great depth of chocolate-coloured and red sandstones and marls of the plain of Strathmore (*c* in Fig. 81). Nevertheless a few hundred feet up in these sedimentary deposits we meet with yet one further thin sheet of lava—the last known eruption of the long volcanic history of this district.

Before quitting the Ochil range I may refer to the evidence there obtainable as to the horizontal extent of separate sheets of lava. The western end of this range affords great facilities for following out individual beds of andesite along the bare terraced front of the great escarpment. Thus, the easily recognizable porphyrite which caps King's Seat Hill, above Tillicoultry (see Fig. 68), can be traced winding along the hill-slopes until it descends to the plain, and is then lost under the great fault, at the foot of Dumyat—a distance of more than six miles. There is, therefore, no difficulty in supposing that from the Ochil line of vents streams of lava should have rolled along the floor of the lake across to the base of the Highland slopes, 10 or 12 miles distant. We cannot tell, of course, whether any buried vents lie below the plain of Strathmore, but certainly no unquestionable trace of vents has yet been found among the crystalline rocks along the borders of the Highlands.[2]

[1] *Palæontographical Society*, vols. xxi. (1867) and xxiii. (1869).

[2] Allusion has already been made to the possible connection of the younger Highland granites

Reference has already been made to the comparative scarcity of sills in this region, and to the occurrence of the acid group of Lintrathen porphyry and the more basic sheets between the Firth of Tay and Forfar. This scarcity no doubt arises in part from the extent to which the rocks that underlie the volcanic series are concealed. Yet it is noteworthy that along the coast-section of these rocks near Stonehaven hardly any intrusive sheets are to be seen.

3. *The Arran and Cantyre Centre*

It is unfortunate that the Ochil chain should be broken across and buried under younger formations at the very place where some of the most interesting vents in the whole area of the Old Red Sandstone might have been looked for.[1] We have to pass westwards across the Firth of Clyde to the Isle of Arran before we again meet with rocks of the same age and character.

In the course of the recent work of the Geological Survey in that island, Mr. W. Gunn has discovered that the Lower Old Red Sandstone includes some interstratified volcanic rocks on the north side of North Glen Sannox, and he has supplied me with the following notes regarding them. "The area in which the volcanic intercalations occur is much faulted and only a part of it has been mapped in detail, but the position of the inter-bedded igneous rocks is quite clear. The Old Red Sandstone here consists of three distinct members, the lowest of which is made up of coarse, well-rounded conglomerates, alternating with sandstones and purple mudstones. Above this, and apparently unconformable to it, is a middle series of light coloured conglomerates and sandstones, the pebbles in which are mainly of quartz. Finally comes an upper series of red sandstones and conglomerates, which occupy nearly the whole of the coast section, and it is this series which has generally been taken as the typical Old Red Sandstone of the island. The volcanic series is intercalated between the middle and upper divisions given above, and may be seen in several places on the hillside between the shepherd's house at North Sannox and Laggan. It consists mainly of old lava-beds of a dull reddish or purplish colour, often soft, and in places much decomposed. It seems basic in character. A specimen from near the Fallen Rocks, examined by Mr. Teall, was found to be too much altered for precise determination, but was probably a basalt originally. These rocks do not occur on the coast."

In the southern extremity of Cantyre some important relics of the volcanic rocks of the Lower Old Red Sandstone have been recently detected and mapped for the Geological Survey by Mr. R. G. Symes.[2]

with the volcanic series of the north-eastern part of Lake Caledonia; also to the occurrence of isolated masses of breccia piercing the crystalline schists near Loch Lomond (*ante*, p. 272).

[1] The Ochil area is not the only example of the abrupt termination of a volcanic band near its centre owing to faults or overlaps. The sudden disappearance of the Pentland lavas and tuffs on the northern side of the Braid Hills is another striking illustration.

[2] The late Prof. James Nicol published in 1852 an account of the geology of the southern portion of Cantyre. He grouped all the igneous rocks of the district as one series, which he

This division of the system has been ascertained by him to be extensively developed to the south of Campbeltown, and to include some small but interesting remains of the volcanic action which was so marked a feature in the areas of Lake Caledonia, lying further to the east. To the student of volcanic geology, indeed, this small tract at the extreme southern end of Argyllshire has a peculiar interest, for in no other part of the British Isles have the phenomena of the eruptive vents of the Lower Old Red Sandstone been more admirably laid bare. Not only are there necks in the interior like that represented in Fig. 82; but others have been dissected by the

FIG. 82.—View of Cnoc Garbh, Southend, Campbeltown. A volcanic neck of Lower Old Red Sandstone age, about 400 yards wide in its longer diameter.

waves along the southern shore, and their relations to the deposits of fragmentary material showered over the bottom of the lake have been more or less clearly exposed.

At Keil Point, a little to the east of the most southerly headland of the Mull of Cantyre, some reddish and purplish highly felspathic sandstones (*a* in Fig. 83) dipping towards the east are found to pass upward into coarse volcanic breccias (*b*), which, followed eastwards, lose almost all trace of stratification, and are then abruptly succeeded by a neck of coarse agglomerate (*c*) measuring 25 yards from north to south, where its limits can be seen, and at least 12 yards from west to east. It is hardly possible to distinguish between the breccias to the west and the agglomerate of the

regarded as later than the Coal-formation and possibly of the same age as those of the north-east of Ireland. He made no distinction between the Lower Old Red Sandstone and the younger unconformable conglomerates (*Quart. Journ. Geol. Soc.* vol. viii. (1852), p. 406).

neck, except by the rude bedding of the former which pass down into the well-bedded sandstones.

The agglomerate is a thoroughly volcanic rock. The materials consist chiefly of angular blocks of a pale purplish or lilac highly porphyritic mica-porphyrite, with large white felspars and hexagonal tables of black mica. These blocks might sometimes be mistaken for slags from their cavernous, weathered surfaces, but this rough aspect is found on examination to be due to the decay of their felspars.

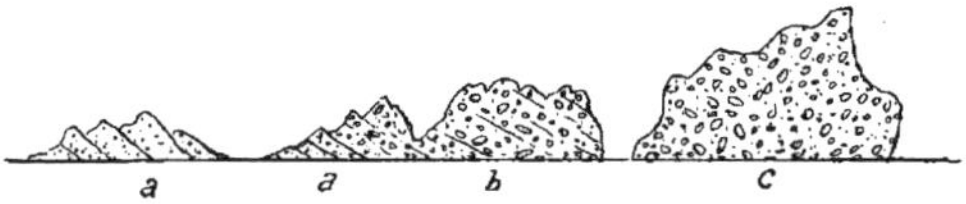

FIG. 83.—Section of volcanic series on beach, Southend, Campbeltown.

a, Fine reddish and purplish highly felspathic sandstones, largely composed of porphyry-debris and passing up into coarse breccias; *b*, volcanic breccias, coarse and only rudely stratified, formed of blocks of porphyry, sandstone fine tuff and andesite, together with water-worn quartzite pebbles derived from some conglomerate; *c*, coarse unstratified agglomerate forming a neck.

Perhaps the most singular feature among the contents of this neck is the number of well-rounded and smoothed pebbles and boulders of quartzite. These are dispersed at random through the mass, and are often placed on end. There can be no doubt that they are water-worn stones, but the contrast of their smooth surfaces and rounded forms with the rough angular blocks of igneous material is so striking as to lead at once to the conclusion that they cannot have acquired their water-worn character in the deposit where they now lie. Their positions and their occurrence with ejected volcanic blocks suggest that they too were discharged by volcanic explosions. They so exactly resemble the quartzite boulders and pebbles in the neighbouring Old Red Conglomerates that there can be little hesitation in regarding them as derived from these conglomerates. They seem to me to have come from a lower part of the Old Red Sandstone, which was shattered by volcanic energy either before the conglomerates were firmly consolidated or afterwards by such violent explosions as served to separate the pebbles from the matrix of the rock.

There occur also in the agglomerate blocks of fine tuff and ashy sandstone sometimes four feet long, and often stuck on end, showing that the deposits of earlier eruptions were broken up during the drilling of this little vent.

A few hundred yards further east a larger neck rises on the beach, immediately to the south of the old Celtic chapel of St. Columba. It consists also of exceedingly coarse agglomerate, with andesite blocks three and four yards in diameter. It is about 125 yards broad from east to west, on which sides it is seen to be flanked by coarse volcanic breccias and conglomerates, resembling in composition the materials of the neck, but showing an increasingly definite stratification as they are traced eastward in the ascending succession of deposits. Following the section in still the same easterly direction along the coast, we find that bands of fine felspathic sandstone, marking probably intervals of quiescence, are again and again succeeded by coarse brecciated conglomerates of igneous materials, which may be inferred to have been due to a renewal of violent eruptions. By degrees the evidence of stratification and of attrition among the volcanic materials becomes more pronounced as the ascending section is followed; blocks of andesite, even 18 inches or two feet in diameter, assume well-rolled, rounded, water-worn forms, like the pebbles of quartzite associated with them, and eventually the strata return to the usual aspect of the conglomerates of the district.

I have never seen anywhere better proofs of volcanic explosions, contemporaneous with a group of strata, and of the distribution of volcanic fragmentary material round the vents. A further point of much interest is the additional evidence furnished by this shore-section of considerable wave-action during the accumulation of the coarse conglomerates. To give to blocks of porphyrite two feet in diameter a smoothed and rounded form must have required the action of water in considerable agitation.

4. *The Ulster Centres*

From the volcanic breccias and conglomerates of the Mull of Cantyre to the coast of Antrim in a straight line is a distance of little more than twenty miles. On a clear day the Old Red Sandstone of Cross Slieve, and the range of cliffs in which it abruptly descends to the sea between Cushendall and Cushendun, can be distinctly seen from the Argyllshire shore. The geologist who passes from the Scottish to the Irish sections cannot fail to be impressed with the resemblance of the rocks in the two countries, and with the persistence of the types of conglomerate in Lake Caledonia.

A picturesque section has been laid bare between the Coastguard Station south of Cushendall and Cushendun Bay.[1] At the south side of the little inlet of Cushendall, a compact dull quartz-porphyry is exposed in crags along the shore. This rock ranges in colour from dark brown and purple to pale-green and buff. Its texture also varies, as well as the proportion of its felspar-crystals and quartz-blebs. Some parts have a cavernous structure, like that of an amygdaloid, the small globular cavities being filled with green decomposition products.

The stratigraphical relations of this rock are not quite clear, but it is certainly older than the Old Red conglomerates which lie to the north of it, for these are largely made up of its fragments. The matrix of these detrital masses consists mainly of the comminuted debris of the porphyry. The pebbles include all the varieties of that rock, and are tolerably well-rounded. There is no distinct evidence of volcanic action among these conglomerates. They resemble, however, many of the conglomerates in the Midland Valley of Scotland, which, as in the case of those on the Forfarshire and Kincardineshire coast, are in great part made of the detritus.

[1] For descriptions of this district see J. Bryce, *Proc. Geol. Soc.* i. (1834) p. 396, v. (1837) p. 69; J. Kelly, *Proc. Roy. Irish Acad.* x. (1868), p. 239. The area is contained in Sheet 14 of the Geological Survey of Ireland, and was mapped by Mr. A. M'Henry and described by him in the accompanying Explanatory Memoir (1886), pp. 12, 25.

of andesitic lavas. The Cushendall rocks become coarser as they are traced northwards into lower members of the series, while at the same time the proportion of porphyry-debris in their constitution diminishes, and materials from the metamorphic series take its place. Thus at Cushendun the percentage of quartz-pebbles rises to 70 or 80. These blocks, of all sizes up to two feet or more in diameter, are admirably rounded and smoothed, like those in the Stonehaven section and those among the conglomerates at the south end of Cantyre. Fragments of the porphyry, however, still continue to appear, and the matrix shows an admixture of the finer detritus of that rock. I may remark in passing that no conglomerates of the Old Red Sandstone show more strikingly than these at Cushendun the effects of mechanical crushing subsequent to deposition and consolidation. In many parts of the rock it is hardly possible to find a rounded block that has not been fractured. Some of them, indeed, may be seen cut into half a dozen slices, which have been pushed over each other under the strain of strong lateral or vertical pressure.

In the interior of the country, after passing over the broad Tertiary basaltic plateau of Antrim, we come upon a large area of Lower Old Red Sandstone in Tyrone. It stretches from Pomeroy to Loch Erne, a distance of about 30 miles, and is about 12 miles broad. In lithological character the strata of this tract exactly resemble parts of the deposits of Lake Caledonia in Central Scotland. They include also a volcanic series which, down to the smallest points of detail, may be paralleled in the sister island.[1] This interesting westward prolongation of the volcanic record consists of a number of outlying patches confined to the eastern part of the district.

The largest of these patches lies to the south of Pomeroy, where it forms a line of hills about four miles long, and covers an area of some five square miles. The rocks consist of successive sheets of andesite-lavas. These, as a rule, are not markedly cellular, though they include some characteristic amygdaloids. A distingushing feature of some of the sheets is their remarkably well-developed flow-structure. Thus on Sentry Box, at the north-western end of the ridge, the fissility resulting from this structure so perfectly divides the rock into parallel flags that the material might easily be mistaken for a bedded rock. Where this structure has been produced in a cellular lava, the cavities have been drawn out and flattened in the direction of flow.

I have not observed true tuffs in any of the sections traversed by me in this district. But the conglomerates furnish abundant evidence of the contemporaneous outpouring of the lavas. Thus, in a brook a little west of Reclain, five miles south of Pomeroy, the section shown in Fig. 84 may be seen. At the base lies a coarse conglomerate (*a*) largely composed of andesite debris, the stones being here, as elsewhere in the district, well rounded. Then comes a series of green and reddish highly-felspathic sandstones (*b*), followed by an exceedingly coarse conglomerate (*c*), formed mainly of the debris of andesites, especially lumps of slag. Some of the stones

[1] This area of Old Red Sandstone is represented on Sheets 33, 34, 45 and 46 of the Geological Survey of Ireland, and the igneous rocks are described in the Memoirs on Sheets 33 (1886, p. 17) by Mr. J. R. Kilroe, and 34 (1878, p. 16) by Mr. J. Nolan.

measure 18 inches in diameter, and all are well water-worn. Immediately over this mass of detritus lies the lowest sheet of andesite-lava (*d*).

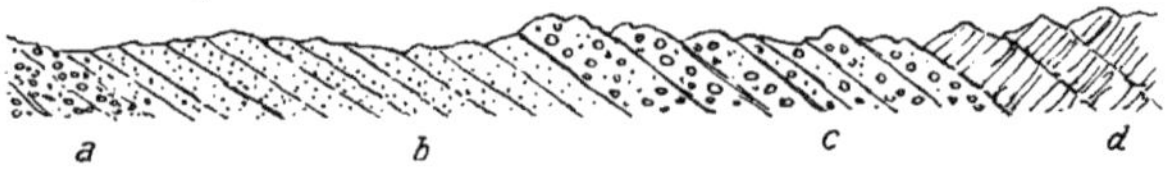

FIG. 84.—Section of the base of the volcanic series, Reclain, five miles south of Pomeroy.

Some sections visible in the neighbourhood of Omagh afford further evidence of volcanic action at the time of the deposition of the Old Red Sandstone of this region. At Farm Hill, a little to the east of the town, felspathic sandstones and breccias enclose angular and subangular pieces of various andesites, and occasionally even pieces of tuff. Near these strata a decayed andesite occurs in the bed of a stream, and a fresher variety is quarried at Farm Hill. A little further south another variety of andesite is exposed in two quarries at Recarson Meeting-House—a fine granular purplish-grey rock, with abundantly-diffused hæmatite pseudomorphs, probably after a pyroxene, and sometimes strongly amygdaloidal.

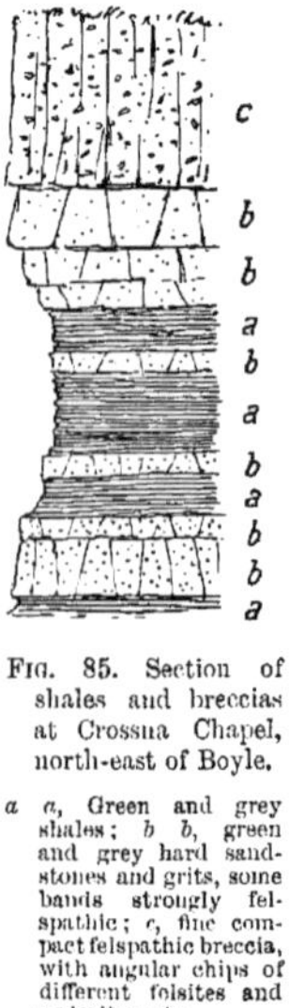

FIG. 85. Section of shales and breccias at Crossna Chapel, north-east of Boyle.

a a, Green and grey shales; *b b*, green and grey hard sandstones and grits, some bands strongly felspathic; *c*, fine compact felspathic breccia, with angular chips of different felsites and andesites, etc.

There can thus be no doubt that this region of Ulster included several centres of volcanic activity during the deposition of the red sandstones and conglomerates, and that the lavas and volcanic conglomerates belonged to precisely the same types as those of the same geological age which occur so abundantly in Scotland.

Further south-west, near Boyle, in the county of Roscommon, certain curious felspathic breccias in the Old Red Sandstone have been mapped as "felstone."[1] So far as I have been able to examine them, however, they are entirely of fragmental origin. They contain pieces of andesitic and felsitic rocks, with fragments of devitrified glass, which undoubtedly point to the occurrence of volcanic eruptions during their deposition, though no tuffs and lavas appear to crop out in the narrow strip of the formation there exposed.

The accompanying section (Fig. 85) may be seen on the hills to the north-east of Boyle. Where quarried on the roadside to the north of Boyle, the series of deposits here represented contains a bed of coarse and exceedingly compact breccia, similar to that just referred to, but containing angular and subangular fragments six or eight inches long. The joints of these compact strata are remarkably sharp and clean cut, so that where the fragmentary character is not very distinct the rocks might easily be mistaken on casual inspection for felsites.

[1] See Sheet 66 Geological Survey of Ireland, and Explanation to that sheet (1878), p. 15. The rocks were previously described by Jukes and Foot, *Journ. Roy. Geol. Soc. Ireland*, vol. i. (1866), p. 249.

CHAPTER XX

VOLCANOES OF THE LOWER OLD RED SANDSTONE OF "LAKE CALEDONIA"—*continued*

The Southern Chain—The Pentland Volcano—The Biggar Centre—The Duneaton Centre—The Ayrshire Volcanoes.

WE have now to note the leading features of the groups of volcanic rocks distributed along the southern line of vents already described. At least four different centres of eruption may be observed on that line. Their mutual limits are, on the whole, better seen than those of the northern line, for from the north-eastern to the south-western end of the volcanic belt the Old Red Sandstone and rocks of older date are almost continuously exposed at the surface. The encroaching areas of Carboniferous formations in Lanarkshire and Ayrshire interrupt but do not entirely conceal the volcanic tracts.

II. THE SOUTHERN CHAIN OF VOLCANOES IN "LAKE CALEDONIA"

5. *The Pentland Volcano*

Beginning at the north-east end of the line we first come upon the classic area of the Pentland Hills, for the study of which the geologist is prepared by the admirable description of Charles Maclaren,[1] and the earlier geognostical papers of Jameson.[2] The area mapped in detail is represented in Sheet 32 of the Geological Survey of Scotland, published in 1859, and described in the Memoir accompanying that sheet.

When in these early days I surveyed this ground I found it extremely difficult to understand. Being then myself but a beginner in geology, and the study of old volcanic rocks not having yet advanced much beyond its elementary stage, I failed to disentangle the puzzle. Not until after more than twenty years, largely spent in the investigation of volcanic rocks elsewhere, had I an opportunity of resurveying the ground and bringing to its

[1] *A Sketch of the Geology of Fife and the Lothians*, 1839. The detailed descriptions in this work are accompanied with a map and two plates of sections. In the map all the volcanic rocks are represented by one colour. In the sections the bedding of the rocks is shown, and an indication is given of the succession of their chief varieties.

[2] See specially *Mem. Wernerian Soc.* vol. ii. ; also MacKnight in vol. i. The account of the Pentland Hills by Hay Cunningham in vol. vii. (1838) is clear but brief.

renewed study a wider knowledge of the subject. A new edition of the map was issued in 1892, and I shall here embody in my summary the chief results obtained in the course of this revision.

The most obvious features in the Pentland area are the marked development of the volcanic rocks at the north end of the chain, their rapid diminution and disappearance towards the south-west, the abrupt truncation of the bedded masses by the line of craggy declivity which forms the northern termination of the hills, and lastly, the continuation of the volcanic series northward in a totally different form in the lower eminences of the Braid Hills.

The length of the whole volcanic tract is about eleven miles; its breadth at the widest northern part is four miles, but from that maximum it dwindles southwards and dies out in seven miles. Its western side is in large measure flanked by the unconformable overlap of the Upper Old Red Sandstone and Lower Carboniferous formations, though in some places the base of the volcanic series is seen. The eastern boundary is chiefly formed by a large fault which brings down the Carboniferous rocks against the volcanic ridge. At the northern end, this ridge plunges unconformably under the Upper Old Red Sandstone of the southern outskirts of Edinburgh.

The bedded aspect of the truncated end of the Pentland chain, as seen from the north, has been already alluded to (p. 281). The rocks dip to the south-east, hence the lower members of the series are to be found along the north-west side of the hills.

It will be noticed from the Geological Survey map that the volcanic rocks of the main body of the Pentland Hills are arranged in alternations of

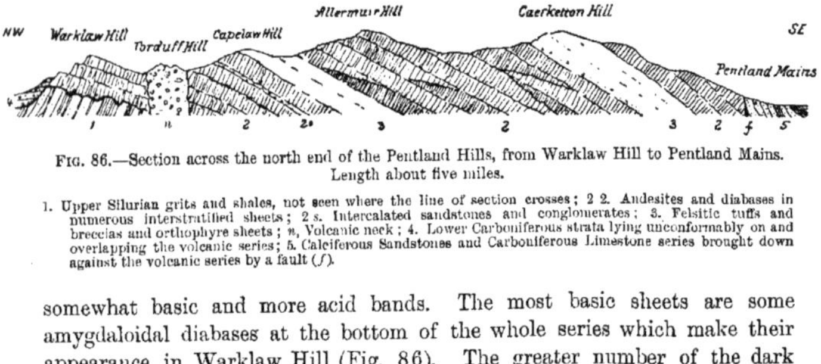

FIG. 86.—Section across the north end of the Pentland Hills, from Warklaw Hill to Pentland Mains. Length about five miles.

1. Upper Silurian grits and shales, not seen where the line of section crosses; 2 2. Andesites and diabases in numerous interstratified sheets; 2 s. Intercalated sandstones and conglomerates; 3. Felsitic tuffs and breccias and orthophyre sheets; n, Volcanic neck; 4. Lower Carboniferous strata lying unconformably on and overlapping the volcanic series; 5. Calciferous Sandstones and Carboniferous Limestone series brought down against the volcanic series by a fault (*f*).

somewhat basic and more acid bands. The most basic sheets are some amygdaloidal diabases at the bottom of the whole series which make their appearance in Warklaw Hill (Fig. 86). The greater number of the dark lavas are varieties of andesite, sometimes tolerably compact, sometimes highly cellular and amygdaloidal. But interstratified with these are thick sheets of what used to be called "claystone," a term which here comprised decayed felsites (orthophyres), and also felsitic tuffs and breccias. The remarkably acid nature of some of these rocks has been already pointed out.

The total thickness of the volcanic series at the north end of the hills is about 7000 feet, but as neither the top nor the bottom is there visible, it may be considerably greater. At these maximum dimensions the rocks form the high scarped front of the Pentland Hills, which rises into so prominent a feature in the southern landscape of Edinburgh. A series of transverse sections across the chain from north to south will illustrate its structure and history. These I shall here describe, reserving for subsequent consideration the great vent of the Braid Hills.

A section taken through the north end of the chain, where the maximum depth of volcanic material is exposed, presents the arrangement represented in Fig. 86. It will be seen that the base of the series is here concealed by the unconformable overlap of the Lower Carboniferous rocks on the west side, while the top is cut off by the great fault which on the east side brings down the Midlothian Coalfield.

The Lower Carboniferous conglomerates (4) creep over the edge and up the slopes of the volcanic series of the Pentland Hills. They contain abundant

FIG. 87.—View of the lava-escarpments of Warklaw Hill, Pentland chain, from the north-west.

pebbles of the lavas, and were evidently laid down along a shore from which the Pentland rocks rose steeply into land. Though the actual base of the lavas is not seen here, two miles further to the south highly-inclined Upper Silurian shales and mudstones are found emerging unconformably from under the volcanic pile, and similar strata probably underlie Warklaw Hill as indicated in the figure. The Upper Silurian strata pass up into a lower group of the Lower Old Red Sandstone, which has also been covered unconformably by the volcanic series. In these underlying deposits we have evidence of the pre-volcanic accumulations of the lake, which were broken up and tilted at the beginning of the volcanic eruptions.

The lowest lavas, consisting of well-marked beds of diabase (2), present their escarpments to the north-west and dip into the rising ground, as sketched in Fig. 87. Their characters have been already noticed in the general petrography of the Old Red Sandstone volcanic rocks. Dark solid compact portions of them pass rapidly into coarsely cellular slag, especially along the upper and under parts of the several sheets. No tuff has been noticed between these basic flows, but here and there thin lenticular layers of sandstone, lying in hollows of the lava-sheets, are connected with vertical or highly-inclined ramifying veins of similar material, with the plains of stratification passing across the breadth of the veins. These features are an exact reproduction of those above described in Forfarshire and Kincardineshire. The amygdales consist of chalcedony, crystallized quartz and calcite.

Torduff Hill, which rises to the east of Warklaw, consists of a mass of coarse volcanic breccia or agglomerate (*n*), markedly felsitic in its materials. It probably forms a neck marking a small volcanic vent, like some others at the north end of the chain to be afterwards referred to.

In the lower part of Capelaw Hill, the next eminence in an easterly direction, bedded andesites, with an intercalated band of sandstone and conglomerate (2*s*), appear and pass under rocks of so decomposing a kind that no good sections of them are to be found. The hill is covered with grass, but among the rubbish of the screes pieces of felsite-like rocks and breccias may be observed. Some of these blocks show an alternation of layers of felsitic breccia with a fine felsite-like material which may be a tuff. These rocks, conspicuous by the light colours of their screes, alternate further up with other dark andesitic lavas, and run south-westward for about five miles.

Beyond Capelaw Hill, upon a band of these pale rocks, comes a thick group of sheets of dark andesite, which form the main mass of Allermuir Hill. They are well seen from the south side and likewise from the north, dipping towards the south-east at angles of from 35° to 40°, and weathering along the crest of the hills into a succession of scars and slopes which show the bedded character of the lavas.

At Caerketton Hill another band of pale material forms the conspicuous craggy face so familiar in the aspect of the Pentland Hills as seen from Edinburgh. This band consists of pale felsitic breccia, and amorphous, compact, much-decayed rock, regarding which it is difficult to decide whether it should be considered as a fine felsitic tuff, or as a decomposed felsite. The band is better seen when traced southwards. The light colour of its screes makes it easily followed by the eye even from a distance along the hill-tops and declivities.

On the next hill to the south-west, known as Castlelaw Hill, this pale band of rock is exposed in a few crags and quarries, and its debris, protruding through the scanty herbage, slips down the slopes. On its north side the screes display the same felsitic breccias and compact, decayed felsitic rocks, occasionally showing a structure like the flow-structure of rhyolite. The breccia which projects in blocks from the summit of the hill has been quarried immediately below the crest on the south side, where it overlies a thin intercalated band of a dull, much-decomposed porphyry.

The breccias are composed almost entirely of thoroughly acid rock-fragments, as may be judged from the percentage of silica shown to occur in them. These fragments vary from the finest lapilli up to angular pieces several inches long. They not infrequently display a fine and extremely beautiful flow-structure. It is thus quite certain that there are acid breccias intercalated among the more basic lavas of the northern Pentlands, and that among the constituents of these breccias are fragments of felsite or perhaps even lithoid rhyolite.

We may therefore be prepared to find that actual outflows of felsitic lava accompanied the discharge of these highly-siliceous tuffs. Unfortunately the manner in which the rocks decay and conceal themselves under their own debris makes it difficult to separate the undoubtedly fragmental bands from those which may be true lavas. But an occasional opening, and here and there a scattered loose block, serve to indicate that the two groups of rock certainly do coexist in this pale band, which can be followed through the chain for upwards of six miles until it is cut off by the eastern boundary fault.

At the south-west end of Castlelaw Hill, where a quarry has been opened above the Kirk Burn, blocks of felsite may be observed showing flow-structure on a large scale. The bands of varied devitrification are sometimes a quarter of an inch broad, and weather out in lighter and darker tints. Some of them have retained their felsitic texture better than others, which have become more thoroughly kaolinized. That these are not deceptive layers of different texture in fine tuffs is made quite clear by some characteristic rhyolitic structures. The bands are not quite parallel, but, on the contrary, are developed lenticularly, and may be observed to be occasionally puckered, and to be even bent back and folded over as in ordinary rhyolites. There is no contortion to be observed among the stratified tuffs of the hills. This irregularity in the layers is obviously original, and can only be due to the flow of a moving lava.

On the east side of Castlelaw Hill, as shown in Fig. 86, dull reddish andesites overlie the pale belt of felsitic rocks. Their lower bands are marked by the presence of well-formed crystals of a dark green mica. Their central and higher portions consist of porphyrites of the prevalent type, both compact and vesicular. These lavas continue as far as any rock can be seen. Beyond the boundary fault, the Burdiehouse Limestone and oil-shales of the Lower Carboniferous series are met with, inclined at high angles against the hills. It is impossible to say how much of the volcanic series has here been removed from sight by the dislocation.

If now we move three miles further to the south-west and take a second section across the Pentland Hills, it will be found to expose the arrangement of rocks represented in Fig. 88. At the western end the Upper Old Red Sandstones (4) and Lower Carboniferous series (5) are seen lying unconformably on the upturned edges of the Upper Silurian shales (1). North Black Hill consists of a large intrusive sheet of pale felsite (F) that has broken through the Silurian strata and has in places thrust itself between them and the conglomerates of the Lower Old Red Sandstone which lie unconformably upon them. In the neighbouring Logan Burn, at the bottom of the Habbie's Howe Waterfall, the felsite can be seen injected into the conglomerate. The felsitic sill of North Black Hill runs for a mile and a half along the western base of the volcanic series, and has a breadth of about half a mile. It is the only important intrusive mass in the Pentland Hills.

To the south of the Silurian shales that lie against the southern flank of North Black Hill, pale felsitic tuffs (3) occur, which are a continuation of those already referred to as running southwards from Capelaw Hill. Above them a series of andesites (2), with intercalated bands of tuff, sandstone and conglomerate (2*s*), occupy the bottom of the Logan valley and part of the slopes on both sides. In the thickest band of tuffs, which is well-

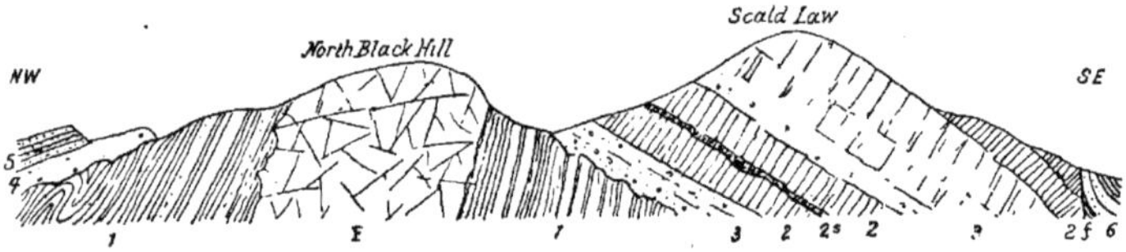

Fig. 88.—Section across the Pentland Hills through North Black Hill and Scald Law (length about three miles).

exposed along the road by the side of the Loganlee Reservoir, a group of well-bedded strata occurs from less than an inch to a foot or more in thickness. Generally they are pale in colour, and are made up of white felsitic detritus, but with a sprinkling of dull purplish-red fragments, and occasional larger rounded pieces of different andesites. Some of the rocks might be called felspathic sandstones. Other bands in the group are dark purplish-red in tint, and consist mainly of andesitic debris, with a dusting of white felsitic grains and fragments. There would thus seem to have been showers both of felsitic and of andesitic ashes and lapilli.

The dark lavas that overlie the tuffs are likewise well displayed along the same road-section. They vary rapidly from extremely compact homogeneous dark blue rocks, that weather with a greenish crust, to coarse, slaggy masses and amygdaloids.

These more basic lavas are a continuation of those of Allermuir Hill, and, as at that locality, they plunge here also under the same band of white tuffs, breccias and felsites (3), which has been referred to as stretching

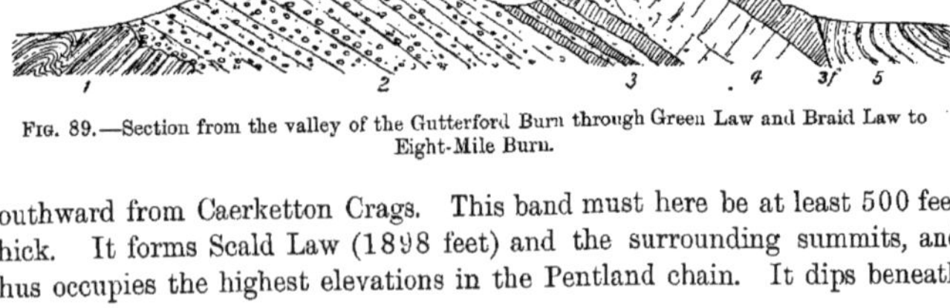

Fig. 89.—Section from the valley of the Gutterford Burn through Green Law and Braid Law to Eight-Mile Burn.

southward from Caerketton Crags. This band must here be at least 500 feet thick. It forms Scald Law (1898 feet) and the surrounding summits, and thus occupies the highest elevations in the Pentland chain. It dips beneath the uppermost group of andesites, which, as before, are here truncated by the eastern fault (*f*), the Calciferous Sandstones and Carboniferous Limestone series (6) being thrown against them.

A third section (Fig. 89), taken two miles still further south, shows a remarkable attenuation of the volcanic series, and the appearance of a thick group of conglomerates (2) lying conformably below that series, but resting on the upturned edges of the upper Silurian shales (1). The thick Allermuir porphyrites are here reduced to a few thin beds (3) intercalated among the conglomerates and sandstones, amidst which the whole volcanic series dies out southward. A detailed section of the rocks exposed on the western front of Braid Law shows the following succession:—

White felsitic rocks of Braid Law (4 in Fig. 89).
Coarse conglomerate passing down into sandstone. About 20 feet visible.
Dark andesite, 4 feet.
Parting of yellow felspathic grit, 8 or 10 inches.
Andesite, 10 feet.
Hard felspathic grit, 6 feet.
Dark green amygdaloidal andesite, 2 feet.
Yellow felspathic sandstone and grit, 2 feet.
Dark green amygdaloidal andesite, 6 feet.
Felspathic grit and red and brown sandstone, 4 feet.
Dark andesite, perhaps 6 or 8 feet.
Great conglomerate with alternating courses of sandstone, rapidly increasing in thickness southwards.

Above these dwindling representatives of the northern andesitic lavas comes the continuation of the white band of tuffs and breccias of Caerketton and Scald Law (4), which in turn dips under the highest group of andesites. The Carboniferous strata (5) are brought in by the fault (*f*). In little more than two miles beyond this line of section the volcanic series disappears, and the Old Red Sandstone for a brief space consists only of sedimentary deposits.

Besides the remarkable alternation of basic and acid ejections, there is a further notable feature in the geology of the Pentland Hills. This volcanic centre presents us with one of the most remarkable vents anywhere to be seen among the volcanic rocks of Britain. The full significance of this feature may best be perceived if we advance along the hills from their south-western end. As has now been made clear, the volcanic materials which begin about the line of the North Esk near Carlops rapidly augment in thickness until, in a distance of not more than seven miles, they attain a thickness of about 7000 feet, and then form the great scarped front of the hills that look over Edinburgh. But at the base of that wall their continuity abruptly ceases. The lower ground, which extends thence to the southern suburbs of Edinburgh, and includes the group of the Braid Hills, is occupied by another and more complex group of rocks in which the parallelism and persistence so marked in the Pentland chain entirely disappear.

This abrupt truncation of the bedded lavas and tuffs marks approximately the southern margin of a large vent from which at least some, if not most, of these rocks were probably ejected. The size of this vent cannot be precisely ascertained on account of the unconformable overspread of Lower Carboniferous strata. But that it must have been a large and important volcanic orifice may be inferred from the fact that the visible area of the materials that fill it up measures two miles from north-east to south-west, and a mile and a half from south-east to north-west, thus including a space of rather more than two square miles. Its original limits towards the north and south can be traced by help of the bedded lavas that partially surround it, but on the two other sides they are concealed by the younger formations. We shall probably not over-estimate the original area of the vent if we state it at about four square miles.

The materials that now fill this imporant orifice consist mainly of "claystones," like those of the Pentland series—dull rocks, meagre to the touch, varying in texture from the rough porous aspect of a sinter through stages of increasing firmness till they become almost felsitic, and ranging in colour from a dark purple-red, through shades of lilac and yellow, to nearly white,

FIG. 90.—Section across the north end of the Pentland Hills, and the southern edge of the Braid Hill vent. Length about two miles.

1 1. Andesites; 2. Fine tuffs, etc., of the Braid Hill vent; 3 3 3. Agglomerate in lateral necks with felsitic intrusions (4).

but often strikingly mottled. A more or less laminar structure is often to be observed among them, indicating a dip in various directions (but especially towards the north) and at considerable angles. Throughout this exceedingly fine-grained material, lines of small lapilli may occasionally be detected, also bands of breccia, consisting of broken-up tuff of the same character, and of fine "hornstone" and felsite, with delicate flow-structure. Exhibiting on the whole so little structure, this tract may be regarded as consisting largely of fine volcanic dust derived from the explosion of felsitic or orthophyric lavas. Some portions indeed are not improbably composed of decayed felsites, like those which present so many difficulties to the geologist who would try to trace their course among the other lavas and tuffs of the Pentland chain. Various veins, dykes and small bosses of felsite, andesite and even more basic material, such as fine dolerite, have been intruded into the general body of the mass.

On the outskirts of the main vent some subordinate necks may be observed (3, 3 in Fig. 90), perhaps, like Torduff Hill, already noticed (Fig. 86), marking lateral eruptions from the flanks of the great cone. Three of these occur in a line more than half a mile long, possibly indicating a fissure on the side of the old volcano, running in a south-westerly direction from the southern edge of the vent. The smallest of them measures about 500 feet in diameter; the largest is oblong in shape, its shorter diameter being about 500 feet, and its longer about 1000 feet. The materials that fill these lateral vents are coarse agglomerates, traversed by veins and irregular intrusions of a fine horny or flinty felsite.

From the acid character of most of the rocks that now fill the wide vent of the Braid Hills it may be inferred that at least the last eruptions from it consisted chiefly of acid tuffs and lavas. The upper portion of the volcanic series being everywhere concealed, there are no means left to verify this inference from an examination of the ejected material. It may be remarked, however, that the pale yellow sandstones which lie on the east side of the fault and are exposed in the Lyne Water above West Linton are in great measure composed of fine felsitic material.[1] They certainly belong to a higher horizon than the most southerly lavas of the Pentland Hills, and if they have not derived their volcanic detritus from the Biggar volcanic area, it may be assumed that they obtained it from the vent of the Braid Hills. In any case they show that after the lavas of the southern end of the Pentland Hills were buried, acid volcanic detritus continued to be abundantly distributed over this part of the floor of Lake Caledonia.

6. *The Biggar Centre*[2]

Another distinct group of volcanoes had its centre about 25 miles south-westward from the Braid vent, and on the same line as those of the Pentland Hills. In no part of the basin can the isolation of the different volcanic clusters be so impressively observed as in the area to the south-west of these hills. On the one hand, the lavas and tuffs from the Braid vent die out, and on the other, as we follow the conglomerates south-westwards, a new volcanic series immediately makes its appearance.

The space between the last extremity of the Pentland lavas and the beginning of the Biggar series does not exceed some 500 yards. It will be remembered that the lower half of the Pentland volcanic series dies out long before it reaches the southern end of the hills, and that it is by lavas on the horizon of some of the dark andesites of Allermuir Hill that the volcanic band is finally prolonged to its extreme southern limit. The most northerly extension of the Biggar lavas lies somewhere on the same general platform. But whereas, at the north end of the Pentland chain, the volcanic sheets rest on the edges of the Upper Silurian shales, at the south end, several hundred feet of coarse conglomerate and sandstone intervene between the Silurian shales and the porphyrites. So rapidly does the bulk of these sedimentary formations increase that in the course of two miles they must be 3000 feet in thickness below the most northerly of the Biggar lavas just

[1] Explanation to Sheet 24 of the Geological Survey of Scotland, pp. 10, 12.

[2] This area is included in Sheets 23 and 24 of the Geological Survey of Scotland. It was mapped and described by myself. (Explanations of Sheets 23 and 24.) Various parts of it have been referred to by earlier writers, particularly Maclaren, *Geology of Fife*, etc., p. 176.

referred to. But after that point, when they cross the Lyne Water, they begin to be more and more interstratified with thin sheets of andesite. These lavas, the beginning of the Biggar series, soon number nine or ten distinct bands, and so quickly do they usurp the place of the sedimentary materials that in a distance of not more than twelve miles they form, where traversed by the river Clyde, the whole breadth of the visible tract of Old Red Sandstone, to the exclusion of the conglomerates.

Unfortunately, soon after the lavas make their appearance at the north end they are in great measure overlapped unconformably by the red sandstones at the base of the Carboniferous system, but where the Medwin Water has cut through this covering, they can be seen here and there underneath on their southerly course.

A section through the northern end of the Biggar series, where the successive lavas are dying out northwards among the conglomerates, shows the structure given in Fig. 91. The sedimentary strata consist largely of

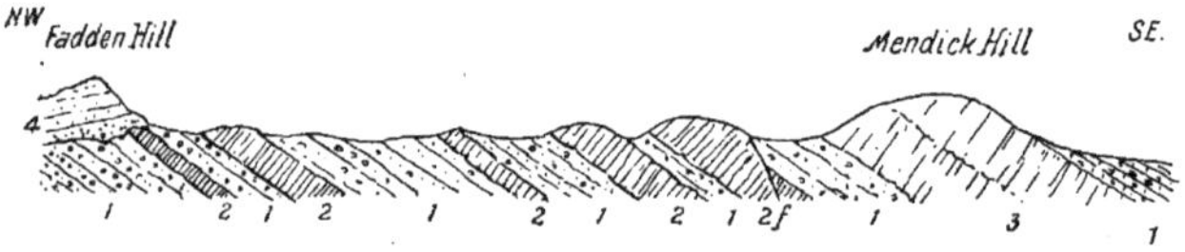

FIG. 91.—Section across the northern end of the Biggar volcanic group, from Fadden Hill to beyond Mendick Hill.

1. Conglomerates and sandstones; 2. Lavas, the lowest being an olivine-diabase or basalt, the main mass being andesites; 3. Felsites and tuffs; 4. Upper Old Red Sandstone. *f*, Fault.

debris of andesite, and the lavas include dark red or purple andesites and also pale felsites, both having the same characters as those of the Pentland Hills.

In one important respect the volcanic series in the northern part of the Biggar area differs from that of the Pentland Hills, for whereas the uppermost parts of the latter are concealed by faults which bring down the Carboniferous strata against the base of the hills, the lavas at the north end of the Biggar district pass conformably under a thick group of Lower Old Red conglomerates and sandstones. We thus learn that here the volcanic eruptions ceased long before the close of the deposition of the Lower Old Red Sandstone. The overlying sedimentary series is disposed in a long synclinal trough, corresponding in direction with the general north-easterly strike of the volcanic rocks which reappear from under the sandstones and conglomerates along its south-eastern border, where they are abruptly truncated by the fault (*f*, Fig. 92), which brings them against the flanks of the Silurian Uplands. It is interesting to note that by this dislocation the lavas of the Lower Old Red Sandstone are placed almost in immediate contact with those of the Lower Silurian series, which appear here on the crests of numerous anticlinal folds that are obliquely cut off by the fault.

There is yet another feature of interest in the northern part of the Biggar volcanic centre. While the lowest visible lava is an olivine-diabase not unlike parts of the Warklaw group of the Pentland Hills, those which occur above it are partly andesites and partly orthoclase-felsites. The latter form, among the hills near Dolphinton, an important group which reaches its greatest development in the Black Mount (1689 feet). These rocks cover a breadth of more than a mile of ground, and probably attain a thickness of not less than 2000 feet. They so closely resemble in their general characters the corresponding rocks of the Pentland Hills that a brief description of them may suffice. As in that chain of hills, they are so prone to decomposition that they are in large part concealed under a covering of their own debris and of herbage, though their fragments form abundant screes, and numerous projecting knobs of rock suffice to show the main features of the lavas and their accompaniments.

The felsites weather into pale yellow and greyish "claystones," but where fresher sections can be procured they often show darker tints of lilac and purple. They are close-grained, sometimes flinty, generally porphyritic with scattered highly-kaolinized white felspars, but without quartz, often presenting beautiful flow-structure, and not infrequently showing a brecciated appearance, which in the usual weathered blocks is hardly to be distinguished from the breccia of interstratified tuffs.

A locality where some of these features may be satisfactorily examined is a dry ravine in the farm of Bank, on the south-east side of the Black Mount. Here the felsite possesses such a perfectly developed flow-structure as to split into slabs which, dipping S.E. at about 25°, might deceive the observer into the belief that it is a sedimentary rock. A fresh fracture shows the laminæ of flow, many of which are as thin as sheets of paper, to be lilac in colour, some of the more decomposed layers assuming tints of grey. The felspars and micas are arranged with their long axes parallel to the lines of flow. The rock is not vesicular, but it breaks up here and there into the brecciated condition just referred to. Below the sheet which displays the most perfect flow-structure, what is probably a true volcanic breccia makes its appearance. It consists of angular fragments of a similar lilac felsite, of all sizes up to pieces two or three inches in length, cemented in a matrix of the same material stained reddish-brown. In this breccia the stones show little or no flow-structure.

Above the group of felsites and felsitic breccias, grey andesites make their appearance, like some of those in the Pentland Hills. They are sometimes extraordinarily vesicular, the vesicles in the body of the rock being filled with calcite, agate, etc. Such lavas must have been originally sheets of rough slag. The elongated steam-vesicles have been partly filled up with micaceous sand and fine red mud that were washed into crannies of the lava in direct communication with the overlying water. It is evident that in the northern part of the Biggar centre the succession of volcanic events followed closely the order observable in the Pentland Hills, but on a feebler scale. We may suppose that the lower diabases and andesites are the equivalents of those of Warklaw and Allermuir, that the felsites and breccias were contemporaneous with those of Capelaw, Caerketton and Castlelaw, and that the last andesites made their appearance together with those which form the highest lavas of the Pentland chain.

A section across the southern end of the Biggar volcanic belt shows less diversity of structure (Fig. 92). The lavas (3) are there found to flatten out and to spread unconformably over the older part of the Lower Old Red Sandstone (2), which, as already stated, passes down into the Upper Silurian shales. A few intercalations of conglomerate, mainly made up of volcanic detritus, are here and there to be detected among these lavas. But the chocolate sandstones and conglomerates that lie unconformably below

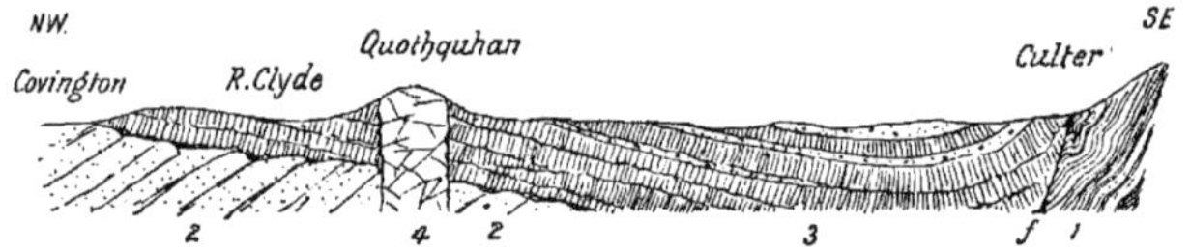

FIG. 92.—Section across the southern part of the Biggar volcanic group from Covington to Culter.

1. Lower Silurian strata; 2. Lower Old Red Sandstone (pre-volcanic group); 3. Andesite lavas with intercalated sandstones and conglomerates; 4. Felsite neck. *f*, The boundary-fault on northern edge of Southern Uplands.

them contain no such detritus, for they belong to the pre-volcanic part of the history of Lake Caledonia, and were here locally upraised, perhaps as an accompanimen. of the terrestrial disturbances that preceded or attended the first outburst of volcanic energy. Followed south-westwards, the stratigraphical break in the Lower Old Red Sandstone disappears, and, as will be shown in the account of the Duneaton centre, a continuous succession can there be traced from the Upper Silurian shales up into the volcanic series.

An interesting feature in this district is the felsitic boss of Quothquan already alluded to (p. 288) as rising up through the andesites, and possibly marking one of the vents of the district. It is one of a number of felsitic intrusions in this neighbourhood, of which the most important is Tinto.

A third section taken across Tinto, from Thankerton Moor on the north side to Lamington on the south, will serve further to illustrate the great

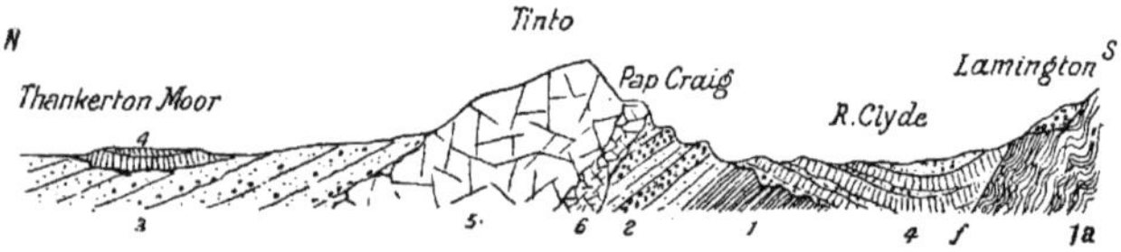

FIG. 93.—Section from Thankerton Moor across Tinto to Lamington.

1a. Lower Silurian; 1. Upper Silurian strata; 2. Lower Old Red Sandstone with two marked bands of conglomerate; 3. Lower Old Sandstone (pre-volcanic chocolate sandstones); 4. Andesite lavas with sandstones, conglomerates and tuffs lying unconformably on No. 3; 5. Felsite sill of Tinto with the smaller sill of the Pap Craig (6). *f*, Fault bounding the Silurian uplands on the north. A small patch of the unconformable Lower Old Red conglomerate is seen on the south side of the fault.

unconformability in the Lower Old Red Sandstone of this district, and to show the relation of the largest felsitic intrusion to the surrounding rocks (Fig. 93). The conglomerates and sandstones that appear on the south slopes of Tinto lie near the base of the Old Red Sandstone, and if we could bore among the overlying andesites we should probably meet with the Upper Silurian shales among or conformably beneath the red passage-beds, as in the Lesmahagow district.

The andesitic lavas creep over the upturned denuded edges of these strata and sweep round the flanks of Tinto. This conspicuous hill reaches a height of 2335 feet above the sea, and consists of the felsitic rocks already described (p. 278). Seen from many points of view it rises as a graceful cone, distinguished from all the other eminences around it by the pinkish colour of its screes. In reality it forms a continuous ridge which runs in an east and west direction for about five and a half miles, with a breadth of about a mile. Some part at least, and possibly the whole of this oblong mass, is in the form of a sill or laccolite which dips towards the north. Conglomerates and sandstones plunge under it on the southern side, and similar sandstones overlie it on the north. If there be a neck in this mass, as one might infer from the shape of the hill, its precise limits are concealed. The rock does not break through the andesites, and may belong to an earlier period of eruptivity than the lavas immediately around it. There were other, though smaller, vents in the immediate neighbourhood. Besides the cone of Quothquan just referred to, another may be marked by the felsite boss which overlooks the village of Douglas, four miles to the south-west of the Tinto ridge, while a third rises into a low rounded hill close to the village of Symington.

The lavas spread out again to the south-west of Tinto in a group of hills, until they are interrupted by a fault which brings in the Douglas coal-field.[1] This dislocation abruptly terminates the Biggar volcanic band in a south-westerly direction, after extending for a length of 26 miles, with a breadth of sometimes as much as five miles.

7. *The Duneaton Centre*

Among the high bleak muirlands on the confines of the three counties of Lanark, Ayr and Dumfries, traversed by the Duneaton Water, a distinct volcanic area may be traced.[2] Its boundaries, however, cannot be satisfactorily fixed. It is overspread with Carboniferous rocks both to the north-east and south-west, so that its rocks are only visible along a strip about seven miles long and two miles broad. On the north-western side its lower members are seen lying interstratified among the sandstones and conglomerates which thence pass down conformably into the Upper Silurian series (Fig. 94). But although we thus get below the volcanic series we meet with no vents or sills among the lower rocks. On the south-east side the highest lavas and tuffs are overlain by some 5000 feet of red sandstones and conglomerates (2, 3), which completely bury all traces of the volcanic history.

[1] See Explanation to Sheet 23 of the Geological Survey of Scotland (1873), p. 15. This ground was mapped and described by Mr. B. N. Peach.

[2] This area was mapped by Mr. B. N. Peach in Sheet 15 of the Geological Survey of Scotland, and is described by him in the accompanying Memoir.

The volcanic series in this limited district reaches an estimated thickness of 4000 feet, built up of purple and green slaggy andesites, dark heavy diabases (melaphyres) and tuffs, with abundant interstratification of sandstone, especially towards the base. One of its chief features of interest is the manner in which it exhibits, better, perhaps, than can be found in any of the other volcanic areas, the frequent and rapid alternations of lavas and tuffs with sandstone and conglomerate. In this part of the region the volcanic discharges were obviously frequent and intermittent, while at the same time the transport and deposition of sediment were continuous. This sediment consisted largely, indeed, of volcanic detritus mixed with ordinary sand and silt. That these conditions of sedimentation were not wholly inimical to animal life is shown by the occasional occurrence of worm-burrows in the ashy sandstones.[1]

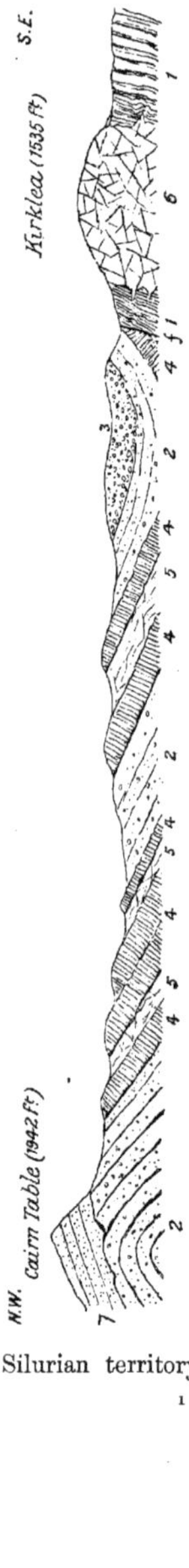

FIG. 94.—Section across the Duneaton volcanic district from the head of the Duneaton Water to Kirklea Hill.
1. Silurian strata; 2. Lower Old Red Sandstone and conglomerate; 3. Coarse conglomerate; 4. Andesite-lavas; 5. Stratified tuffs; 6. Spango granite; 7. Upper Old Red Sandstone.

The thick accumulation of sandstones and conglomerates above the main mass of lavas has been derived almost wholly from the waste of the volcanic rocks (3). Blocks of andesite, well rounded and often from six to twelve inches in diameter, may be seen in the remarkable band of coarse conglomerate which runs as a nearly continuous ridge from the Nith to the Clyde—a distance of more than twenty miles. Nothing impresses the geologist more, as he wanders over this district, than the evidence of the prodigious waste which the volcanic series underwent before it was finally buried. Some part of the detritus may have been supplied, indeed, by occasional discharges of fragmental matter, as has already been suggested in the case of the Ochil and Montrose conglomerates. But the nature of the pebbles in these masses of ancient shingle shows them to be not bombs, but pieces worn away from sheets of lava.

That the lavas underlie these piles of detritus and extend southwards even up to the very edge of the Silurian Uplands is shown by the rise of a number of successive beds from under the trough into which the conglomerate has been thrown. These lavas, however, are almost immediately cut off by the great boundary fault (*f*) which flanks the Silurian territory. That they are not met with now to the south-east of

[1] Memoir on Sheet 15 Geol. Surv. Scotland (1871), p. 22.

the dislocation, where they must once have lain, is an evidence of the great denudation which the district has undergone. Fig. 94, which gives a section across the broadest part of the area, from the edge of the Muirkirk coal-field to the Silurian uplands, shows the general structure of the ground.

No satisfactory evidence regarding the position of any of the vents of the period has been met with in this district. The rocks to the south of the boundary-fault are older than the Old Red Sandstone, and as they must have been for some distance overspread by the conglomerates, sandstones and volcanic series, we might hope to find somewhere among them traces of necks or bosses. The only mass of eruptive rock in that part of the district is the tract of Spango granite which has been already referred to in connection with the subject of the vents and granite protrusions of Old Red Sandstone time. This mass, about four miles long and two miles broad, rises through Silurian strata, and by means of the boundary fault is brought against the higher group of conglomerates and sandstones. The Silurian shales and sandstones around the granite have undergone contact-metamorphism, becoming highly micaceous and schistose. The ascent of this granite must have taken place between the upheaval and contortion of the Upper Silurian strata, and the deposition of the higher parts of the Lower Old Red Sandstone of this region. Its date might thus come within the limits of the volcanic period. But one must frankly own that there is no positive evidence to connect its production with the volcanic history.

8. *The Ayrshire Group of Vents*

The original limits of the volcanic districts in the remaining portion of the Old Red Sandstone area on the mainland of Scotland, from the valley of the Nith to the Firth of Clyde, can only be vaguely indicated.[1] There is a difficulty in ascertaining the south-western termination of the Duneaton area, and in deciding whether the lavas and tuffs of Corsincone in Nithsdale should be assigned to that district or be placed with those further to the south-west. Between Corsincone and the next visible volcanic rocks of the Lower Old Red Sandstone there intervenes a space of six miles, along which, owing to the effect of the great fault that flanks the north-western margin of the Southern Uplands, the Carboniferous Limestone and even the Coal-measures are brought against the Silurian formations, every intermediate series of rocks being there cut out. It may therefore be, on the whole, better to include all the volcanic rocks on the left side of the Nith as part of the Duneaton series. There will still remain a tract of five miles of blank intermediate ground before we enter upon the volcanic rocks of Ayrshire.

[1] The mapping of the Old Red Sandstone volcanic areas of Ayrshire for the Geological Survey was thus distributed :—The district east of Dalmellington was surveyed by Mr. B. N. Peach, that between Dalmellington and Straiton by Prof. James Geikie, and all from the line of the Girvan Valley south of Straiton westward to the sea by myself. The ground is embraced in Sheets 8, 13 and 14 of the Map of Scotland, and is described in accompanying Explanations.

Owing to complicated faults, extensive unconformable overlaps of the Carboniferous formations, and enormous denudation, the volcanic tracts of Old Red Sandstone age in Ayrshire have been reduced to mere scattered patches, the true relations of which are not always easily discoverable. One of these isolated areas flanks the Silurian Uplands as a belt from a mile to a mile and a half in breadth and about six miles long, but with its limits everywhere defined by faults. A second much more diversified district extends for about ten miles to the south-west of Dalmellington. It too forms a belt, averaging about four miles in breadth, but presenting a singularly complicated geological structure. Owing to faults, curvatures and denudation, the volcanic rocks have there been isolated into a number of detached portions, between some of which the older parts of the Old Red Sandstone, and even the Silurian rocks, have been laid bare, while between others the ground is overspread with Carboniferous strata. A third unbroken area forms the Brown Carrick Hills, south of the town of Ayr, and is of special interest from the fact that its rocks have been exposed along a range of sea-cliffs and of beach-sections for a distance of nearly four miles. Other detached tracts of volcanic rocks are displayed on the shore at Turnberry and Port Garrick, on the hills between Mochrum and the vale of the Girvan, and on the low ground between Dalrymple and Kirkmichael.

The isolation of these various outliers and separated districts is probably not entirely due to the effects of subsequent geological revolutions. More probably some of the areas were always independent of each other, and their igneous rocks were discharged from distinct volcanic centres. We may conjecture that one of these centres lay somewhere in the neighbourhood of New Cumnock, for the lavas between that town and Dalmellington appear to diminish in thickness and number as they are traced south-westward. Another vent, or more probably a group of vents, may have stood on the site of the present hills to the right and left of the Girvan Valley, south of the village of Straiton. A third probably rose somewhere between Dailly and Crosshill, and poured out the lavas of the ridges between Maybole and the Dailly coal-field. The important centre of eruption that produced the thick and extensive lavas of the Brown Carrick Hills may be concealed under these hills, or may have stood somewhere to the west of Maybole. Still another vent, perhaps now under the sea, appears to be indicated by the porphyrites of the coast-section between Turnberry and Culzean Bay.

Owing to the complicated structure of the ground, several important points in the history of the Old Red Sandstone of this region have not been established beyond dispute. In particular, the unconformability which undoubtedly exists in that system in the south-west of Ayrshire has not been traced far enough eastwards to determine whether it affects the volcanic belt east of Dalmellington, or whether the break took place before or after the eruption of that belt. West of Dalmellington it clearly separates a higher group of sandstones, conglomerates and volcanic rocks from everything older than themselves. The structure is similar to that in the Pentland Hills, a marked disturbance having taken place here as well as there after a considerable portion of the Lower Old Red Sandstone had been deposited. These earlier strata were upraised, and on their denuded ends another group of sandstones and conglomerates was laid down, followed by an extensive eruption of volcanic materials.

It is the upper unconformable series that requires to be considered here, as it includes all the volcanic rocks of the Old Red Sandstone lying to the west of the meridian of Dalmellington. The position of these rocks on their underlying conglomerates is admirably exposed among the hills between the valleys of the Doon and the Girvan, as well as on Bennan Hill to the south of Straiton. The andesites rise in a craggy escarpment crowning long green slopes that more or less conceal the conglomerates and sandstones below.

Along the coast-sections the structure of the volcanic rocks may be most advantageously studied. The shore from the Heads of Ayr to Culzean Castle affords a fine series of exposures, where every feature in the succession of the lavas may be observed. Still more instructive, perhaps, is the mile and a half of beach between Turnberry Bay and Douglaston, of which I shall here give a condensed account, for comparison with the coast-sections of Kincardineshire and Forfarshire already described.

The special feature of this part of the Ayrshire coast-line is the number of distinct andesite sheets which can be discriminated by means of the thin layers of sandstone and sandy tuff that intervene between them. In the short space of a mile and a half somewhere about thirty sheets can be recognized, each marking a separate outflow of lava. It was in this section that I first observed the sandstone-veinings which have been described in previous pages, and nowhere are they more clearly developed. Almost every successive stream of andesite has been more or less fissured in cooling, and its rents and irregular cavernous hollows have been filled with fine sand silted in from above. The connection may often be observed between these sandstone partitions or patches and the bed of the same material, which overspread the surface of the lava at the time that the fissures were being filled up.

The andesites of the Turnberry shore are of the usual dark purplish-red to green colours, more or less compact in the centre and vesicular towards the top and bottom. They display with great clearness the large empty spaces that were apt to be formed in such viscous slaggy lavas as they moved along the lake-bottom. These spaces, afterwards filled with fine sand, now appear as irregular enclosures of hard green sandstone embedded in the andesite. The example shown in Fig. 95 may be seen in one of the lavas at John o' Groat's Port.

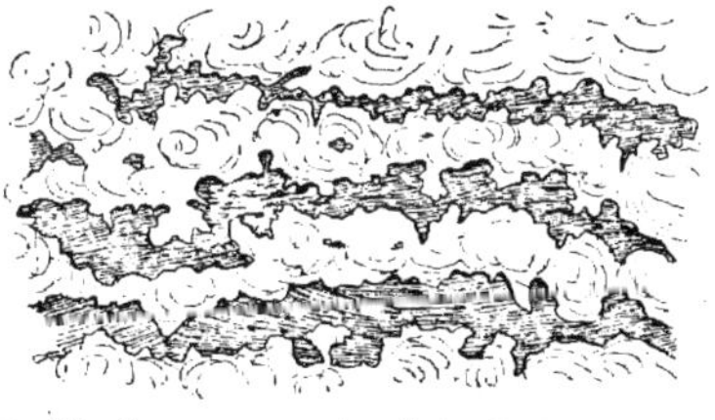

FIG. 95.—Cavernous spaces in andesite, filled in with sandstone, John o' Groat's Port, Turnberry, Ayrshire.

From the arrangement of the veins of sandstone it is evident that irregularly divergent, but often more or less stellate, fissures opened in the lavas as they cooled. Sometimes, indeed, the molten rock appears to have broken up into a shattered mass of fragments, as must often have happened when lavas were poured over the lake-floor. What may be an instance of this effect is to be seen on the cliff under the walls of Turnberry Castle, whence the annexed sketch (Fig. 96) was taken. The lower andesite (*a*) is highly amygdaloidal towards the top, and is traversed in all directions with irregular veins and nests of sandstone which can be traced upward to the bed (*b*), consisting of sandstone, but so full of lumps or slags of amygdaloidal andesite that one is here and there puzzled whether to regard it as a sedimentary deposit, or as the upper layer of clinkers of a lava-stream strewn with sand. Above this fragmentary layer lies another bed of andesite (*c*) of a coarsely amygdaloidal structure, which encloses patches of the underlying sandstone. It passes upward, in a space of from four to six feet, into a mass of angular scoriaceous fragments (*d*) of all sizes up to blocks 18 inches in length cemented in a vein-stuff of calcite, chalcedony and quartz. This brecciated structure ascends for about 13 or 14 feet, and is then succeeded by a greenish compact andesite (*e*), which further north becomes amygdaloidal and much veined with sandstone, passing into a breccia of lava fragments and sandstone.

FIG. 96.—Section of andesites, Turnberry Castle, Ayrshire.

The remarkable brecciated band (*d*) in this cliff, though 13 or 14 feet

FIG. 97.—Lenticular form of a brecciated andesite (shown in Fig. 96), Turnberry, Ayrshire.

in the centre, immediately thins out on either side, until in the course of a few yards it completely disappears and allows the lavas *c* and *e* to come together, as shown in Fig. 97. We may suppose that this section reveals

MAP OF THE
VOLCANIC DISTRICTS
LOWER OLD RED SANDSTONE
"LAKE CALEDONIA"

the structure of the terminal portion of a highly viscous lava which was shattered into fragments as it moved along under water.

No clear evidence of the sites of any of the volcanic vents has yet been detected in the Old Red Sandstone of Ayrshire. Possibly some of the numerous felsitic bosses to the south-west of Dalmellington may partly mark their positions. But the sills connected with the volcanic series are well exposed in the 12 miles of hilly ground between Dalmellington and Barr. Two groups of intrusive sheets may there be seen. The most numerous consist of pale or dark-pink felsite, often full of crystals of mica. They form prominent hills, such as Turgeny, Knockskae and Garleffin Fell. The second group comprises various diabase-sheets which have been intruded near the base of the red sandstones and conglomerates, over a distance of seven miles on the north side of the Stinchar Valley above Barr. They attain their greatest development on Jedburgh Hill, where they form a series of successive sills, the largest of which unite northwards into one thick mass and die out southward among the sandstones and conglomerates.

CHAPTER XXI

VOLCANOES OF THE LOWER OLD RED SANDSTONE OF THE CHEVIOT HILLS, LORNE, "LAKE ORCADIE" AND KILLARNEY

THE CHEVIOT AND BERWICKSHIRE DISTRICT

In the south-east of Scotland, and extending thence into the north of England, the remains of several distinct volcanic centres of the Lower Old Red Sandstone may still be recognized. Of these the largest and most interesting forms the mass of the Cheviot Hills; a second has been partially dissected by the sea along the coast south from St. Abb's Head; while possibly relics of others may survive in detached bosses of eruptive rock which rise through the Silurian formations of Berwickshire. The water-basin in which these volcanic groups were active was named by me "Lake Cheviot,"[1] to distinguish it from the other basins of the same geological period (Map I.).

The volcanic rocks of the Cheviot Hills, though their limits have been reduced by faults, unconformable overlap of younger formations and severe denudation, still cover about 230 square miles of ground, and rise to a height of 2676 feet above the sea. As they have been mapped in detail by the Geological Survey, both on the English and the Scottish sides of the Border, their structure is now known.[2] No good horizontal section, however, has yet been constructed to show this structure—a deficiency which, it is hoped, may before long be supplied.

[1] *Trans. Roy. Soc.* Edin. xxviii. (1878), p. 354.

[2] The Geology of the Cheviot Hills is comprised in Sheets 108 N.E., 109 N.W., and 110 S.W. of the Geological Survey of England and Wales, and in Sheets 17, 18 and 26 of the Geological Survey of Scotland. For descriptive accounts the Memoirs to some of these Sheets may be consulted, particularly "Geology of the Cheviot Hills" (English side), by C. T. Clough (*Mem. Geol. Surv.* 1888); "Geology of Otterburn and Elsdon," by H. Miller and C. T. Clough (*Mem. Geol. Surv.* 1887); "Geology of Part of Northumberland between Wooler and Coldstream," by W. Gunn and C. T. Clough, with Petrographical Notes by W. W. Watts (*Mem. Geol. Surv.* 1895). Other descriptions have been published by Professor James Geikie, *Good Words*, vol. xvii. (1876), reprinted in *Fragments of Earth-lore* (1893), and by Prof. Lebour, *Outlines of the Geology of Northumberland*, 2nd edit. 1886. For the petrography of the rocks consult Mr. J. J. H. Teall, *Geol. Mag.* 1883, pp. 100, 145, 252, 344; 1884, p. 226; 1885, p. 106; *Proc. Geol. Assoc.* ix. (1886) p. 575; and his *British Petrography*, 1888; Dr. J. Petersen, *Mikroskopische und chemische Untersuchungen am Enstatitporphyrit aus den Cheviot Hills*, Inaugural Dissertation, Kiel, 1884.

This volcanic pile, consisting mainly of bedded andesites which rest unconformably on the upturned edges of Wenlock shales and grits, presents a most typical display of the lavas of the Lower Old Red Sandstone. These rocks range from vitreous or resinous pitchstone-like varieties to coarsely porphyritic forms, on the one hand, and to highly vesicular and amygdaloidal kinds, on the other. Analyses of some of these rocks, and an account of their petrography, have already been given.

The lavas are often separated by thin partings of tuff, and their upper surfaces show the fissured character with sandstone infillings, so characteristic among the lavas of "Lake Caledonia."[1] Tuffs form a very subordinate part of the whole volcanic series. One of the most important bands is a thick mass at the base of the series, lying immediately on the highly inclined Silurian shales. The fragments are generally of a fine-grained purple mica-andesite, often two or three feet and sometimes at least five feet long. For a few feet near the bottom of this mass of tuff, pieces of Silurian shale an inch in length may be noticed. Mr. Clough remarks that distinct bedding is not usual among the tuffs. Though no doubt most of the fragmental materials really lie intercalated between successive lava-streams, yet some of the isolated patches of coarse volcanic breccia may mark the sites of eruptive vents. One such probable neck has been mapped on the Scottish side between Cocklawfoot, at the head of the Bowmont Water, and King's Seat, while others may perhaps occur among the detached patches that have been observed on the Northumbrian side. No thick conglomerates or sandstones have been noticed in the Cheviot District. The volcanic eruptions appear to have usually succeeded each other without the spread of any notable amount of ordinary detritus over the floor of the water-basin. It is difficult to estimate the total thickness of volcanic material here piled up, but it probably amounts to several thousand feet. The top of the series is not visible, having been partly removed by denudation and partly buried under the Carboniferous formations.

It will thus be seen that the Cheviot area stands apart from the other volcanic districts of the Lower Old Red Sandstone in the great relative thickness of its accumulated lavas, the comparative thinness of its tuffs, and the absence of the thick intercalations of coarse conglomerate so abundantly developed among the volcanic series all over Central Scotland. But there is yet another characteristic in which this area is pre-eminently conspicuous. In the heart of the andesites lies a core of augite-granitite, around which these rocks are traversed with dykes.

This interesting granitic boss rises into the highest summit of the whole Cheviot range, and covers an area of rather more than 20 square miles. While its petrographical characters have been described by Mr. Teall, its boundary has been mapped by Mr. Clough, who found the line difficult to trace, owing partly to the prevalent covering of peat, and partly to the jagged and irregular junction caused by the protrusion of dykes from and into the boss. He obtained evidence that the granite has broken through

[1] Clough, *Geology of the Cheviot Hills*, p. 15.

the bedded andesites, and that it is in turn traversed by dykes composed of a material indistinguishable from that of some of the flows. He therefore considered that it is essentially of the same age as the rest of the volcanic series, and "not improbably the deep-seated source of it."[1] Mr. Teall also, from a chemical and microscopical examination of the rocks, drew a similar conclusion.[2]

The andesites around the granite have undergone contact-metamorphism, but the nature and extent of the change have not yet been studied. There occur around the granite many dykes of felsite and quartz-felsite, to the petrographical character of which reference has already been made. But the most abundant and remarkable dykes of the district are those of a reddish mica-porphyrite, of which Mr. Clough has mapped no fewer than forty, besides those in the granitic area. He has called attention to the significant manner in which all the dykes of the district tend to point in a general way to the great core of granite, as if that were the nucleus from which they had radiated.[3]

The central granite of the Cheviot Hills, with its peripheral dykes, has no accompanying agglomerates nor any decided proof that it ever communicated with the surface. When, however, we consider its petrographical and chemical constitution, its position as a core among the bedded lavas, and the intimate way in which it is linked with these rocks by the network of dykes, we are, I think, justified in accepting the inference that it belongs to the volcanic series. It possesses some curious and interesting features in common with the great granophyre bosses of Tertiary age in the Inner Hebrides. Like these it has no visible accompaniment of superficial discharges. Yet it may have ascended by means of some central vent or group of vents which, offering to it a weak part of the crust, allowed it to communicate with the surface and give rise to the outflow of lavas and fragmental ejections. In any case, it affords us a most interesting and instructive insight into one of the deeper-seated ducts of a volcanic region, and the relation of a volcanic focus to the ascent of the granitic magma.

About twenty miles to the north of the Cheviot Hills, and separated from them by the Carboniferous and Upper Old Red Sandstones which spread across the broad plain of the Merse, a group of volcanic rocks has been laid open in a singularly instructive manner along the coast of Berwickshire, between the village of Eyemouth and the promontory of St. Abb's Head. Not only the actual vents, but the lavas and tuffs connected with them, have there been admirably dissected by the forces of denudation.

That this volcanic area was quite distinct from that of the Cheviot Hills may be inferred from its coarse agglomerates, and from the fact that when the rocks are followed inland in a south-westerly direction, that is, towards the Cheviot area, they are found to diminish in thickness and to disappear among the ordinary sediments. For the same reason we may

[1] *Op. cit.* p. 24. [2] *Geol. Mag.* 1885, p. 106. [3] *Op. cit.* pp. 26-28.

regard the area as independent of any vents which may have risen further west about Cockburn Law and the Dirrington Laws. Unfortunately, however, only a small part of the area comes into view, the rest of it lying beneath the waters of the North Sea.[1]

Of the several vents dissected along this coast-line, one may be seen at Eyemouth, filled with a very coarse tumultuous agglomerate of andesite fragments embedded in a compact felspathic matrix, through which are scattered broken crystals of felspar, and imperfect tabular crystals of black mica. Another of similar character is exposed for more than a mile and a half along the shore at Coldingham. It contains blocks, sometimes more than a yard in diameter, of different varieties of andesite, and, as at Eyemouth, is much invaded by veins and bosses of intrusive andesite.

To the north of Coldingham, a series of bedded volcanic rocks which form the picturesque headland of St. Abb's Head, are, according to the estimate of Professor James Geikie, about 1000 to 1200 feet thick, but neither their bottom nor their top is seen. The same observer found them to consist of three groups of andesite sheets separated and overlain by

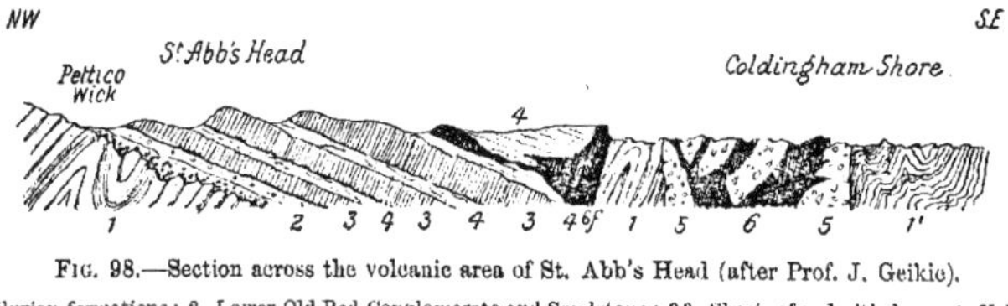

FIG. 98.—Section across the volcanic area of St. Abb's Head (after Prof. J. Geikie).

1 1. Silurian formations; 2. Lower Old Red Conglomerate and Sandstone; 3 3. Sheets of andesitic lava; 4. Volcanic tuffs, largely composed of scoriæ in the higher parts; 5. Volcanic agglomerate of neck on shore; 6. Intrusive andesites. *f*, Fault.

bedded tuffs. The lowest lavas have their base concealed under the sea, and are covered by a thick band of coarse agglomeratic tuff, above which lies the second group of andesites, about 250 feet thick. An intercalation of various tuffs from 40 to 50 feet thick then succeeds, followed by the third lava-group, 250 or 300 feet in depth. The highest member of the series is a mass of bedded tuffs some 400 feet thick.

The andesites lie in beds varying from about 15 to about 50 feet or more in thickness. They are fine-grained, purplish-blue, or greyish-blue, often reddish rocks, of the usual type. Generally rather close-grained, they are not as a rule very porphyritic, but often highly scoriaceous and amygdaloidal, especially towards the top and bottom of each bed. The more slaggy portions are sometimes so filled in with fine tuff that the rock might be mistaken for one of fragmental origin.

The bedded tuffs are usually well stratified deposits. The most import-

[1] This area lies in Sheet 34 Geological Survey of Scotland, and was described by myself in the Memoir to accompany that Sheet ("Geology of Eastern Berwickshire," 1864, p. 20). More recently the shore between St. Abb's Head and Coldingham has been remapped by Professor James Geikie who has also studied the microscopic character of the rocks, *Proc. Roy. Soc. Edin.* xiv. (1887).

ant band of them is that which forms the highest member of the volcanic series. It consists of successive beds that vary from fine red mudstones up to volcanic breccias with blocks one foot or more in diameter. The materials have been derived from the explosion of andesitic lavas. Most of the lapilli are vesicular or amygdaloidal, and many of them have evidently come from vitreous scoriaceous lavas. Professor Geikie remarks that "from their highly vesicular character, they might well have floated in water at the time of their ejection—they are in short mere cinders." He could detect no trace of ordinary sediment in the matrix, the whole material being thoroughly volcanic in origin.

The lavas, tuffs and agglomerates have been abundantly invaded by intrusive rocks, chiefly andesites.[1]

The agglomerates of this Berwickshire coast extend for a short way inland from the Coldingham and Eyemouth vents, but the fragmental material soon becomes finer and more water-rolled, and assumes a distinctly stratified structure, as it is gradually and increasingly interleaved with layers of ordinary sediment. Hence in passing towards the south-west, away from the coast-line, we are obviously receding from the vents of eruption and entering into the usual non-volcanic deposits of the time. That these deposits belong to the Lower Old Red Sandstone was first ascertained during the progress of the Geological Survey in this district by the discovery of abundant plant-remains in the form of linear grass-like strips, and also pieces of *Pterygotus* in some of the green shales interstratified among fine tuffs and ashy sandstones.[2] Before the volcanic detritus disappears from the strata as they are followed in a south-westerly direction, the whole series is unconformably overlain by the Upper Old Red Sandstone. The lower division of the formation is not again seen until it rises from under the southern margin of the plain of the Merse into the Cheviot Hills.

About ten miles to the south-west of the large Coldingham neck the great boss of Cockburn Law and Stoneshiel Hill rises out of the Silurian rocks.[3] Five miles still further in the same direction the group of the beautiful cones of Dirrington (Fig. 70) overlooks the wide Merse of Berwickshire,[4] and six miles to the north of these hills, in the very heart of Lammermuir, lies the solitary boss of the Priestlaw granite.[5] To these protrusions of igneous material reference has already been made as possible volcanic vents connected with the eruptions of the Lower Old Red Sandstone. As regards their age they must certainly be younger than the Llandovery rocks which they disrupt, and older than the Upper Old Red Sandstone, of which the conglomerates, largely made from their debris, lie on them unconformably. It seems therefore probable that these great bosses may form a part of the

[1] See Prof. J. Geikie, *op. cit.*

[2] "Geology of Eastern Berwickshire," *Mem. Geol. Surv. Scotland* (1864), pp. 26, 27, 57.

[3] See "The Geology of Eastern Berwickshire" (Sheet 34), *Mem. Geol. Surv. Scotland* (1864), p. 29.

[4] These hills are chiefly represented in Sheet 25. But see "The Geology of East Lothian," *Mem. Geol. Surv. Scotland* (1866), p. 26.

[5] "Geology of East Lothian," *Mem. Geol. Surv. Scotland*, p. 15, and authorities there cited.

volcanic history of the Lower Old Red Sandstone period. But no positive proof has yet been obtained that any one of them was the site of an eruptive vent, and no trace has been detected around them of any lavas or tuffs which might have proceeded from them.

"THE LAKE OF LORNE"

The basin of Lorne has not yet been carefully examined and described, though various writers have referred to different parts of it (Map I.). My own observations have been too few to enable me to give an adequate account of it. The volcanic sheets of this area form a notable feature in the scenery of the West Highlands, for they are the materials out of which the remarkable terraced hills have been carved, which stretch from Loch Melfort to Loch Creran (Fig. 99), and which reappear in picturesque outliers among the mountains traversed by Glen Coe. Between the ancient schists that form the foundation-rocks of this district and the base of the volcanic series, lies a group of sedimentary strata which in the western part of the district must be 500 feet thick. This group consists of exceedingly coarse breccias at the bottom, above which come massive conglomerates, volcanic grits or

FIG. 99.—View of terraced andesite hills resting on massive conglomerate, south of Oban.

tuffs, fine sandstones and courses of shale. While the basement-breccias are composed mainly of detritus of the underlying schists, including blocks six feet long, they pass up into coarse conglomerates made up almost entirely of fragments of different lavas (andesites, diabases, etc.), and more than 100 feet thick, which often show little or no trace of stratification, but break up into large quadrangular blocks by means of joints which cut across the imbedded boulders. These volcanic conglomerates form some of the more conspicuous features of the coast to the south and north of Oban, and are well exposed in the Isle of Kerrera. They offer many points of resemblance to those of Lake Caledonia, but no certain proof has been noted that they belong to the Lower Old Red Sandstone. They have obviously been derived from lava-sheets that were exposed to strong breaker-action, which at the same time transported and rounded blocks of granite, schist and other crystalline rocks derived from the adjacent hills. During the intervals of quieter sedimentation indicated by the fine sandstones and shales, volcanic explosions continued, as may be seen by the occurrence of occasional large bombs which have fallen upon and pressed down the fine ashy silt that was gathering on the bottom.

It would seem from the characters of some of the strata in this sedimentary series that over the area of deposit portions of the shallower waters were occasionally laid bare to the sun and air. Among the conglomerates there lie certain bands of reddish sandy, ripple-marked, sun-cracked and rain-pitted shales and fine sandstones. Such accumulations, indicative of the ultimate exposure of fine sediment that silted up hollows in the great banks of coarse shingle, may be noticed at the south end of the Island of Kerrera, on at least two horizons which are separated from each other by thick masses of conglomerate and fine felspathic grit. We may infer, therefore, that the fine littoral mud, which gathered during pauses in the heaping up of the coarse gravel and shingle, was occasionally laid dry. Similar strata may be observed behind Oban, where the alternation of exceedingly fine sediment and granular ashy bands is well exhibited.

But the explosions that gave rise to the volcanic materials so largely represented in these lower conglomerates, were merely preliminary to those which led to the outflow of the great sheets of lava that now constitute so large a part of the hills of Lorne. In the few traverses which I have made across different parts of this district I have noted the general resemblance of the lavas to those of the Lower Old Red Sandstone of the Midland Valley of Scotland, their bedded character, and the occurrence of occasional layers of stratified material between them. The prominent features of these rocks, and their relation to the volcanic conglomerates below them, and to the underlying slates and schists are well displayed on Beinn Lora at the mouth of Loch Etive (Fig. 100). There the black slates of the district are unconformably covered by the coarse volcanic conglomerate, formed chiefly of blocks of andesite, cemented in a hard matrix of similar composition. About 150 or 200 feet of this material underlie the great escarpment of the lavas, which here rise in successive beds to the top of the hill, 1000 feet above its base.

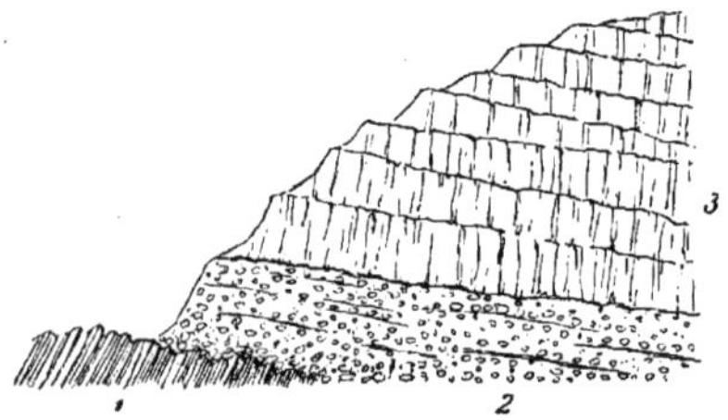

FIG. 100.—Section of lava-escarpment at Beinn Lora, north side of mouth of Loch Etive, Argyllshire.

1. Phyllites; 2. Thick conglomerate; 3. Successive sheets of andesite.

On the south side of Loch Etive the base of the volcanic series, with its underlying conglomerate, may be followed westward to Oban and thence southward to Loch Feochan. The lavas cover most of the ground from the western shore eastwards to near Loch Awe. But this area is still very imperfectly known. The Geological Survey, however, has now advanced to this part of the country, so that we shall before long be in possession of more detailed information regarding the character and sequence of its volcanic history and the geological age of the eruptions.

Mr. H. Kynaston, who has begun the mapping of the eastern portion of the district, finds that there, as further west, the bottom of the volcanic series is generally a breccia or conglomerate. He has met with two leading types among the lavas, the more abundant being strongly vesicular, the other more compact. He has observed also numerous dykes and sills of intrusive porphyrite, trending in a general N.N.E. and S.S.W. direction, and pointing towards the great granite mass of Ben Cruachan.[1]

Mr. R. G. Symes has traced the volcanic series to the north and south of Oban. While visiting with him part of his ground, I was much struck with the evidence of an intrusive mass at the base of the volcanic series in the Sound of Kerrera. A prominent feature on the east side of the channel, known as Dun Uabairtich and 270 feet high, consists of andesite which appears to combine both a central boss and a sill. The rock breaks through the black slates and the overlying conglomerates and sandstones, and has wedged itself into the unconformable junction between the two formations. It is beautifully columnar on its sea-covered face, some of the columns being 120 feet or more in length, and gently curved.

"LAKE ORCADIE"

We now cross the whole breadth of the Scottish Highlands in order to peruse the records of another of the great detached water-basins of the Old Red Sandstone, which for the sake of brevity of reference I have named and described as "Lake Orcadie" (Map I.). This area has its southern limits along the base of the hills that enclose the wide Moray Firth. It spreads northward over the Orkney and much of the Shetland Islands, but its boundaries in that direction are lost under the sea. In the extensive sheet of water which spread over all that northern region the peculiar Caithness Flags, with their associated sandstones and conglomerates, were deposited to a total depth of 16,000 feet. A sigillaroid and lycopodiaceous vegetation flourished on the surrounding land, together with ferns, *Psilophyton* and conifers. The waters teemed with fishes of which many genera and species have now been described. The remains of these creatures lie crowded upon each other in the flagstones in such a manner as to indicate that from time to time vast quantities of fish were suddenly killed. Not impossibly, these destructions may have been connected with the volcanic activity which has now to be described.

In the year 1878 I called attention to the evidence for the existence of contemporaneous volcanic rocks in the Old Red Sandstone north of the range of the Grampians, and specially noted three localities where this evidence could be seen—Strathbogie, Buckie and Shetland.[2] Since that time Messrs.

[1] *Ann. Report Geol. Surv.* (1895), p. 29 of reprint.
[2] *Trans. Roy. Soc. Edin.* xxviii. (1878).

B. N. Peach and J. Horne have added a fourth centre in the Orkney Islands. At present, therefore, we are acquainted with the records of four separate groups of volcanic vents in Lake Orcadie.

The southern margin of this water-basin appears to have indented the land with long fjord-like inlets. One of these now forms the vale of Strathbogie, which runs into the hills for a distance of fully 20 miles beyond what seems to have been the general trend of the coast-line. In this valley I found a bed of dark vesicular diabase intercalated among the red sandstones and high above the base of the formation, as exposed on the west side of the valley near Burn of Craig. On the east side a similar band has since been mapped for the Geological Survey by Mr. L. Hinxman who has traced it for some miles down the Strath.[1] This latter band, as shown in Fig. 101, lies not far above the bottom of the Old Red Sandstone of this district, and is thus probably distinct from the Craig outcrop. There would thus appear to be evidence of two separate outflows of basic lava in this fjord of the Old Red Sandstone period.

No vestige has here been found of any vent, nor is the lava accompanied with tuff. The eruptions took place some time after the earlier sediments

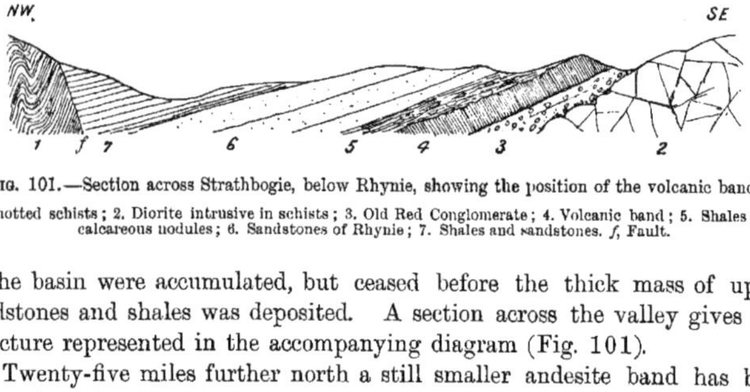

FIG. 101.—Section across Strathbogie, below Rhynie, showing the position of the volcanic band.

1. Knotted schists; 2. Diorite intrusive in schists; 3. Old Red Conglomerate; 4. Volcanic band; 5. Shales with calcareous nodules; 6. Sandstones of Rhynie; 7. Shales and sandstones. *f*, Fault.

of the basin were accumulated, but ceased before the thick mass of upper sandstones and shales was deposited. A section across the valley gives the structure represented in the accompanying diagram (Fig. 101).

Twenty-five miles further north a still smaller andesite band has been detected by Mr. J. Grant Wilson among the sandstones and conglomerates near Buckie.[2] It is a truly contemporaneous flow, for pebbles of it occur in the overlying strata. But again it forms only a solitary bed, and no trace of any accompanying tuff has been met with, nor of the vent from which it came. Both this vent and that of Strathbogie must have been situated near the southern coast-line of the lake.

At a distance of some 90 miles northward from these Moray Firth vents another volcanic district lies in the very heart of the Orkney Islands.[3] The lavas which were there ejected occur at the south-eastern corner of the island of Shapinshay, where they are regularly bedded with the flagstones. They consist of dark green olivine-diabases with highly

[1] See Sheet 76 of the Geological Survey of Scotland.

[2] See Sheet 95 of the Geological Survey of Scotland and *Trans. Roy. Soc. Edin.* vol. xxviii. (1878), p. 435.

[3] Messrs. B. N. Peach and J. Horne, *Proc. Roy. Phys. Soc. Edin.* vol. v. (1879), p. 80.

amygdaloidal and vesicular upper surfaces. Their thickness cannot be ascertained, as their base is not seen, but they have been cut by the sea into trenches which show them to exceed 30 feet in depth. The position of the vent from which they came has not been ascertained. Neither here nor in the Moray Firth area do any sills accompany the interbedded sheets, and in both cases the volcanic action would seem to have been of a feeble and short-lived character.

Much more important were the volcanoes that broke out nearly 100 miles still further north, where the Mainland of Shetland now lies. I shall never forget the pleasure with which I first recognized the traces of these eruptions, and found near the most northerly limits of the British Isles proofs of volcanic activity in the Lower Old Red Sandstone. Since my observations were published,[1] Mr. Peach, who accompanied me in Shetland, has returned to the district, and, in concert with his colleague Mr. Horne, has extended our knowledge of the subject.[2] The chief vent or vents lay towards the west and north-west of the Mainland and North Mavine; others of a less active and persistent type were blown out some 25 miles to the east, where the islands of Bressay and Noss now stand. In the western district streams of slaggy andesite and diabase with showers of fine tuff and coarse agglomerate were ejected, until the total accumulation reached a thickness of not less than 500 feet. The volcanic eruptions took place contemporaneously with the deposition of the red sandstones, for the lavas and tuffs are intercalated in these strata. The lavas and volcanic conglomerates are traceable from the southern coast of Papa Stour across St. Magnus' Bay to the western headlands of Esha Ness, a distance of more than 14 miles. They have been cut by the Atlantic into a picturesque range of cliffs, which exhibit in some places, as at the singular sea-stalk of Doreholm, rough banks of andesitic lava with the conglomerate deposited against and over them, and in other places, as along the cliffs of Esha Ness, sheets of lava overlying the conglomerates.

No trace of any vents has been found in the western and chief volcanic district, but in Noss Sound a group of small necks occurs, filled with a coarse agglomerate composed of pieces of sandstone, flagstone and shale. Messrs. Peach and Horne infer that these little orifices never discharged any streams of lava. More probably they were opened by explosions which only gave forth vapours and fragmentary discharges, such as a band of tuff which is intercalated among the flagstones in their neighbourhood.

But one of the most striking features of the volcanic phenomena of this remote region is the relative size and number of the sills and dykes which here as elsewhere mark the latest phases of subterranean activity. Messrs. Peach and Horne have shown us that three great sheets of acid rocks (granites and spherulitic felsites, to which reference has already been made, p. 292) have been injected among the sandstones and basic lavas, that abundant veins of granite, quartz-felsite and rhyolite radiate from these acid sills, and that the latest phase of igneous action in this region was the

[1] *Trans. Roy. Soc. Edin.* vol. xxviii. (1878), p. 418. [2] *Ibid.* vol. xxxii. (1884), p. 359.

intrusion of a series of bosses and dykes of basic rocks (diabases) which traverse the sills.

THE KILLARNEY DISTRICT

In the south of Ireland the Upper Silurian strata are followed upwards conformably by the great series of red sandstones and conglomerates known as the "Dingle Beds." Lithologically these rocks present the closest resemblance to the Lower Old Red Sandstone of Central Scotland. They occupy a similar stratigraphical position, and though they have not yielded any palæontological data for comparison, there can, I think, be no hesitation in classing them with the Scottish Lower Old Red Sandstone, and regarding them as having been deposited under similar geographical conditions. They offer one feature of special interest for the purpose of the present inquiry, since they contain a well-marked group of contemporaneous volcanic rocks, including nodular felsites, like those so characteristic of the Silurian period.

The area where this remote and isolated volcanic group is best developed forms a range of high rugged ground along the northern front of the hills that stretch eastward from the Lakes of Killarney. Their general distribution is shown on Sheets 184 and 185 of the Geological Survey of Ireland;[1] though I may again remark that petrography has made great strides during the thirty years and more that have passed since these maps and their accompanying Memoirs were published, and that, were the district to be surveyed now, probably a considerable tract of ground coloured as ash would be marked as felsite. At the same time the existence of both these rocks here cannot be gainsaid.

The felsite was long ago brought into notice by Dr. Haughton, who published an analysis of it.[2] It is also referred to by Mr. Teall for its spherulitic structure.[3] Seen on the ground it appears as a pale greenish-grey close-grained rock, sometimes exhibiting flow-structure in a remarkably clear manner, the laminae of devitrification following each other in wavy lines, sometimes twisted and delicately puckered or frilled, as in some schists. Portions of the rock are strongly nodular, the nodules varying in size from less than a pea to that of a hen's egg.

The close resemblance of this rock to many of the Lower Silurian nodular felsites of Wales cannot but strike the geologist. It presents analogies also to the Upper Silurian felsites of Dingle. But its chief interest arises from the geological horizon on which it occurs. Lying in the so-called "Dingle-Beds," which may be regarded as the equivalents of the Lower Old Red Sandstone of England and Scotland, it is, so far as my observations go, the only example of such a nodular felsite of later date than the Silurian period. We recognize in it a survival, as it were, of the peculiar Silurian type of acid lava, the last preceding eruption

[1] See the Memoir (by J. B. Jukes and G. V. Du Noyer) on Sheet 184, p. 15. Other volcanic rocks have been mapped at Valentia Harbour in the Dingle Beds, but these I have not had an opportunity of personally examining.

[2] *Trans. Roy. Irish Acad.* vol. xxiii. (1859), p. 615.

[3] *British Petrography*, p. 349.

of which took place not many miles to the west, in the Dingle promontory. But elsewhere this type does not appear to have survived the end of the Silurian period.

The detrital rocks accompanying the felsite, in the district east of Killarney, vary from such closed-grained felsitic material as cannot readily be distinguished from the felsite itself to unmistakable felsitic breccias. Even in the finest parts of them, occasional rounded quartz-pebbles may be detected, while here and there a reddish shaly band, or a layer of fine pebbly conglomerate with quartz-pebbles an inch in length, shows at once the bedding and the dip. Mr. W. W. Watts, who, with Mr. A. M'Henry of the Irish Staff of the Geological Survey, accompanied me over this ground, found that a microscopic examination of the slides which were prepared from the specimens we collected completely confirmed the conclusions reached from inspection of the rocks in the field.[1] He detected among the angular grains slightly damaged crystals of felspar, chiefly orthoclase. Many portions of these felspathic grits much resemble the detrital Cambrian rocks which in the Vale of Llanberis have been made out of the pale felsite of that locality.

[1] Mr. Watts also examined the microscopic structure of the felsite of Benaun More. He found that the spherulites appear to have a micropegmatitic structure, owing to the intergrowth of quartz and felspar. In some parts of the rock the spherulites, from ·02 to ·01 inch in diameter, are surrounded by exceedingly minute green needles, possibly of hornblende, while inside some of them are small quartz-grains. Larger porphyritic felspars occur outside the spherulites, some being of plagioclase, but most of orthoclase. The spherulitic structure is not so well developed near the felspars. A few of the large nodules are hollow and lined with crystals, while some of them show a finely concentric lamination like the successive layers of an agate.

CHAPTER XXII

VOLCANOES OF THE UPPER OLD RED SANDSTONE—THE SOUTH-WEST OF IRELAND, THE NORTH OF SCOTLAND

IN the northern half of Britain, where the Old Red Sandstone is so well displayed, the two great divisions into which this series of sedimentary deposits is there divisible are separated from each other by a strongly marked unconformability. The interval of time represented by this break must have been of long duration, for it witnessed the effacement of the old water-basins, the folding, fracture, and elevation of their thick sedimentary and volcanic accumulations, and the removal by denudation of, in some places, several thousand feet of these rocks. The Upper Old Red Sandstone, consisting so largely as it does of red sandstones and conglomerates, indicates the return or persistence of geographical conditions not unlike those that marked the deposition of the lower subdivision. But in one important respect its history differs greatly from that which I have sketched for the older part of the system. Though the Upper Old Red Sandstone is well developed across the southern districts of Scotland from the Ochil to the Cheviot Hills, and appears in scattered areas over so much of England and Wales, no trace has ever been there detected in it of any contemporaneously erupted volcanic rocks. The topographical changes which preceded its deposition must have involved no inconsiderable amount of subterranean disturbance, yet the volcanic energy, which had died out so completely long before the close of the time of the Lower Old Red Sandstone, does not appear to have been rekindled until the beginning of the Carboniferous period.

Two widely separated tracts in the British Isles have yielded traces of contemporaneous volcanic rocks in the Upper Old Red Sandstone. One of these lies in the south-west of Ireland, the other in the far north of Scotland.

THE SOUTH-WEST OF IRELAND

The Irish locality is situated a few miles to the south of the town of Limerick, where the Carboniferous Limestone has been thrown into long folds, and where, along the anticlines, strips of the underlying red sandstones have been brought up to the surface. Two such ridges of Upper Old Red Sandstone bear, each on its crest, a small but interesting relic of volcanic activity[1] (Map I.).

The more northerly ridge rises in the conical eminence of Knockfeerina to a height of 949 feet above the sea. Even from a distance the resemblance of this hill (Fig. 102) to many of the Carboniferous necks of Scotland at once attracts the eye of the geologist. The resemblance is found to hold still more closely when the internal structure of the ground is examined. The cone consists mainly of a coarse agglomerate, with blocks generally somewhat rounded and varying in size up to two feet in length. The most prominent of these, on the lower eastern slopes, are pieces of a fine flinty felsite weathering white, but there also occur fragments of grit and baked shale. The matrix is dull-green in colour, and among its ingredients are

FIG. 102.—View of Knockfeerina, Limerick, from the north-east—a volcanic neck of Upper Old Red Sandstone age.

abundant small lapilli of a finely vesicular andesite or diabase. These more basic ingredients increase in number towards the top of the eminence, where much of the agglomerate is almost wholly made up of them. No marked dip is observable over most of the hill, the rock appearing as a tumultuous agglomerate, though here and there, particularly near the top and on the south side, a rude bedding may be detected dipping outwards. On the west side the agglomerate is flanked with yellow sandstone baked into quartzite, so that the line of junction there between the two rocks not improbably gives us the actual wall of the vent. The induration of the surrounding sandstones is a familiar feature among the Carboniferous vents. Some intrusive dark flinty rock traverses the agglomerate near the top on the north side.

Retiring eastwards from the cone, the observer finds evidence of the intercalation of tuff among the surrounding Upper Old Red Sandstone. At the east end of the village of Knockfeerina a red nodular tuff, with rounded pieces of andesite, grit and sandstone, sometimes 12 inches long, is seen to dip under yellow, grey and red sandstones and shales, while other shales

[1] See Sheet 153 of the Geological Survey of Ireland, and Explanation to that Sheet (1861), by Messrs. G. H. Kinahan and J. O'Kelly. The account of the ground above given is from notes which I made during a personal visit.

and sandstones underlie this tuff and crop out between it and the agglomerate. There is thus evidence of the intercalation of volcanic tuff in the Upper Old Red Sandstone of this district. And there seems no reason to doubt that the tuff was ejected from the adjoining vent of Knockfeerina.

On the next ridge of Old Red Sandstone, which runs parallel to that of Knockfeerina at a distance of little more than a mile to the south, another mass of volcanic material rises into a prominence at Ballinleeny. On the north side it consists of agglomerate like that just described, and is flanked by sandstone baked into quartzite. Here again we probably see the edge of a volcanic funnel. There may possibly be more than one vent in this area. But well-bedded tuffs can be observed to dip away from the centre and to pass under sandstones and shales which are full of fine ashy material. Gradations can be traced from the tuff into ordinary sediment. In this instance, therefore, there is additional proof of contemporaneous volcanic action in the Upper Old Red Sandstone. There can be no uncertainty as to the horizon of the strata in which these records have been preserved, for they dip conformably under the shales and limestones at the base of the Carboniferous Limestone series. They are said to have yielded the characteristic fern *Palæopteris* of Kiltorcan.[1]

THE NORTH OF SCOTLAND

The only district in England or Scotland wherein traces of volcanic action during the time of the Upper Old Red Sandstone have been observed lies far to the north among the Orkney Islands, near the centre of the scattered outliers which I have united as parts of the deposits of "Lake Orcadie"[2] (Map I.). The thick group of yellow and red sandstones which form most of the high island of Hoy, and which, there can be little doubt, are correctly referred to the Upper Old Red Sandstone, rest with a marked unconformability on the edges of the Caithness flagstones (Fig. 103). At the base of these pale sandstones, and regularly interstratified with them, lies a band of lavas and tuffs which can be traced from the base of the rounded hills to the edge of the cliffs at the Cam, along the face of which it runs as a conspicuous feature, gradually sloping to a lower level, till it reaches the sea. At the Cam of Hoy it is about 200 feet thick, and consists of three or more sheets of andesite. The upper 50 or 60 feet show a strongly slaggy structure, the central portion is rudely columnar, and the lower part presents a kind of horizontal jointing or bedding. There can be no question that this rock is not a sill but a group of contemporaneous lava-flows. Beneath it, and lying across the edges of the flagstones below, there

[1] There may be some other examples of Upper Old Red Sandstone volcanic rocks in Ireland which I have not yet been able personally to examine. On the maps of the Geological Survey (Sheet 198, and Explanation, pp. 8, 17) contemporaneous rocks of this age are marked as occurring at Cod's Head and Dursey Island, on the south side of the mouth of the Kenmare estuary.

[2] First noticed in *Geol. Mag.* February 1878; and *Trans. Roy. Soc. Edin.* xxviii. (1878), p. 411.

is a zone of dull-red, fine-grained tuff, banded with seams of hard red and yellow sandstone. This tuff zone dies out to the eastward of the Cam.

A few miles south of the Cam the volcanic zone appears again as the platform on which the picturesque natural obelisk of the Old Man of Hoy stands. Here the lava runs out as a promontory from the base of the cliff, and on this projection the Old Man has been left isolated from the main precipice. The cliffs of Hoy are traversed by numerous small faults which have shifted the volcanic zone. But on the great face of rock behind the Old Man there appears to be a second volcanic zone lying several hundred feet above that just described. It is probably this upper zone which emerges from under the hills a mile and a half to the south at Black Ness in the bay of Rackwick. A good section is there visible, which is represented in Fig. 104. The ordinary red and yellow sandstones (*a*) appear from under the volcanic rocks at this locality, and stretch southwards to the most southerly headland of Hoy.

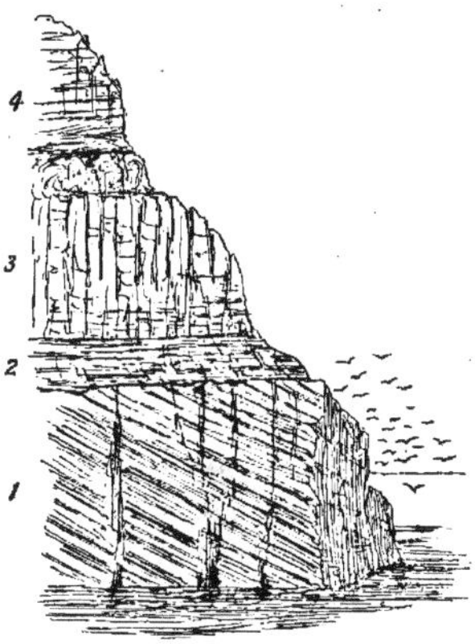

FIG. 103.—Section of the volcanic zone in the Upper Old Red Sandstone, Cam of Hoy, Orkney.

1. Caithness flagstones; 2. Dull-red tuff and bands of sandstone; 3. Lava zone in three bands; 4. Yellow and red sandstone.

The lowest volcanic band in the section is one of red sandy well-bedded tuffs (*b*). Some of the layers are coarse and almost agglomeratic, while others are fine marly and sandy, with dispersed bombs, blocks and lapilli of diabase and andesite. Hard ribs of a sandy tufaceous material also occur. These fragmental deposits are immediately overlain by a dark-blue rudely prismatic diabase with slaggy top (*c*). It is about 150 feet thick at its thickest part, but rapidly thins away in a westerly direction. It passes under a zone of red tuff (*d*) like that below, and above this highest member of the volcanic group comes the great overlying pile of yellow and reddish sandstone of Hoy. Followed

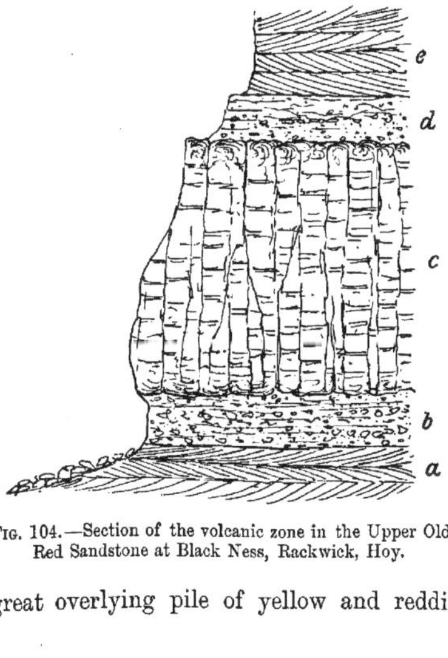

FIG. 104.—Section of the volcanic zone in the Upper Old Red Sandstone at Black Ness, Rackwick, Hoy.

westward for a short distance, the whole volcanic zone is found to die out and the sandstones below and above it then come together.

The interest of this little volcanic centre in Hoy is heightened by the fact that the progress of denudation has revealed some of the vents belonging to it. On the low ground to the east of the Cam, and immediately to the north of the volcanic escarpment, the flagstones which there emerge from under the base of the unconformable upper sandstones are pierced by three volcanic necks which we may with little hesitation recognize as marking the sites of vents from which this series of lavas and tuffs was discharged (Fig. 105). The largest of them forms a conspicuous

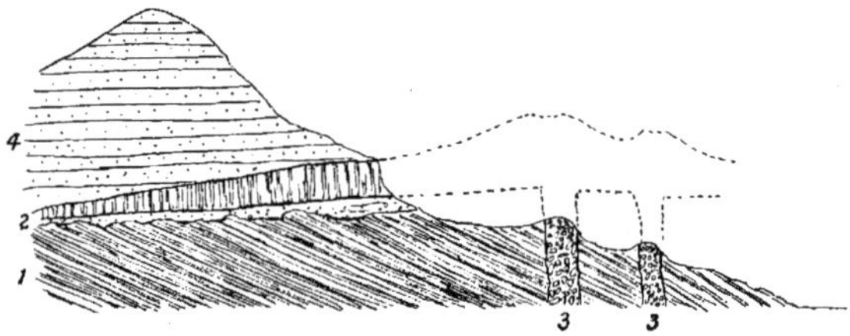

Fig. 105.—Section across the volcanic band and its associated necks, Hoy, Orkney.

1. Caithness flagstones; 2. Volcanic band lying on red sandstones and conglomerates and dying out eastwards; 3 3. Two vents between the base of the hills and the sea; their connection with the volcanic band is shown by dotted lines; 4. Overlying mass of Upper Old Red Sandstone forming the hills of Hoy.

hill about 450 feet high, the smallest is only a few yards in diameter, and rises from the surface of a flagstone ridge. They are filled with a coarse, dull-green, volcanic agglomerate, made up of fragments of the lavas with pieces of hardened yellow sandstone and flagstone. Around the chief vent the flagstones through which it has been opened have been greatly hardened and blistered. The most easterly vent, which has been laid bare on the beach at Bring, due east of Hoy Hill, can be seen to pierce the flagstones, which, although their general dip is westerly at from 10° to 15°, yet at their junction with the agglomerate are bent in towards the neck, and are otherwise much jumbled and disturbed.

On the northern coast of Caithness I have described a remarkable volcanic vent about 300 feet in diameter, which rises through the uppermost group of the Caithness flagstones. It is filled with a coarse agglomerate consisting of a dull-greenish diabase paste crowded with blocks of diabase, sometimes three feet in diameter, and others of red sandstone, flagstone, limestone, gneiss and lumps of black cleavable augite (Fig. 106).[1] The sandstones around it present the usual disrupted, indurated and jointed character, and are traversed by a small diabase dyke close to the western margin of the neck. Another similar neck has since been found by the officers of the Geological Survey on the same coast.

[1] See *Trans. Roy. Soc. Edin.* xxviii. (1878), p. 405; also p. 482 of the same volume for an account of the cleavable augite.

That these volcanic orifices were active about the same time with those in the opposite island of Hoy may be legitimately inferred.

These northern volcanoes made their appearance in a district where during the preceding Lower Old Red Sandstone period there had been several widely separated groups of active volcanic vents. So far as the fragmentary nature of the geological evidence permits an opinion to be

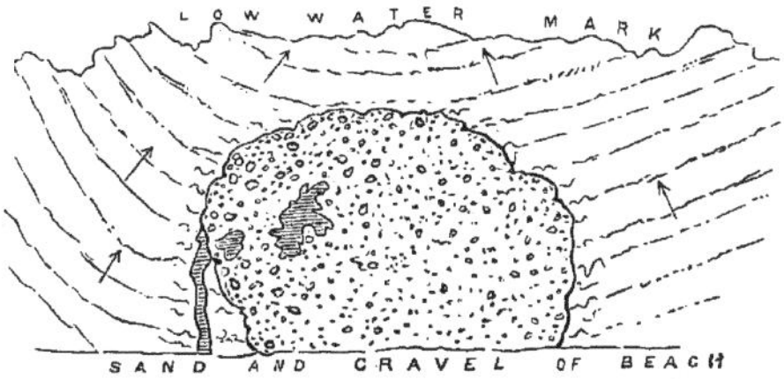

FIG. 106.—Ground-plan of volcanic neck piercing the Caithness Flagstone series on the beach near John o' Groat's House.

formed, they seem to have broken out at the beginning, or at least at an early stage, of the deposition of the Upper Old Red Sandstone, and to have become entirely extinct after the lavas of Hoy were poured forth. No higher platform of volcanic materials has been met with in that region. With these brief and limited Orcadian explosions the long record of Old Red Sandstone volcanic activity in the area of the British Isles comes to an end.[1]

[1] There appear to be traces of volcanic eruptions contemporaneous with the Upper Old Red Sandstone of Berwickshire, but as they merely formed a prelude to the great volcanic activity of Carboniferous time, they are included in the account of the Carboniferous plateau of Berwickshire in Chapters xxiv. and xxv.

BOOK VI

THE CARBONIFEROUS VOLCANOES

CHAPTER XXIII

THE CARBONIFEROUS SYSTEM OF BRITAIN AND ITS VOLCANIC RECORDS

Geography and Scenery of the Carboniferous Period—Range of Volcanic Eruptions during that time—I. The Carboniferous Volcanoes of Scotland—Distribution, Arrangement and Local Characters of the Carboniferous System in Scotland—Sketch of the Work of previous Observers in this Subject.

WITHIN the area of the British Isles, the geological record is comparatively full and continuous from the base of the Upper Old Red Sandstone to the top of the Coal-measures. We learn from it that the local basins of deposit in which the later portion of the Old Red Sandstone was accumulated sank steadily in a wide general subsidence, that allowed the clear sea of the Carboniferous Limestone ultimately to spread for some 700 miles from the west coast of Ireland into Westphalia. Over the centre of England this Carboniferous Mediterranean had a breadth of at least 150 miles, gradually shallowing northwards in the direction of land in Scotland and Northern Ireland. The gentle sinking of the floor of the basin continued until more than 6000 feet of sediment, chiefly composed of the remains of crinoids, corals and other marine organisms, had been piled up in the deeper parts. Traces of the southern margin of this sea, or at least of a long insular ridge that rose out of its waters, are to be seen in the protuberances of older rocks which appear at intervals from under the Coal-measures and later formations between the borders of Wales and the heart of Leicestershire, and of which the crags of Charnwood Forest are among the few peaks that still remain visible. To the south of this ridge, open sea extended far southward and westward over the site of the Mendip Hills and the uplands of South Wales.

The Carboniferous period, as chronicled by its sedimentary deposits, was a time of slow submergence and quiet sedimentation, terrestrial and marine conditions alternating along the margins of the sinking land, according as the rate of depression surpassed or fell short of that of the deposition of sediment. There is no trace of any general disturbance among the strata, such as would be marked by an important and widely extended unconformability. But many indications may be observed that the rate of subsidence did not continue uniform, if, indeed, the downward movement was not locally arrested, and even exchanged for a movement in the opposite direction. It is difficult, for instance, to believe the ancient ridge of the Midlands to have been so lofty that even the prolonged subsidence required for the accumulation of the whole Carboniferous system was insufficient to carry its highest crests below the level of the coal-jungles. More probably the depression reached its maximum along certain lines or bands running in a general north-easterly direction, the intervals between these lines sinking less, or possibly even undergoing some measure of uplift. One of the subsiding tracts, that of the wide lowlands of Central Scotland, was flanked on the south by a ridge which, while its north-eastern portion was buried under the Upper Old Red Sandstone and Lower Carboniferous rocks, remained above water towards the south-west, and does not appear to have been wholly submerged there even at the close of the Carboniferous period.

So abundant and varied are the sedimentary formations of Carboniferous time, and so fully have they preserved remains of the contemporary plants and animals, that it is not difficult to realise in some measure the general aspect of the scenery of the time, and the succession of changes which it underwent from the beginning to the end of the period. The land was green with a luxuriant if somewhat monotonous vegetation. Large pine trees flourished on the drier uplands. The lower grounds nourished dense groves of cycads or plants allied to them, which rose as slim trees twenty or thirty feet high, with long hard green leaves and catkins that grew into berries. The swamps and wetter lands bore a rank growth of various gigantic kinds of club-moss, equisetaceous reeds and ferns.

Nor was the hum of insect-life absent from these forests. Ancestral types of cockroaches, mayflies and beetles lived there. Scorpions swarmed along the margins of the shallow waters, for their remains, washed away with the decayed vegetation among which they harboured, are now found in abundance throughout many of the dark shales.

The waters were haunted by numerous kinds of fish quite distinct from those of the Old Red Sandstone. In the lagoons, shoals of small ganoids lived on the cyprids that peopled the bottom, and they were in turn preyed on by larger ganoids with massive armature of bone. Now and then a shark from the opener sea would find its way into these more inland waters. The highest types of animal life yet known to have existed at this time were various amphibians of the extinct order of Labyrinthodonts.

The open sea, too, teemed with life. Wide tracts of its floor supported a thick growth of crinoids whose jointed stems, piled over each other generation after generation, grew into masses of limestone many hundreds of feet in thickness. Corals of various kinds lived singly and in colonies, here and there even growing into reefs. Foraminifera, sponges, sea-urchins, brachiopods, gasteropods, lamellibranchs and cephalopods, in many genera and species, mingled their remains with the dead crinoids and corals to furnish materials for the wide and thick accumulation of Carboniferous Limestone.

Looking broadly at the history of the Carboniferous period, and bearing in mind the evidence of prolonged depression already referred to, we can recognize in it three great eras. During the first, the wide clear sea of the Carboniferous Limestone spread over the centre and south of Britain, interrupted here and there by islands that rose from long ridges whereby the sea-floor was divided into separate basins. Next came a time of lessened depression, when the sea-bottom was overspread with sand, mud and gravel, and was even in part silted up, as has been chronicled in the Millstone Grit. The third stage brings before us the jungles of the Coal-measures, when the former sea-floor became a series of shallow lagoons where, as in the mangrove-swamps of our own time, a terrestrial vegetation sprang up and mingled its remains with those of marine shells and fishes.

Such a state of balance among the geological forces as is indicated by the stratigraphy of the Carboniferous system would not prepare us for the discovery of the relics of any serious display of contemporary volcanic activity. And, indeed, throughout the Carboniferous rocks of Western Europe there is for the most part little trace of contemporaneous volcanic eruptions. Yet striking evidence exists that, along the western borders of the continental area, in France as well as over much of Britain, which had for so many previous geological ages been the theatre of subterranean activity, the older half of Carboniferous time witnessed an abundant, though less stupendous and prolonged, renewal of volcanic energy.

From the very commencement of the Carboniferous period to the epoch when the Coal-measures began to be accumulated, the area of the British Isles continued to be a scene of active volcanism. In the course of that prolonged interval of geological time the vents shifted their positions, and gradually grew less energetic, but there does not appear to have been any protracted section of the interval when the subterranean activity became everywhere entirely quiescent.

The geologist who traces, from older to younger formations, the progress of some persistent operation of nature, observes the evidence gradually to increase in amount and clearness as it is furnished by successively later parts of the record. He finds that the older rocks have generally been so dislocated and folded, and are often so widely covered by younger formations, that the evidence which they no doubt actually contain may be difficult to decipher, or may be altogether concealed from view. In following, for instance, the progress of volcanic action, he is impressed, as he passes from the older to the younger Palæozoic chronicles, by the striking contrast between the fulness and legibility of the Carboniferous records and the comparative meagreness and obscurity of those of the earlier periods. The Carboniferous rocks have undergone far less disturbance than the Cambrian and Silurian formations; while over wide tracts, where their volcanic chapters are fullest and most interesting, they lie at the surface, and can thus be subjected to the closest scrutiny. Hence the remains of the volcanic phenomena of the later Palæozoic periods present a curiously modern aspect, when contrasted with the fragmentary and antique look of those of older date.

The history of volcanic action during the Carboniferous period in Britain is almost wholly comprised in the records of the earlier half of that period, that is, during the long interval represented by the Carboniferous Limestone series and the Millstone Grit. It was chiefly in the northern part of the region that volcanic activity manifested itself. In Scotland there is the chronicle of a long succession of eruptions across the district of the central and southern counties, from the very beginning of Carboniferous time down to the epoch when the Coal-measures began to be accumulated. In England, on the other hand, the traces of Carboniferous volcanoes are confined within a limited range in the Carboniferous Limestone, while in Ireland they appear to be likewise restricted to the same lower division of the system. During the whole of the vast interval represented by the Coal-measures volcanic energy, so far as at present known, was entirely dormant over the region of the British Isles.

These general statements will be more clearly grasped from the accompanying table, which shows the various sections into which the Carboniferous system of Britain has been divided, and also, by black vertical lines, the range of volcanic intercalations in each of the three kingdoms.

		England.	Scotland.	Ireland.
Coal-measures.	Upper Red Sandstones with *Spirorbis*-limestone.			
	Middle or chief coal-bearing measures.			
	Gannister group.			
Millstone Grit.	Grits, flagstones and shales with thin coals.			
Carboniferous Limestone.	Yoredale group of shales and grits with limestones.			
	Thick (Scaur or Main) Limestone of England, with sandstones and coals in Scotland.			
	Lower Limestone Shale (Calciferous Sandstones of Scotland)			

Such being the general range in time of the Carboniferous volcanic phenomena, it may be convenient, in this preliminary survey, to take note of the general distribution of the volcanic districts over the British Isles, as in this way we may best realise the extent and grouping of the eruptions, which will then be considered in further detail (see Map I.).

Not only were the Carboniferous volcanoes most abundant and persistent in Scotland, but they attained there a variety and development which give their remains an altogether exceptional interest in the study of volcanic geology. They were distributed over the wide central valley, from the south of Cantyre to beyond the mouth of the estuary of the Forth. On the southern side of the Silurian Uplands, they were likewise numerous and active. There is thus no considerable tract of Lower Carboniferous rocks in Scotland which does not furnish its evidence of contemporaneous volcanic action.

Although some portions of the Scottish Carboniferous igneous rocks run for a short distance into England, it is remarkable that, when these at last die out southwards, no other relics of contemporaneous volcanic energy take their place. Along the Pennine chain, from the Border into the heart of England, though natural sections are abundant, no trace of included volcanic rocks appears until we reach Derbyshire. The whole of that wide interval of 150 miles, so far as the present evidence goes, remained during Carboniferous time entirely free from any volcanic eruption. But from the picturesque country of the Peak southwards, the sea-floor of the Carboniferous Limestone, in what is now the heart of England, was dotted with vents whence the sheets of "toadstone" were ejected, which have so long been a familiar feature in English geology. Beyond this limited volcanic district the Carboniferous formations of the south-west of England remain, on the whole, devoid of contemporaneous volcanic intercalations, traces of Carboniferous volcanic action having been recognized only in West Somerset and Devonshire. In the Mendip district and in the ridges of limestone near Weston-super-Mare bands of cellular lava and tuff have been observed. To the west of Dartmoor, Brent Tor and some of the surrounding igneous masses may mark the positions of eruptive vents during an early part of the Carboniferous period.

At the south end of the Isle of Man relics remain of a group of vents among the Carboniferous limestones. Passing across to Ireland, where these limestones attain so great a thickness and cover so large a proportion of the surface of the island, we search in vain for any continuation of the abundant and varied volcanic phenomena of Central Scotland. So far as observation has yet gone, only two widely separated areas of Carboniferous volcanic rocks are known to occur in Ireland.[1] One of these shows that a little group of vents probably rose from the floor of the Carboniferous Limestone sea, near Philipstown, in King's County. The other lies far to the west in the Golden Vale of Limerick, where a more important series of vents poured out successive streams of lava with showers of ashes, from an early part of the Carboniferous period up to about the beginning of the time of the Coal-measures.

The total area within which the volcanic eruptions of Carboniferous time took place was thus less than that over which the volcanoes of the

[1] The supposed Carboniferous volcanic rocks of Bearhaven on the coast of Cork are noticed on p. 49, vol. ii.

Lower Old Red Sandstone were distributed, yet they were scattered across the larger part of the site of the British Isles. From the vents of Fife to those of Limerick is a distance of above 300 miles; from the latter eastward to those of Devonshire is an interval of 250 miles; while the space between the Devonshire volcanoes and those of Fife is about 400 miles. In this triangular space volcanic action manifested itself at each of the apices, to a slight extent along the centre of the eastern side, but with much the greatest vigour throughout the northern part of the area.

Since the volcanic phenomena of Carboniferous time are exhibited on a much more extensive scale in Scotland than in any other region of the world yet studied, it will be desirable to describe that area in considerable detail. The other tracts in Britain where volcanic rocks of the same age occur need not be so fully treated, except where they help to a better comprehension of the general volcanic history.

It is in the southern half of Scotland that the Carboniferous system is developed (Map IV.). A line drawn from Machrihanish Bay, near the Mull of Cantyre, north-eastward across Arran and Bute to the south end of Loch Lomond, and thence eastward by Bridge of Allan, Kinross and Cupar to St. Andrews Bay, forms the northern limit of this system. South of that line Carboniferous volcanic intercalations are to be met with in nearly every county across into the borders of Northumberland.

That we may follow intelligently the remarkably varied volcanic history of this region, it is desirable to begin by taking note of the nature and sequence of the sedimentary formations among which the volcanic rocks are intercalated, for these serve to bring before us the general conditions of the geography of the period. The subjoined table exhibits the subdivisions into which the Carboniferous system in Scotland has been grouped:—

Coal-measures.	Upper Red Sandstone group, nearly devoid of coal-seams. Coal-bearing, white, yellow and grey sandstones, dark shales and ironstones (Upper Coal series).
Millstone Grit.	Thick white and reddish sandstones and grits.
Carboniferous Limestone series.	Sandstones, shales, fireclays, coal-seams, ironstones and three seams of marine limestone, of which the uppermost is known as the Castlecary seam, the second as the Calmy or Arden, and the lowest as the Index (Lower Coal series). Bands of marine limestone intercalated among sandstones, shales and some coal-seams. A thick band of limestone lying at or near the bottom of the group, traceable all over Central Scotland, is known as the Hurlet or Main Limestone. Some higher and thinner seams are called Hosie's (see Fig. 155).

Calciferous Sandstones.[1]	In the basin of the Firth of Forth, below the Hurlet Limestone, comes a varied series of white and yellow sandstones, black shales (oil-shales), cyprid shales and limestones (Burdiehouse), and occasional coal-seams (Houston), having a total depth of about 3000 feet. This local group abounds in fossil plants, entomostraca and ganoid fishes. It passes down into the Cement-stone group, which, however, is feebly developed in this district, unless it is partly represented by the sandstones, shales, limestones and coals just mentioned.
	Cement-stone group consisting of red, blue and green marls and shales, red and grey sandstones, and thin bands of cement-stone: fossils scarce.
	Reddish and grey sandstones and shales, with occasional plant-remains, passing down into the deep red (sometimes yellow) sandstones, red marls and cornstones of the Upper Old Red Sandstone.

From this table the gradual geographical evolution of the Carboniferous period in Scotland may be gleaned. We observe that at the beginning, the conditions under which the Old Red Sandstone had been accumulated still in part continued. The great lacustrine basins of the Lower Old Red Sandstone had indeed been effaced, and their sites were occupied by comparatively shallow areas of fresh or brackish water in which the Upper Old Red Sandstone was laid down. Their conglomerates and sandstones had been uplifted and fractured. Their vast ranges of volcanic material, after being deeply buried under sediment, had been once more laid bare, and extended as ridges of land, separating the pools and lagoons which they supplied with sand and silt. This singular topography had not been entirely effaced at the beginning of the Carboniferous period, for we find that many of the ridges which bounded the basins of the Upper Old Red Sandstone remained as land until they sank beneath the waters in which the earliest Carboniferous strata accumulated. Thus, while no trace of an unconformability has yet been detected at the top of the Upper Old Red Sandstone, there is often a strong overlap of the succeeding deposits. At the south end of the Pentland Hills, for example, the Upper Old Red Sandstone attains a thickness of 1000 feet, but only three miles further south it entirely disappears, together with all the overlying mass of Calciferous Sandstones, and the Carboniferous Limestone then rests directly on the Lower Old Red Sandstone. Again, at the north end of the same chain the upper division of the Old Red Sandstone dies out against the lower, which is eventually overlapped by the Calciferous Sandstones.

The change from the physical conditions of the Scottish Old Red Sandstone to those of the Carboniferous system was no doubt gradual and slow. The peculiar red sandy sediment continued to be laid down in basins that were apparently being gradually widened by access of water from the open sea. Yet it would seem that in Scotland these basins still for a long time continued saline or, from some other cause, unfavourable to life; for the red, blue and green shales or marls, and occasional impure limestones or cement-stones and gypseous layers, which were deposited in them, are in general unfossiliferous, though drifted plants from the neighbouring land are here

1 The Calciferous Sandstones are the stratigraphical equivalents of the Limestone Shale and lower portion of the Carboniferous Limestone of England.

and there common enough. The sediments of these early Carboniferous waters are met with all over the southern half of Scotland, but in very unequal development, and constitute what is known as the "Cement-stone Group."

It was while these strata were in course of deposition that the earliest Carboniferous volcanoes broke into eruption. In some localities a thickness of several hundred feet of the Cement-stone group underlies the lowest lavas. In other places the lavas occur in and rest on the Upper Old Red Sandstone and have the Cement-stone group wholly above them; while in yet other districts the volcanic rocks seem entirely to take the place of that group. So vigorous was the earliest display of volcanic action in Carboniferous times that from the borders of Northumberland to the uplands of Galloway, and from the slopes of the Lammermuirs to Stirlingshire and thence across the estuary of the Clyde to Cantyre, innumerable vents were opened and large bodies of lava and ashes were ejected.

The Cement-stone group, save where succeeded by volcanic intercalations, passes up conformably into the lowest crinoidal limestones of the Carboniferous Limestone series. In the basin of the Firth of Forth, however, the cement-stones, feebly represented there, are overlain by a remarkable assemblage of white sandstones, black carbonaceous shales, or "oil-shales," cyprid limestones, occasional marine limestones and thin seams of coal, the whole having a thickness of more than 3000 feet. These strata, unlike the typical Cement-stone group, abound in fossils both vegetable and animal. They prove that, over the area of the Forth, the insalubrious basins wherein the red and green sediments of the Cement-stone group were laid down, gave place to opener and clearer water with occasional access of the sea. The peculiar lagoon-conditions which favoured the formation of coal were thus developed in Central Scotland earlier than elsewhere in Britain. We shall see in later pages that these conditions were accompanied by a fresh outbreak of volcanic activity, in a phase less vigorous but more enduring and extensive than that of the first Carboniferous eruptions.

The Carboniferous Limestone sea over the site of the southern half of Scotland appears never to have reached the depth which it attained in England and Ireland. To the north of it lay the land from which large quantities of sand and mud were carried into it, as shown by the deep accumulations of sandstone and shale, which far surpass in thickness the few comparatively thin marine limestones intercalated in them. There is thus a striking contrast between the thick masses of limestone in central and south-western England and their dwindled representatives in the north. Another marked difference between the Scottish and English developments of this formation is to be noticed in the abundant proof that the comparatively shallow waters of the northern basin were plentifully dotted over with active volcanoes. The eruptions were especially vigorous and prolonged in the basin of the Firth of Forth. They continued at intervals, even after the peculiar geographical conditions of the Carboniferous Limestone had ceased. But they had died out by the time of the beginning of the Coal-measures.

Owing to the number and variety of the natural sections, the Carboniferous volcanic rocks of Scotland have been the subject of numerous observations and descriptions, from the early days of geology down to the present time. The mere enumeration of the titles of the various publications regarding them would make a long list. These rocks formed the subject of some of Hutton's early observations, and furnished him with facts from which he established the igneous origin of "whinstone."[1] They supplied Playfair with numerous apt illustrations in support of Hutton's views, and he seems to have made himself thoroughly familiar with them.[2] In the hands of Sir James Hall they became the groundwork of those remarkable experiments on the fusion of whinstone which may be said to have laid the foundation of experimental geology.[3] In the controversies of the Neptunian and Plutonian schools these rocks were frequently appealed to by each side in confirmation of its dogmas. The appointment in 1804 of Jameson to the Chair of Natural History in the Edinburgh University gave increased impetus to the study of the igneous rocks of Scotland. Though he did not himself publish much regarding them, we know that he was constantly in the habit of conducting his class to the hills, ravines and quarries around Edinburgh, and that the views which he taught were imbibed and extended by his pupils.[4] Among the early writers the names of Allan,[5] Townson,[6] Lord Greenock,[7] and Ami Boué,[8] deserve especial mention.

The first broad general sketch of the Carboniferous igneous rocks of a large district of the country was that given by Hay Cunningham in his valuable essay on the geology of the Lothians.[9] He separated them into two series, the Felspathic, including "porphyry" and "clinkstone," and the Augitic or Trap rocks. To these he added "Trap-tufa," which he considered to be identical in origin with modern volcanic tuff. It was the eruptive character of the igneous rocks on which he specially dwelt, showing by numerous sections the effects which the protrusion of the molten masses have had upon the surrounding rocks. He did not attempt to separate the intrusive from the interstratified sheets, nor to form a chronological arrangement of the whole.

Still more important was the sketch given by Maclaren, in his classic *Geology of Fife and the Lothians.*[10] This author clearly recognized that many of the igneous rocks were thrown out contemporaneously with the strata among which they now lie. He constantly sought for analogies among modern volcanic phenomena, and presented the Carboniferous igneous rocks of the Lothians not as so many petrographical varieties, but as monu-

1 Hutton's *Theory of the Earth*, vol. i. p. 155 *et seq.*

2 Playfair's *Illustrations of the Huttonian Theory*, § 255 *et seq*

3 *Trans. Roy. Soc. Edin.* (1805), vol. v. p. 43.

4 *Mem. Wern. Soc.* ii. 178, 618 ; iii. 25 ; *Edin. Phil. Journ.* i. 138, 352 ; xv. 386.

5 *Trans. Roy. Soc. Edin.* (1811), vi. p. 405.

6 *Tracts and Observations in Natural History and Physiology*, 8vo, Lond. 1799.

7 *Trans. Roy. Soc. Edin.* (1833), xiii. pp. 39, 107.

8 *Essai géologique sur l'Écosse.* Paris ; no date, probably 1820.

9 *Mem. Wern. Soc.* vii. p. 1. Published separately, 1838.

10 Small 8vo, Edin. 1838, first partly published as articles in the *Scotsman* newspaper. A second edition, which was little more than a reprint of the first, appeared in 1866.

ments of different phases of volcanic action previous to the formation of the Coal-measures. His detailed descriptions of Arthur Seat and the rocks immediately around Edinburgh, which alone the work was originally intended to embrace, may be cited as models of exact and luminous research. The portions referring to the rest of the basin of the Forth did not profess to be more than a mere sketch of the subject.

Various papers of more local interest, to some of which allusion will be made in the sequel, appeared during the next quarter of a century. But no systematic study of the volcanic phenomena of any part of Scotland was resumed until the extension in 1854 of the Geological Survey to the north of the Tweed by A. C. Ramsay. The volcanic rocks of the Lothians and Fife were mapped by Mr. H. H. Howell and myself. The maps of that district began to be published in the year 1859, and the Memoirs two years later. In 1861, in a chronological grouping of the whole of the volcanic phenomena of Scotland, I gave an outline of the Carboniferous eruptions.[1] By degrees the detailed mapping of the Geological Survey was pushed across the whole of the rest of the south of Scotland, and the Carboniferous volcanic rocks of each area were then for the first time carefully traced and assigned to their various stratigraphical horizons. In the following pages reference will be given to the more important features of the Survey maps and Memoirs. In the year 1879, availing myself of the large amount of information which my own traverses and the work of the Survey had enabled me to acquire, I published a Memoir on the geology and petrography of the volcanic rocks of the basin of the Firth of Forth;[2] and lastly, in my Presidential Address to the Geological Society in 1892, I gave a summary of all that had then been ascertained on the subject of the volcanic rocks of Carboniferous time in the British Isles.[3]

Two well-marked types of volcanic accumulations are recognizable in the British Isles, which may be conveniently termed Plateaux and Puys.

1. Plateaux.—In this type, the volcanic materials were discharged over wide tracts of country, so that they now form broad tablelands or ranges of hills, reaching sometimes an extent of many hundreds of square miles and a thickness of more than 1000 feet. Plateaux of this character occur within the British area only in Scotland, where they are the predominant phase of volcanic intercalations in the Carboniferous system.

It is noteworthy that the Carboniferous plateaux appeared during a well-marked interval of geological time. The earliest examples of them date from the close of the Upper Old Red Sandstone. They were all in vigorous eruption during the time of the Calciferous Sandstones, but in no case did they survive into that of the Hurlet and later limestones. They are thus eminently characteristic of the earliest portion of the Carboniferous period.

2. Puys.—In this type, the ejections were often confined to the discharge of a small amount of fragmentary materials from a single solitary

[1] *Trans. Roy. Soc. Edin.* vol. xxii.

[2] *Ibid.* vol. xxix. (1879), p. 437.

[3] *Quart. Journ. Geol. Soc.* xlviii. (1892), p. 104. This summary, with additional details and illustrations, is embodied in the text.

vent, and even where the vents were more numerous and the outpourings of lava and showers of ash more copious, the ejected material usually covered only a small area round the centres of eruption. Occasionally streams of basic lava and accumulations of tuff were piled up into long ridges. Volcanoes of this character were specially abundant in the basin of the Firth of Forth, and more sparingly in Ayrshire and Roxburghshire. They form the persistent type throughout the rest of the British Isles.

The Puys also occupy a well-defined stratigraphical position. They did not begin until some of the volcanic plateaux had become extinct. From the top of the Cement-stone group up into the Carboniferous Limestone series, their lavas and tuffs are met with on many platforms, but none occur above that series save in Ayrshire, where some of the eruptions appear to have been as late as about the beginning of the Coal-measures.

Arranged in tabular form the stratigraphical and geographical distribution of the two great volcanic types of the Carboniferous system in Scotland will be more easily followed. I have therefore drawn up the accompanying scheme:—

[TABLE

		Plateau-type.					Puy-type.						
		Ayrshire and Renfrewshire.	Stirlingshire.	Midlothian.	East Lothian.	Berwick & Roxburghshire.	Ayrshire and Renfrewshire.	Stirlingshire.	West Lothian.	Midlothian.	East Lothian.	Fife.	Berwick & Roxburghshire.
Coal Measures		...	...	...	...	...	...	...	...	...	...	...	...
Millstone Grit		...	...	...	...	...	│	...	...	...	...	...	...
Carboniferous Limestone Series.	Castlecary Limestone	...	...	...	...	...	│	...	...	...	...	...	...
	Calmy ,,	...	...	...	...	...	│	...	...	...	...	...	...
	Index ,,	...	...	...	...	...	│	...	│	...	...	│	...
	Hurlet ,,	...	...	...	...	...	...	...	│	...	...	│	│
Calciferous Sandstone Series.	Burdiehouse Limestone and Oil-shale Group	│	│	... │	│	...	...	...	│	│	...	│	│
	Cement-stone Group	...	...	...	...	│	...	...	...	...	...	...	...
	Red Sandstones passing down into Upper Old Red Sandstone	...	...	...	...	│	...	...	...	...	...	...	...

CHAPTER XXIV

CARBONIFEROUS VOLCANIC PLATEAUX OF SCOTLAND

I. The Plateau-type restricted to Scotland—i. Distribution in the Different Areas of Eruption—ii. Nature of the Materials erupted.

In the division of the Plateaux I group all the more copious eruptions during the Carboniferous period, when the fragmentary materials generally formed but a small part of the discharges, but when the lavas were poured out so abundantly and frequently as to form lava-fields sometimes more than 2000 square miles in area, and to build up piles of volcanic material sometimes upwards of 3000 feet in thickness. As already remarked, this phase of volcanic action, especially characteristic of the earlier part of the Carboniferous period across the south of Scotland, but not found elsewhere in the same system in Britain, preceded the type of the Puys. Its eruptions extended from about the close of the Old Red Sandstone period through that section of Carboniferous time which was marked by the deposition of the Calciferous Sandstones, but they entirely ceased before the accumulation of the Main or Hurlet Limestone, at the base of the Carboniferous Limestone Series of Scotland. Its stratigraphical limits, however, are not everywhere the same. In the eastern part of the region, the lavas appear to be intercalated with, and certainly lie directly upon, the Upper Old Red Sandstone containing scales of *Bothriolepis* and other characteristic fishes, and they are covered by the Cement-stone group of the Calciferous Sandstones. In the western district a considerable thickness of Carboniferous strata sometimes underlies the volcanic sheets. On the other hand, the type of the Puys, although it appeared in Fife, Linlithgowshire and Midlothian during the time of the Calciferous Sandstones, attained its chief development during that of the Carboniferous Limestone, and did not finally die out in Ayrshire until the beginning of the deposition of the Coal-measures.

i. DISTRIBUTION OF THE PLATEAUX

Notwithstanding the effects of many powerful faults and extensive denudation, the general position of the Plateaux and their independence of each other can still be traced. They are entirely confined, as I have said, to the southern half of Scotland (see Map IV.). In noting their situations we are once more brought face to face with the remarkable fact, so strikingly manifested in the geological history of Britain, that volcanic action has been apt to recur again and again in or near to the same areas. The Carboniferous volcanic plateaux were poured out from vents, some of which not impossibly rose among the extinct vents of the Old Red Sandstone. Another fact, to which also I have already alluded as partially recognizable in the records of Old Red Sandstone volcanism, now becomes increasingly evident—the tendency of volcanic vents to be opened along lines of valley rather than over tracts of hill. The vents that supplied the materials of the largest of the Carboniferous volcanic plateaux broke forth, like the Old Red Sandstone volcanoes, along the broad Midland Valley of Scotland, between the ridge of the Highlands on the north and that of the Southern Uplands on the south. Others appeared in the long hollow between the southern side of these uplands, and the Cheviot Hills and hills of the Lake District. It is not a question of the rise of volcanic vents merely along lines of fault, but over broad tracts of low ground rather than on the surrounding or neighbouring heights. It can easily be shown that this distribution is not the result of better preservation in the valleys and greater denudation from the higher grounds, for, as has been already remarked in regard to the volcanoes of the Old Red Sandstone, these higher grounds are singularly free from traces of necks which, had any vents ever existed there, would certainly have remained as memorials of them. The following summary of the position and extent of the Plateaux will afford some idea of their general characters:—

1. THE CLYDE PLATEAU.—The chief plateau rises into one of the most conspicuous features in the scenery of Central Scotland. Beginning at Stir-

FIG. 107.—View of the escarpment of the Clyde Plateau in the Little Cumbrae, from the south-west.

ling, it forms the table-land of the Fintry, Kilsyth, Campsie and Kilpatrick Hills, stretching westwards to the Clyde near Dumbarton. It rises again on the south side of that river, sweeping southwards into the hilly moorlands which range from Greenock to Ardrossan, and spreading eastwards along the high watershed between Renfrewshire, Ayrshire, and Lanarkshire to Galston and Strathavon. But it is not confined to the mainland, for its prolongation can be traced down the broad expanse of the Firth of Clyde by the islands of Cumbrae to the southern end of Bute, and thence by the east of Arran to Campbeltown in Cantyre. Its visible remnants thus extend for more than 100 miles from north-east to south-west, with a width of some thirty-five miles in the broadest part. We shall probably not exaggerate if we estimate the original extent of this great volcanic area as not less than between 2000 and 3000 square miles.

It is in this tract that the phenomena of the plateaux are most admirably displayed. Ranges of lofty escarpments reveal the succession of the several eruptions, and the lower ground in front of these escarpments presents to us, as the result of stupendous denudation, many of the vents from which the materials of the plateau were ejected, while in the western portion of the area admirable coast-sections lay bare to view the minutest details of structure.[1]

It will be seen from the map (No. IV.), that the Clyde plateau extends in a general north-east and south-west direction. It is inclined on the whole towards the east, where, when not interrupted by faults, its highest lavas and tuffs may be seen to pass under the Carboniferous Limestone series. Its greatest elevations are thus towards its escarpment, which, commencing above the plains of the Forth a little to the west of Stirling, extends as a striking feature to the Clyde above Dumbarton. On the south side of the great estuary the escarpment again stretches in a noble range of terraced slopes for many miles into Ayrshire. It is well developed in the Little Cumbrae Island (Fig. 107), and in the south of Bute, where its successive platforms of lava mount in terraces and green slopes above the Firth. Even as far as the southern coast of Cantyre the characteristic plateau scenery reappears in the outliers which there cap the hills and descend the slopes (Fig. 108).

While the escarpment side of this plateau is comparatively unfaulted, so that the order of succession of the lavas and their superposition in the sedimentary rocks can be distinctly seen, the eastern or dip side is almost everywhere dislocated. Innumerable local ruptures have taken place, allowing the limestone series to subside, and giving to the margin of the volcanic area a remarkably notched appearance. To the effects of this faulting may be attributed the way in which the plateau has been separated into detached blocks with intervening younger strata. Thus a complex series of dislocations brings in a long strip of Carboniferous Limestone which extends from Johnston to Ardrossan, while another series lets in the limestone that runs from Barrhead to near Dalry. In each of these instances, the

[1] This plateau is represented in Sheets 12, 21, 29, 30, 31 and 39 of the Geological Survey, and is described in the accompanying Memoirs as far as published. The eastern part of the Campsie Hills was surveyed by Mr. B. N. Peach, the western part by Mr. R. L. Jack, who also mapped the rest of the plateau to the Clyde, and a portion of the high ground of Renfrewshire and Ayrshire; the rest of the area, south to Ardrossan, was surveyed by myself. The tract from Stewarton to Strathavon was surveyed by Mr. James Geikie, the Cumbraes and Bute by Mr. W. Gunn, and southern Cantyre by Mr. R. G. Symes. The Campsie Hills have been partly described by Mr. John Young in the first volume of the *Transactions of the Glasgow Geological Society*. The occurrence of plants in the tuffs of the east coast of Arran was discovered by Mr. E. Wunsch. The Campbeltown igneous rocks were described by J. Nicol, *Quart. Journ. Geol. Soc.* viii. (1852), p. 406. See also J. Bryce's *Arran and Clydesdale*.

continuity of the volcanic plateau is interrupted. To the same cause we owe the occasional reappearance of a portion of the plateau beyond the limits of the main mass, as for instance in the detached area which occurs in the valley of the Garnock above Kilwinning.

Denudation has likewise come into play, not only in reducing the area of the plateau, but in isolating portions of it into outliers, with or without the assistance of faults. The site of the Cumbraes and Bute was no doubt at one time covered with a continuous sheet of volcanic material, and there appears to be no reason for refusing to believe that this sheet formed part of that which caps the opposite uplands of Ayrshire. From the southern end

FIG. 108.—View of the edge of the Volcanic Plateau south of Campbeltown, Argyllshire.

The uppermost of the three zones is the volcanic series with its lava-ridges. The central band is the Upper Old Red Sandstone, lying conformably beneath the lavas, with its cornstone which has been quarried. The lowest band, tinted dark, is the Lower Old Red Sandstone, on which the other rocks rest unconformably.

of Bute it is only about seven miles across to the shore of Arran near Corrie, where the lavas and tuffs reappear. They are so poorly represented there, however, that we are evidently not far from the limit of the plateau in that direction. So vast has been the denudation of the region that it is now impossible to determine whether the volcanic ejections of Campbeltown, which occupy the same geological platform as those of Arran, Bute and Ayrshire, were also actually continuous with them. But as the distance between the denuded fragments of the volcanic series in Arran and in Cantyre is only about 20 miles it is not improbable that this continuity existed, and thus that the volcanic accumulations reached at least as far as the southern end of Argyllshire, where they now slip under the sea.

2. THE EAST LOTHIAN OR GARLETON PLATEAU.—Some 50 miles to the

Fig. 109.—View of North Berwick Law from the east, a trachyte neck marking one of the chief vents of the Garleton Plateau. (From a photograph.)

This illustration and Figs. 119, 133 and 136 are from photographs taken by Mr. Robert Lunn for the Geological Survey.

east of the Clyde volcanic district, and entirely independent of it, lies the plateau of the Garleton Hills in East Lothian, which, as its limits towards the east and north have been reduced by denudation, and towards the west are hidden under the Carboniferous Limestone series of Haddington, covers now an area of not more than about 60 square miles.[1] That the eruptions from this area did not extend far to the north is shown by the absence of all trace of them among the Lower Carboniferous rocks of Fife. A relic of them occurs above Borthwick, in Midlothian, about twelve miles to the south-west of the nearest margin of the plateau. The area over which the lavas and tuffs were discharged may not have exceeded 150 square miles. Small though this plateau is, it possesses much interest from the remarkable variety of petrographical character in its lavas, from the size and composition of its necks, and from the picturesque coast-line where its details have been admirably dissected by the waves. In many respects it stands by itself as an exception to the general type of the other plateaux.

FIG. 110.—The Bass Rock, a trachytic neck belonging to the Garleton plateau, from the shore at Canty Bay.

From its proximity to Edinburgh this volcanic area has been often studied and described. The memoirs of Hay Cunningham and Maclaren gave the fullest account of it until its structure was mapped by the Geological Survey. Its scenery differs from that of the other plateaux chiefly in the absence of the terraced contour which in them is so characteristic. The

[1] This plateau is represented in Sheets 33 and 41 of the Geological Survey of Scotland, and is described in the Explanation to accompany Sheet 33.

peculiar lavas of the Garleton Hills form irregularly-uneven ground, rising to not more than 600 feet above the sea. They slope gradually down to the coast, where a succession of fine sections of the volcanic series has been laid bare for a distance of altogether about ten miles. Nowhere, indeed, can the phenomena of the plateau-tuffs and their association with the Carboniferous strata be so well studied as along the coast-line from North Berwick to Dunbar. Among the necks of this plateau distinguished for their size, conspicuous prominence and component materials, the most important are those that form the conical eminences of North Berwick Law (Fig. 109), Traprain Law (Fig. 133), and the Bass Rock (Fig. 110).

3. THE MIDLOTHIAN PLATEAU.—On the same general stratigraphical horizon as the other volcanic plateaux, a narrow band of lavas and tuffs can be followed from the eastern outskirts of the city of Edinburgh into Lanarkshire, a distance of about 23 miles. It is not continuously visible, often disappearing altogether, and varying much in thickness and composition.

FIG. 111.—Corston Hill—a fragment of the Midlothian Plateau, seen from the north. The volcanic rocks form a cake on the top, the slopes lying across the edges of the Calciferous Sandstones.

This volcanic tract, which may be conveniently termed the Midlothian Plateau, is the smallest and most fragmentary of all the series. Its most easterly outliers form Arthur Seat and Calton Hill at Edinburgh.[1] Three miles to the south-west a third detached portion is known as Craiglockhart Hill. After another interval of ten miles, the largest remaining fragment forms the prominent ridge of Corston Hill (Fig. 111), whence a discontinuous narrow strip may be traced nearly as far as the River Clyde.

The well-known Arthur Seat and Calton Hill have been fully described by Maclaren, and have been the subject of numerous observations by other geologists.[2] They have been likewise mapped in detail on a large scale by

[1] I formerly classed these eminences with the Puys, but I am now of opinion that they ought rather to be regarded as fragments of a long and somewhat narrow plateau. Their basic lavas and overlying sheets of porphyrite repeat the usual sequence of the plateaux, which is not met with among the Puys. But, as will be pointed out in the sequel, Arthur Seat in long subsequent time became again the site of a volcanic vent.

[2] Maclaren's *Geology of Fife and the Lothians*, 1839, pp. 1-67 ; and Hay Cunningham, *Mem. Wer. Soc.* vii. pp. 51-62. The plateau is represented in Sheets 24 and 32 of the Geological Survey, and Arthur Seat and Calton Hill will be found on Sheet 2 of the Geological Survey map of Edinburghshire on the scale of 6 inches to a mile.

the Geological Survey, and have been described in the Survey Memoirs. The rest of the plateau to the south-west is much less familiar.

In Fig. 112 the great escarpment which descends from the right towards the centre is the sill of Salisbury Crags. The long dark crag (Long Row) rising between the two valleys is the lowest of the interstratified lavas. The slope that rises above it has been cut out of well-bedded tuffs, on which lie the basalts and andesites in successive sheets that form all the eastern or left side of the hill. The rocks around the summit belong to a much later period of volcanic eruption, and are referred to in Chapter xxxi.

The rocks of this plateau are comparatively limited in thickness, and have a much more restricted vertical range than those of other districts. At Arthur Seat and Corston Hill they begin above the cement-stones and cease in a low part of the great group of white sandstones and dark shales which form the upper half of the Calciferous Sandstones of Midlothian. They do not ascend as high as the Burdiehouse Limestone, which to the west of Corston Hill is seen to come on above them. One of their most

FIG. 112.—View of Arthur Seat from Calton Hill to the north.

remarkable features is the manner in which they diminish to a single thin bed and then die out altogether, reappearing again in a similar attenuated form on the same horizon. This impersistence is well seen in the south-western part of the area, between Buteland, in the parish of Currie, and Crosswood, in the parish of Mid-Calder. The lowest more basic band may there be traced at intervals for many miles without the overlying andesitic group. Yet that andesites followed the basalts, as in other plateaux, is well shown by large remnants of these less basic lavas left in Arthur Seat and Calton Hill. On the extreme southern margin of the area also a thin band of porphyrite with a group of overlying tuffs is seen above the red sandstones near Dunsyre.[1] The eruptions over the site of this plateau seem to have been much more local and limited than in the other plateaux. They appear to have gathered chiefly around two centres of activity, one of which lay about the position of Edinburgh, the other in the neighbourhood of Corston Hill. It is worthy of remark that this tract of volcanic material flanks the much older range of lavas and tuffs of the Pentland Hills and

[1] *Explanation, Geol. Surv. Scotland*, Sheet 24, p. 13 (1869).

wraps round the south-western end of this range, thus furnishing another illustration of the renewal of volcanic activity in the same region during successive geological periods.

4. THE BERWICKSHIRE PLATEAU.—Another and entirely disconnected area occurs in the broad plain or Merse of the lower portion of the valley of the Tweed.[1] The northern limit of its volcanic tuff occurs in the River Whitadder above Duns, whence the erupted materials rapidly widen and thicken towards the south-west by Stitchell and Kelso, until they die out against the flanks of the Cheviot Hills. The eastern extension of the area is lost beneath the Cement-stone group which covers the Merse down to the sea. Its western boundary must once have reached far beyond its present limits, for the low Silurian ground in that direction is dotted over with scattered vents to a distance of ten miles or more from the present outcrop of the bedded lavas, extensive denudation having cleared away the erupted materials and exposed the volcanic pipes over many square miles of country. Among the more prominent of these old vents are the Eildon Hills, Minto Crags and Rubers Law, as well as many other eminences familiar in Border story.

The bedded volcanic rocks of this area form a marked feature in the topography and geology of the district. They rise above the plain of the Merse as a band of undulating hills, of which the eminence crowned by Hume Castle, about 600 feet above the sea, is the most conspicuous height. In the geological structure of this part of Scotland they are mainly interposed between the Upper Old Red Sandstone and the base of the Carboniferous system, which they thus serve to divide from each other. But their lowest sheets appear to be in some places intercalated in the Old Red Sandstone, so that their eruption probably began before the beginning of the Carboniferous period. They form a band that curves round the end of the great Carboniferous trough at Kelso and skirts the northern edge of the andesites of the Lower Old Red Sandstone in the Cheviot Hills.

5. THE SOLWAY PLATEAU.—The last plateau, that of the Solway basin, though its present visible eastern limits approach those reached by the lavas from the Berwickshire area, was quite distinct, and had its chief vents at some distance towards the south-west.[2] On the north-western flanks of the Cheviot Hills, the Upper Old Red Sandstone is overlain by the lowest Carboniferous strata, without the intercalation of any volcanic zone, so that there must have been some intermediate ground that escaped being flooded with lava from the vents of the Merse on the one hand, and of the Solway on the other. The Solway lavas form a much thinner group than those of Berwickshire. From the wild moorland between the sources of the

[1] This plateau is shown on Sheets 17, 25, 26 and 33 of the Geological Survey Map of Scotland. It was chiefly mapped by Prof. James Geikie and Mr. B. N. Peach.

[2] For a delineation of the distribution and structure of this plateau see Sheets, 5, 6, 10, 11 and 17 of the Geological Survey of Scotland. In the upper part of Liddesdale, Ewesdale and Tarras it was mapped by Mr. B. N. Peach; in lower Liddesdale and Eskdale by Mr. R. L. Jack and Mr. J. S. Grant Wilson; from Langholm to the Annan by Mr. H. Skae; and in Kirkcudbright by Mr. John Horne.

Liddell and the Rule Water, they run in a narrow and much-faulted band south-westward across Eskdale and the foot of Annandale, and are traceable in occasional patches on the farther side of the Nith along the southern flanks of Criffel, even as far as Torrorie on the coast of Kirkcudbright—a total distance of about 45 miles. It is probable that this long outcrop presents merely the northern edge of a volcanic platform which is mainly buried under the Carboniferous rocks of the Solway basin. Yet it exhibits many of the chief characters of the other plateaux, and even occasionally rivals them in the dignity of the escarpments which mark its progress through the lonely uplands between the head of Liddesdale and the Ewes Water (Figs. 113, 142).

FIG. 113.—View of Arkleton Fell, part of the Solway Plateau, from the south-west.

The lower slopes below the single bird, round to the left side of the sketch, are on the Upper Old Red Sandstone; the line of crag below the two birds marks the volcanic group above which lies an outlier of the Calciferous Sandstone series, forming the upper part of the hill (three birds). The knobs under the four birds are bosses of andesite.

The plateaux of the Merse and the Solway illustrate in a striking manner the distribution of the volcanic eruptions along valleys and low plains. The vents from which the lavas and tuffs proceeded are chiefly to be found on the lower grounds, though these bedded volcanic rocks rise to a height of 1712 feet (the Pikes) to the west of the Cheviot Hills. Between the Silurian uplands of Selkirkshire and Berwickshire on the north and the ridge of the Cheviot Hills on the south, the broad plain was dotted with volcanic vents and flooded with lava, while to the south-west the corresponding hollow between the uplands of Dumfries and Galloway on the one side, and those of Cumberland on the other, was similarly overspread. The significance of these facts will be more apparent when the grouping of the vents has been described. We shall then also be better able to realize the validity of the inference that the present plateaux are mere fragments of what they originally were, wide areas having been removed from the one side of them by denudation, and having been concealed on the other under later portions of the Carboniferous system.

The same two plateaux likewise supply further illustrations of the outflow of similar volcanic materials in the same locality at widely separated intervals of time. They may be traced up to and round the margin of the great pile of andesites of Lower Old Red Sandstone age forming the Cheviot Hills.

ii. NATURE OF THE MATERIALS ERUPTED

The volcanic materials characteristic of the plateau-type of eruptions consist mainly of lavas in successive sheets, but include also various tuffs in frequent thin courses, and less commonly in thick local accumulations. The lavas are chiefly andesites in the altered condition of porphyrites. They vary a good deal in the relative proportions of silica. Some of them are decidedly basic and take the form of dolerites and olivine-basalts. With these rocks are occasionally associated "ultra-basic" varieties, where the felspar almost disappears and the material consists mainly of ferro-magnesian minerals. The more basic rocks are generally found towards the bottom of the volcanic series, where they appear as the oldest flows. In the Garleton Hills lavas of a much more acid nature are met with—true sanidine-trachytes, which overlie the porphyrites and basalts of the earlier eruptions.

No adequate investigation has yet been made of the chemical and microscopic characters of these various rocks, regarded as a great volcanic series belonging to a definite geological age, though many of the individual rocks and the petrography of different districts have been more or less fully described. I cannot here enter into much detail on the subject, but must content myself with such a summary as will convey some idea of the general composition and structure of this very interesting volcanic series.

(*a*) AUGITE-OLIVINE ROCKS (PICRITES and LIMBURGITES).—Towards the bottom of the plateaux there are found here and there sheets of "ultra-basic" material, some of which appear to be bedded with the other rocks and to have flowed out as surface-lavas, though it may be impossible to prove that they are not sills. Thus at Whitelaw Hill, on the south side of the Garleton Hills, a dark heavy rock is found to contain hardly any felspar, but to be made up mainly of olivine and augite. Dr. Hatch has published a description and drawing of this rock, together with the following analysis by Mr. Player:[1]—

Silica	40·2
Titanic oxide	2·9
Alumina	12·0
Ferric oxide	4·0
Ferrous oxide	10·4
Lime	10·4
Magnesia	11·9
Potash	0·8
Soda	2·7
Loss by ignition	3·4
	99·5

Spec. grav. 3·03.

[1] *Trans. Roy. Soc. Edin.* vol. xxxvii. (1893), p. 116.

(*b*) Dolerites and Basalts.[1]—These rocks are found both as interstratified lavas and as intrusive masses. In the former condition they take a conspicuous place among the sheets of the plateaux, but especially in the lower parts of the series. They are dark, often black, usually more or less porphyritic, with large felspars, frequently also large crystals of augite or olivine, and may be described as porphyritic olivine-dolerites and olivine-basalts, more rarely as olivine-free dolerites and basalts. Their groundmass consists of short laths or microlites of felspar (probably labradorite) and granules or small crystals of augite and magnetite, with sometimes a little fibrous brown mica. The large porphyritic felspars are striped (probably labradorite), the augites are frequently chloritized, and the olivines are generally more or less serpentinized. But in some cases all these minerals are as fresh as in a recent basalt. The rocks are sometimes beautifully columnar, as at Arthur Seat.

Of these basic lavas conspicuous examples may be seen at Arthur Seat, Calton Hill and Craiglockhart Hill. The eastern part of Arthur Seat, known as Whinny Hill, furnishes examples of olivine-dolerites of the Jedburgh type (p. 418). The beautiful basalt of Craiglockhart with its large porphyritic olivines and augites has afforded a distinct type of Carboniferous basalt (p. 418). The same type occurs on the Calton Hill in the cliff below the gaol. Similar basic lavas are especially abundant and remarkable in the Clyde plateau near Campbeltown in Argyllshire, and at the south end of Bute and in the Cumbraes, where they are associated with an interesting series of dykes and sills. But even where, as in the Garleton Hills, the lavas are for the most part somewhat acid in composition, those first poured out, which form the lowest band, include some typical olivine-basalts, of which a characteristic example occurs at Kippie Law at the base of the Garleton plateau (p. 418). It has been described by Dr. Hatch as exhibiting under the microscope porphyritic crystals of felspar and olivine lying in a groundmass composed of lath-shaped felspars, granular olivine and magnetite, and microlitic augite. The olivine, originally the most abundant constituent, has been converted into a fibrous aggregate of serpentine. All the minerals are more or less idiomorphic, but especially the augite, which crowds the groundmass in delicately-shaped prisms, most of which are terminated at both ends by faces of the hemi-pyramid. The analysis of this rock is given in the accompanying table of analyses of Garleton basalts. The Kippie Law type of basalt was recognized by Dr. Hatch among the Geological Survey collections from other districts, as in the intrusive bosses of Neides Law and Bonchester near Jedburgh, and from the Campsie plateau a mile and a half north of Lennoxtown.[2]

At Hailes Castle, in the Garleton plateau, the lower basic lavas include another olivine-basalt somewhat more felspathic than that just described, and

[1] A general classification of the whole series of Scottish Carboniferous dolerites and basalts, including both the plateau and puy examples, will be given in the account of the rocks of the puys in Chapter XXVI. (p. 418).

[2] *Trans. Roy. Soc. Edin.* vol. xxxvii. (1893), pp. 117-119.

at Markle quarry the rock is still more felspathic and contains the olivine only in small sporadic grains. The composition of these basic rocks of the Garleton plateau is shown in the subjoined table of analyses by Mr. J. S. Grant Wilson:—

	SiO_2	Al_2O_3	Fe_2O_3	FeO	MnO	CaO	MgO	K_2O	Na_2O	H_2O	Total
Kippie Law, specific gravity 2·8	46·01	19·19	5·91	6·75	0·19	8·68	6·81	1·20	3·27	3·07	101·08
Hailes Castle, specific gravity 2·76	49·07	19·43	10·58	2·35	0·32	7·87	4·36	0·98	3·31	2·26	100·53
Markle Quarry, specific gravity 2·7	49·54	22·23	9·55	1·12	0·08	7·19	2·80	1·81	4·56	2·42	101·30

Olivine-dolerites are more especially developed in the district around Jedburgh, where they form some of the most prominent bosses, such as Dunian and Black Law. They show a subophitic groundmass, with inconspicuous porphyritic crystals, among which those of olivine are more prominent than the felspars (p. 418).

(*c*) ANDESITES (PORPHYRITES).—These are the most abundant lavas of the plateaux. They occur in every district, and usually form the main constituents of the pile of volcanic material. They vary in colour from a pale pinkish grey, through many shades of red, purple, brown and yellow, to sometimes a dark green or nearly black rock. Their texture ranges from almost semi-vitreous, through different degrees of compactness, to open, cellular, slaggy masses. Generally through their base porphyritic felspars are abundantly disseminated, sometimes in large, flat, tabular forms, like those of the Lower Old Red Sandstone already referred to. The amygdaloidal kernels consist of calcite, zeolites, chalcedony or quartz. It is from the amygdaloids on either side of the Clyde that the fine examples of zeolites have been chiefly obtained for which the south of Scotland has long been famed. Occasionally, as at the south end of Bute, the andesitic lavas display a marked columnar structure.

Under the microscope these rocks present the usual fine felted aggregate of felspar microlites, with granules or crystals of magnetite and sometimes pyroxene. The porphyritic felspars, often large and well defined, generally contain inclusions of the groundmass. Occasionally some of the large porphyritic constituents are augite, or pseudomorphs after that mineral. The alteration of the rocks has oxidized some of the iron-ore and given rise to the prevalent purplish and reddish tints.

(*d*) TRACHYTES.—Some of the most remarkable lavas to be found in any of the plateaux are those which constitute a large part of the Garleton Hills. They overlie the lower andesite and basalt platform, which surrounds them as a narrow belt, while they occupy the central and much the largest part of the area. They have been included among the porphyrites, but are pale rocks, generally with a yellowish crust, presenting when quite fresh a grey, compact, felsitic base with large porphyritic crystals of unstriped felspar.

A number of specimens selected as illustrative of the different varieties have been analyzed and the results are stated in the subjoined table.[1] The specific gravity of the rocks is about 2·6.

	SiO_2	Al_2O_3	Fe_2O_3	FeO	MnO	CaO	MgO	K_2O	Na_2O	H_2O	Total
Pepper Craig . .	62·61	18·17	0·32	4·25	0·21	2·58	0·74	4·02	6·49	0·80	100·19
Kae Heughs . .	61·35	16·88	0·41	5·01	0·26	2·39	0·44	6·12	5·26	1·70	99·82
Hopetoun Monument	62·50	18·51	4·39		...	2·00	0·61	6·31	3·44	2·10	99·86
Phantassie . .	59·50	18·25	4·81	2·34	...	2·10	0·70	6·30	5·03	1·60	100·63
Bangley Quarry .	58·50	21·12	4·68	...	...	3·70	0·93	5·84	3·90	2·00	100·67

The microscopic characters of these rocks, as worked out by Dr. Hatch, show them to be well-marked and wonderfully fresh sanidine-trachytes. Some of them are porphyritic, with large crystals of perfectly unaltered sanidine, sometimes also oligoclase. Small but well-formed crystals of yellowish-green augite, in addition to the porphyritic felspars, are imbedded in a fine groundmass composed chiefly of microlites of sanidine, but with granules of augite and magnetite plentifully interspersed, and occasionally prisms of apatite. There is a group in which the porphyritic felspars are scarce or absent. In these there is little or no ferro-magnesian constituent. Other trachytes, rather less basic than the augite-bearing varieties here referred to, occur as bosses in the Garleton Hills district, and are referred to in the following section (*e*).[2]

(*e*) Rocks of the Necks.—In the necks connected with the plateaux other types of massive rock are to be found. Among these perhaps the most frequent are trachytes, grey to pink in colour, but apt to weather yellow, exceedingly compact, sparingly porphyritic, and with a peculiar platy structure and waxy lustre. Rocks of this character also appear as sills and dykes. Other varieties that occur in similar positions are more basic in composition, including dark, coarse, granular diabases. In the Jedburgh district the most frequent rocks are beautiful varieties of olivine-dolerite and olivine-basalt, which form most of the prominent hills of the neighbourhood. These bosses are sometimes associated with agglomerates as at Rubers Law.

In the Garleton Hills district, some of the necks present another petrographical type which directly connects them with the remarkable lavas of the higher part of that plateau. Thus the rock of Traprain Law was ascertained by Dr. Hatch to be a true phonolite. In its general platy structure and sonorous ring under the hammer it reminds one of typical phonolites. Under the microscope the rock is found to consist mainly of small lath-shaped crystals of sanidine arranged in a marked minute flow-structure, but with few porphyritic crystals. It contains small crystals and ophitic patches

[1] The first two analyses are by Mr. J. S. Grant Wilson, the last two by Mr. A. Dick jun., and that from Hopetoun Monument by Mr. G. Barrow. *Trans. Roy. Soc. Edin.* vol. xxxvii. p. 122.

[2] For fuller petrographical details consult Dr. Hatch's paper above cited.

of a light green soda-augite, with practically no magnesia in it. A small quantity of iron-ore and isolated granules of apatite are also present, together with patches of nepheline which, though generally decomposed and replaced with zeolitic products, occasionally display six- and four-sided crystal-contours. An analysis of the Traprain phonolite by Mr. Player is subjoined:—[1]

Silica	56·8
Titanic acid	0·5
Alumina	19·7
Ferric oxide	2·2
Ferrous oxide	3·5
Manganous oxide	0·2
Lime	2·2
Magnesia	0·4
Soda	4·3
Potash	7·1
Loss by ignition	2·5
	99·4

Spec. grav. 2·588

The neck of North Berwick Law was found by Dr. Hatch to be a trachyte, showing a plexus of lath-shaped sanidines that diminish in size to minute microlites, but with no porphyritic or ferro-magnesian constituent. The Bass Rock, though its geological relations are concealed by the sea, is in all probability another neck of this district. It is likewise a mass of trachyte, composed almost entirely of lath-shaped crystals of sanidine, with no ferro-magnesian constituent, but a good deal of iron ore. It shows none of the large porphyritic felspars so characteristic of the Garleton Hills lavas, but it closely resembles the non-porphyritic varieties, particularly the lavas of Score Hill, Pencraig, Lock Pit Hill, and Craigie Hill.[2]

(*f*) TUFFS.—The fragmentary ejections of the plateaux vary in texture from the finest-grained tuffs to coarse agglomerates.[3] As they have been derived from the explosion of andesite-lavas, they consist mainly of the debris of these rocks. They are often deep red in colour, as for example those of Dunbar, but are most frequently greenish. They have a granular texture, due to the small lapilli of various porphyrites imbedded in a fine dust of the same material. Grains of quartz, frequently to be detected even in the finer tuffs, may either have been ejected from the volcanic vents, or may have been grains of sand in the ordinary sediment of the sea-bottom. Both at the base and at the top of the plateau-series, the tuffs are interstratified with and blend into sandstones and shales, so that specimens may be collected showing a gradual passage from volcanic into non-volcanic detritus. In many of the tuffs of the necks fragments of sandstone and

[1] *Trans. Roy. Soc. Edin.* vol. xxxvii. p. 125.

[2] The composition of the rocks of North Berwick Law and the Bass closely resembles that of the trachytic lavas of the plateau. For analyses, see Dr. Hatch's Paper, *ibid.* pp. 123, 124.

[3] For accounts of these rocks, see Explanation of Sheet 33 *Geol. Surv. Scot.* p. 32; Sheet 22, pp. 11-14; Sheet 31 pp. 14-17.

other stratified rocks occur, representing the strata through which the vents were drilled. In the tuffs of the Eaglesham district pieces of grey and pink granite have been met with which, if they are portions of an old granite mass below, must have come from a great depth.[1] In the coarser tuffs and agglomerates a larger variety of lava-form rocks is to be found than can be seen among the bedded lavas of the Plateaux. They include felsites and quartz-porphyries, and more rarely basic lavas (diabases, etc.).

[1] Explanation of Sheet 22 *Geol. Surv. Scot.* p. 12.

CHAPTER XXV

GEOLOGICAL STRUCTURE OF THE CARBONIFEROUS VOLCANIC PLATEAUX OF SCOTLAND

1. Bedded Lavas and Tuffs; Upper Limits and Original Areas and Slopes of the Plateaux; 2. Vents; Necks of Agglomerate and Tuff; Necks of Massive Rock; Composite Necks; 3. Dykes and Sills; 4. Close of the Plateau-eruptions.

THE structure of the various plateaux presents a general similarity, with many local variations. Each plateau is built up entirely, or almost entirely, of sheets of volcanic material, the intercalations of ordinary sedimentary layers being, for the most part, few and unimportant, and usually occurring either towards the base or the top of the volcanic series, though at a few localities inter-stratifications of shale and sandstone, marking pauses in the eruptions, occur throughout that series. The vents of eruption are in some instances still to be recognized on the plateaux themselves. More usually they occur on the lower ground flanking the volcanic escarpments, where they have been laid bare by denudation. Dykes, though seldom abundant, are associated with the plateaux, while the sills which may mark the latest manifestations of volcanic energy, though not developed on so large a scale as among the Cambrian and Silurian volcanoes, can nevertheless be distinctly recognized.

It is a question of some interest to determine the geological date of the commencement of the plateau-eruptions by fixing the precise stratigraphical horizon on which the base of the volcanic series rests. I have already referred to the fact that this base does not always lie on the same platform among the Lower Carboniferous formations. In Berwickshire, as above mentioned, the earliest eruptions appear to have taken place before the close of the Upper Old Red Sandstone period. These are the earliest of the whole series. In Cantyre, the lowest lavas and tuffs come directly upon the sandstones, marls and cornstones of the Upper Old Red Sandstone. In Stirlingshire, Renfrewshire and Ayrshire several hundred feet of the Cement-stone group are sometimes interposed between the bottom of the volcanic rocks and the top of the Old Red Sandstone. This divergence doubtless indicates that the eruptions began earlier in some districts than in others. But there were also probably unequal terrestrial movements preceding, and perhaps accompanying, the volcanic outbursts. In the case of the Clyde plateau, for example, if we examine its base in the neighbourhood of Fintry, we find that it lies upon some 500 feet of Carboniferous white sandstone, red and green marls and cement-stones, which rest on the Upper Old Red Sandstone. Yet only eight miles to the eastward, this considerable mass of strata disappears, and the bottom of the lavas comes down upon the red sandstones. Five miles still further in the same direction the volcanic masses likewise die out, and then the Carboniferous Limestone series is found at Abbey Craig to lie, with scarcely any representative of the Cement-stone group, on the Upper Old Red Sandstone (Fig. 114). Again, to the south-west of Fintry, the zone of cement-stones below the volcanic series continues to vary considerably in thickness and sometimes almost to disappear, while in Ayrshire the lavas lie immediately on the red sandstones.

These irregularities, not improbably indicative of inequalities of sub-

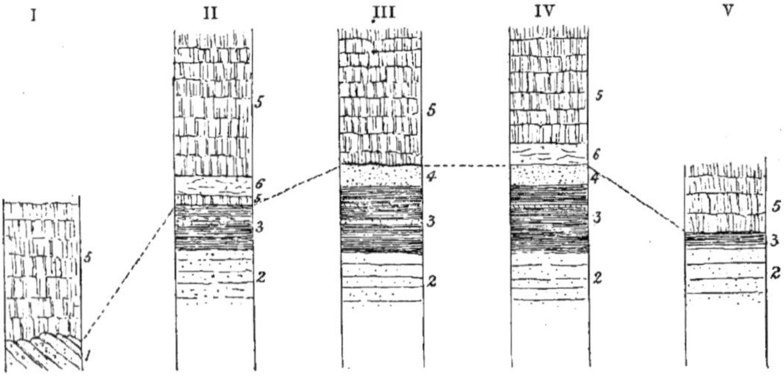

FIG. 114.—Vertical sections of the escarpment of the Clyde plateau from north-east to south-west.

I. Section at the east end of the Campsie Hills, four miles west from Stirling. II. Section above Glins, six miles west from No. I. III. Section at Strathblane Hill, eight miles further south-west. IV. Section at Lang Craig, east from Dumbarton, eight miles south-west from No. III. V. Section above Fort Matilda, Greenock, eleven miles from the previous section and on the south side of the Clyde.

1. Lower Old Red Sandstone; 2. Upper Old Red Sandstone; 3. Carboniferous shales, sandstones and cement-stones, (the "Ballagan beds"); 4. Thick white sandstone which comes in above the Ballagan beds; 5. Andesite lava-sheets; 6. Interstratified tuffs. The dotted lines connect the base of the volcanic series.

sidence and of deposition, may have been connected with the subterranean disturbances which culminated in the abundant outbreak of volcanic action. But though the volcanic rocks of the plateaux may be traced overlapping the underlying strata, no evidence has anywhere been detected of an unconformability between them and the Lower Carboniferous or Upper Old Red Sandstone series.

1. BEDDED LAVAS AND TUFFS

The successive sheets of lava in a plateau usually form thin and wide-spread beds which are only occasionally separated by intercalations of tuff or of red marl. In this, as well as in other respects, they present much resemblance to the lavas of the Tertiary plateaux of Antrim and the Inner Hebrides. They are generally marked off from each other by the slaggy upper and under portions of the successive flows, and this structure gives a distinctly bedded aspect to the escarpments, as in the Campsie and Largs Hills, or still more conspicuously in Little Cumbrae (Fig. 107) and the southern end of Bute. Considerable diversity of structure may be noticed among these sheets. Some present a compact jointed centre passing up and down into the slaggy material just referred to; others have assumed a vesicular character throughout, the vesicles being often elongated in the direction of flow. Where, as usually occurs, the vesicular is replaced by the amygdaloidal structure, some of the rocks have long been famous for the minerals found in their cavities. The beautiful zeolites of the Kilpatrick and Renfrewshire Hills, for example, may be found in every large mineralogical collection in the country. Well-developed columnar structure occasionally appears among the lavas of the plateaux, but chiefly, so far as I have observed, in the lower or more basic group, as in the basalts along the east side of the Dry Dam at Arthur Seat.

In each plateau the lavas may be observed to thicken in one direction, or more usually towards more than one, and this increase no doubt indicates in which quarters the chief centres of discharge lay. Thus in the Clyde plateau, several areas of maximum development may be detected. In the Kilpatrick Hills the total thickness of lavas and tuffs exceeds 3000 feet (Fig. 120). Above Largs it is more than 1500 feet, rapidly thinning away towards the south. The continuation of the plateau far to the north-east in the Campsie Fells reveals a thickness of about 1000 feet of lavas at Kilsyth, which become thicker further west, but eastward rapidly diminish in collective bulk, until in about twelve or thirteen miles they disappear altogether, and then, as already remarked, the Calciferous Sandstone series closes up without any volcanic intercalation.

In the Solway plateau, the lavas attain a maximum development about Birrenswark, whence they diminish in bulk towards the north-east and south-west. The Berwickshire plateau reaches its thickest mass about Stitchill, whence it rapidly thins away towards the north-east, until at a distance of some twelve miles it disappears altogether, the last trace of it in that direction being a band of tuff which dies out in the Calciferous Sandstones to the north of Duns.

In the Midlothian Plateau, the development of the volcanic series is more irregular than in any of the others. As already remarked, there appear to have been at least two chief centres of discharge in this region, one at Edinburgh and one some fourteen miles to the south-west. At the former, the volcanic materials attain in Arthur Seat and Calton Hill a thickness of about 1100 feet. In Craiglockhart Hill, three miles distant, they are still about 600 feet thick. But beyond that eminence they cease to be traceable for about eight miles, either because they entirely die out, or because their dwindling outcrops are concealed under superficial deposits. As we approach the south-western centre of eruption around Corston Hill a new volcanic group begins and soon increases in bulk.

A distinguishing feature of the plateaux is found in the difference between the lavas that were first erupted and those which followed them. The earlier eruptions, as above remarked, were generally basic, sometimes highly so. Thus at Arthur Seat the thick series of lavas which form the eastern part of the hill have at their base several sheets of columnar basalt, over which come the andesites that make up the main mass of the erupted material. In the Calton Hill the same sequence may be observed. Underneath the andesites of Campbeltown comes a well-marked and persistent band of olivine-dolerite. Still more basic are some portions of the earliest lavas of the Garleton plateau where, as already stated, rocks present themselves composed mainly of olivine and augite.

It is worthy of notice that where the lavas of a plateau diminish greatly in thickness or become impersistent, the lowest basic group may continue while the overlying andesites disappear. This feature has been already mentioned as well seen in the Midlothian plateau. The thick group of andesites in Arthur Seat and Calton Hill is not to be found in the next volcanic eminence, Craiglockhart Hill; but the basalts with their underlying tuffs continue. In the south-western tract from Harper Rig to Hare Law in Lanarkshire, the thin lava-band, which can be found only at intervals along the line of outcrop of the volcanic series for about nine miles, is a dolerite often highly slaggy in structure. Again, at Corrie in Arran, the lavas which appear upon the shore, apparently at the extreme western limits of the Clyde plateau, are basic rocks.

But whether or not the lowest and more basic lavas appear in any plateau, the main mass of the molten material erupted has usually consisted of varieties of andesite. The successive discharges of these intermediate lavas have flowed out in sheets, some of which must have been little more than heaps of clinkers and scoriæ, while others were more fluid and rolled along with a ropy or slaggy surface. Occasionally the upper part of an andesite shows the reddened and decomposed character that suggests some degree of disintegration or weathering before the next lava-stream buried it. The intervals between successive outflows of these lavas are not, as a rule, defined by any marked breaks or by the intercalation of other material. In general, the plateaux are mainly built up of successive sheets of lava which have followed each other at intervals sufficiently short to prevent the accumulation of much detritus between them. Thus the Campsie Hills have the upper 600 feet of their mass formed of admirably-well-defined sheets of andesite, separated sometimes by thin partings of tuff, but more usually only by the slaggy vesicular surfaces between successive flows.

Where the lavas consisted of trachytes they were apt to assume more irregular forms. Of this tendency the rocks of the Garleton Hills supply an excellent example. As already stated, their lumpy character gives to these hills an outline which offers strong contrast to the ordinary symmetrical terraced contours of the andesitic plateaux.

Although tuffs play, on the whole, a comparatively unimportant part among the constituents of the plateaux, they attain in a few localities an exceptionally great development, and even where they occur only as thin partings between the successive lava-flows, they are always interesting memorials of the volcanic activity of a district. In many portions of the plateaux, the lowest members of the volcanic series are tuffs and agglomerates, showing that the eruptions often began with the discharge of fragmentary materials. Thus in the Midlothian plateau at Arthur Seat, though the lowest interbedded volcanic sheet is a dolerite, it is immediately followed by a series of bedded tuffs, before the main mass of the lavas of that hill

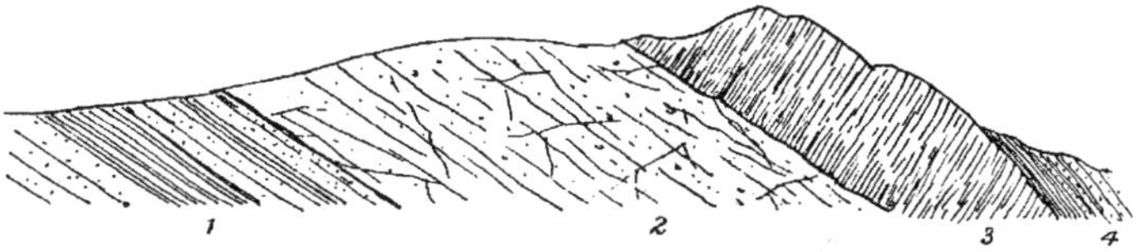

FIG. 115.—Section of Craiglockhart Hill, Edinburgh.

1. Red sandstones and clays; 2. Green stratified tuffs; 3. Columnar basalt; 4. Dark shales, ironstones and sandstones, with plants.

make their appearance. At Craiglockhart Hill, three miles distant (Fig. 115) this lowest lava is absent, and a group of tuffs about 300 feet thick rests immediately on the red Carboniferous sandstones and shales, and is overlain by sheets of columnar basalt. The scoriaceous bottom of the latter rock may here and there be seen to have cut out parts of the tuff as it rolled over the still unconsolidated material. In the same district, a few miles further to the south-west, some interesting sections of the Midlothian plateau are laid bare in the streams which descend from the western slopes of the Pentland Hills. I

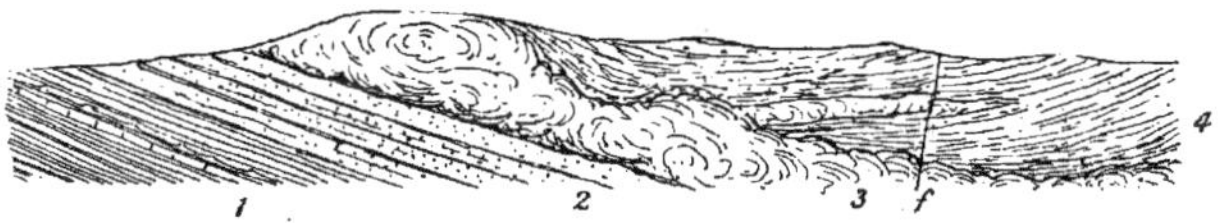

FIG. 116.—Section of the bottom of the Midlothian Plateau, Linnhouse Water above Mid-Calder Oilworks.

1. Shales and cement-stones; 2. Sandstones; 3. Highly vesicular lava; 4. Tuffs and sandstone bands. *f*, Fault.

may cite, in particular, those exposed in the course of the Linnhouse Water. At the railway viaduct near the foot of Corston Hill, a good section is displayed of the Cement-stone group—thick reddish, purplish, and greenish-blue marly shales or clays, with thin ribs and bands of cement-stone and grey compact cyprid-limestone, as well as lenticular seams and thicker beds of grey shaly sandstone, sometimes full of ripple-marks and sun-cracks. These strata, which exactly reproduce the typical lithological characters of the Cement-stone group of Stirlingshire (Ballagan Beds), Ayrshire and Berwickshire, are surmounted by a group of reddish, yellow and brown sandstones, sometimes pebbly and containing a band of conglomerate.

Among the stones in this band, pieces of the radiolarian cherts of the Lower Silurian series of the Southern Uplands are conspicuous, likewise pieces of andesite which may have come from the neighbouring Pentland Hills.

Above these strata lie the lavas of Corston Hill. These are highly vesicular in some parts, and include bands of tuff which are well exposed further down the same stream, immediately above the railway bridge near the Mid-Calder oilworks (Fig. 116). There the lavas, though much decomposed, show a highly vesicular structure with a rugged upper surface, in the hollows and over the prominences of which fine flaky and sandy tuffs have been deposited, while thin seams of vesicular lava are intercalated among these strata.

The upper part of the same plateau, as exposed in the course of the Murieston Water, contains evidence that the last eruptions consisted of tuff. The highly slaggy lava (1 in Fig. 117) is there surmounted by a thick mass of grey and greenish-white well-bedded granular tuff (2) in-

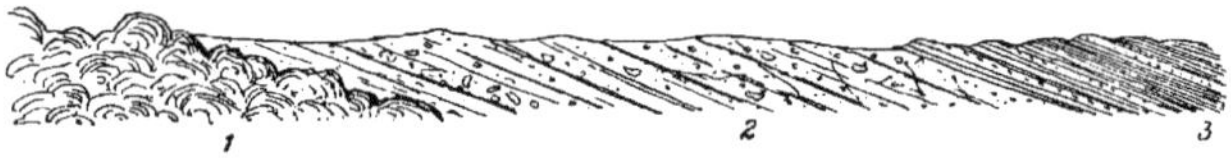

FIG. 117.—Section of the top of the Midlothian Plateau in the Murieston Water.

cluding occasional lumps of the basic lava, and passing up into black shale (3). But that the volcanic eruptions continued during the accumulation of the shale is proved by the intercalation of thin partings and thicker layers of tuff in the black sediment. A short way higher up the Burdiehouse Limestone comes in.

The great lava-escarpment of the Kilpatrick Hills rests on a continuous band of tuff which is thickest towards the west, near the group of vents above Dumbarton, while it thins away eastward and disappears in Strathblane, the lavas then forming the base of the volcanic series. But perhaps the most remarkable group of basal tuffs is that which underlies the lavas of the Garleton plateau, to which further reference will be immediately made.

Extensive accumulations of tuff form in one or two localities a large proportion of the thickness of the whole volcanic series of a plateau. Thus in the north-eastern part of Ayrshire, between Eaglesham and the valley of the Irvine, the lavas die out for a space and give place to tuffs. During the discharge of the fragmentary materials over that ground no lava seems to have flowed out for a long period. Ordinary sediment, however, mingled with the volcanic detritus, and there were even pauses in the eruptions when layers of ironstone were deposited, together with thin impure limestone that inclosed shells of *Productus giganteus*.[1]

In some of the plateaux, particularly within the older part of the volcanic series, intercalations of ordinary sediment among the tuffs and lavas show that eruptions occurred only occasionally, and that during the long intervals

[1] Explanation of Sheet 22 *Geol. Surv. Scotland*, p. 12.

between them the deposition of sand and mud went on as before. Thus the lower 400 feet of the Campsie Fells are built up of slaggy andesites and thick beds of fine-grained stratified tuff, with bands of red, green and grey clays and cement-stone and a zone of white sandstone. The Calton Hill at Edinburgh (Fig. 118) affords an excellent illustration of the interstratifica-

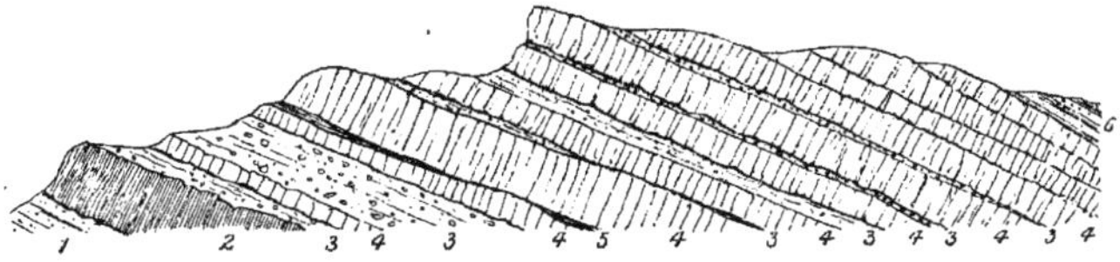

FIG. 118.—Section of Calton Hill, Edinburgh.

1. Lower Carboniferous sandstones; 2. Basic lava at the bottom of the volcanic series; 3. Tuff often interstratified with sandstones and shales; 4. Sheets of andesite-lava frequently separated by layers of tuff; 5. Shale passing into tuff; 6. White sandstone and black carbonaceous shales overlying the volcanic series.

tion both of tuffs and ordinary sediments among the successive outflows of lava. In the total thickness of about 1100 feet of volcanic material in this hill, at least eight intervals in the discharge of the lavas are marked by the intercalation of as many bands of nodular tuff, together with seams of shale and sandstone more or less charged with volcanic detritus. The highest lava is immediately covered by the white sandstones and black shales of the Calciferous Sandstone series.

The tuffs, as might be expected, are coarsest in texture and thickest in mass where they approach most nearly to some of the vents of eruption, and, on the other hand, become finer as they recede from these. As a rule, they are distinctly stratified, and consist of layers varying in the size of their component lapilli. Here and there, near the centres of discharge, the bedding becomes hardly traceable or disappears, and the fragmentary materials take the form of agglomerate.

In the admirable range of coast-cliffs which extend from North Berwick to Dunbar, we learn that above the red sandstones at the base of the Carboniferous system, a thick pile of volcanic ashes was accumulated by numerous discharges from vents in the immediate neighbourhood. Some of the explosions were so vigorous that blocks of different lavas, sometimes a yard or more in length, were thrown out and heaped up in irregular mounds and hollows. Others discharged exceedingly fine dust, and between these two extremes every degree of coarseness of material may be recognized.

As an illustration of the remarkable alternation of coarse and fine materials, according to the varying intensity of the volcanic paroxysm, Fig. 119 is here introduced. It represents a portion of the tuff-cliffs east of Tantallon Castle, and shows at the bottom fine well-stratified tuff, over which a shower of large blocks of lava has fallen. Fine detritus is seen to cover the deposits of this shower, and successive discharges of large stones may be noticed higher up on more or less well-defined horizons.

The space over which this pyroclastic material can now be traced, large though it is, does not represent the whole of the original area included within the range of the discharges of ash and stones, for much has been removed by denudation. During pauses of various length between the eruptions, waves and currents washed down the heaps of volcanic material and distributed ordinary sediment over the bottom of the water. Hence, abundantly interstratified in some parts of the tuff, seams of sandstone, blue and green shale, cement-stone and limestone occur. One thick band of limestone may be traced from near Tynningham House to Whittinghame, a distance of about four miles; another patch appears near Rockville House; and a third at Rhodes, near North Berwick. No fossils have been noticed in these limestones. The calcareous matter, together sometimes with silica, appears to have been supplied, at least in part, by springs, which may have been connected with the volcanic phenomena of the district. The North Berwick limestone, in particular, has the peculiar carious wavy structure with minute mamillated interstices so common among sinters. It contains grains of pyrites, flakes of white kaolin, which probably represent decayed prisms or tufts of natrolite, and cavities lined with dog-tooth spar. Some portions give out a strongly foetid odour when freshly broken.

After the tuffs of the Garleton plateau had accumulated to a depth of perhaps 200 feet or more, lavas began to be poured out. First came basic outflows (olivine-basalts with picrites) and andesites (porphyrites), which form a thin but continuous sheet all over the area. These were succeeded by the series of trachytes which distinguish this area. Although the observer remarks the absence there of the usual terraced arrangement, yet from some points of view, particularly from the westward, a succession of low escarpments and longer dip-slopes can be detected among the trachytes of the Garleton Hills, while there can be no doubt that, in spite of their irregular lumpy contours, these lavas lie as a great cake above the lower platform of more basic flows (Fig. 10). There is evidence that during the emission of the trachytes occasional eruptions of andesite took place. Not the least striking and interesting feature of this plateau is the size and distribution of its necks, to which reference will be made in the sequel.

The latest eruption in the Garleton area had ceased and the cones and lava sheets had probably been buried under sediment before the commencement of the deposition of the Hurlet or thick Main Limestone of the Carboniferous Limestone series which lies immediately to the west of the plateau.

The tuffs of the plateaux are seldom fossiliferous, probably for the same reason that fossils are scarce in the Cement-stone group which the plateau volcanic rocks overspread and with which they are interstratified. Occasional stems and other fragments of vegetation occur in the plateau-tuffs, as in those of North Berwick, where I have found a decayed coniferous trunk three feet in length. The green tuff at the base of the volcanic group of Arthur Seat contains abundant macerated plant-remains, together with scales of *Rhizodus* and other fishes. In some places the plants are represented by trunks or roots, which appear to remain in their positions of

FIG. 119.—Cliff of tuff and agglomerate, east side of Oxroad Bay, a little east from Tantallon Castle, East Lothian.

growth. A remarkable instance of this nature occurs in some bands of tuff in the volcanic group of the east coast of the Isle of Arran, first brought to notice by Mr. E. Wunsch,[1] and of which the plants have been so fully investigated by Professor Williamson.[2]

Plant-remains also occasionally occur in the stratified layers intercalated among the lavas and tuffs of the plateaux. Some of the best examples of their occurrence are to be found in the shales and tuffs interstratified among the enormous pile of volcanic material near Bowling. Not only does abundant vegetable debris occur distributed through the detrital strata in the volcanic series at that locality, but it is even aggregated into thin seams of coal which have been examined and described by various observers.[3] It may

FIG. 120.—Section across part of the Clyde Plateau to the west of Bowling (reduced from Sheet 6 of the Horizontal Sections of the Geological Survey of Scotland).

1. "Ballagan Beds"; 2. White sandstone; 3. Tuffs, 600 feet thick, with a thin sheet of andesite; 4. Andesite sheets, 500 feet; 5. Stratified tuffs with thin coals, shales, fireclays and plant-remains, 500 or 600 feet; 6 6. A series of andesite-lavas, about 1500 feet thick, enclosing a thin coal-seam at *; 7. Stratified tuffs, 200 feet; 8. Shales with plants and coaly seams, 150 feet; 9. Base of another andesite series, which must be some hundreds of feet thick. 10 and 11. Necks of agglomerate.

be remarked that the plant remains thus found intercalated in the volcanic series, especially when they have been entombed in tuff, have often had their internal structure admirably preserved, the organic tissues having been delicately replaced by calcite or other petrifying medium. The remarkably perfect structure of some of these plants has been demonstrated by Professor Williamson, especially in the case of the Arran deposit just referred to. Mr. John Young has also found the structure well preserved among the *Sigillariæ* and *Stigmariæ* that occur in the stratified intercalations between the lavas near Bowling.

Upper Limits and Original Areas and Slopes of the Plateaux.—Where the highest members of the volcanic series can be seen passing conformably under the overlying Carboniferous strata they are frequently found to be mainly composed of fine tuffs, the last feeble efforts of the plateau-volcanoes having consisted in the discharge of showers of ashes. These materials were mingled with a gradually increasing proportion of ordinary mechanical sediment, which finally overspread and buried the volcanic tracts of ground, as these slowly sank in the general subsidence of the region. The characteristic corals, crinoids and shells of the Carboniferous Limestone begin to appear in these ashy sediments. There is thus an insensible passage from volcanic detritus into fossiliferous shales and limestones.

[1] *Trans. Geol. Soc. Glasgow*, vol. ii. (1867) p. 97. [2] *Phil. Trans.* 1871-1883.

[3] See in particular J. Young, *Trans. Geol. Soc. Glasgow*, vol. iv. (1874) p. 123.

Examples of this gradation may be seen in many natural sections along the flanks of the Ayrshire plateau from above Kilbirnie to Strathavon.

It is still possible to fix in some quarters the limits beyond which neither the lavas nor the tuffs extended, and thus partially to map out the original areas of the plateaux. For example, in certain directions the Carboniferous formations can be followed continuously downward below the Main Limestone, without the intervention of any volcanic material, or with only a slight intermixture of fine volcanic lapilli, such as might have been carried by a strong wind from some neighbouring active vents. By this kind of evidence and by the proved thinning-out of the materials of the plateau, we can demonstrate that in the north of Ayrshire the southern limits of the great volcanic bank did not pass beyond a line drawn from near Ardrossan to Galston. We can show, too, that the lavas of the Campsie Fells ended off about a mile beyond Stirling before they reached the line of the Ochil heights, and that the *coulées* which flowed from the Solway vents did not quite join with those from the Berwickshire volcanoes.

Moreover, evidence enough remains to enable us to form a tolerably clear conception of the original average slopes of the surface of some of

FIG. 121.—Diagram illustrating the thinning away southwards of the lavas of the Clyde Plateau between Largs and Ardrossan. Length about 10 miles.

1. Upper Old Red Sandstone; 2. Sandstones, shales, etc., with "Ballagan Beds"; 3. Tuffs; 4. Andesite lavas; 5. Carboniferous Limestone series.

the plateaux. Thus in the great escarpment above Largs and the high ground eastward to Kilbirnie the volcanic series, as already stated, must be at least 1500 feet thick. This thick mass of lavas and tuffs thins away southwards and probably disappears a short distance south from Ardrossan in a space of about ten miles (Fig. 121). The original southward slope of the plateau would thus appear to have been about 1 in 35. Again, the northward slope of the same plateau may be estimated from observations in the Campsie Fells. We have seen that above Kilsyth the total depth of the volcanic sheets is about 1000 feet, while to the westward it is much thicker. From the top of the Meikle Bin (1870 feet) above Kilsyth north-eastwards to Causewayhead, where the whole volcanic series has died out, is a distance of 12 miles, so that the slope of the surface of erupted materials on this side was about 1 in 63 (Fig. 122).

Judging from the sections exposed along the faces of the escarpments, we may infer that the volcanic sheets had a tolerably uniform surface which sloped gently away from the chief vents, but with local inequalities according to the iregularities of the lava-streams that were heaped up round the vents and flowed outward in different directions and to various distances from them. At the beginning, these flat volcanic domes were certainly subaqueous. While they were being formed, continuous subsidence appears to have been in progress. But the great thickness of the volcanic accumulations, as in the Kilpatrick and Renfrewshire areas, and the paucity of ordinary sedimentary strata among them, make it not improbable that at least their higher parts rose above the water. Where this was the case there may have been considerable degradation of the lava-banks before these were reduced or were by subsidence submerged beneath the water-level. Evidence

FIG. 122.—Diagram illustrating the thinning away eastwards of the lavas of the Clyde Plateau in the Fintry Hills. Length about 12 miles.

1. Upper Old Red Sandstone; 2. White sandstone, blue shales and cement-stones ("Ballagan Beds"); 3. Andesite sheet, about 100 feet thick; 4. Tuffs (250 feet), with an included band of ashy sandstone containing plant-remains; 5. Andesite lavas; 6. Carboniferous Limestone series, which to the east lies immediately on the Upper Old Red Sandstone.

of this waste is probably to be recognized in the bands of conglomerate, occasionally of considerable thickness, which, particularly in some parts of Ayrshire, intervene between the top of the volcanic group and the Hurlet Limestone. As I shall have occasion to point out further on, there seems to be some amount of evidence in favour of the view that a considerable interval of time elapsed between the close of the plateau-eruptions and the date of that widespread depression which led to the deposition of the Hurlet Limestone over the whole of Central Scotland. If such an interval did occur it would include a prolonged abrasion of any projecting parts of the plateaux, and the production and deposition of volcanic conglomerate.

2. VENTS

We have now to consider the external forms, internal contents and distribution of the vents from which the material of the plateaux was discharged. In the Carboniferous system these interesting relics of former volcanoes are far more distinctly defined and better preserved than in older geological formations. Moreover, in Scotland, they are laid bare to greater advantage, both inland and along the sea-coast, and may indeed be studied there as typical illustrations of this kind of geological structure.

In external form the necks connected both with the plateaux and the puys generally rise from the surrounding ground as isolated, rounded, conical or dome-shaped prominences, their details of contour depending mainly upon the materials of which they consist. When these materials are of agglomerate, tuff or other readily disintegrated rock, the surface of the domes is generally smooth and grass-covered. Where, on the other hand, they consist wholly or in part of dolerite, basalt, diabase, andesite, trachyte or other

FIG. 123.—View of the two necks Dumgoyn and Dumfoyn, Stirlingshire, taken from the south.

These two necks form a conspicuous feature in front of and below the lava plateau, a portion of which is shown on the right hand. The ground-plan of the same necks is shown in Fig. 124.

crystalline rock, they present more irregular rocky outlines. Illustrations

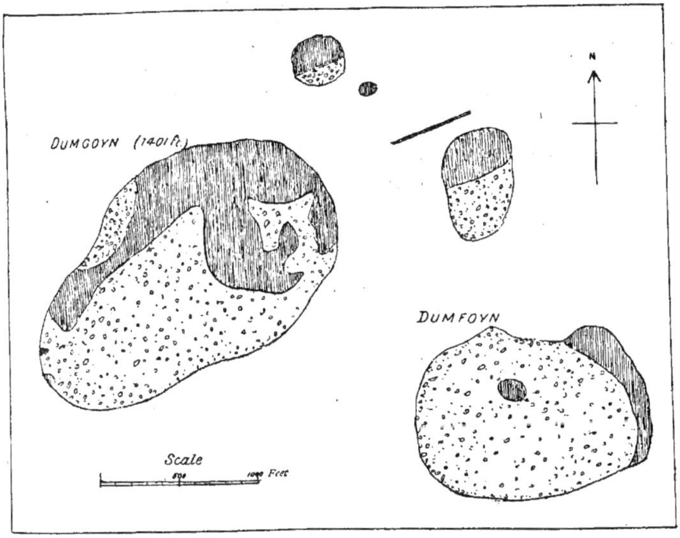

FIG. 124.—Ground-plan of Plateau-vents near Strathblane, Stirlingshire, on the scale of 6 inches to a mile.

of some of those varying forms are given in Figs. 23 and 123. In rare instances the vent is marked at the surface not by a hill but by a hollow, as in the great neck in the heart of the Campsie Fells (Fig. 128).

As regards their ground-plan, which affords a cross-section of the original

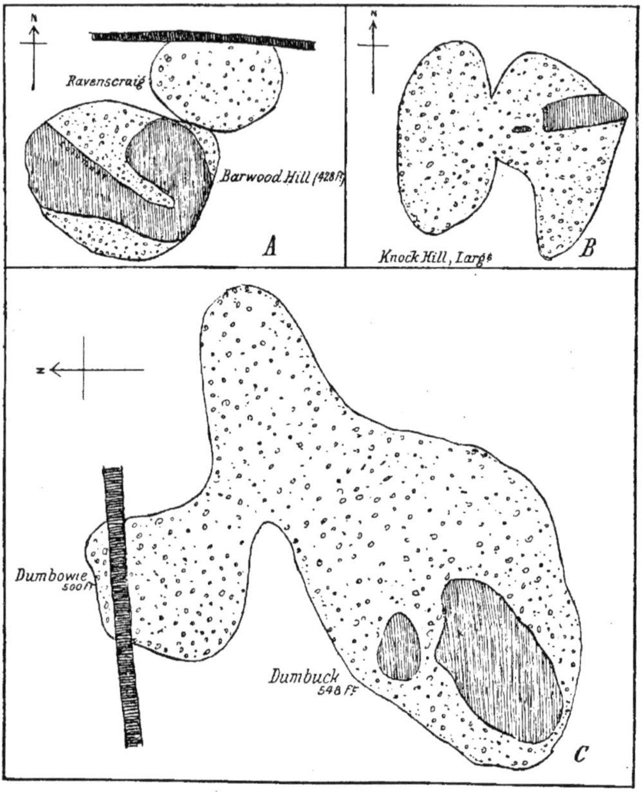

Fig. 125.—Ground-plans of double and triple necks in the Plateau series, on the scale of 6 inches to a mile.

A. Barwood Hill and Ravenscraig, east of Dumbarton, double vent. B. The Knock Hill, Largs, Ayrshire, double vent (see Fig. 28). C. Dumbowie and Dumbuck, east of Dumbarton, triple vent.

volcanic funnel, the plateau-vents present considerable variety. The simplest cases are those in which the form is approximately circular or somewhat elliptical. Here the outline corresponds to the cross-section of a single and normal orifice. Some examples of this simple type are given in Fig. 124, which represents a group of vents on the edge of the Clyde plateau near Strathblane. The two larger necks here shown are the same which appear in the view in Fig. 123.[1] Where two vents have been successively opened close to each other, or where the same vent has shifted its position, the ground-plan may be greatly modified. In some instances the double funnel can be distinctly traced. Thus in the conspicuous Knock Hill above Largs in Ayrshire (Fig. 125, B) there are two conjoined necks, and such appears to be also the structure shown by the ground-plan of the neck of Barwood Hill and Raven's Craig, east of Dumbarton (Fig. 125, A).[2] But more complex forms occur which point to a still larger number of coalescing necks. A group of hills to the east of Dumbarton gives the ground-plan shown in C, Fig. 125, where traces may be detected of three separate vents. Still more irregular are long narrow dyke-like masses of tuff or agglomerate which have probably risen along lines of fissure (Fig. 22, No. 1). The most striking example of these, however, occur in association with the puys and will be described in later pages.

Connected with their ground-plan is the relative size of the plateau-vents. On the whole they are larger than those of the puy series. The simple circular or elliptical type presents the smallest necks, some of them not exceeding 100 feet in diameter. The more complex forms are generally also of larger dimensions. By much the largest vent or connected group of vents is that which lies among the uplands of Misty Law in the heart of the Renfrewshire part of the Clyde plateau, where a connected mass of tuff and agglomerate now occupies a space of about 4 miles in length by 2½ miles in breadth (Fig. 129). It has not been found possible, however, to trace the boundaries of the separate vents of this tract, nor to distinguish the material of the necks from that which surrounds them. Another large mass which from its shape may be conjectured to represent more than one vent is the great tract north of Melrose, which measures 8800 by 4200 feet.[3]

The distribution of the necks can best be understood from the maps of the Geological Survey, where they have been carefully indicated. As might have been expected, they are not found outside the original limits within which it may be reasonably inferred that the lavas and tuffs were erupted. They occur most abundantly and attain their largest size in and around the districts where the plateaux are most extensively developed. No doubt a

[1] The illustrations in Figs. 124 and 125 are taken from the field-maps of the Geological Survey on the scale of 6 inches to a mile. The ground represented in Fig. 124 was mapped by Mr. R. L. Jack.

[2] These ground-plans are likewise taken from the field-maps of the Geological Survey. A and C were mapped by Mr. Jack, B by myself. The shaded parts are intrusive andesites and dolerites; the dark bars in A and C being dolerite dykes of much later date than the necks. The dotted portions mark tuff and agglomerate.

[3] The following measurements are, like those in the text, taken from the field-maps of the Geological Survey. Carewood Rig, on the borders of Roxburghshire and Dumfriesshire, 7000 × 2400 feet; the great vent in the middle of the Campsie Fells, 5200 × 2600; Black Law, between Bedrule and Jedburgh, 3400 × 1600; Dumgoyn, Strathblane, 2300 × 1300; Rubers Law, 1500 × 1000; Minto Hill (south), 2300 × 1650; Minto Hill (north), 1500 × 1100; Doughnot Hill, Kilpatrick range, 1000 × 700; four of the smallest agglomerate vents along the northern escarpment of the Clyde plateau between Strathblane and Fintry, 500 × 450, 450 × 400, 250 × 100, 200 × 200; Pike Law, Arkleton, Tarras Water, 500 × 500; Harwood, Stonedge, 5 miles S.E. from Hawick, 500 × 300; Arkleton Burn, Dumfriesshire, 400 × 100; Dalbate Burn, 250 × 120.

large number of them are concealed under these plateaux. A few appear at the surface among the lavas and tuffs, but by far the largest number now visible have been revealed by denudation, the escarpments having been cut back so as to lay bare the underlying rocks through which the necks rise. Thus, along the flanks of the great escarpment that extends from near Stirling by Fintry and Strathblane to Dumbarton, more than two dozen of agglomerate necks may be counted in a distance of about sixteen miles, while if the necks of lava-form material are included, the number of vents must be about fifty. Nowhere in Scotland do such necks form a more conspicuous feature in the scenery as well as the geology than they do between Fintry and Strathblane, where, standing out as bold isolated hills in front of the escarpments, their conical and rounded outlines present a striking contrast to the terraced escarpments behind them. I would especially refer again to the two remarkable cones of Dumfoyn and Dumgoyn above Strathblane (Figs. 123, 124, 127). Along the west front of the hills between Gourock and Ardrossan seventeen agglomerate-vents occur in a distance of sixteen miles. In Roxburghshire a group of large agglomerate-necks is dotted over the Silurian country around Melrose and Selkirk[1] (see Fig. 130).

From the evidence of these necks it is plain that the volcanic materials

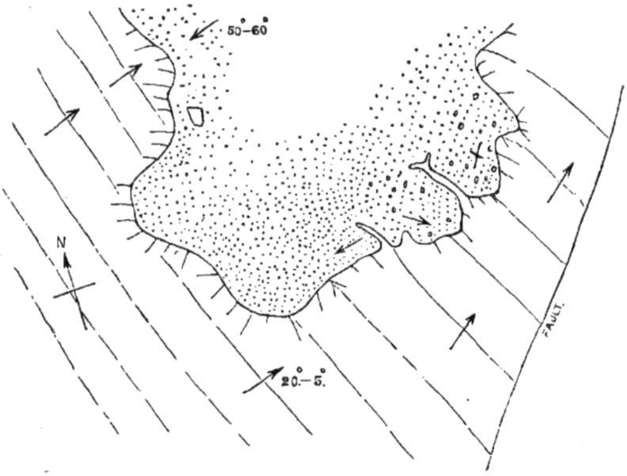

FIG. 126.—Ground-plan of tuff-neck, shore east of Dunbar.

The surrounding rocks are sandstones, which are much hardened round the vent in the zone marked by the short divergent lines. The arrows mark the direction of dip. See "Geology of East Lothian," *Mem. Geol. Survey*, p. 44.

of the plateaux must in each case have been supplied not from great central orifices, but from abundant vents standing sometimes singly, with intervening spaces of several miles, often in groups of four or five within a single square mile.

[1] In this region and farther southward, besides the plateau-eruptions, a later group of puys is to be seen, and it is difficult to discriminate between the necks belonging to the two groups. Those which lie to the east are probably connected with the plateaux, those to the west with the puys. The latter are referred to on p. 475.

In the interior of the country, it is seldom possible to examine the actual junction of necks with the rocks through which they rise, the boundary-line being usually obscured by debris or herbage. On the coast, the vents of the plateaux have not been bared by the sea so fully as in the case of the much younger series of the the east of Fife to be described in later pages. But where the East Lothian plateau touches the shore, the waves have laid bare a number of its minor vents, which have thus been dissected in ground-plan on the beach. As an illustration of these vents an example is given in Fig. 126, from the shore east of Dunbar. Here the sandstones, which are inclined in an easterly direction at 20° to 25°, are pierced by an irregular mass of tuff. It is observable that in this instance long tongue-like projections of the sandstones protrude into the neck; more frequently the material of a neck sends veins or dykes into the surrounding walls. A volcanic chimney would seem to have been often much shattered and fissured in the course of the volcanic explosions, and the fragmentary material has fallen or been injected into the rents thus caused. As a rule, the rocks immediately around the Carboniferous necks are more or less indurated, as in this instance from the Dunbar shore.

The materials which have filled up the vents connected with the plateau-eruptions generally consist of (*a*) agglomerates or tuffs, but occasionally of (*b*) some kind of lava, and frequently (*c*) of both these kinds of rock combined.

(*a*) *Necks of Agglomerate or Tuff.*—These materials vary greatly in the nature and relative proportions of their constituents. Usually the included blocks and lapilli are pieces of andesite, diabase, basalt or other lava, like the rocks of the plateaux. But with these occur also fragments probably detached from the sides of the funnels through which the explosions took place, such as pieces of greywacke, sandstone, limestone and shale. Considerable induration may be observed among these non-volcanic ingredients. In some cases, as in that of the occurrence of pieces of granite referred to on p. 382, the stones have probably been brought up from some considerable depth. In others it is easy to see that the blocks have slipped down from some higher group of strata now removed from the surrounding surface by denudation. Some striking illustrations of this feature will be cited from necks of the puy-series in the south of Roxburghshire (p. 476).

The lava blocks in the tuffs and agglomerates are usually rounded or sub-angular. Pear-shaped blocks, or flattened discs, or hollow spherical balls are hardly ever to be observed, though I have noticed a few examples in the tuffs of Dunbar. A frequent character of the blocks is that of roughly rounded, highly amygdaloidal pieces of lava, the cellular structure being specially developed in the interior, and the cells on the outside being often much drawn out round the circumference of the mass. Such blocks were probably torn from the cavernous, partially consolidated, or at least rather viscous, top of a lava column. Most of the stones, however, suggest that they were produced by the explosion of already solidified lava, and were somewhat rounded by attrition in their ascent and descent The vents filled with such materials must have been the scene of prolonged and intermittent activity; successive paroxysms resulting in the clearing out of the hardened lava column in the throat of the volcano, and in the rise of fresh lava, with abundant ejection of dust and lapilli.

Necks formed entirely of agglomerate are abundant among the vents connected with the plateaux. As examples of them I may refer to the series already mentioned as fronting the escarpment of the Clyde plateau from

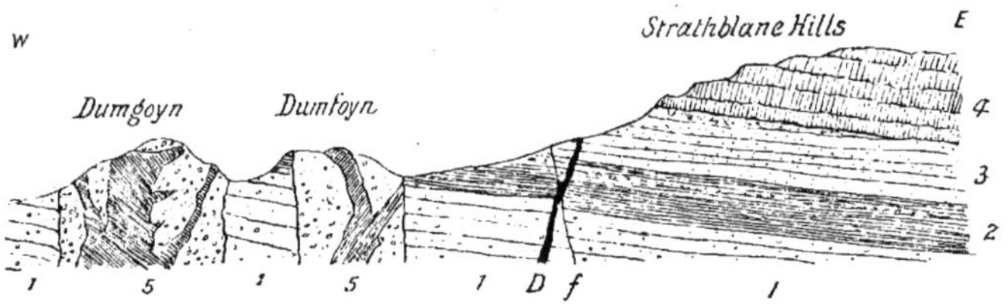

FIG. 127.—Section across the vents Dumgoyn and Dumfoyn, and the edge of the Clyde plateau above Strathblane, Stirlingshire.

1. Upper Old Red Sandstone; 2. Shales, cement-stones and sandstones ("Ballagan beds"); 3. White sandstone; 4. Andesite lavas; 5. Agglomerate (shown by the dotted portions), traversed by intrusive diabase. *f*, Fault. D. Late dolerite dyke.

Fintry to Largs. Another interesting group rises through the Silurian and Old Red Sandstone rocks to the west of the escarpment of the Berwickshire plateau, that near Melrose forming one of the largest in Scotland.

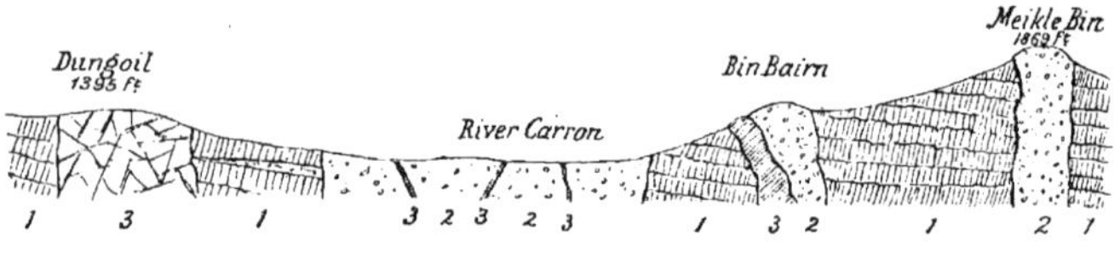

FIG. 128.—Section through the large vent of the Campsie Hills.

1. Andesite lavas; 2. Agglomerate and tuff; 3. Trachytic and andesitic intrusive rocks.

Illustrations of the varying structure of these vents are given in the accompanying figures. In Fig. 127, a section is drawn through the two necks

FIG. 129.—Diagrammatic section across the central vent of the Clyde plateau in Renfrewshire.

1. Andesite lavas; 2. Agglomerates and fine tuffs often much altered; 3. Dykes of trachytic and andesitic rocks; 4. Later dykes of dolerite and basalt.

Dumgoyn and Dumfoyn, which have already been shown in outline and in ground-plan. The relation of these two vents to the neighbouring plateau to the right can here be seen. Fig. 128 gives a section taken through the great vent of the Campsie Hills, with the minor adjacent necks of Dungoil, Bin Bairn, and the Meikle Bin.

The diagram in Fig 129 is meant to convey in a general way what appears to be the structure of the central vent of the Renfrewshire plateaux, to be afterwards referred to. But, as already mentioned, the limits of the various rocks are too much obscured to allow an accurate delineation to be given of their areas and relations to each other. The Berwickshire plateau supplies abundant interesting examples of tuff necks which rise through the Old Red Sandstone many miles distant from the edge of the lavas. This structure is shown in Fig. 130.

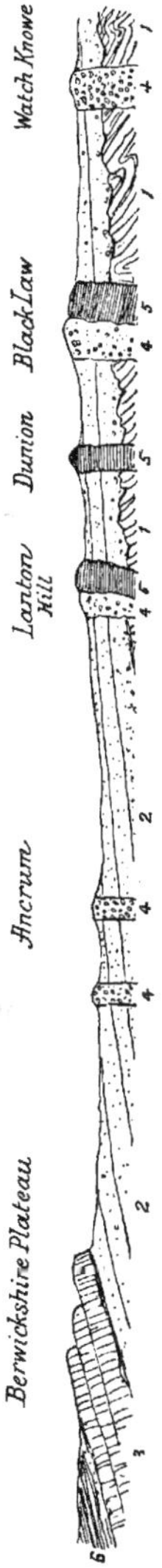

FIG. 130.—Section across Southern Berwickshire to show the relation of the volcanic plateau to the vents lying south from it.
1. Upper Silurian strata; 2. Upper Old Red Sandstone; 3. The volcanic plateau; 4. Agglomerate and tuff of the vents; 5. Basalt and dolerite; 6. Lower Carboniferous strata.

Indications may occasionally be observed of an agglomerate vent having been first occupied by one kind of material and then, after being in great measure cleared out by explosions, having been subsequently filled up with another. As an example of this structure I may cite again the double neck of the Knock Hill a little to the north of Largs, of which the outline is shown in Fig. 23, and the ground-plan in Fig. 125, B. This hill rises from the red sandstone slopes that front the great Ayrshire plateau and forms a conspicuous cone the top of which is rather more than 700 feet above the sea. Its summit commands a remarkably extensive and interesting panorama of the scenery of the Clyde, but to the geologist perhaps the most striking feature in the landscape is the range of terraced hills behind, mounting up into the great vents of the Renfrewshire uplands. On these declivities the successive lava-streams that have built up the plateau can be seen piled over each other for a thickness of more than 1000 feet, and presenting their escarpments as parallel lines of brown crag with green slopes between.

The Knock has had its upper part artificially dressed, for lines of trench have been cut out of its rocks by some early race that converted the summit of the hill into a strongly intrenched camp. From the apex of the cone the ground falls rapidly westward into a hollow, beyond which rises a lower rounded ridge of similar materials. It is possible that this western ridge may really form part of the main hill, but the grass-covered ground does not afford sufficient exposures of the rocks to settle this point. From the contours of the surface, it may be inferred that there are two closely adjacent vents, and that the western and lower eminence is the older of the two. This hill or ridge consists partly of a coarse agglomerate, and partly of veins and irregular protrusions of a dark, compact, slightly cellular lava. The stones in the fragmental rock are different olivine-basalts, or other basic lavas, and sandstones. The paste is rough, loose and granular. The sandstone fragments are much indurated and sometimes bleached.

The Knock itself is formed mainly of a remarkably coarse and strikingly volcanic agglomerate. Round the outside, and particularly on the south-east, the rock is finer in texture, compact, and gravelly, or like a mudstone, with few or no imbedded blocks, dull-green to red in colour, and breaking with a clean fracture which shows angular lapilli of various basalts or diabases. At the southern end of the neck, where the surrounding red sandstone can be seen within a few feet of the tuff, the latter is bright red in colour, and contains much debris of red sandstone and marl. Possibly this finer tuff, which is traceable as an irregular band round the outside of the neck, may mark an older infilling of the vent than the agglomerate of the centre ; but there is no sharp line to be drawn between the two, though a hollow can sometimes be traced on the surface where they join.

The agglomerate of this locality is one of the most characteristic among the plateau-necks of the Clyde region. Its blocks sometimes measure from two to three feet in diameter. They consist almost wholly of a dark crystalline porphyritic olivine-basalt. These blocks are subangular in form, often with clean-fractured surfaces. Though occasionally slightly cellular, they are never slaggy so far as I could see, nor are any true scoriae to be noticed among them. The blocks suggest that they were derived from the disruption of an already solidified mass of lava. The agglomerate is entirely without any trace of stratification.

Through this tumultuous accumulation of volcanic debris some irregular veins of olivine-basalt, sometimes glassy in structure, have been injected, and reach nearly to the summit of the hill. This intrusive material resembles generally some of the dark intrusive masses in the Dumbartonshire necks. Like these, it exhibits a tendency to assume a more or less distinctly columnar structure, its columns having the same characteristic wavy sides and irregular curvature. The intrusive rocks in the two eminences of the Knock may be paralleled among the stones in the agglomerate. The neck on its north-eastern side rises steeply from the red sandstones which it pierces, but which, although they are much jointed and broken, are not sensibly indurated. Unfortunately the actual junction of the igneous and sedimentary rocks is concealed under herbage.

As a rule, the fragmental materials of the plateau-necks are quite unstratified. Their included blocks, distributed irregularly through the mass, have evidently undergone little or no assortment after they fell back into the vents. Occasionally, however, a more or less distinct bedding of the agglomerate or tuff may be observed, the layers having a tendency to dip inward into the centre. One of the most conspicuous examples of this structure is to be found in the hill of Dumbuck, to the east of Dumbarton. This neck, which forms so prominent a feature in the landscape, presents a precipitous face towards the south, and allows the disposition of its component materials to be there seen. The agglomerate consists of a succession of rudely stratified beds of coarser and finer detritus, which on both sides are inclined towards the centre, where a plug of fine-grained olivine-basalt has risen and spread out into a columnar sheet above (Fig. 131). In general form this basalt resembles such intrusions as that of Largo Law, to be afterwards described (Fig. 226), where what may have been the hollow or bottom of the crater is filled with basalt.

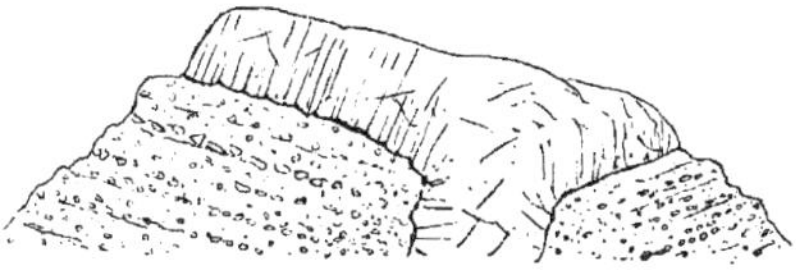

FIG. 131.—Section of south end of Dumbuck Hill. East of Dumbarton.

(*b*) *Necks of Andesite, Trachyte, Dolerite, Diabase, or other massive Rock.*—When the vents have been filled by the uprise of some molten rock, it is generally, as we have seen, of a more acid character than the ordinary lavas of the plateaux. Frequently it consists of some variety of trachyte or andesite, commonly of a dull yellow or grey tint and waxy lustre. Good examples may be seen among the remarkable group of necks on either side of the valley north of the village of Strathblane and in those above Bowling. The three great necks in East Lothian, already alluded to,—Traprain Law

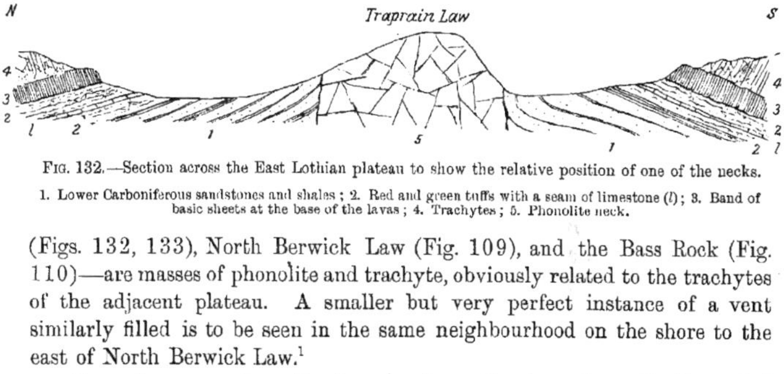

FIG. 132.—Section across the East Lothian plateau to show the relative position of one of the necks.
1. Lower Carboniferous sandstones and shales; 2. Red and green tuffs with a seam of limestone (*l*); 3. Band of basic sheets at the base of the lavas; 4. Trachytes; 5. Phonolite neck.

(Figs. 132, 133), North Berwick Law (Fig. 109), and the Bass Rock (Fig. 110)—are masses of phonolite and trachyte, obviously related to the trachytes of the adjacent plateau. A smaller but very perfect instance of a vent similarly filled is to be seen in the same neighbourhood on the shore to the east of North Berwick Law.[1]

Examples occur where the funnels of eruption have been finally sealed up by the rise of more basic material, and this has happened even in a district where most of the lava-form necks consist of trachyte or some other intermediate lava. Thus, in the Campsie Fells, several such bosses appear, of which the most conspicuous forms the hill of Dungoil (1396 feet, Fig. 128). Further west, among the Kilpatrick Hills, bosses of this kind are still more numerous. The group of bosses near Ancrum and Jedburgh is mainly made up of olivine-dolerites and olivine-basalts (Fig. 130). This more basic composition of itself suggests that these bosses may be connected rather with the puy- than with the plateau-eruptions.

[1] See "Geology of East Lothian," *Geological Survey Memoir*, p. 40.

(*c*) *Necks of Composite Character.*—In not a few examples, the vents have been filled with agglomerate which has been pierced by a plug or veins of lava-form material. Many illustrations of this composite structure may be observed along the west front of the great escarpments from Fintry to Ardrossan (see Figs. 124, 125, 127 and 128). In that region the intruded rock is often a dull yellowish or grey trachytic or andesitic material. Olivine-basalt is the chief rock intruded in the vents in the Dumbarton district. Among the Roxburghshire vents, where the injected material is commonly olivine-basalt or dolerite, it occasionally happens, as in Rubers Law, that the uprise of the lava has almost entirely cleared out or concealed the agglomerate, and in some of the bosses, where no agglomerate is now to be seen, the basalt may have taken its place (Fig. 130).

The largest and most interesting vents connected with this type of Carboniferous volcano, are those which occur within the limits of the plateaux, where they are still surrounded with lavas and tuffs that probably came out of them. Of these by far the most extensive and remarkable lies among the high moorlands of Renfrewshire between Largs and Lochwinnoch, where the ground rises to more than 1700 feet above the sea (see Fig. 129). This area, as already remarked, is unfortunately much obscured with drift and peat, so that the limits of its rocks cannot be so satisfactorily traced as might be desired. I think it probable that several successive vents have here been opened close to each other, but their erupted ashes probably cannot be distinguished. Over a space measuring about four miles in length by two and a half in breadth, the rocks exposed at the surface are fine tuffs, breccias and coarse agglomerates, largely made up of trachytic, andesitic or felsitic material, and pierced by innumerable protrusions of various andesitic, trachytic or felsitic rocks in bosses and veins, as well as also by dykes of a more basic kind, such as dolerites and basalts. Some of the tuffs present a curiously indurated condition; and they are frequently much decayed at the surface.[1] Another large mass of tuff and agglomerate lies a little to the south-west of the main area.

After the explosions ceased, by which the vents were opened and the cones of debris were heaped up, heated vapours would in many cases, as in modern volcanoes, continue for a long while to ascend in the funnels. The experiments of Daubrée on the effects of water and vapour upon silicates under great pressure and at a low red heat, have shown how great may be the lithological changes thereby superinduced. It is improbable that where a mass of tuff and lava, lying deep within a volcanic vent, was thoroughly permeated with constantly ascending heated vapours, it should escape some kind of change. I am inclined to attribute to this cause the frequent conversion of the sandstones round the walls of the vents into quartzite. The most remarkable example of metamorphism within a vent which I have observed among the plateaux, occurs in the heart of the Campsie Fells, where, instead of forming a prominence, the neck is marked

[1] This tract of ground was mapped for the Geological Survey by Mr. R. L. Jack, now in charge of the Geological Survey of Queensland. See Sheet 31, *Geological Survey of Scotland*.

FIG. 133.—View of Traprain Law from the south, a phonolite neck of the Garleton Plateau.

by a great hollow, measuring about a mile in length and half a mile in breadth (Fig. 128).[1] It is occupied mainly by a coarse tumultuous agglomerate, like that of other necks in the same district, but with a matrix rather more indurated, and assuming in certain parts a crystalline texture, so as to be at first sight hardly distinguishable from some of the surrounding andesites. Even in this altered condition, however, its included fragments may be recognized, particularly blocks of sandstone which have been hardened into quartzite. Numerous small veins of pink and yellow trachyte traverse the agglomerate, and are found also cutting the bedded andesites that encircle it.

3. DYKES AND SILLS

Intrusive masses both in the form of dykes and of sills are of frequent occurrence in connection with the Carboniferous volcanic plateaux. From the variety of their component materials it may be inferred that these rocks belong to different ages of intrusion.

DYKES.—The great majority of the Dykes consist of trachyte or of andesite, resembling in lithological characters the material of the necks and doubtless connected with its uprise. There occur also dykes of diabase, basalt or dolerite. Some of the latter, especially those which run for many miles, cutting every rock in the districts in which they occur, and crossing large faults without deviation, are certainly long posterior to the plateau volcanic period. Whether the small inconstant dykes of more basic composition, found in the same districts with the trachytes, are to be looked upon as part of the volcanic phenomena of the plateaux, is a question to which at present no definite answer can be given. I shall have occasion to show that in the next volcanic period the lavas that flowed from the puys are more basic than most of those of the plateaux, and that they are associated with more basic dykes and sills. In Roxburghshire, where it is so difficult to distinguish between the denuded vents of the two periods, the dark heavy olivine-basalts and dolerites of the bosses may possibly belong rather to the later than to the earlier volcanic episode. And if that be their true age, the dykes of similar material may be connected with them. At the same time it must be remembered that the earliest eruptions of the plateaux were markedly basic, that many vents in the plateaux are pierced by basic intrusions, and that basic dykes may have been associated with the uprise of the same magma.

The dykes occur in considerable numbers and in two distinct positions, though these may be closely related to each other: 1st, among the rocks outside and beneath the plateau-lavas, or cutting these lavas; and 2nd, in and around the vents.

1. Among the rocks which emerge from under the Carboniferous volcanic plateaux, dykes are sometimes to be observed in considerable numbers. They may be compared to the far more extensive series con-

[1] See Explanation to Sheet 31, *Geological Survey of Scotland*, par. 21 (1878).

nected with the Tertiary basalt-plateaux, like which they may have had a close relation to the actual building up of the successive sheets of andesite, trachyte and basalt that were erupted at the surface. They are particularly well developed in the Clyde plateau, where by extensive denudation they have been admirably exposed. I would especially refer to those that traverse the tract of red sandstones which underlie the volcanic series along the flanks of the great escarpments from Fintry to Strathblane and Dumbarton, and between Gourock and Ardrossan. These dykes have been dissected by the sea along both sides of the estuary of the Clyde and in the islands of Cumbrae. In these islands and in Bute they have recently been mapped in great detail for the Geological Survey by my colleague, Mr. W. Gunn, who has supplied me with notes of his observations on the subject, from which the following summary is compiled.

"There are at least four distinct groups of intrusive rocks in the Greater Cumbrae. The oldest of these is trachytic in character, and occurs both as dykes and sheets, which run generally in the same E.N.E. direction. The rock is usually pinkish in colour, sometimes grey or purplish. A specimen from the dyke of the Hawk's Nest, north of Farland Point, analyzed by Mr. Teall, was found to contain 11 per cent of alkalies, principally potash, while the percentages of lime and iron were very low. Sometimes these rocks are fine in grain with only a few porphyritic orthoclase crystals, though numerous small crystals of this mineral are found with the aid of the microscope. These red trachyte dykes are almost confined to the Upper Old Red Sandstone, rarely entering the overlying white Calciferous Sandstones, and never invading the plateau-lavas. They are therefore probably of early Carboniferous age.

"The next group follows the same general direction, but clearly traverses the trachytes, and must therefore be of later date. The dykes of this group are the most numerous of the whole, the greater part of the island being intersected by them. In the north-east corner about 40 of them may be counted in half a mile of coast-line, some being of large size. All of them which can be clearly made out are porphyritic olivine-basalts of the type of the Lion's Haunch at Arthur's Seat. They are generally grey in colour and finer at the edges than in the centre, which is often coarsely porphyritic and amygdaloidal. Olivine seems always characteristic, but has often been replaced by hæmatite or calcite. In Bute a good many dykes have been mapped to the north of Kilchattan Bay resembling this basalt series of Cumbrae, and running in the same direction. But they appear to be all porphyritic andesites. The second group of dykes, though it cuts the first and is thus proved to be later in date, is nevertheless confined within the same stratigraphical limits. It may thus belong nearly to the same period of intrusion.

"The dykes of the third group are dolerites without olivine, and follow on the whole an east and west direction. They cut both of the two foregoing sets of dykes, and likewise the lavas of the plateau. They must thus belong to a far later period of intrusion. They may be connected with other dykes and sills on the mainland, which traverse the Coal-measures, and would thus be not older than late Carboniferous or Permian time.

"The fourth group of dykes intersects all the others, and is probably of Tertiary age. The prevalent direction of these dykes in the Cumbraes is N.N.W." The Tertiary dykes are more fully described in Chapters xxxiv. xxxv.

The great group of tuffs which underlies the lavas of the East Lothian plateau is traversed by numerous dykes and sills, of which many good examples may be seen in the coast-cliffs of North Berwick. Among these rocks are beautiful olivine-basalts with singularly fresh olivine, as on the shore at North Berwick. Some of them are still more basic, as in the case of a limburgite intrusion at the Gin Head, Tantallon Castle.

2. In the necks, dykes are sometimes abundant, and they may be

FIG. 134.—Veins and dykes traversing the agglomerate and tuff of the great Renfrewshire vent.

observed occasionally to traverse the surrounding lavas. They consist of similar materials to those found outside the plateaux. Some of the larger necks are intersected by a network of dykes and veins. The great vent or group of vents among the uplands of Renfrewshire, already described (Fig. 129), furnishes some admirable examples of this characteristic volcanic feature. An illustration from that locality forms the subject of Fig. 134. The agglomerate which fills the large hollow among the Campsie Hills may be quoted as another illustration (Fig. 128). Further instances will be found in some of the sections given in preceding pages (see Figs. 124, 125, 127). The general aspect of a dyke in the volcanic series is shown in Fig. 135.

The SILLS associated with the plateau-type of Carboniferous volcanic action form a less prominent feature than they do among the earlier

FIG. 135.—"The Yellow Man," a dyke in volcanic tuff and conglomerate on the shore a little east of North Berwick.

Palæozoic formations or in the puy-type which succeeded them. They consist in general of short lenticular sheets of andesite or trachyte, like the necks and dykes in proximity to which they commonly appear. The best area for the study of them is the ground which stretches out from the base of the great escarpments of the Campsie, Kilpatrick and Ayrshire Hills (Fig. 136), where, among the agglomerate-vents and abundant dykes, intrusive sheets have likewise been injected between the bedding-planes of the red sandstones. But these sheets are of comparatively trifling dimensions. Very few of them reach a mile in length, the great majority falling far short of that size. In the Cumbraes and in Bute, Mr. Gunn has observed

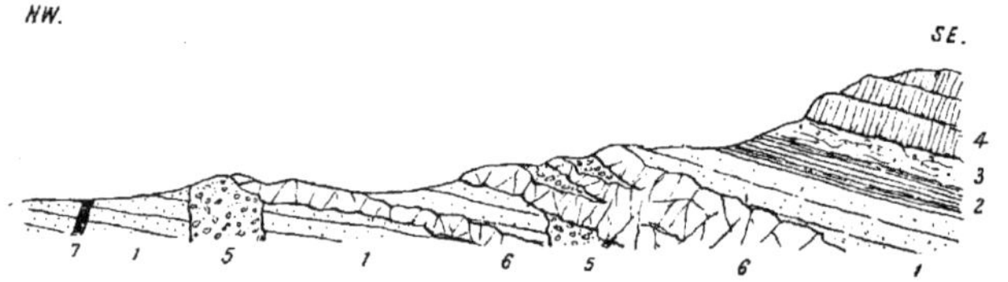

Fig. 136.—Trachytic sills, Knockvadie, Kilpatrick Hills.

a, Upper Old Red Sandstone; 2. "Ballagan Beds"; 3. Tuffs; 4. Lavas of the Plateau; 5. Agglomerate of necks; 6. Trachyte sills; 7. Dolerite dyke (? Tertiary).

that the trachytic, olivine-basalt and dolerite dykes are apt to pass into intrusive sheets. That the sills, as well as the dykes and bosses of the same material, are not of older date than the lavas of the plateaux is proved by the manner in which they pierce these lavas, especially towards the bottom of the series. The general absence of basic sills, when we consider how thick a mass of these rocks has sometimes been poured out in the plateaux, is not a little remarkable. Only in the basin of the Firth of Forth do we encounter thick basic sills near the plateaux, such, for instance, as Salisbury Crags at Edinburgh. But it is doubtful whether they ought not rather to be classed with the sills of the puys, to be afterwards described.

Close of the Plateau-eruptions

The relative geological date when the eruptions of each plateau ceased can fortunately be determined with much more precision than the time of their beginning. The Hurlet Limestone, so well known as the lowest thick calcareous seam in the Carboniferous Limestone series, of which it is generally taken as the base, can be identified over the whole of Central Scotland, and thus forms an excellent stratigraphical horizon, from which the upward termination of the volcanic sheets underneath it can be measured.

When the volcanic episode of the plateau-eruptions came to an end, such banks or cones as rose above the level of the shallow sea which then overspread Central Scotland were brought beneath the water, as I have already remarked, either by prolonged denudation or more probably in large part by the continued subsidence of the region. The downward movement may possibly for a time have been accelerated, especially in some districts. Thus the Hurlet Limestone, though usually not more than five or six feet thick, increases locally to a much greater thickness. At Petersfield, near Bathgate, for example, it is between 70 and 80 feet in depth, while at Beith, in North Ayrshire, it increases to 100 feet (Fig. 137), which is the thickest mass of Carboniferous Limestone known to exist in Scotland. At

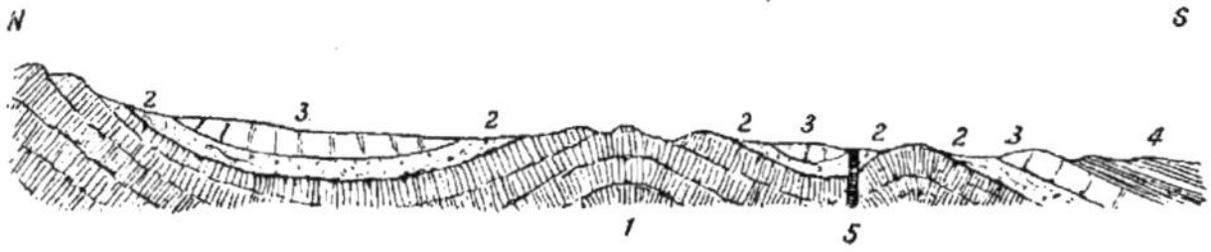

FIG. 137.—Section across the edge of the Clyde plateau, south-east of Beith.

1 Plateau-lavas; 2. Tuffs and volcanic conglomerates; 3. Hurlet Limestone; 4. Coal-bearing strata above the limestone; 5. Dolerite dyke.

both of these localities the limestone lies upon a series of volcanic rocks, and we may perhaps infer that the subsidence advanced there somewhat more rapidly or to a greater extent, so as to form hollows in which the limestone could gather to an abnormal depth. The water would appear to have become for a time tolerably free from mechanical sediment. The limestone is hence comparatively pure, and is extensively quarried all over the country for industrial purposes. It is a crinoidal rock, abounding in many species of corals, brachiopods, lamellibranchs, and gasteropods, with trilobites, cephalopods, and fishes.

A variable thickness of strata intervenes between the top of the volcanic series and the Main Limestone. Sometimes these deposits consist in large measure of a mixture of ordinary sandy and muddy material with the washed-down tuff of the cones, and probably with volcanic dust and lapilli

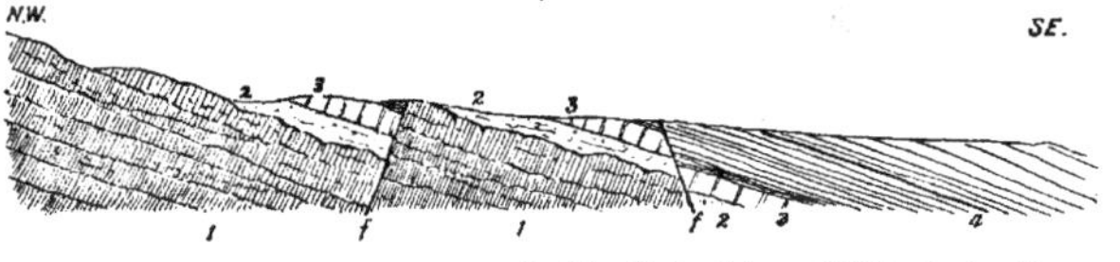

FIG. 138.—Section across the upper part of the Clyde plateau at Kilbirnie, Ayrshire.

1 1. Plateau-lavas; 2 2. Tuffs; 3 3. Hurlet Limestone; 4 Black-band Ironstone. *ff.* Faults.

thrown out by the latest eruptions. Thus along the flank of the hills from Barrhead to Strathavon, yellow and green ashy sandstones, grits and conglomerates are succeeded by ordinary sandstones, black shales and ironstones, while here and there true volcanic tuff and conglomerate make their appearance.[1] Further west, in the Kilbirnie district, the limestone lies directly on the tuffs that rest upon the andesites (Figs. 137, 138).

[1] Explanation of Sheet 22, *Geol. Surv. Scotland*, p. 12.

But perhaps the most striking contrast between adjacent localities in regard to the distance between the limestone and the top of the volcanic series is to be observed along the southern front of the Campsie Fells. In spite of the abundant faults which have there so broken up the regular sequence of the rocks, we can see that at Banton and Burnhead the lime-

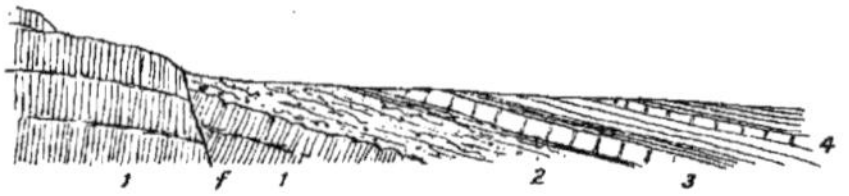

FIG. 139.—Section across the upper surface of the Clyde volcanic plateau, Burnhead, north-west of Kilsyth.

1. Lavas of the plateau; 2. Tuffs; 3. Hurlet Limestone; 4. Hosie's Limestone. *f*, Fault.

stone lies almost immediately on the volcanic series (Fig. 139). But a little to the westward, sandstones, conglomerates, shales and thin limestones begin to intervene between the volcanic series and the Hurlet Limestone and swell out so rapidly that on Craigmaddie Muir and South Hill of Campsie, only some five miles off, they must form a total thickness of not

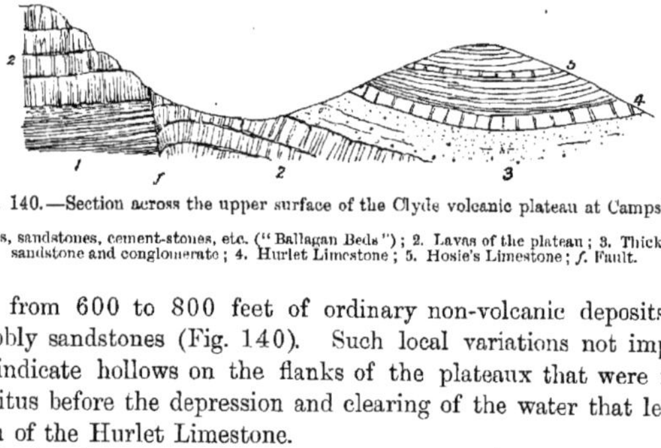

FIG. 140.—Section across the upper surface of the Clyde volcanic plateau at Campsie.

1. Shales, sandstones, cement-stones, etc. ("Ballagan Beds"); 2. Lavas of the plateau; 3. Thick white sandstone and conglomerate; 4. Hurlet Limestone; 5. Hosie's Limestone; *f*. Fault.

less than from 600 to 800 feet of ordinary non-volcanic deposits, chiefly thick pebbly sandstones (Fig. 140). Such local variations not improbably serve to indicate hollows on the flanks of the plateaux that were filled up with detritus before the depression and clearing of the water that led to the deposition of the Hurlet Limestone.

I have already remarked that the eruptions of the plateau period lasted

FIG. 141.—Section across western edge of the Garleton plateau.

1. Trachyte lavas of the plateau; 2. Calciferous Sandstones; 3. Hurlet Limestone.

longer in the western than in the eastern parts of the region. In the Garleton district, where the peculiar viscous trachytic lavas probably gave rise to a more uneven surface or more prominent cones than was usual among the andesitic plateaux, the eruptions ceased some time before the deposition of the Hurlet Limestone. As the area sank, the successive zones of the Calciferous Sandstones crept over the flanks of the trachytes, until at last they had completely buried these rocks before the limestone spread over the area (Fig. 141). In consequence, probably, of the uneven surface of this plateau, there is here a strong overlap of the higher part of the Calciferous Sandstones. On the west side of the volcanic area there can hardly be more than some 200 feet of strata between the top of the trachytic series and the limestone, while on the south side there must be greatly more than that thickness. This structure probably indicates that the Garleton volcanoes became extinct after having piled up a mass of tuffs and lavas to such a height that its summits were not submerged until the area had subsided 800 or 1000 feet in the waters, over the floor of which the Calciferous Sandstones were laid down. Hence, in spite of the proximity of the lavas to the limestone, there may have been a vast interval of time between their respective epochs, as has been already suggested with regard to other plateaux. This subject will be again referred to in discussing the relative chronology of the plateaux and puys.

In the Berwickshire and Solway districts, the extinction of the plateau-vents appears to have taken place at a still earlier part of the Carboniferous period, for there the andesites, while they rest on the Upper Old Red Sandstone, are covered with at least the higher group of the Calciferous Sandstones (Fig. 142). The equivalent of the Hurlet Limestone of Central Scotland must lie many hundred feet above them.

The submergence of the plateaux, and their entombment under the thick Carboniferous Limestone series, did not mark the close of volcanic activity in Central Scotland during Carboniferous time. The plateau-type of eruption ceased and was not repeated, but a new type arose, to which I would now call the reader's attention.

FIG. 142.—Section across the Solway plateau from Birrenswark to Kirtlebridge.
1. Upper Silurian strata; 2. Upper Old Red Sandstone; 3. Plateau-lavas; 4. Calciferous Sandstones and Carboniferous Limestone series; 5. Trias.

CHAPTER XXVI

THE CARBONIFEROUS PUYS OF SCOTLAND

i. General Character and Distribution of the Puys; ii. Nature of the Materials Erupted—Lavas Ejected at the Surface—Intrusive Sheets—Necks and Dykes—Tuffs.

i. GENERAL CHARACTER AND DISTRIBUTION

AFTER the beginning of the Carboniferous Limestone period, when eruptions of the plateau-type had generally ceased, volcanic activity showed itself over the area of the British Isles in a different guise both as regards the nature of its products and the manner and scale of their discharge. Instead of widely extended lava-sheets and tuffs, piled above each other sometimes to a thickness of many hundred feet, and stretching over hundreds of square miles, we have now to study the records of another phase of volcanism, where scattered groups and rows of *Puys*, or small volcanic cones, threw out in most instances merely tuffs, and these often only in trifling quantity, though here and there their vents also poured forth lavas and gradually piled up volcanic ridges which, in a few cases, almost rivalled some of the plateaux. The evidence for these less vigorous manifestations of volcanic activity is furnished (1) by layers of tuff and sheets of basaltic-lavas intercalated among the strata that were being deposited at the time of the eruptions, (2) by necks of tuff, agglomerate, or different lava-form rocks that mark the positions of the orifices of discharge, and (3) by sills, bosses, and dykes that indicate the subterranean efforts of the volcanoes. The comparatively small thickness of the accumulations usually formed by these vents, their extremely local character, the numerous distinct horizons on which they appear, and the intimate way in which they mingle and alternate with the ordinary Carboniferous strata are features which at once arrest the attention of the geologist, presenting, as they do, so striking a contrast to those of the plateaux.

From the clear intercalation of these volcanic materials on successive platforms of the Carboniferous system, the limits of geological time within which they were erupted can be fixed with considerable precision. It may be said that, in a broad sense, they coincided with the period of the Carboniferous Limestone, and certainly it was during the deposition of that formation that the eruptions which produced them reached their greatest vigour and widest extent. Here and there in Scotland evidence may be found that the phase of the Puys began during that earlier section of Carboniferous time recorded by the Calciferous Sandstones. This is markedly the case in Liddesdale and the neighbouring territory. Over the western part of Midlothian also, the eastern portion of West Lothian, and the southern margin of Fife, abundant traces occur of puy-eruptions during the deposition of the Calciferous Sandstones. Elsewhere in Central Scotland there is no evidence of the vents having been opened until after the deposition of the Hurlet Limestone, which, as we have seen, may conveniently be taken as the base of the Scottish Carboniferous Limestone series. The volcanoes remained active in West Lothian until near the close of the time represented by that series; but in Ayrshire they continued in eruption until the beginning of the accumulation of the Coal-measures. These western examples of the puy-type are, so far as I am aware, the latest known in Britain.

Whether or not the earliest puy-eruptions began before the latest plateau-lavas and tuffs were accumulated is a question that cannot be readily answered. It will be remembered that in the basin of the Firth of Forth a thickness of more than 3000 feet of sedimentary strata, including the Burdiehouse Limestone and numerous oil-shales as well as thin coal-seams, lies above the red and green marls, shales, sandstones and cement-stones of the Calciferous Sandstone series. This remarkable assemblage of strata is absent in the western parts of the country, where the top of the Clyde volcanic plateau is almost immediately overlain by the Hurlet Limestone. If we were to judge of the sequence of events merely from the stratigraphy, as expressed in such sections as Figs. 137, 138, 139 and 140, we might naturally infer that as no trace of any break occurs at the top of the Clyde plateau, the tuffs shading upward there into the limestone series, no important pause in sedimentation took place, but that the last volcanic eruptions were soon succeeded by the conditions that led to the deposition of the widespread encrinite-limestones. If this inference were well founded it would follow that while the plateau-eruptions in the west lasted till the time of the Hurlet Limestone, those in the east ceased long before that time and were succeeded by the puys of Fife and the Lothians. There would thus be an overlap of the two phases of volcanic action.

I am inclined to believe, however, that in spite of the superposition of the Hurlet Limestone almost immediately upon the volcanic rocks of the Clyde plateau, and the absence of any trace of a break in the process of sedimentation, a long interval nevertheless elapsed between the last eruptions and the deposit of that limestone. The Campsie section (Fig. 140) shows us how rapidly a thick mass of strata can come in along that horizon. The volcanic ridges may have remained partly unsubmerged for such time as was required for the subsidence of the Forth basin and the deposit of the thick Calciferous Sandstone series there, and their summits may only have finally sunk under the sea not long before the Hurlet Limestone grew as a continuous floor of calcareous material over the whole area of central Scotland. In these circumstances, the puy-eruptions of that basin would be long subsequent to the eruptions of the Clyde plateau, as they certainly were to those of the plateaux of Midlothian and the Garleton Hills.

In tracing the geographical distribution of the puy-eruptions we are first impressed with the force of the evidence for their extremely local and restricted character (Map IV.). Thus in the area of the basin of the Firth of Forth, which may be regarded as the typical region in Britain for the study of this form of Carboniferous volcano, traces of them are abundant to the west of the line of the Pentland Hills. To the east of that line, however, not a vestige of puy-eruptions, save a few sills of uncertain relationhip, can be detected, though the same series of stratigraphical horizons is well developed on both sides of the Lothian coal-field. Again, to the westward of the Forth basin over the area of Stirlingshire, Lanarkshire and Renfrewshire lying to the north of the great volcanic plateau, no record of puy-eruptions has been noticed. Immediately to the south of that plateau, however, these eruptions were numerous in the north of Ayrshire. Yet the rest of the Carboniferous area in that large county has supplied no relics of these eruptions save at one locality—the Heads of Ayr. Lastly, while no trace of any younger display of volcanic activity occurs in the Merse of Berwickshire, east of the plateau series of that district, the ground immediately to the west abounds in puys, and contains likewise extensive sheets of tuff and beds of basic lavas connected with these vents.

Another fact which at once attracts notice in Scotland is the way in which the puy-vents have generally avoided the areas of the plateaux, though they sometimes approach them closely. As a rule, it is possible to distinguish the tuffs and agglomerates which have filled up these vents from those that mark the sites of the eruptive orifices of the plateaux. There are, no doubt, some instances, as in Liddesdale, where puys have appeared on the sites of the older lavas, but these are exceptional collocations.[1] On the other hand, many examples may be found where puys have risen in the interspace between the limits of the eruptions of two plateau-areas. Thus the tract between the Clyde plateau-eruptions on the west and those of the Garleton Hills on the east was dotted over with puys. Again, the southern margin of the Clyde plateau in Ayrshire, from Dalry to Galston is flanked with puys and long sheets of their lavas and tuffs.[2]

ii. NATURE OF THE MATERIALS ERUPTED

A. The Lava-form Rocks

We have now to consider the nature of the materials erupted by the volcanic activity of the puys. The geologist who passes from the study

[1] A means of definitely placing some of these vents in the series of puy-eruptions is stated further on, at p. 476.

[2] Reference may again be made here to the remarkable similarity between the Scottish Carboniferous puy-vents and those of older Tertiary time in the Swabian Alps so fully described by

of the plateau lavas to those of the puys at once remarks the prevalent more basic character of the latter. The great majority of them are basalts, generally olivine-bearing, in the various types embraced in the table on the following page. The olivine-free dolerites are generally found as intrusive bosses, sills and dykes. Such more acid rocks as andesites occur only rarely, and still more seldom are quartziferous masses met with in some of the bosses.

DOLERITES and BASALTS.—The great majority of the lava-form rocks connected with the puys are basic in composition, and belong to the family of the Dolerites and Basalts. They graduate, on the one hand, into ultra-basic rocks such as limburgite and picrite, and on the other, into compounds that approach andesites or trachytes in composition. A large series of specimens from Central Scotland was studied a few years ago by Dr. Hatch, who proposed a petrographical classification of the rocks, and arranged them in a number of types which he named after localities where they are well developed.[1] More recently the rocks have again been subjected to microscopic investigation by my colleague Mr. Watts, who, confirming generally Dr. Hatch's discriminations, has made some modifications of them. He has furnished me with a revised classification (p. 418), based on purely petrographical considerations. The doleritic and basaltic series may be grouped into two divisions, one with, and the other without, olivine, and each division may be further separated into a dolerite group, which presents an ophitic or subophitic structure, and a basalt-group in which the groundmass is made up of felspar and granular augite, and possesses the "intersertal structure" of Rosenbusch, or consists of idiomorphic augite embedded in felspar substance. The term "sub-ophitic" is employed by Mr. Watts "to imply that the augite grains are neither very large nor very continuous, optically, and that they rarely contain entire felspar-crystals imbedded in them, merely the ends of a group of these crystals as a rule penetrating into them."

Transitional forms occur between many of the following types by the increase or diminution in the relative proportions of the constituents. Thus it is not easy to draw a line between *2b* and *2c*; the latter again shades into *2d* and *2e* by the decrease of the felspar.

Mr. Watts has further observed that the rocks containing no olivine offer greater difficulties in classification than those in which that mineral is present. "The very distinction," he remarks, "between dolerites and basalts is less marked, the types are much less sharply distinguished, and decomposition and masking of the structure are more common. While using the term Dolerite for such rocks as have a sub-ophitic structure, I have extended it to those rocks in which evidence exists that a great part of the crystallization took place under intratelluric conditions. Although

Professor Branco in the work already cited p. 46. Denudation in that region has bared the cones and exposed the structure of the necks which, down to even minute details, repeat the phenomena of Carboniferous and Permian time in Scotland.

[1] This classification was given in my Presidential Address to the Geological Society, 1892, *Quart. Journ. Geol. Soc.* vol. xlviii. p. 129. See Report of Geological Survey for 1896.

not quite holo-crystalline, the crystals of felspar, augite and magnetite are large and the structure coarse-grained, while the groundmass is confined to comparatively small interstitial patches. In these rocks there is usually no one dominant porphyritic ingredient."

I. The Olivine-bearing Series

1. *Olivine-Dolerites*

1*a*.	Porphyritic elements inconspicuous, olivine being the principal, and felspar of secondary importance ; groundmass sub-ophitic.	Jedburgh Type.
1*b*.	Strongly porphyritic ; felspar-phenocrysts large ; olivine smaller ; groundmass sub-ophitic.	Kilsyth Type.
1*c*.	Porphyritic olivine, but not felspar ; groundmass sub-ophitic.	Gallaston Type.

2. *Olivine-Basalts*

2*a*.	Porphyritic olivine, augite and felspar ; groundmass of felspar-laths imbedded in granules of augite.	Lion's Haunch Type. (See Fig. 207.)
2*b*.	Porphyritic olivine and augite ; groundmass of felspar-laths imbedded in granules of augite. More rarely the groundmass is made of idiomorphic augite imbedded in felspar-substance.	Craiglockhart Type.
2*c*.	Porphyritic olivine abundant, augite much less common, and felspar very rare or absent ; groundmass with granular or idiomorphic augite (one of the most common types).	Dalmeny Type.
2*d*.	Porphyritic olivine more common than augite ; groundmass of granules of augite set amongst lath-like felspars which are much fewer in number than in 2*c*.	Picrite Type.
2*e*.	Porphyritic olivine more common than augite ; groundmass of idiomorphic augite imbedded in felspathic material which is not abundant.	Limburgite Type.
2*f*.	Porphyritic olivine and felspar, without augite ; groundmass of granular or idiomorphic augite, with lath-shaped felspars.	Kippie Law Type. (For analysis see p. 379).

II. The Non-olivine-bearing Series

3. *Olivine-free Dolerites*

Felspar, augite, magnetite in coarse-grained aggregate usually ophitic or sub-ophitic ; groundmass not plentiful.

3*a*.	Groundmass absent	Ophitic Type.
3*b*.	Groundmass micropegmatitic . . .	Ratho Type.
3*c*.	Groundmass an unstriated felspar (not orthoclase) and occasionally some interstitial altered glass or a little quartz.	Burntisland Sill Type.

4. *Doleritic Basalts*

Felspar, augite and magnetite in coarse-grained aggregate ; groundmass rather more plentiful and often in large patches.

4*a*.	Felspar and augite, related sub-ophitically where together, but augite showing crystalline contours in contact with the groundmass ; some interstitial quartz and unstriated felspar.	Bowden Hill Type.

5. *Basalts*

Finer-grained rocks, generally with a porphyritic ingredient and much scattered interstitial matter in the groundmass.

5*a*. Porphyritic felspar, and occasionally a little augite; groundmass of granular augite, felspar needles and magnetite with some interstitial matter.	Binny Craig Type.
5*b*. Porphyritic felspars not conspicuous and small; the rock mainly made up of a mesh of fine felspar-laths set amongst granular augite, magnetite and base.	Tholeiite Type.
5*c*. Similar to the last but even finer-grained, and with the base in a cryptocrystalline condition.	Cryptocrystalline Type.

Taking first the superficial lavas, I know of only one locality where picrite occurs in such a position that it may be included among the surface outflows. This is the quarry near Blackburn, to the east of Bathgate, where I originally observed it.[1] The rock occurs there on the line of the basalt-flows from the Bathgate Hills, and I mapped it as one of them before the microscope revealed the remarkable composition of the mass. I still believe it to be a lava like the "leckstone" described on p. 443, though the other known examples of this rock in the basin of the Firth of Forth are intrusive sheets. The rock locally known as "leckstone" or "lakestone" has long been quarried for the purpose of constructing the soles of bakers' ovens, as it stands a considerable temperature without cracking. Its microscopic structure is now well known. As exposed in Blackburn quarry, an interesting difference is observable between the lower and upper parts of the sheet. The lower portion is a picrite, with abundant serpentinized olivine, large crystals of augite, and a considerable amount of ores. The upper portion, on the other hand, has plagioclase as its most abundant definite mineral, with a minor quantity of minute prisms of augite and of iron-ores, and scattered crystals of olivine. Here, within the compass of a few yards and in one continuous mass of rock, we have a transition from a variety of olivine-basalt into a picrite.

The great majority of the puy lavas belong to the olivine-bearing series. A few of them are dolerites, but most are true basalts of the Dalmeny type, of which typical examples may be seen at the Kirkton quarries, Bathgate, and in the coast section between Pettycur and Kinghorn. Occasionally they present transitions towards picrite, as in the sheet overlying the lowest limestone at Kirkton, and in the lowest lava of King Alexander's Crag, Burntisland. These puy lavas exhibit considerable variety of structure as seen in the field. Some are solid, compact, black rocks, not infrequently columnar and weathering into spheroidal exfoliating forms. Others are somewhat granular in texture, acquiring green and brown tints by weathering, often showing amygdaloidal kernels, and even passing into well-marked amygdaloids. Many of them exhibit a slaggy structure at their upper and under surfaces (Figs. 153, 170, 171). These external differences are an

[1] *Trans. Roy. Soc. Edin.* vol. xxix. (1879) p. 506.

index to the corresponding variations in composition and microscopic structure enumerated in the foregoing tabular arrangement.

As a rule, the basic rocks which occur intrusively in connection with the puys, especially where they form a considerable mass, have assumed a much more coarsely crystalline texture than those of similar composition which have been poured out at the surface. They are generally dolerites rather than basalts. But with this obvious distinction, the two groups have so much in common, that the geologist who passes from the study of the subterranean phenomena of the Plateaux to that of the corresponding phenomena of the Puys is at once impressed with the close relationship between the material which, in the case of the puys, has consolidated above ground, and that which has been injected below. There is no such contrast between them, for example, as that between the basic and intermediate lavas of the plateaux and the more acid intrusions associated with them.

By far the largest number of the basic sills, bosses and dykes associated with the puys are somewhat coarsely crystalline dolerites without olivine. They include, however, olivine-dolerites and basalts, and even some extremely basic compounds. Of these last, a typical example is supplied by the now well-known picrite of Inchcolm, in the Firth of Forth, which occurs as an intrusive sheet among the Lower Carboniferous Sandstones.[1] In recent years one or two other picrite-sills have been observed in the same district. An interesting example has been described from a railway cutting between Edinburgh and Cramond where the rock invades and alters shales. More detailed reference to it will be made in the account of the sills connected with the puys. Another instance of the occurrence of this rock is in a railway cutting immediately to the west of Burntisland where it has been intruded among the Calciferous Sandstones below the Burdiehouse Limestone.

Rocks approaching limburgite occur among the sills and bosses which pierce the Carboniferous Limestone series of Fife between Cowdenbeath and Inverkeithing. One of these is found at Pitandrew, near Fordel Castle. Dr. Hatch observed that it consists of "numerous porphyritic crystals of olivine, with a few grains of augite and an occasional small lath-shaped crystal of felspar imbedded in a groundmass which is composed principally of idiomorphic augite microlites, small crystals of a brown mica, granules of magnetite and prisms of apatite. In addition, there is a considerable amount of interstitial matter, which is partly colourless glass, and partly shows a slight reaction between crossed nicols." Another example of the same type of rock occurs as a plug or boss in the tuff-vent of the Hill of Beath, and a further display of the limburgite type is to be seen in Dunearn Hill near Burntisland.

Although olivine-basalts of the Dalmeny type are most frequently met with as interstratified lavas, they also occur as bosses and sills. The typical example from Dalmeny is itself intrusive. Other illustrations are to be found in the Castle Rock of Edinburgh and in the sheets near Crossgates and Blairadam in Fife. The presence or absence of olivine, however, may

[1] *Trans. Roy. Soc. Edin.* vol. xxix. (1879) p. 506. Teall, *British Petrography*, p. 94.

sometimes be a mere accident of cooling or otherwise. I have shown that in the same mass of rock at Blackburn a gradation can be traced from a rock largely composed of altered olivine into one consisting mainly of felspar with but little olivine, and another example occurs in the picrite-sill between Edinburgh and Cramond. Dr. Stecher has ascertained that the marginal portions of the sills in the basin of the Firth of Forth, which cooled first and rapidly, and may be taken, therefore, to indicate the mineral composition of the rock at the time of extrusion, are often rich in olivine, while that mineral may be hardly or not at all discernible in the main body of the rock.[1]

Of the ordinary and characteristic dolerites without olivine which constitute most of the intrusive masses, the various types enumerated in the tabular arrangement are abundantly developed in Central Scotland. Thus the normal ophitic type is displayed by the uppermost sill of the Burntisland series, and by the rock which forms the plug of the Binns Hill neck in Linlithgowshire. The Ratho type is well seen in the large sill at Ratho, likewise in the extensive intrusive sheets in the west of Linlithgowshire as at Muckraw and Carribber. The Burntisland sill type is shown by the lower sills of Burntisland and by some others in the same region, especially by that of Colinswell, and by another on the shore east from the Poorhouse, near Kinghorn. The great boss among the Bathgate Hills likewise displays it. The Bowden Hill type occurs in well-marked development at Bowden Hill, three miles south-west of Linlithgow, and in the massive sill at St. Margaret's, west from North Queensferry.

The non-olivine-bearing basalts are found in various bosses and sheets in the basin of the Firth of Forth. Thus the Binny Craig type occurs in the prominent and picturesque sill from which it is named, likewise among the intrusive sheets near Kirkcaldy, in Fife. Sometimes the same mass of rock displays more than one type of structure, as in the case of the great Galabraes neck among the Bathgate Hills wherein both the Tholeiite and Burntisland sill types may be recognized.

Some of the sills in West Lothian, as I pointed out many years ago, contain bitumen and give off a bituminous odour when freshly broken. They have been injected into bituminous shales or coal-seams.[2]

2. ANDESITES.—Rocks referable to this series appear to have been of rare occurrence among the puy-eruptions. Examples of them containing as much as 60 per cent of silica occur among the lavas of the Limerick basin. Some of the necks and what may be sills in the same district likewise consist of them.

3. TRACHYTES AND QUARTZ-BEARING ROCKS.—Acid rocks, as I have already said, are extremely rare among the puy-eruptions. The only important examples known to me are those around the Limerick basin, where they rise apparently in old vents and form conspicuous rounded or conical hills. These rocks have been examined microscopically by Mr. W. W.

[1] Dr. Stecher, *Tschermak's Mineralog. Mittheil.* vol. ix. (1887) p. 193. *Proc. Roy. Soc. Edin.* vol. xv. (1888) p. 162.

[2] *Geol. Survey Memoir on Geology of Edinburgh* (Sheet 32, Scotland), p. 46.

Watts. One of the most interesting varieties, which occurs at the Standing Stone near Oola, was found by him to show quartz enclosing ophitically the felspars which, with well-terminated prisms, project into it. Further west, near Knockaunavoher, another boss occurs with conspicuous quartz. These rocks have much in common with trachytes but have a wholly crystalline structure. They will be described in the account of the Limerick basin.

B. Tuffs

The fragmental rocks connected with the puy-eruptions form a well-marked group, easily distinguishable, for the most part, from the tuffs of the plateaux. They vary from exceedingly fine compacted dust or volcanic mud, through various stages of increasing coarseness of texture, to basalt-conglomerates and tumultuous agglomerates.

The fragmentary material found in the necks of the puys is generally an agglomerate of a dull dirty-green colour. The matrix ranges from a fine compact volcanic mud to a thoroughly granular detritus, and sometimes shows a spheroidal concentric structure in weathering. In this matrix the lapilli are distributed with great irregularity and in constantly varying proportions. They consist in large measure of a pale yellowish-green, sometimes pale grey, very basic, finely vesicular, devitrified glass, which is generally much decomposed and cuts easily with the knife. This highly basic substance is a kind of palagonite. So minute are its vesicles that under the microscope a thin slice may present a delicate lace-like network of connected walls, the palagonite occupying much less space than the vesicles. The material has been a finely frothed-up pumice.

Besides this generally distributed basic pumice, the stones in the agglomerate of the necks likewise include fragments of older volcanic grits or tuffs, blocks of basalt or diabase, as well as pieces of the Carboniferous strata of the district, especially shale, sandstone and limestone. Not infrequently also, they comprise angular blocks of fossil wood.

The materials which fill the necks are generally much coarser than those that form intercalated beds. But while in numerous cases huge blocks of basalt and large masses of sandstone, shale, limestone, ironstone or other strata may be seen wrapped up in a matrix of coarse basalt-tuff, in not a few instances the material in the necks may be observed to consist of a tuff quite as fine as that of the interstratified bands. Such necks appear to mark the sites of tuff-cones where only fine ashes and lapilli were ejected, and where, after sometimes a brief and feeble period of activity, the orifice became extinct.

The bedded tuffs interstratified with the ordinary Carboniferous strata do not essentially differ in composition from the material of the necks. They are basalt- (diabase-) tuffs and basalt- (diabase-) conglomerates, usually dull green in colour and granular in texture, the lapilli consisting in great measure of various more or less decayed basalts, but containing the same highly vesicular basic glass or pumice above referred to. They are mainly to be distinguished by their conspicuous stratification, and especially by their rapid alternations of coarser and finer material, by the intercalation of shales, limestones, sandstones or ironstones in them, and by the insensible gradations by which they pass both vertically and laterally into ordinary sediments. Occasional large blocks or bombs, indicating some paroxysm of explosion, may be observed even among the finer tuffs, shales and other strata, which round the sides of these masses have had their layers bent down by the fall of heavy blocks.[1] Many of the bedded tuffs contain fossils, such as crinoids, corals, brachiopods, fish-teeth or macerated fragments of land-plants. Coal-seams also are occasionally interstratified among them.

Of the finer kinds, the best example is furnished by a remarkable group of "green and red marls" which lie above a seam of coal (Houston Coal) in the Calciferous Sandstones of West Lothian.[2] These strata, which differ much from any of the rocks with which they are associated, are exceedingly fine in grain, dull sage-green and brownish or chocolate-red in colour, not well laminated like the shales, but breaking under the influence of weathering into angular fragments, sometimes with a conchoidal fracture. They look like indurated mud. Mr. H. M. Cadell, who has recently re-examined them in connection with a revision of the Geological Survey Map (Sheet 32) has found them passing into ordinary granular tuff.

Palagonitic-tuff is of frequent occurrence. It is met with in the Firth of Forth district,[3] and Mr. Watts has detected fragments of palagonite among the tuffs of the Limerick basin.

1 *Ante*, p. 36, and Figs. 15 and 151. See also *Geol. Mag.* i. (1864), p. 22; *Trans. Roy. Soc. Edin.* vol. xxix. (1879) p. 515.

2 Memoir on Sheet 32 *Geol. Surv. Scotland* (1861), p. 42. The stratigraphical position of these "Houston Marls," as they are locally called, is indicated in Fig. 155.

3 *Trans. Roy. Soc. Edin.* vol. xxix. (1819) p. 515.

CHAPTER XXVII

GEOLOGICAL STRUCTURE OF THE CARBONIFEROUS PUYS OF SCOTLAND

1. Vents: Relation of the Necks to the Rocks through which they rise—Evidence of the probable Subærial Character of some of the Cones or Puys of Tuff—Entombment of the Volcanic Cones and their Relation to the Superficial Ejections. 2. Bedded Tuffs and Lavas—Effects of Subsequent Dislocations. 3. Sills, Bosses and Dykes.

THE puy-type of volcanic hill differs widely in one respect from those which we have hitherto been considering. In the earlier epochs of volcanism within the British area, it is the masses of material discharged from the vent, rather than the vents themselves which arrest attention. Indeed, so copiously have these masses been erupted that the vents are often buried, or their positions have been rendered doubtful, by the uprise in and around them of sills and bosses of molten rock. But among the Carboniferous puys the vent is often the only record that remains of the volcanic activity. In some cases we know that it never ejected any igneous material to the surface. In others, though it may be filled with volcanic agglomerate or tuff, there is no record of any shower of such detritus having been discharged from it. In yet a third class of examples, we see that lava rose in the vent, but no evidence remains as to whether or not it ever flowed out above ground. Other cases occur where beds of lava or of tuff, or of both together, have been intercalated in a group of strata, but with no trace now visible of the vent from which they came. The most complete chronicle, preserving at once a record of the outflow of lava, of the showering forth of ashes and bombs, and of the necks that mark the vents of eruption, is only to be found in some of the districts.

I shall therefore, in the present instance, reverse the order of arrangement followed in the previous chapters, and treat first of the vents, then of the materials emitted from them, and lastly of the sills and dykes.

i. VENTS

A large number of vents rise through the Carboniferous rocks of Scotland. Some of these are not associated with any interbedded volcanic material, so that their geological age cannot be more precisely defined than by saying that they must be later than the particular formations which they pierce. Some of them, as I shall endeavour to show, are in all probability of Permian age. But many, from their position with reference to the nearest intercalated lavas and tuffs, are to be regarded as almost certainly belonging to the Carboniferous period. Those which are immediately surrounded by sheets of lava and tuff, similar in character to the materials in the vents themselves, may without hesitation be connected with these sheets as marking the orifices of discharge.

The vents of the puys are in general much less than those of the plateaux. Their smallest examples measure only a few yards in diameter, their largest seldom much exceed half a mile.[1]

The dislocations of the Carboniferous system are probably on the whole later than its volcanic phenomena. It is at least certain that the lavas and tuffs of the puys have been extensively faulted, like the surrounding sedimentary strata, and the vents seldom show any apparent relation to faults. It may sometimes be observed, however, that the vents are arranged in lines suggestive of fissures underneath. A remarkable instance of the linear distribution is furnished by the chain of necks which extends from the Vale of the Tweed at Melrose south-westwards across the watershed and down Liddesdale. The most notable part of this line lies among the uplands to the east of the Old Mosspaul Inn at the head of the Ewes Water. A string of masses of agglomerate has there solidified in a fissure among the Silurian greywackes and shales, running in a north-easterly direction for several miles. The largest connected mass of agglomerate is 4700 feet long, and from 350 to 600 feet broad (see No. 1 in Fig. 22). That this curious vent, or connected line of vents along a great fissure, belongs to the puy-eruptions of Liddesdale is shown by the abundant fragments of yellow sandstone and cement-stone which occur in the agglomerate.[2]

Most frequently the vents are distributed irregularly in groups. As examples of this arrangement I may cite those of the west of Fife, of West Lothian and of the north of Ayrshire.

A convenient classification of the vents may be made by dividing them into four groups according to the nature of the material that now fills them: 1st, Necks of non-volcanic debris; 2nd, Necks of tuff and agglomerate; 3rd, Necks of similar materials, but with a central plug of basalt; 4th, Bosses of basalt or other lava, without agglomerate or tuff.

1 The following measurements of necks belonging to the puy-eruptions in different parts of Scotland are taken from the 6-inch field-maps of the Geological Survey:—Saline Hill, Fife, 6000 × 4000 feet; Binn of Burntisland, 3500 × 1500; Hill of Beath, Fife, 2900 × 1550; Binns Hill, Linlithgowshire, 4800 × 2200; Tor Hill, Ecclesmachan, Linlithgowshire, 1900 × 1000 (Fig. 155); Great Moor, near Maiden Pap, Roxburghshire, 2600 × 2400; Tinnis Hill, Liddesdale, 1500 × 1000; Roan Fell, Liddesdale, 300 × 200; Hadsgarth Burn, Liddesdale, 250 × 200; Dalbate Burn, 250 × 120. In some cases, especially in those of the larger necks, it is probable that the tuff belongs to more than one funnel. Thus the Binn of Burntisland almost certainly includes two necks, a smaller one to the west and a much larger one to the east. Saline Hill may also conceal more than one vent. But in the continuous mass of tuff at the surface it is at present impossible to determine precisely the number and boundaries of the several orifices.

2 These facts were ascertained by Mr. Peach in mapping the ground for the Geological Survey. See Sheet 17, Scotland.

1. *Necks of Non-volcanic Debris.*—In a few instances the orifices of eruption have been filled up entirely with non-volcanic debris. They have served as funnels for the discharge of explosive vapours only, without the expulsion of any solid volcanic materials. At least no trace of fragmentary lavas is met with in them, nor are any beds of tuff or lava intercalated among the surrounding strata. Some interesting examples of this kind were laid bare in the open ironstone-workings near Carluke in Lanarkshire. They were circular in ground-plan, descended vertically into the strata, and were somewhat wider at the top of the quarry than at the bottom. They were filled with angular pieces of Carboniferous sandstone, shale, limestone, ironstone and other rocks, these materials being rudely arranged with a dip towards the centre of the neck, where the blocks were largest in size. Though no fragments of igneous rocks were observed among the debris, a few string-like veins of "white trap," or altered basalt, were seen to traverse the agglomerate here and there. The necks and the strata surrounding them were highly impregnated with pyrites and sulphate of lime.[1]

A vent of the same nature, but on a much larger scale, has been mapped by Mr. Peach in the south of Perthshire, near East Grange, where

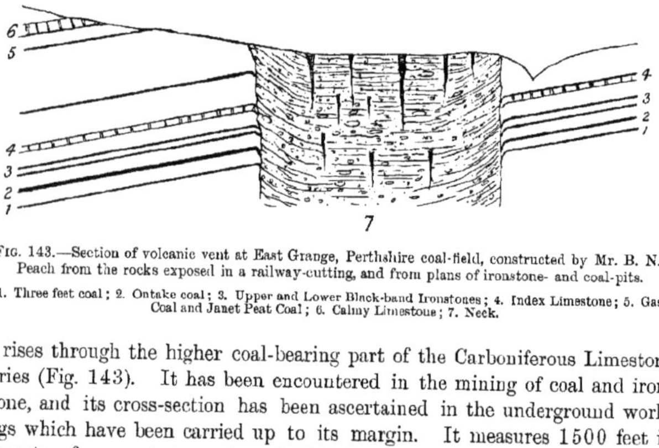

FIG. 143.—Section of volcanic vent at East Grange, Perthshire coal-field, constructed by Mr. B. N. Peach from the rocks exposed in a railway-cutting, and from plans of ironstone- and coal-pits.

1. Three feet coal; 2. Ontake coal; 3. Upper and Lower Black-band Ironstones; 4. Index Limestone; 5. Gas Coal and Janet Peat Coal; 6. Calmy Limestone; 7. Neck.

it rises through the higher coal-bearing part of the Carboniferous Limestone series (Fig. 143). It has been encountered in the mining of coal and ironstone, and its cross-section has been ascertained in the underground workings which have been carried up to its margin. It measures 1500 feet in diameter from east to west and 2000 feet from north to south. It does not appear ever to have emitted any ashes or lava. Mr. Peach found it filled with dark sandy crumbling clays, full of fragments of sandstone, shale and coal. These sediments are arranged in layers that dip in the same general direction as the strata surrounding the vent. They contain abundant calcareous nodules of all sizes from that of a hazel-nut up to concretions 18 feet in diameter. The clays likewise include many of the common shells and crinoids of the Carboniferous Limestone sea, and the same fossils are enclosed in the nodules. A remarkable feature in this vent is the occur-

[1] Prof. Jas. Geikie, *Mem. Geol. Surv. Scotland*, Explanation of Sheet 23, p. 39.

rence of abundant vertical rents, which have been filled partly with the same material that forms the nodules, and partly with sandstone.

The formation of the neck took place after the deposition of the Index Limestone, and probably about the time of the accumulation of the next limestone, which lies immediately to the west somewhat higher in the series. It would appear that the eruption which produced this funnel gave forth only gaseous explosions, and occurred on the sea-floor; that the low crater-walls were washed down to such an extent that the sea entered and carried some of its characteristic organisms into the lagoon or *maar* within; further, that as the silt gathered inside, successive subsidences occurred, whereby the sediment was rent by cracks into which sand and calcareous mud were washed from above.[1]

Many necks occur wherein non-volcanic materials, though not forming the whole of the agglomerate, make up by far the larger part, with only a slight admixture of volcanic tuff between them. Among the Burntisland necks of Fife, for instance, abundant fragments of the well-marked cyprid limestone and shale may be observed, while at Niddry in Linlithgowshire blocks several yards in length, and consisting of different layers of shale and cement-stone still adhering to each other, may be seen imbedded at all angles in the tuff.

Where only the debris of non-volcanic rocks occupies a vent, we may infer that the volcanic action was limited to the explosion of steam, whereby the rocks were dislocated, and an orifice communicating with the surface was drilled through them, and that while no true volcanic rock in such a case appeared, the pipe was filled up to perhaps not far from the surface by the falling back of the shattered detritus. A little greater intensity or farther prolongation of the volcanic action would bring the column of lava up the funnel, and allow its upper part to be blown out as dust and lapilli; while still more vigorous activity would be marked by the rise of the lava into rents of the cone or its actual outflow at the surface. Every gradation in this scale of progress may be detected among the Carboniferous volcanoes of the basin of the Firth of Forth.

2. *Necks of Tuff and Agglomerate.*—The majority of the necks connected with the puys consist of tuff or agglomerate. Externally they generally appear as smooth rounded grassy hills that rise disconnected from other eminences. In some districts their materials consist of a greenish granular often stratified tuff, enclosing rounded balls of various basic lavas and pieces of sandstone, shale, limestone or other strata through which they have been drilled. This is their usual character in the Forth region. But in some cases, the tuff becomes a coarse agglomerate, made up partly of large blocks of basalt and other volcanic rocks and partly of the sedimentary strata around them, of which large masses, many cubic yards in bulk, may be seen. Among the enclosed fragments it is not unusual to find pieces of older stratified tuff. These resemble in general petrographical character parts of the tuff among which they are imbedded. Sometimes they have been

[1] The vent is shown in Sheet 39, *Geol. Surv. Scotland.*

Fig. 144.—View of the Binn of Burntisland—a volcanic neck of agglomerate.
(This illustration and Figs. 145, 152, 153, 164, 166 and 168 are from photographs taken by Mr. Robert Lunn for the Geological Survey.)

derived from previous tuffs which, interstratified among the sedimentary strata, had been broken up by the opening of a new vent. But probably in most cases they should be regarded as portions of the volcanic debris which, having solidified inside the crater, was blown out in fragments by subsequent explosions. In a modern volcano a considerable amount of stratified tuff may be formed inside the crater. The ashes and stones thrown out during a period of activity fall not only on the outer slopes of the cone, but on the steep inner declivities of the crater, where they arrange themselves in beds that dip at high angles towards the crater bottom. This feature is well seen in some of the extinct cones in the Neapolitan district. In some of the Scottish puys the tuff is stratified and has tumbled down into a highly inclined or vertical position (Fig. 145).

As a good illustration of the variety and relative proportions of the ejected blocks in the green tuff of the puy-vents, I may cite the following table of percentages which I took many years ago in the tuff which rises through the Cement-stone group on the beach at the Heads of Ayr.

Diabase and basalt	57	per cent.
Older tuff	3	"
Andesite (probably from Old Red Sandstone volcanic series of the neighbourhood	14	"
Limestone (cement-stone, etc.)	20	"
Shale	3	"
Sandstone	2	"
Fossil wood	1	"
	100	

While many examples might be cited where no molten rock of any kind has risen in the vents, or where at least all the visible materials are of a fragmentary character, yet small veins and dykes of basalt have not infrequently been injected into the tuff or agglomerate. These seldom run far, and usually present a more or less tortuous course. Thus, on the south front of the Binn of Burntisland (Figs. 166, 168) a number of basalt-dykes, which vary in breadth from five or six feet to scarcely so many inches, bifurcate and rapidly disappear in the tuff, one of them ascending tortuously to near the top of the cliff. They at once recall the appearance of the well-known dykes in the great crater wall of Somma.

Though not by any means the largest or most perfect of the vents in the basin of the Firth of Forth, the Binn of Burntisland, of which a view is given in Fig. 144, may be cited in illustration of their general characters. It presents in detail some of the most strikingly volcanic aspects of scenery anywhere to be seen in that region. Consisting of a dull green granular volcanic tuff, it rises abruptly out of the Lower Carboniferous formations to a height of 631 feet above the sea. The southern edge of this neck has been so extensively denuded, that it presents steep craggy slopes and rugged precipices, which descend from the very summit of the cone to the plain below—a vertical distance of nearly 500 feet. Here and there the action of atmospheric waste has hollowed out huge crater-like chasms in the crumbling tuff. Standing in one of these, the geologist can realize what must have been the aspect of the interior of these ancient Carboniferous volcanic cones. The scene at once reminds him of the crater-walls of a modern or not long extinct volcano. The dull-green rudely stratified tuff rises around in verdureless crumbling sheets of naked rock, roughened by the innumerable blocks of lava, which form so conspicuous an element in the composition of the mass. The ribs or veins of columnar basalt run up the declivities as black shattered walls. The frosts and rains of many centuries have restored to the tuff its original loose gravelly character. It disintegrates rapidly, and rolls down the slopes in long grey lines of volcanic sand, precisely as it no doubt did at the time of its ejection, when it fell on the outer and inner declivities of the original cone. Some of these features may be partly realized from Fig. 145, which represents a portion of the south front of the hill. Sections of this neck are given in Figs. 149 and 159.

(3) *Necks of Tuff or Agglomerate with a Central Plug of Basalt or other Lava.*—It has often happened that, after the explosions in a vent have begun to decrease in vigour, or have at last ceased, lava has risen in the chimney and finally sealed it up. In such cases the main mass of the rock may consist of tuff or agglomerate, which the enfeebled volcanic activity has been unable to expel from the orifice, while a plug of basalt, dolerite, or even more basic material, of much smaller dimensions, may have risen up the pipe in the centre or towards one side. Binns Hill, West Lothian, the Beath and Saline Hills of Fife, and Tinnis Hill in Liddesdale are good examples of this structure. (See Figs. 26, 148, 149 and 174).

(4) *Necks of Basalt, Dolerite, etc.*—In other cases no fragmental material is present in the vent, or possibly traces of it may be seen here and there adhering to the walls of the funnel, the prevailing rock being some form of lava. Necks of this kind are much less frequent in the puy- than in the plateau-type. But examples may be found in several districts. The most striking with which I am acquainted are those which form so picturesque a group of isolated cones around the volcanic basin of Limerick, to be afterwards described (Figs. 195, 196). The vents there have been filled by the uprise of much more acid rocks than the lavas of the basin, for, as I have already stated, they include even quartziferous trachytes. In the basin of the Firth of Forth some prominent bosses of basalt probably mark the sites of former vents, such as Dunearn Hill in Fife, the Castle Rock of Edinburgh, and Galabraes Hill near Bathgate. Some striking vents which occur in the Jedburgh district, in the debateable land between the plateau series on the east and the puy-series on the west, show the nearly complete usurpation of the funnel by basalt, but with portions of the tuff still remaining visible.

Relation of the Necks to the Rocks through which they rise.—A remarkable feature among the Carboniferous and Permian vents of central Scotland is presented by the effect which has been produced on the strata immediately surrounding them. In the interior of the country this effect is often

FIG. 145.—View of part of the cliffs of vertical agglomerate, Binn of Burntisland.

concealed by herbage, but where the rocks have been laid bare by the sea it may be most instructively studied. In such shore-sections, a singular change of dip is often observable among the strata round the edge of a vent. No matter what may be the normal inclination at the locality, the beds are bent sharply down towards the wall of the neck, and are frequently placed on end. This structure (shown in Figs. 24, 143, 147, 148 and 149) is precisely the reverse of what might have been anticipated, and can hardly be due to upward volcanic explosions. It is frequently associated with considerable metamorphism in the disturbed strata. Shales are converted into porcellanite or various jaspery rocks, according to their composition. Sandstones pass into quartzite, with its characteristic lustrous fracture. It is common to find vents surrounded with a ring of this contact-metamorphism, which, from the hardness and frequently vertical or highly inclined bedding of its strata, stands up prominently on the beach (as in Figs. 126 and 210), and serves to mark the position of the necks from a distance.

I have not been able to find an altogether satisfactory explanation of this inward dip of the strata around vents. Taking it in connection with the metamorphism, I am inclined to believe that it arose after the close of the long-continued volcanic action which had hardened the rocks around the volcanic pipe, and as the result of some kind of subsidence within the vent. The outpouring of so much tuff and lava as escaped from many of the volcanoes would doubtless often be apt to produce cavities underneath them, and on the decay of volcanic energy there might be a tendency in the solid or cavernous column filling up the funnel, to settle down by mere gravitation. So firmly, however, did much of it cohere to the sides of the pipe, that if it sank at all, it could hardly fail to drag down a portion of these sides. So general is this evidence of downward movement in all the volcanic districts of Scotland where the necks have been adequately exposed, that the structure may be regarded as normal to these volcanic vents. It has been observed among the shore-sections of the volcanoes of the Auckland district, New Zealand. Mr. C. Heaphy, in an interesting paper upon that district, gives a drawing of a crater and lava-stream abutting on the edge of a cliff where the strata bend down towards the point of eruption, as in the numerous cases in Scotland.[1]

Evidence for the probable subaerial Character of some of the Cones or Puys of Tuff.—From the stratigraphical data furnished by the basin of the Firth of Forth, it is certain that this region, during a great part of the Carboniferous period, existed as a wide shallow lagoon, sometimes overspread with sea-water deep enough to allow of the growth of corals, crinoids, and brachiopods; at other times, shoaled to such an extent with sand and mud as to be covered with wide jungles of a lepidodendroid and calamitoid vegetation. As volcanic action went on interruptedly during a vast section of that period, the vents, though generally submarine, may occasionally have been subærial. Indeed, we may suppose that the same vent might begin as a subaqueous orifice and continue to eject volcanic materials, until, as these

[1] *Quart. Journ. Geol. Soc.* 1860, vol. xvi. p. 245.

rose above the level of the water, the vent became subærial. An instance of a submarine vent has been cited from the Perthshire coal-field (p. 426).

Among the evidence which may be collected to show that some Carboniferous volcanoes probably rose as insular cones of tuff above the surrounding waters, the structure of the tuff in many necks may be cited, for it suggests subærial rather than subaqueous stratification. The way in which the stones, large and small, are grouped together in lenticular seams may be paralleled on the slopes of many a modern volcano. Another indication of this mode of origin is supplied by the traces of wood to be met with in some of the tuff-necks. The vents of Fife and Linlithgowshire contain these traces sometimes in great abundance. The specimens are always angular fragments, and are frequently encrusted with calcite.[1] Sometimes they present the glossy fracture and clear ligneous structure shown by sticks of well-made wood charcoal. In a neck at St. Magdalen's, near Linlithgow, the wood fragments occur as numerous black chips. So far as can be ascertained from the slices already prepared for the microscope, the wood is always coniferous. These woody fragments seldom occur in the interstratified tuffs or in the associated strata where *Stigmaria, Lepidodendron*, etc., are common. They are specially characteristic of the necks and adjacent tuffs. The parent trees may have grown on the volcanic cones, which as dry insular spots would support a different vegetation from the club-mosses and reeds of the surrounding swamps. As the fragments occur in the tuffs which, on the grounds already stated, may be held to have been deposited within the crater, they seem to point to intervals of volcanic quiescence, when the dormant or extinct craters were filled with a terrestrial flora, as Vesuvius was between the years 1500 and 1631, when no eruptions took place. Some of the cones, such as Saline Hill and the Binn of Burntisland, may have risen several hundred feet above the water. Clothed with dark pine woods, they would form a notable feature in the otherwise monotonous scenery of central Scotland during the Carboniferous period.

Entombment of the Volcanic Cones and their relation to the bedded Lavas and Tuffs.—From the facts above detailed, it is evident that in most cases the necks represent, as it were, the mere denuded stumps of the volcanoes. As the puys took their rise in areas which, on the whole, were undergoing a movement of subsidence, they were eventually submerged and buried under sedimentary accumulations. Their loose ashes would be apt to be washed down and strewn over the sea bottom, so that only the lower and inner part of a cone might remain. We can hardly hope to discover any of the actual craters among these volcanic relics. The cones having been submerged and buried under many hundred feet of sediment, their present position at the surface is due to subsequent elevation and prolonged denudation. It is obvious that there must still be many buried cones which the progress of denudation has not yet reached. Some of these have been

[1] The largest I have observed is a portion of a stem about two feet long and six inches broad, in the (Permian ?) neck below St. Monan's church.

revealed in the course of mining operations. Valuable seams of coal, ironstone and oil-shale in the Scottish Carboniferous Limestone and Calciferous Sandstone series are extensively worked, and in the underground operations many illustrations of former volcanic action have been met with. The most remarkable instances of the discovery of buried volcanoes have occurred in the Dalry coal-field in the north of Ayrshire. In one pit-shaft about a mile and a half to the south-west of the village of Dalry, a thickness of 115 fathoms of tuff was passed through, and in another pit 90 fathoms of similar tuff were sunk into before the position of the black-band ironstone of that mineral field was reached by driving levels through the tuff into the sedimentary strata outside of it. Only a short distance from these thick piles of tuff, their place is entirely taken up by the ordinary sedimentary strata of the district. The working-plans of the mines show the tuff to occur in irregular patches and strips, between which the ironstone is workable. From these data we perceive that the shafts have in some cases been sunk directly upon the tops of puys of tuff, which were, in one case, nearly 700 feet, and in another instance, 540 high[1] (Fig. 146).

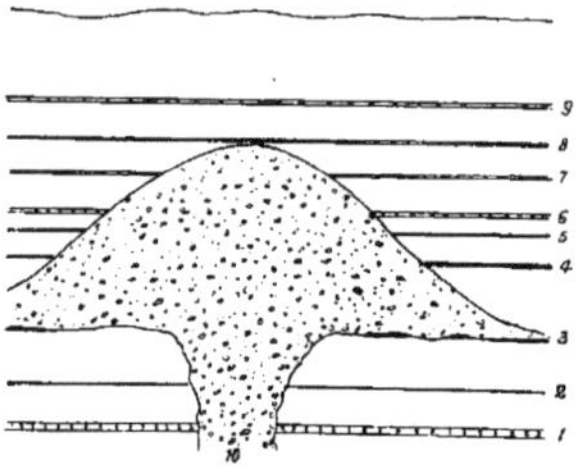

FIG. 146.—Diagram of buried volcanic cone near Dalry, Ayrshire. Constructed from information obtained in mining operations.

1. Hurlet Limestone. 2. Clayband Ironstone. 3. Black-band Ironstone. 4. Borestone Coal. 5. Wee Coal. 6. Highfield Limestone. 7 and 8. Thin Limestones. 9. Linn Limestone. 10. Volcanic neck and cone of tuff.

It is obvious that from the condition of a completely buried and concealed cone every stage may be expected to occur up to the deeply worn-down neck representing merely the stump of the volcanic column. The subjoined diagram (Fig. 147) may serve to illustrate this process of gradual re-emergence.

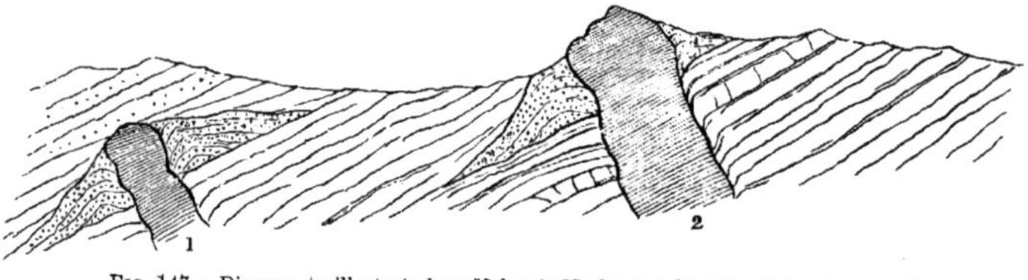

FIG. 147.—Diagram to illustrate how Volcanic Necks may be concealed and exposed.

1, Neck, still buried under the succeeding sedimentary accumulations; 2, Neck uncovered and denuded.

When, in the progress of denudation, a volcanic cone began to show itself from under the cover of removed strata, it would still for a time maintain its connection with the sheets of tuff or of lava which, when active, it had erupted. A number of examples of this structure may be observed in the basin of the Firth of Forth, where the degradation of the surface has

[1] Explanation of Sheet 22, *Geol. Surv. of Scotland*, p. 16.

not yet proceeded so far as to isolate the column of agglomerate or tuff from the sheets of tuff that were strewn around the old volcano. In such cases, the actual limits of the vent are still more or less concealed, or at least no sharp line can be drawn between the vent and its ejections. As an illustration of this connection of a volcanic pipe with the materials ejected from it over the surrounding country I would cite Saline Hill in the west of Fife. That eminence rises to a height of 1178 feet above the sea, out of a band of tuff which can be traced across the country for fully three miles. Numerous sections in the water-courses show that this tuff is regularly interbedded in the Carboniferous Limestone series, so that the relative geological date of its eruption can be precisely fixed. On the south of Saline Hill, coal and ironstone, worked under the tuff, prove that this portion of the mass belongs to the general sheet of loose ashes and dust, extending outwards from the original cone over the floor of the sheet of water in which the Carboniferous Limestone series of strata was being deposited. But the central portion of the hill is occupied by one or more volcanic pipes. A section across the eminence from north-west to south-east would probably show the structure represented in Fig. 148. Immediately to the east of the Saline Hill lies another eminence, known as the Knock Hill, which marks the site of another eruptive vent. A coal-seam (the Little Parrot or Gas Coal) is worked along its southern base, and is found to plunge down steeply towards the volcanic rocks. This seam, however, is not the same as that worked under the Saline Hill, but lies some 600 feet below it. Probably the whole of the Knock Hill occupies the place of a former vent.

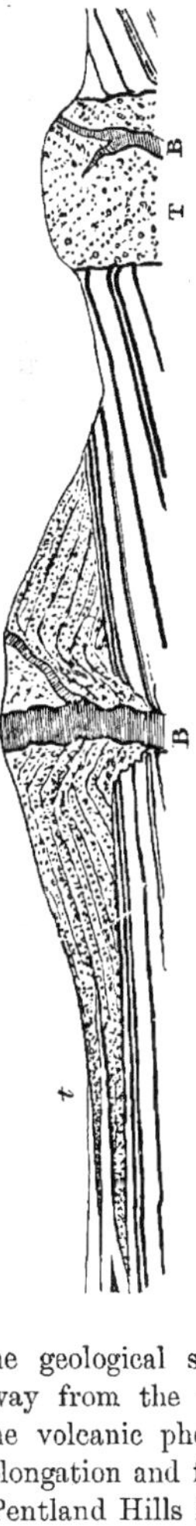

FIG. 148.—Section across the Saline Hills, Fife.

The thick parallel black lines mark the position of seams of coal and ironstone, some of which are worked under Saline Hill. T, Tuff of the necks; *t*, Tuff at a little distance from the cone, interstratified with the ordinary sedimentary beds; B, Basalt. The larger eminence is Saline Hill, the lower is Knock Hill.

A further stage of decay and denudation brings before us the entire severance of the volcanic column from the materials that were ejected from it. An excellent example of this isolation of the neck in the midst of surrounding masses of tuff and lava which proceeded from it is presented by the Binn of Burntisland, to which I have already alluded. A section across that eminence gives the geological structure represented in Fig. 149. The dip of the rocks away from the volcanic pipe at this locality has been produced long after the volcanic phenomena had ceased. The arch here shown is really the prolongation and final disappearance of the great anticlinal fold of which the Pentland Hills form the axis on the opposite side of the Firth. But if we restore the rocks to a horizontal, or approximately horizontal position, we find the Binn of Burntisland rising among them in one or more necks, which doubtless mark centres of volcanic activity in that district. A series of smaller neck-like eminences runs for two miles westward.

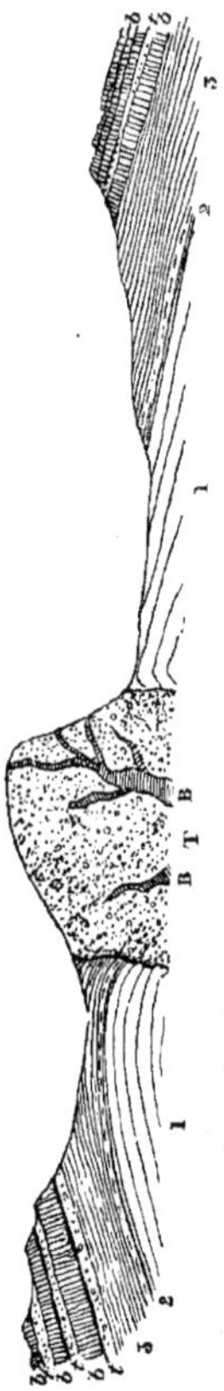

FIG. 149.—Section across the Binn of Burntisland, in an East and West direction. 1, Sandstones · 2, Limestone (Burdiehouse) · 3, Shales, etc.; *b*, *b*, Interstratified basalts; *t t*, Bedded tuff, etc.; T, Tuff of the great neck of Burntisland; B, Basalt veins.

Striking as the forms of many of the necks are, and much as their present conical forms resemble those of active and extinct volcanoes, the evidence of extensive denudation proves that these contours are not the original outlines of the Carboniferous vents, but are in every case the result of prolonged waste. What we now see is a section of the volcanic chimney, and the conical form is due to the way in which the materials filling the chimney have yielded to the forces of denudation.

ii. BEDDED TUFFS AND LAVAS

During at least the earlier part of the period of the puys, in some districts or from certain vents, such as those of East Fife, Western Midlothian, Eastern Linlithgowshire, Northern Ayrshire, Heads of Ayr and Lower Eskdale, only fine tuff seems to have been thrown out, which we now find intercalated among the surrounding strata. These eruptions, neither so vigorous nor so long-continued as those of the plateaux, never gave forth such thick and widespread sheets of fragmentary materials as those associated with the plateaux in East Lothian and the north-east of Ayrshire. A single discharge of ashes seems in many cases to have been the sole achievement of one of those little volcanoes; at least only one thin band of tuff may be discoverable to mark its activity.

The tuff of these solitary bands is seldom coarse in texture. It usually consists of the ordinary dull green paste, with dust and lapilli of basic pumice. The local variations in the tuffs of the puys generally arise mainly from differences in the composition, size and numbers of the included ejected blocks. Generally the most abundant stones are pieces of different diabases, or basalts; then come fragments from the surrounding Carboniferous strata, from older tuffs and rarely from rocks of much deeper-seated origin.

Now and then the eruptions of tuff have consisted of extremely fine volcanic dust, which, mingling with water, took the form of a compact mudstone, as in the case of the Houston Marls (p. 423), which remind one of a volcanic mud. But in most localities the discharge of tuff, though for a time it may have completely obscured the ordinary contemporaneous sedimentation, was intermittent, so that in the intervals between successive showers of detritus, the deposition of non-volcanic sediment went on as usual. Hence it is that bands of tuff, whether they lie among lavas or among sedimentary formations, are apt to contain interstratifications of sandstone, shale, limestone or other detrital deposit, and to pass insensibly into these. The extremely gentle gradation from volcanic into non-volcanic sediment, and the occasional re-appearance of thin partings of tuff bring vividly before the mind the slow dying out of volcanic energy among the Carboniferous lagoons.

The comparatively quiet character of the volcanic explosions, and the contemporaneous undisturbed deposition of sediment during the earlier part of the puy period, are exemplified in many sections throughout the areas above enumerated, as will be more fully illustrated in subsequent pages. Two typical examples may suffice for this general statement of the characters of the discharges of tuff in the puy-eruptions. In the Linlithgowshire quarry represented in Fig. 150, where about ten feet of strata have been exposed, a black shale (1) of the usual carbonaceous character, so common in the Oil-shale series of this region, may be seen at the bottom of the section. It is covered by a bed of nodular bluish-grey tuff (2) containing black shale fragments. A second black shale (3) is succeeded by a second thin band of fine pale yellowish tuff (4). Black shale (5) again supervenes, containing rounded fragments of tuff, perhaps ejected lapilli, and passing up into a layer of tuff (6). It is evident that we have here a continuous deposit of black shale which was three times interrupted by showers of volcanic dust and stones. At the close of the third interruption, the deposition of the shale was renewed and continued, with sufficient slowness to permit of the segregation of thin seams and nodules of clay ironstone round the decomposing organic remains of the muddy bottom (7). A fourth volcanic interlude now took place, and the floor of the water was once more covered with tuff (8). But the old conditions of deposit were immediately afterwards resumed (9); the muddy bottom was abundantly peopled with ostracod crustaceans, while many fishes, whose coprolites have been left in the mud, haunted the locality. At last, however, a much more serious volcanic explosion took place. A coarse agglomeratic tuff (10), with blocks sometimes nearly three feet in diameter, was then thrown out, and overspread the lagoon.[1]

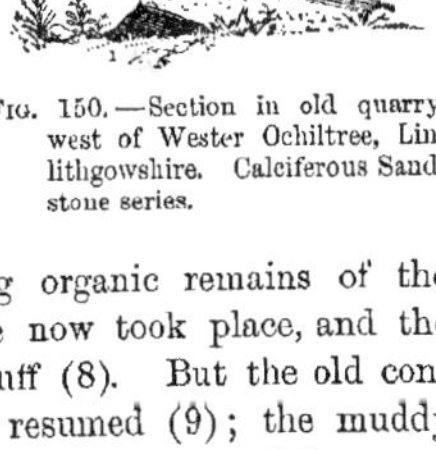

FIG. 150.—Section in old quarry, west of Wester Ochiltree, Linlithgowshire. Calciferous Sandstone series.

The second illustration may be taken from the admirable coast-section between Burntisland and Kinghorn, where the number of intercalations of

[1] See *Geol. Surv. Memoir of Edinburgh*, p. 45. These tuffs are further described on pp. 465 *et seq.*

tuff is very great. Besides thicker well-marked bands, successive innumerable thin layers occur there among the associated zones of sedimentary strata which separate the sheets of basalt. The character of these tuff-seams may be inferred from the following details of less than two feet of rock at Pettycur Point:—

Tuff	1·5	inch.
Limestone	0·2	"
Tuff	0·5	"
Shale	0·2	"
Tuff	0·1	"
Shale and tuff	1·0	"
Shale	0·2	"
Limestone	0·5	"
Shale full of volcanic dust	3·5	"
Shaly limestone	1·5	"
Laminated tufaceous limestone	2·0	"
Limestone in thin bands, with thin laminæ of tuff	0·8	"
Granular tuff	0·6	"
Argillaceous limestone, with diffused tuff	0·9	"
Fine granular tuff	0·7	"
Argillaceous limestone, with diffused tuff	1·5	"
Laminated limestone	0·1	"
Limestone, with parting of granular tuff in middle	0·9	"
Tufaceous shale	2·0	"
Limestone	0·4	"
Shaly tuff	1·25	"
Laminated limestone	0·1	"
Tuff	1·2	"
	21·65	inches.

Such a section as this brings vividly before the mind a long-continued intermittent feeble volcanic action during pauses between successive outbursts of lava. In such intervals of quiescence, the ordinary sediment of the lagoons accumulated, and was mixed up with the debris supplied by occasional showers of volcanic dust. In this Fife volcanic series, thin layers of sandstone, streaked with remains of the Carboniferous vegetation; beds of shale full of cyprid-cases, ganoid scales, and fragmentary ferns; thin beds of limestone, and bands of fire-clay supporting seams of coal, are interleaved with strata of tuff and sheets of basalt. Now and then a sharp discharge of larger stones is seen to have taken place, as in the case of the block many years ago described by me as having fallen and crushed down a still soft bed of coal (Fig. 151).[1]

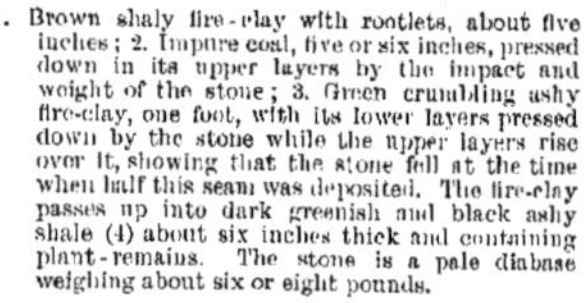

FIG. 151.—Ejected volcanic block in Carboniferous strata, Burntisland.

1. Brown shaly fire-clay with rootlets, about five inches; 2. Impure coal, five or six inches, pressed down in its upper layers by the impact and weight of the stone; 3. Green crumbling ashy fire-clay, one foot, with its lower layers pressed down by the stone while the upper layers rise over it, showing that the stone fell at the time when half this seam was deposited. The fire-clay passes up into dark greenish and black ashy shale (4) about six inches thick and containing plant-remains. The stone is a pale diabase weighing about six or eight pounds.

[1] *Geol. Mag.* vol. i. p. 22. This Fife coast-section is given in full at p. 470.

FIG. 152.—View of volcanic agglomerate becoming finer above. East end of Kingswood Craig, two miles east from Burntisland.

The Fife coast-section from which these details are taken supplies almost endless instances of the varying characters of the pyroclastic materials of the puy-eruptions. The very same cliff, bank or reef will show at one point an accumulation of excessively coarse volcanic debris and at another thin laminæ of the finest dust and lapilli. These rapid gradations are illustrated in Fig. 152, which is taken from the east end of the Kingswood Craig. The lower part of the declivity is a coarse agglomerate which passes upward into finer tuff.

Besides the thin partings and thicker layers of tuff which, intercalated among the sedimentary strata of the Carboniferous system, mark a comparatively feeble and intermittent volcanic activity, we meet in some localities with examples where the puys have piled up much thicker accumulations of fragmentary material without any intercalated streams of lava, or interstratified sandstone, shale or limestone. Thus the widespread Houston marls above described reach a thickness of some 200 feet. The vents of the Saline Hills in Fife covered the sea-floor with volcanic ashes to a depth of several hundred feet. In the north of Ayrshire the first eruptions of the puys have formed a continuous band of fine tuff traceable for some 15 miles, and in places at least 200 feet thick.

Where volcanic energy reached its highest intensity during the time of the puys, not only tuffs but sheets of lava were emitted, which, gathering round the vents, formed cones or long, connected banks and ridges. Of these there are four conspicuous examples in Scotland—the hills of the Burntisland district, the Bathgate Hills, the ground between Dalry and Galston in north Ayrshire, and a broken tract in Liddesdale. Nowhere in the volcanic history of this country have even the minutest details of that history been more admirably preserved than among the materials erupted from puys in these respective districts.

Lava-cones, answering to solitary tuff-cones among the fragmental eruptions, do not appear to have existed, or, like some of those in the great lava-fields of Northern Iceland and Western America, must have been mere small heaps of slag and cinders at the top of the lava-column, which were washed down and effaced during the subsidence and entombment of the volcanic materials. The lavas never occur without traces of fragmentary discharges. Two successive streams of basalt may indeed be found at a given locality without any visible intercalation of tuff, but proofs of the eruption of fragmental material will generally be observed to occur somewhere in the neighbourhood, associated with one or both of them, or with other lavas above or below them.

Where the phenomena of the puys have been most typically developed, lavas and tuffs succeed each other in rapid succession, with numerous or occasional interstratifications of ordinary sediment. Perhaps the most complete and interesting example of this association is to be found on the coast between Burntisland and Kirkcaldy, where, out of a total thickness of rock which may be computed to be between 1500 and 2000 feet, it will probably be a fair estimate to say that the igneous materials constitute four-fifths, or from

FIG. 153.—Alternations of basalt and tuff with shale, etc., Kingswood Craig, Burntisland.

1200 to 1600 feet. The lavas are varieties of basalt ranging in character from a black compact columnar to a dirty green earthy cellular or slaggy rock. Each separate flow may be on the average about 20 or 30 feet in thickness. Columnar and amorphous sheets succeed each other without any interposition of fragmentary material (Fig. 171). But along the junctions of the separate flows layers of red clay, like the bole between the basalts of the Giant's Causeway, may frequently be noticed. The characteristic slaggy aspect of the upper parts of these ancient *coulées* is sometimes remarkably striking. The full details of this most interesting section will be given in later pages (p. 470). But some of its more characteristic external features may be understood from the views which are presented in Figs. 152, 153, 170, 171.

The general bedded character of the volcanic series is well shown in Fig. 153, which represents the alternations of lavas and tuffs in the Kingswood Craig two miles to the east of Burntisland. The harder basalts will be seen to project as bold crags while the tuffs and other stratified deposits between them give rise to grassy slopes and hollows. A nearer view of the alternation of lavas and tuffs with non-volcanic sedimentary deposits is supplied in Fig. 170, which is taken from a part of the Fife coast a little further to the east than the last illustration. Here one of the limestones of the Carboniferous Limestone series is overlain with shale and tuff, which, being easily disintegrated, have been cut away by the waves, leaving the lava above to overhang and fall off in blocks. The columnar structure of some of the basalts of this coast is well brought out in Fig. 171, which shows further how the columns sometimes merge into an amorphous part of the same sheet.

These Fife basalts illustrate admirably the peculiarities of the sheets of lava which are intercalated among the Carboniferous strata. They show how easy it generally is to discriminate between such sheets and intrusive sills. The true lavas are never so largely crystalline, nor spread out in such thick sheets as the sills; they are frequently slaggy and amygdaloidal, especially towards the top and bottom, the central portion being generally more fine-grained and sometimes porphyritic. Where most highly cellular they often decompose into a dull, earthy, dirty-green rock. Where they form a thick mass they are usually composed of different beds of varying texture. Except the differences between the more compact centre and the slaggy layer above and below, the bedded lavas do not present any marked variation in composition or structure within the same sheet. A striking exception to this rule, however, is furnished by the Bathgate "leckstone" already described.[1] This mass forms a continuation of the great basaltic ridge of the Bathgate Hills, and though its exact relations to the surrounding strata are concealed, it appears to be an interbedded and not an intrusive sheet. The remarkable separation of its constituent minerals into an upper, lighter felspathic layer, and a lower, heavier layer, rich in olivine, augite and iron-ores, is a structure which might be more naturally expected to occur in

[1] *Trans. Roy. Soc. Edin.* xxix. (1879) p. 504.

a sill. An instance of its development in an undoubted sill will be described further on. Nevertheless, if we follow the trend of the volcanic band of the Bathgate Hills southward for only two miles beyond the picrite quarry, we find in the Skolie Burn a rock in many respects similar, and quarried for the same purpose of building oven-soles. This "leckstone" is

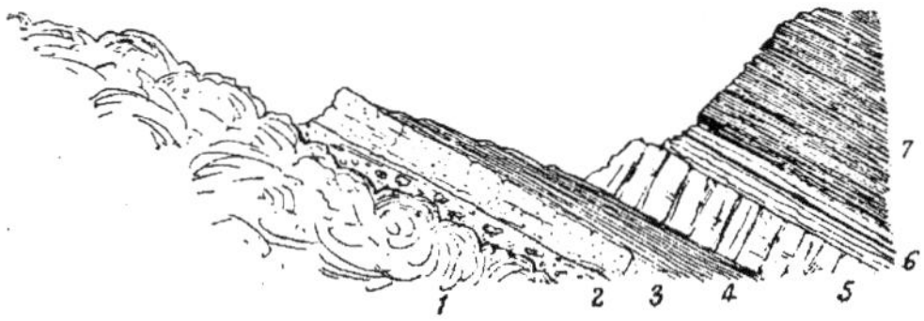

FIG. 154.—Section of the upper surface of a diabase ("leckstone") sheet, Skolie Burn, south-east of Bathgate.

1. Slaggy diabase; 2. Green sandy shale and shaly sandstone containing *Lingulæ*, also pieces of slag from the underlying lava, which are completely wrapped round in the sediment; 3. Yellow calcareous shelly sandstone; 4. Dark shale with *Spiriferæ*, etc.; 5. Bed of blue crinoidal limestone; 6. Clays and thin coal; 7. Black and blue calcareous shales and thin limestones.

there seen to be surmounted by a group of calcareous shales and thin limestones. The section laid bare in the stream is represented in Fig. 154. Immediately above the diabase, which is highly cellular, lies a green felspathic sandstone or shale containing detached fragments of the amygdaloid together with *Lingulæ* and other shells. There seems no reason to doubt that this is a true interstratified lava.[1]

Where the puys attained their greatest development in Scotland, they rose in the shallow lagoons, and here and there from deeper parts of the sea-bottom, until by their successive discharges of lavas and tuffs they gradually built up piles of material, which, in the Linlithgow and Bathgate district, may have been nearly 2000 feet in thickness. It must be remembered, however, that the eruptions took place in a subsiding area, and that even the thickest volcanic ejections, if the downward movement kept pace with the volcanic activity, need not have grown into a lofty volcanic hill. Indeed, largely as the lavas and tuffs bulk in the geology of some parts of Central Scotland, their eruption does not seem to have seriously interfered with the broader physical changes that were in progress over the whole region. Thus the subsidence which led to the spread of a marine and limestone-making fauna over much of Central Scotland included also the volcanic districts. The limestones, formed of crinoids, corals and other marine organisms, extended over the submerged lavas and tuffs, and were even interstratified with them.

While the volcanic materials are found to replace locally the ordinary Carboniferous sedimentary strata, it is interesting in this regard to note that, during pauses in the volcanic activity, while the subsidence doubtless was still going on, some groups of sandstones, shales or limestones extended themselves across the volcanic ridges so as to interpose, on more than one

[1] *Trans. Roy. Soc. Edin.* xxix. (1879), pp. 505-507.

FIG. 155.—Section across the volcanic ridge of the Linlithgow and Bathgate Hills, showing the intercalation of limestones that mark important stratigraphical horizons.

1. Houston Coal; 2. Houston Marls and tuffs; 3. Interstratified sheets of basic lavas with occasional tuffs and intercalations of shale, sandstone, etc.; 4. Tartraven Limestone; 5. Hurlet Limestone with tuffs, shales and sandstones above and below; 6. Wardlaw Limestone; 7. Index Limestone; 8. Highest band of tuff—upward limit of the volcanic series; 9 9. Volcanic necks; 10. Sill of basalt; 11. Levenseat or Castlecary Limestone; 12. Millstone Grit; 13. Base of Coal-measures; 14. Thick doleritic sill; 15. Dolerite dyke (? Tertiary).

platform, a mass of ordinary sediment between the lavas or tuffs already erupted and those of succeeding discharges, and thus to furnish valuable geological chronometers by which to define the stratigraphical horizons of the successive phases of volcanic energy.

The volcanic banks or ridges not improbably emerged as islets out of the water, and were sometimes ten miles or more in length. Their materials were supplied from many separate vents along their surface, but probably never attained to anything approaching the elevation which they would have reached had they been poured out upon a stable platform. This feature in the history of the volcanic ridges is admirably shown by the fact just referred to, that recognizable stratigraphical horizons can sometimes be traced right through the heart of the thickest volcanic accumulations. One of the largest areas of basalts and tuffs connected with the puys is that of the Linlithgow and Bathgate Hills, where, as already remarked, a depth of some 2000 feet of igneous rocks has been piled up. Yet several well-known seams of stone can be traced through it, such as the Hurlet Limestone and the Index Limestone (Fig. 155). Only at the north end, where the volcanic mass is thickest and the surface-exposures of rock are not continuous, has it been impossible to subdivide the mass by mapping intercalations of sedimentary strata across it. It would thus seem that, even where the amplest accumulations gathered round the puys, they formed low flat domes, rather than prominent hills, which, as subsidence went on and the tuff-cones were washed down, gradually sank under water, and were buried under the accumulating silt of the sea-floor.

As a detailed illustration of the manner in which the growth of organically-formed limestones and the deposit of ordinary sediment took place concurrently with the occasional outflow of lava-streams over the sea-bottom, I may cite the section presented in another Linlithgowshire quarry (Fig. 156). At the bottom of the group of strata there exposed, a pale amygdaloidal, somewhat altered basalt (A) marks the upper surface of one of the submarine lavas of the period. Directly over it comes a bed of limestone (B) 15 feet thick, the lower layers of which are made up of a dense growth of the thin-stemmed coral *Lithostrotion irregulare*. The next stratum is a band of dark shale (C) about two feet thick, followed by about the same thickness of an impure limestone with shale seams (D). The conditions for coral and crinoid growth were evidently not favourable, for this argillaceous limestone was eventually arrested first by the deposit of a dark mud, now to be seen in the form of three or four inches of a black pyritous shale (E), and next by the inroad of a large quantity of dark sandy mud and drift vegetation, which has been preserved as a sandy shale (F), containing *Calamites*, *Producti*, ganoid scales and other traces of the life of the time. Finally, a great sheet of lava, represented by the uppermost amygdaloid (G), overspread the area, and sealed up these records of Palæozoic history.[1]

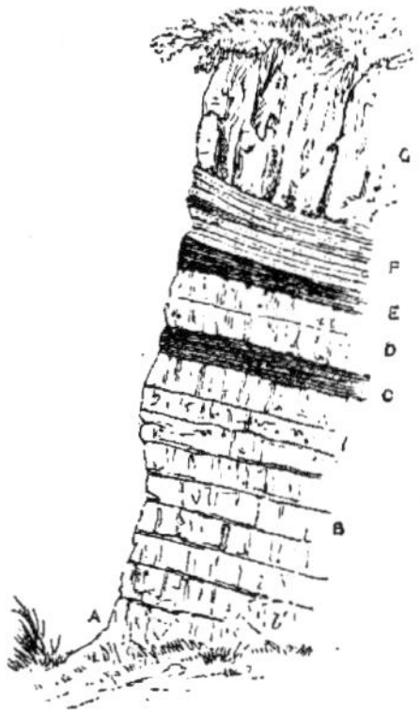

FIG. 156. — Section in Wardlaw Quarry, Linlithgowshire.

Among the phenomena associated with the Carboniferous volcanoes mention may, in conclusion, be made of the evidence for the former existence of thermal springs and saline sublimations or incrustations. Among the plateau-tuffs of North Berwick, as has been already pointed out (p. 390), a fœtid limestone has been quarried, which bears indications of having been deposited by springs, probably in connection with the volcanic action of the district. The lower limestones of Bathgate furnish abundant laminæ of silica interleaved with calcareous matter, the whole probably due to the action of siliceous and calcareous springs connected with the active puys of that district. Some portions of the limestone are full of cellular spaces, lined with chalcedony.[2] A saline water has been met with among the volcanic rocks to the west of Linlithgow, in a bore which was sunk to a depth of 348 feet in these rocks without reaching their bottom. The water that rose from the bore-hole was found to contain as much as 135 grains of chloride of sodium in the gallon. It is not improbable that this salt was originally produced by incrustations on the Carboniferous lavas immediately after their eruption, as has happened so often in recent times at Vesuvius, and that it was then buried under succeeding showers of tuff and streams of lava.[3]

Subsequent Dislocation of Bedded Lavas and Tuffs.—As the interstratified volcanic materials were laid down in sheets at the surface, they necessarily behave like the ordinary sedimentary strata, and have undergone with them the various curvatures and fractures which have occurred since Carbon-

[1] *Geol. Surv. Mem.* "Geology of Edinburgh," p. 58.

[2] *Ibid.* p. 49, *et seq.*

[3] *Proc. Roy. Soc. Edin.* vol. ix. p. 367. Besides chloride of sodium the water contained also chlorides of calcium, magnesium and potassium, carbonates of lime and magnesia, sulphate of lime, and other ingredients in minute proportions.

iferous times. Notwithstanding their volcanic nature, they can be traced and mapped precisely as if they had been limestones or sandstones. This perfect conformability with the associated stratified rocks is strikingly seen in the case of the sheets of lava which lie imbedded in the heart of the great volcanic ridge of Linlithgowshire. The overlying strata having been

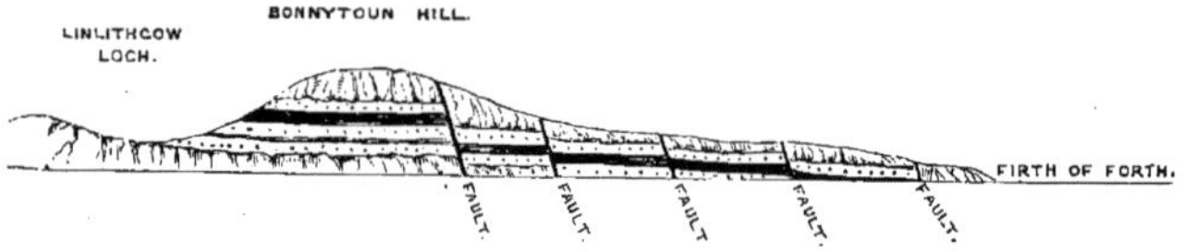

FIG. 157.—Section from Linlithgow Loch to the Firth of Forth.

removed from their surface for some distance, and the ground having been broken by faults, these volcanic rocks might at first be taken for irregular intrusive bosses, but their true character is that shown in Fig. 157, where by a succession of faults, with a throw in the same direction, the upper basalts of Bonnytoun Hill are gradually brought down to the level of the Firth of Forth.

iii. SILLS, BOSSES AND DYKES

One of the characteristic features of Central Scotland is the great number, and often the large size and extraordinary persistence, of the masses of eruptive, more or less basic material, which have been injected among the Carboniferous strata. The precise geological age of these intrusions cannot, of course, be more exactly defined than by stating that they are younger than the rocks which they traverse, though in many cases their association with the necks, lavas and tuffs is such as to show that they must be regarded as part of the Carboniferous volcanic phenomena.

SILLS.—With regard to the sills I have been led, for the following reasons, to connect the great majority of them with the puys, though some are certainly of far later date, while others should possibly be assigned to the plateaux.

In the first place, the sills obviously connected with the plateaux are in great measure intermediate, or even somewhat acid rocks, while those of the puy series are much more basic. It is hardly possible, however, in all cases to decide to which series a particular sill should be assigned. This difficulty is particularly manifest in the western part of Midlothian, where the plateau of that district exhibits such frequent interruption, and where it often consists only of a single basaltic sheet. To the west of it lie the abundant puys with their lavas and tuffs, and between the two volcanic areas numerous sills of dolerite and diabase make their appearance. In the difficulty of deciding to which series these sills should be referred, it will be convenient to consider them with those of the puys.

A remarkable illustration of the contrast in petrographical character between the typical sills of the plateaux and those of the puys is furnished by the chain of the Campsie Fells, where, on the north side, among the Calciferous Sandstones which emerge from under the andesitic lavas of the Clyde plateau, many intrusive sheets and bosses of trachytic material may be seen, while on the southern side come the great basic sills which, from Milngavie by Kilsyth to Stirling, run in the Carboniferous Limestone series (Fig. 158). A similar contrast may be observed in Renfrewshire between the trachytic sills below the plateau-lavas south of Greenock and the basic sills above these lavas in the Carboniferous Limestone series around Johnstone and Paisley.

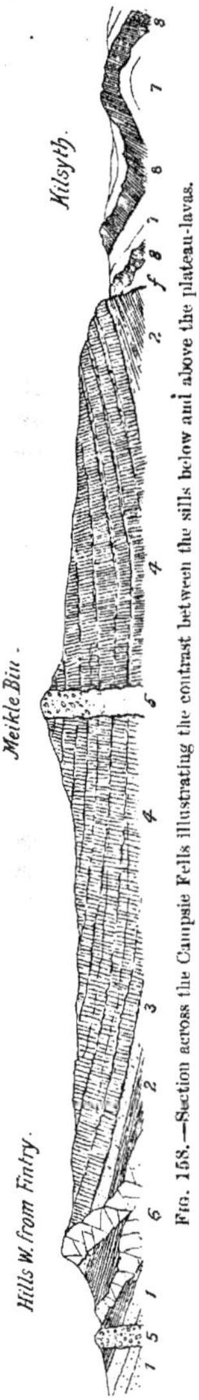

FIG. 158.—Section across the Campsie Fells illustrating the contrast between the sills below and above the plateau-lavas.

1. Upper Old Red Sandstone; 2. "Ballagan Beds"; 3. Tuffs; 4. Lavas of the Campsie district of the Clyde plateau; 5 5. Necks belonging to the plateau volcanic series; 6. Trachytic sills belonging to the plateau; 7. Carboniferous Limestone series; 8. Dolerite sills cutting the Carboniferous Limestone series. *f*, Fault.

In the second place, the more basic sills, as a rule, appear on platforms higher in stratigraphical position than the plateaux, and wherever this is their position there cannot be any hesitation in deciding against their association with the older phase of volcanic activity.

In the third place, the basic sills often occur in obvious connection with the vents or bedded lavas and tuffs of the puy series. A conspicuous example of this dependence is supplied by the intrusive sheets of Burntisland, underlying the basalts and tuffs of that district in the immediate neighbourhood of some of the vents from which these bedded rocks were erupted (Fig. 159).

In the fourth place, even where no visible vents appear now at the surface near the sills, the latter generally occupy horizons within the stratigraphical range indicated by the interbedded volcanic rocks. It must be remembered that all the Carboniferous vents were deeply buried under sedimentary deposits, and that large as is the number of them which has been exposed by denudation, it is probably much smaller than the number still concealed from our view. The sills are to be regarded as deep-seated parts of the volcanic protrusions, and they more especially appear at the surface where the strata between which they were injected crop out from under some of the higher members of the Carboniferous system. Thus the remarkable group of sills between Kilsyth and Stirling (Fig. 158) may quite possibly be connected with a group of vents lying not far to the eastward, but now buried under the higher parts of the Carboniferous Limestone, Millstone Grit and Coal-measures. Again, the great series of sills that gives rise to such a conspicuous range of hills in the north and middle of Fife may have depended for its origin upon the

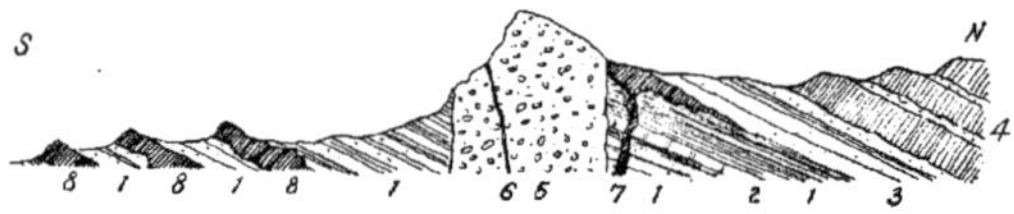

FIG. 159.—Section showing the position of the basic sills in relation to the volcanic series at Burntisland, Fife.

1. Calciferous Sandstone series; 2. Burdiehouse Limestone; 3. Sandstones, shales and tuffs; 4. Basalts and tuffs, with intercalations of sandstone, shale and limestone; 5. Agglomerate of the Binn of Burntisland neck; 6. Basalt dyke; 7. Dyke and sill; 8 8 8. Three sills.

efforts of a line of vents running east and west through the centre of the county, but now buried under the Coal-measures. Some vents, indeed, have been laid bare in that district, such as the conspicuous groups of the Saline Hills and the Hill of Beath, but many more may be concealed under higher Carboniferous strata further east.

In the fifth place, the materials of which the sills consist link them in petrographical character with those that proceeded from the puys. The rocks of the intrusive sheets in West Lothian, Midlothian and Fife are very much what an examination of the bedded lavas of the puys in the same region would lead us to expect. There is, of course, the marked textural difference between masses of molten rock which have cooled very slowly within the crust of the earth and those which have solidified with rapidity at the surface, the sills being for the most part much more coarsely crystalline than the lavas, and more uniform in texture throughout, though generally finer at the margins than at the centre. There is likewise the further contrast arising from differences in the composition of the volcanic magma at widely-separated periods of its extravasation. At the time when the streams of basalt flowed out from the puys its constitution was comparatively basic, in some localities even extremely basic. Any sills dating from that time may be expected to show an equal proportion of bases. But those which were injected at a long subsequent stage in the volcanic period may well have been considerably more acid.

In actual fact the petrographical range of the sills reasonably referable to the puy-eruptions varies from picrite or limburgite to dolerite without olivine. The great majority of these sheets in the basin of the Firth of Forth, where they are chiefly displayed, are dolerites (diabases), sometimes with, but more frequently without, olivine. They include all the more coarsely crystalline rocks of the region, though occasionally they are ordinary close-grained basalts. Their texture may be observed to bear some relation to their mass, so far at least as that, where they occur in beds only two or three feet or yards in thickness, they are almost invariably closer-grained. A cellular or amygdaloidal texture is seldom to be observed among them, and never where they are largely crystalline. This texture is most often to be found in thin sills which have been injected among carbonaceous shales or coals. These intrusive sheets are generally finely cellular, and more or less decayed ("white trap").

Differences of texture may often be observed within short distances in the same sill, and likewise considerable varieties in colour and composition. The most finely crystalline portions are, as usual, those along the junction with the stratified rocks, the most crystalline occurring in the central parts of the mass. A diminution in the size of the crystalline constituents may be traced not only at the base, but also at the top of a sheet, or at any

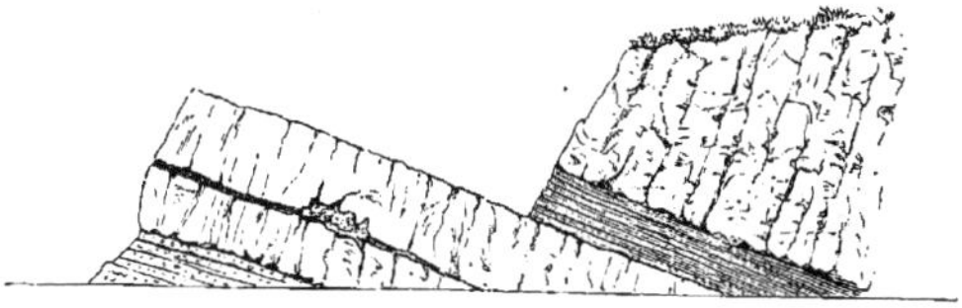

FIG. 160.—Sills between shales and sandstones, Hound Point, Linlithgowshire.

intermediate portion which has come in contact with a large mass of the surrounding rock. A good illustration is supplied by the intrusive sheet at Hound Point (Fig. 160), to the east of South Queensferry, where some layers of shale have been involved in the igneous rock, which becomes remarkably close-grained along the junction.[1] This change in texture and absence of cellular structure form a well-marked distinction between these sheets and those which have flowed out at the surface as true lava-streams.

Some of the larger doleritic sills display a somewhat coarsely crystalline texture in their central portions, and occasionally present a notable micropegmatitic aggregate, which plays the part of interstitial substance enclosing the other minerals. Mr. Teall has referred to the frequent occurrence of this structure in the coarser parts of the Whin Sill of the north of England.[2] It occurs also in a marked degree in the Ratho sill and in some portions of the great doleritic sill of which the crags of Stirling form a part.[3]

But beside the differences in texture, mainly due to varying rates of cooling, the sills sometimes exhibit striking varieties of composition in the same mass of rock. These variations are more especially noticeable among the larger sills, and particularly where the material is most markedly basic. The special type of differentiation, so noticeable in the Bathgate diabase and picrite mass already alluded to, is likewise well exhibited in an intrusive sheet or group of sheets, recently exposed at Barnton, in the cutting of a railway from Edinburgh to Cramond[4] (Fig. 161). The intrusive nature of the

[1] See Hay Cunningham's "Essay," p. 66, and plate ix.; and *Geol. Survey Memoir* on "Geology of Edinburgh," p. 114.

[2] *British Petrography*, p. 208.

[3] Mr. H. W. Monckton. *Quart. Journal Geol. Soc.* vol. li. (1895), p. 482.

[4] This rock has been described by Mr. J. Henderson and Mr. Goodchild, *Trans. Geol. Soc. Edin.* vi. (1893) pp. 297, 301, and by Mr. H. W. Monckton, *Quart. Journ. Geol. Soc.* l. (1894)

several bands of igneous rock which occur here is made quite evident by the alteration they have produced upon the shales with which they have come in contact. It is the uppermost and most extensive of these sills which specially deserves notice, for the differentiation of its constituents. It stretches along the cutting for several hundred yards at an angle of dip of about 15°. At the western or upper part of the mass its actual contact with the superincumbent sedimentary strata is not visible, but as the igneous rock is there a good deal finer in grain than elsewhere, its upper surface cannot be many feet distant. The upper visible portion is a light

FIG. 161.—Section of Sill, Cramond Railway, Barnton, near Edinburgh.

1. Baked shale; 2. Sill of very felspathic dolerite about nine feet thick; 3. Baked shale, eight inches; 4. Dolerite showing chilled fine-grained edge and adhering firmly to the shale below; it rapidly passes up into (5) Picrite with white felspathic veins (6); 7. Junction of picrite and dolerite with a similar vein along the line of contact; 8. Large globular body of dolerite enclosing a mass of picrite.

well-crystallized dolerite with a rudely bedded structure, the planes dipping westwards at 15°. About 20 or 30 feet below the upper visible termination of the mass, the dark ferro-magnesian minerals begin rapidly to increase in relative proportion to the pale felspar, and the rock consequently becomes dark-greenish brown. The change is particularly noticeable in certain bands which run parallel with the general dip. There is no definite line between the pale and dark body of the rock, the two graduating into each other and the darker part becoming deeper in colour, heavier and more decomposing, until it becomes a true typical picrite. Even in this ultra-basic portion the same rude bedding or banding may be observed.

Veins in which felspar predominates over the darker minerals traverse the rock, sometimes parallel with the bedding, sometimes across it. They vary from less than an inch to a foot in width, sometimes dividing and enclosing parts of the surrounding mass. But that they are on the whole contemporaneous with the sill itself, and not long subsequent injections, is shown by the way in which the dark ferro-magnesian minerals project from the picrite into the veins and lock the two together.

But besides these injections, which doubtless represent the last and more acid portions of the magma injected into the basic parts before the final consolidation of the whole, there are to be observed irregular concretionary patches, of similar character to the veins, distributed through the picrite. On the other hand, towards its base the sill becomes a coarse dolerite round which the picrite is wrapped, and which encloses a detached portion of that rock.

It is deserving of note that while the ultra-basic portion descends almost to the very bottom of the sill, the lowest five feet show the same

p. 39. Mr. Goodchild recognized the occurrence of picrite, and Mr. Monckton has described the succession of rocks, and given a diagram of them.

change as occurs at the top of the mass. There the felspar rapidly begins to predominate over the darker minerals, and the dolerite into which the rock passes shows a fine-grained margin adhering firmly to the shales on which it rests. This lower doleritic band, showing as it does the effect of chilling upon its under surface, may be due to more rapid cooling and crystallization, while in the overlying parts the mass remained sufficiently mobile to allow of a separation of the heavier minerals from the felspars, which appear in predominant quantity towards the top. It must be frankly admitted, however, that we are still very ignorant of the causes which led to this separation of ingredients in a few sills, while they were entirely absent or non-efficient in most of them.

The intrusive character of the Carboniferous sills of Central Scotland and their contact-metamorphism have been fully described, and some of them have become, as it were, "household words" in geology.[1] Exposed in so many fine natural sections in the vicinity of Edinburgh, they early attracted the notice of geologists, and furnished a battle-ground on which many a conflict took place between the Plutonist and Neptunist champions at the beginning of the present century.

As the sills frequently lie in even sheets perfectly parallel with the bedding of the strata between which they have been injected, care is required in some cases to establish that they are of intrusive origin. One of the most obvious tests for this purpose is furnished by the alteration they produce among the strata through which they have made their way, whether these lie above or below them. The strata are sometimes crumpled up in such a manner as to indicate considerable pressure. They are occasionally broken into fragments, though this may have been due rather to the effects of gaseous explosions than to the actual protrusion of melted rock. But the most frequent change superinduced upon them is an induration which varies greatly in amount even along the edge of the same intrusive sheet. Sandstones are hardened into quartzite, breaking with a smooth clear glistening fracture. Coals are converted into a soft sooty substance, sometimes into anthracite. Limestones acquire a crystalline saccharoid structure. Shales pass generally into a kind of porcellanite, but occasionally exhibit other types of contact-metamorphism. Thus below the thick picrite sill at Barnton, near Edinburgh, the shales have assumed a finely concretionary structure by the appearance in them of spherical pea-like aggregates.

Another proof of intrusion is to be found in the manner in which sills catch up and completely enclose portions of the overlying strata. The well-known examples on Salisbury Crags (Fig. 162) are paralleled by scores of other instances in different parts of the same region.

Moreover, sills do not always remain on the same horizon; that is,

[1] See, for instance, Maclaren's *Geology of Fife and the Lothians*, 1839; Hay Cunningham's *Essay*, previously cited; *Geological Survey Memoir on the Geology of Edinburgh* (Sheet 32), 1861; Mr. Allport, *Quart. Journ. Geol. Soc.* vol. xxx. (1874) p. 553; Teall, *British Petrography*, p. 187; E. Stecher, *Contacterscheinungen an schottischen Olivindiabasen*, Tschermak's *Mineralog. Mittheil.* vol. ix. (1887) p. 145; *Proc. Roy. Soc. Edin.* vol. xv. (1888) p. 160.

between the same strata. They may be observed to steal across or break through the beds, so as to lie successively between different layers. No more instructive example of this relation on a small scale could be cited than that of the intrusive sheet which has been laid open in the Dodhead Limestone Quarry, near Burntisland. As shown in the accompanying figure (Fig. 163), this rock breaks through the limestone and then spreads out among the overlying shales, across which it passes obliquely.

FIG. 162.—Intrusive dolerite sheet enclosing and sending threads into portions of shale, Salisbury Crags, Edinburgh.

Among the larger sills this transgressive character is seen to be sometimes manifested on a great scale. Thus, along the important belt of intrusive rocks that runs from Kilsyth to Stirling, the Hurlet Limestone lies in one place below, in another above, the invading mass, but in the intervening ground has been engulphed in it. Similar evidence of the widely separate horizons occupied by different parts of the same sill is supplied at Kilsyth, where the intrusive sheet lies about 70 or 80 fathoms below the Index Limestone, while at Croy, in the same neighbourhood, it actually passes above that seam.[1]

Other interesting evidence of the intrusive nature of the Carboniferous dolerite sills of Central Scotland is supplied by the internal modifications which the eruptive rock has undergone by contact with the strata between which it has been thrust. These alterations, though partly visible to the

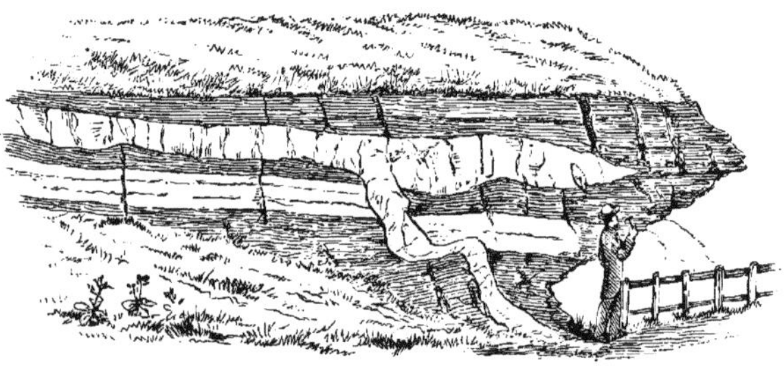

FIG. 163.—Intrusive sheet invading limestone and shale, Dodhead Quarry, near Burntisland.

naked eye, are best studied in thin slices with the aid of the microscope. Tracing the variations of an intrusive dolerite outwards in the direction of the rocks which it has invaded, we perceive change first in the augite. The large crystals and kernels of that mineral grow smaller until they pass into a granulated form like that characteristic of basalts. The large plates and

[1] Explanation of Sheet 31, *Geological Survey of Scotland*, §§ 43 and 83.

amorphous patches of titaniferous iron or magnetite give place to minute particles, which tend to group themselves into long club-shaped bodies. The labradorite continues but little affected, except that its prisms, though as defined, may not be quite so large. The interstitial glassy groundmass remains in much the same condition and relative amount as in the centre of the rock.

Along the line of contact, while the dolerite becomes exceedingly close-grained, its felspar crystals are still quite distinct even up to the very edge. But they become fewer in relative number, and still smaller in size, though an occasional prism two or three millimetres in length may occur. They retain also their sharpness of outline, and their comparative freedom from enclosures of any kind. They tend to range themselves parallel with the surface of the contact-rock. The augite exists as a finely granular pale green substance, which might at first be taken for a glass, but it gives the characteristic action of augite with polarized light. It is intimately mixed through the clear glass of the groundmass, which it far exceeds in quantity. The iron oxides now appear as a fine granular dust, which is frequently aggregated into elongated club-shaped objects, as if round some inner pellucid or translucent microlite. In patches throughout the field, however, the oxides take the form of a geometrically perfect network of interlacing rods. This beautiful structure, described and figured by Zirkel and others,[1] is never to be seen in any of the dolerites, except close to the line of contact with the surrounding rocks. It occurs also in some of the dykes. I have not succeeded in detecting any microlites in the sandstones at the edge of a dolerite sheet, though I have had many slices prepared for the purpose.

Where one of the dolerite sills has invaded sandstone, there is usually a tolerably sharp line of demarcation between the two rocks, though it is seldom easy to procure a hand-specimen showing the actual contact, for the stone is apt to break along the junction-line. Where, however, the rock traversed by the igneous mass is argillaceous shale, we may find a thorough welding of the two substances into each other. In such cases the dolerite at the actual contact becomes a dark opaque rock, which in thin slices under the microscope is found to be formed of a mottled or curdled segregation of exceedingly minute black grains and hairs in a clear glassy matrix, in which the augite and felspar are not individualized. But even in this tachylyte-like rock perfectly formed and very sharply defined crystals of triclinic felspar may be observed ranging themselves as usual parallel to the bounding surfaces of the rock. These characters are well seen in the contact of the intrusive sheet of dolerite with shale and sandstone at Hound Point (Fig. 160).

Another instructive example is furnished by the small threads which proceed from the dolerite of Salisbury Crags, and traverse enclosed fragments of shale (Fig. 162). Some of these miniature dykes are not more than one-eighth of an inch in diameter, and may therefore easily be included, together with part of the surrounding rock, in the field of the microscope. The dolerite in these ramifications assumes an exceedingly fine texture. The felspar is the

[1] *Mikroskopische Beschaffenheit der Mineralien und Gesteine*, p. 273; Vogelsang's *Krystalliten*.

only mineral distinctly formed into definite crystals. It occurs in prisms of an early consolidation, sometimes one-fifth of an inch long, and therefore readily recognizable by the naked eye. These prisms are perfectly shaped, contain abundant twin lamellæ, and show enclosures of the iron of the base. They had been already completely formed at the time of injection; for occasionally they may be observed projecting beyond the wall of the vein into the adjacent shale or sandstone, and they have ranged themselves parallel to the sides of the vein.[1] The black ground, from which these large well-defined crystals stand out prominently, consists of a devitrified glass, rendered dark by the multitude of its enclosed black opaque microlites. These are very minute grains and rudely feathered rods, with a tendency to group themselves here and there into forms like portions of the rhombohedral skeletons of titaniferous iron. So thoroughly fused and liquid has the dolerite been at the time of its injection, that little threads of it, less than $\frac{1}{100}$ of an inch in diameter, consisting of the same dark base, with well-defined felspars, may be seen isolated within the surrounding sedimentary rock. Minute grains and rounded portions of the latter may also be noticed in the marginal parts of the dolerite.

It is thus evident that specimens taken from the edge of an intrusive sheet, where the rock has rapidly chilled and solidified, represent to us an earlier stage in the history of the whole mass than specimens taken from its central portions. In fact, a series of samples collected at short intervals from the outer contact to the inner mass shows, as it were, the successive stages in the consolidation of the molten rock.

From the observations just described, it appears that the triclinic felspars began to assume the shape of large definite crystals before any of the other minerals. These felspars already existed when the molten mass forced its way among the shales, for they can be seen lying with their long axes parallel to the surface of shale, precisely as, in the well-known flow-structures, they behave round a large crystal embedded in the heart of a rock. A few feet from where the consolidation was not so rapid, the iron oxides have grouped themselves into incipient crystalline forms and skeleton crystals; the felspar crystals abundantly occur, and the augite has been left in the finely granular condition. Still further towards the interior of the mass, the normal character of the dolerite is gradually assumed.[2]

[1] The infusibility of the felspar was well shown in some experiments on the rocks of the neighbourhood of Edinburgh, made at my request by Dr. R. S. Marsden, who subjected some of these rocks to fusion at the laboratory of the University of Edinburgh. Microscopic sections were prepared of the products obtained. The basalt of Lion's Haunch is peculiarly instructive. Its large labradorite crystals have resisted the intense white heat which, continued for four hours, has reduced the rest of the minerals to a perfect glass. We can thus well understand how large definite crystals of felspar should have survived or appeared in dykes and veins while the rock was still thoroughly liquid. The glass obtained from the Lion's Haunch rock is of a honey-yellow, and contains translucent tufted microlites. The iron forms beautiful dendritic films in the cracks. Altogether, the glass presents a strong resemblance to the palagonitic substance so abundant among the lapilli in the tuffs of the vents.

[2] For a further and more detailed investigation of the contact phenomena of the Carboniferous doleritic sills of the basin of the Firth of Forth, see the papers by Dr. Stecher, quoted on p. 451.

FIG. 164.—Spheroidal weathering of dolerite sill, quarry east of North Queensferry, Fife.

Where a sill has been injected among carbonaceous shales and coals it has undergone great alteration along the contact, and if the sheet is only a few inches or feet thick, the change extends throughout its whole mass. Black basalts and dolerites, in such circumstances, pass into a substance like a white or pale yellow clay, which at first might be mistaken for some band of fireclay intercalated among the other sediments. But evidence of actual intrusion may usually be found, as where the igneous rock has caught up or broken through the adjacent strata, besides altering them. Such "white traps," as they have been called, have long been familiar in the coal-fields of Scotland and Central England.

As a good illustration of the behaviour of such thin sills among carbonaceous shales I give here a section (Fig. 165) exposed in the old limestone quarry of Hillhouse, south of Linlithgow. At the bottom lies the Hurlet Limestone which has once been extensively mined at this locality. Above it comes a group of black shales which in turn are surmounted by a sheet of beautifully columnar basalt. The shales seem at first sight to include two layers of pale fireclay, each only a few inches in thickness. Closer inspection, however, will show that these two thin intercalations are really sills which, though on the whole parallel with the bedding of the shale, may be seen to cut across it, and even at one place to send a finger into it. The upper example may also be observed to diminish rapidly in thickness in one direction.

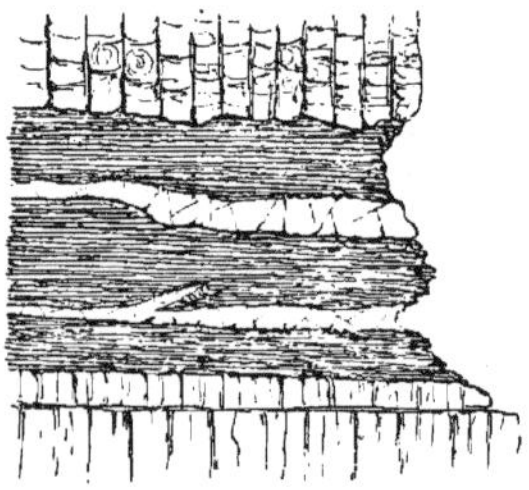

FIG. 165.—Two thin sills of "White Trap" injected into black carbonaceous shale overlying the Hurlet Limestone, Hillhouse Quarry, Linlithgow.

1. Hurlet Limestone; 2. Black shales; 3 3. Two sills of "White Trap"; 4. Columnar Basalts.

The dimensions of the sills vary within tolerably wide limits. Although here and there the injected material dwindles down to an inch or less in thickness, running away even into threads, it more usually forms sheets of considerable depth. The rock of Salisbury Crags, for example, is fully 150 feet thick at its maximum. That of Corstorphine Hill is probably about 350 feet. The great sheet which runs among the lower limestones from Kilsyth by Denny to Stirling has been bored through to a depth of 276 feet, but as the bore started on the rock, and not in overlying strata, some addition may need to be made to that thickness.

The spheroidal weathering so characteristic of basic eruptive rocks is nowhere more characteristically displayed than among the great doleritic sills of the basin of the Firth of Forth. As an illustration of this structure an example is taken here from the large sheet at North Queensferry (Fig. 164).

While one is struck with the great size and extent of some of the sills connected with the puys, as compared with the small and local sheets underneath the plateaux, there is a further fact regarding them that

FIG. 166.—Dyke cutting the agglomerate of a neck. Binn of Burntisland.

deserves remark—their capricious distribution. Their occurrence seems to have little or no relation to the measure of volcanic energy as manifested in superficial eruptions. Thus in the north of Ayrshire, where a long band of lavas and tuffs, pointing to vigorous activity, lies at the top of the Carboniferous Limestone series, and where the strata underneath it are abundantly exposed at the surface, the sills occur as thin and inconstant bands in the central and eastern parts of the district only. The bedded lavas and tuffs at the head of the Slitrig Water have no visible accompaniment of sills. On the other hand, in the Edinburgh and Burntisland districts, the sills bear a large proportion to the amount of bedded lavas and tuffs, while in the Bathgate and Linlithgow district, where the superficial eruptions were especially vigorous and prolonged, the sills are of trifling extent.

It would seem from these facts that the extent to which the crust of the earth round a volcanic orifice is injected with molten rock, in the form of intrusive sheets between the strata, does not depend upon the energy of the volcano as gauged by its superficial outpourings, but on other considerations not quite apparent. Possibly, the more effectively volcanic energy succeeded in expelling materials from the vent, the less opportunity was afforded for subterranean injections. And if the protrusion of the sills took place after the vents were solidly sealed up with agglomerate or lava, it would doubtless often be easier for the impelled magma to open a way for itself laterally between the bedding-planes of the strata than vertically through the thick solid crust. The size and extent of the sills may thus be a record of the intensity of this latest phase of the volcanic eruptions.

Bosses.—The rounded, oval or irregularly shaped masses of igneous rock included under this head are found in some cases to be only denuded domes of sills, as, for example, in the apparently isolated patches in the oil-shale

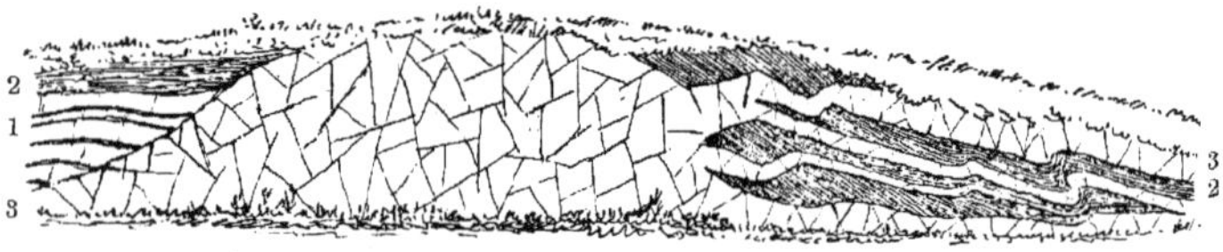

Fig. 167.—Boss of diabase cutting the Burdiehouse Limestone and sending sills and veins into the overlying shales. Railway cutting, West Quarry, East Calder, Midlothian.

1. Burdiehouse Limestone; 2. Shales; 3. Diabase.

district of Linlithgowshire, which have been found to unite under ground. (Compare Fig. 157). In other instances, bosses possibly, or almost certainly, mark the position of volcanic funnels, as at the Castle Rock of Edinburgh, Dunearn Hill, Burntisland, and Galabraes, near Bathgate. But many examples occur which can only be regarded as the exposed ends of irregular bodies of molten material which has been protruded upwards into the

FIG. 168.—Side of columnar basalt-dyke in the same agglomerate as in Fig. 166.

Carboniferous formations. The area between Edinburgh and Linlithgow and the hills of the north of Fife furnish many examples.

The connection between bosses and intrusive sheets is instructively exhibited in a railway cutting to the west of Edinburgh, where the section shown in Fig. 167 may be seen. In the space of a few yards no fewer than four distinct bands of diabase traverse the shale, thickening rapidly in one direction and uniting with a large boss of more coarsely crystalline material. Such connections must exist in all sills, for the material injected as a sheet between stratified formations cannot but be united to some column, dyke or irregular protrusion which descends to the parent magma in the interior. But it is very rarely that the geologist is permitted to see them.

DYKES take a comparatively unimportant place in the eruptive phenomena of the puys. They occur in some numbers, but on a small scale, among the tuff vents, and there they can without much hesitation be

FIG. 169.—Dyke rising through the Hurlet Limestone and its overlying shales. Silvermine Quarry, Linlithgowshire.

set down as part of the phenomena of eruption through these pipes. The Binn of Burntisland, which has been so often referred to in this Chapter, may again be cited as a typical vent for the display of this series of dykes (Figs. 149 and 159). Two additional illustrations from this locality are here given. In Fig. 166 a dyke of compact black basalt is seen on the right hand running up the steep slopes of the agglomerate. Some of these dykes are distinctly columnar, the columns diverging from the walls on each side. Where the encasing agglomerate has been removed by the weather, the side of the dyke presents a reticulated network of prism-ends. A narrow basalt-dyke of this character near the top of the Binn vent is represented in Fig. 168.

But dykes also occur apart from vents and without any apparent relation to these. They are sometimes associated with sills and bosses in such a manner as to suggest that the whole of these forms of injected material belong to one connected series of intrusions. Among the Bathgate Hills, for example, from which I have already cited instances of sills and a boss, the section represented in Fig. 169 occurs. Yet in this same district there is a group of large east and west dykes which cut all the other rocks including the bedded lavas and tuffs, and must be of later date than the highest part of the Coal-measures (Fig. 155).

It is difficult to ascertain the age of the dykes which rise through the Carboniferous formations at a distance from any interbedded sheets of lava and tuff, or from any recognizable vent. The south-east and north-west dykes, increasing in number as they go westward, and attaining a prodigious development in the great volcanic area of Antrim and the Inner Hebrides, are probably of Tertiary date.[1] Others may possibly be Permian, while a certain number may reasonably be looked upon as Carboniferous. In petrographical characters the latter resemble the dolerites and basalts (diabases) of the finer-grained sills.

[1] These are fully described in Chapters xxxiv. and xxxv.

CHAPTER XXVIII

ILLUSTRATIVE EXAMPLES OF THE CARBONIFEROUS PUYS OF SCOTLAND

The Basin of the Firth of Forth—North Ayrshire—Liddesdale.

THOUGH many of the geological details of each of the Scottish districts of Puys have been given in the foregoing pages, it will be of advantage to describe in connected sequence the structure and geological history of a few typical areas. By far the fullest and most varied record of this phase of volcanic activity has been preserved in the basin of the Firth of Forth; but the north of Ayrshire and the district of Liddesdale furnish also many interesting characteristics.

1. BASIN OF THE FIRTH OF FORTH

Reference has already been made to the remarkable peculiarity in the development of the lower part of the Carboniferous system in this district.[1] Elsewhere throughout Scotland the Cement-stone group and the plateau lavas are immediately overlain by the Carboniferous Limestone series. But in the basin of the Firth of Forth a varied succession of strata, more than 3000 feet in thickness, intervenes between the Cement-stones and the Hurlet Limestone. The lower portion of this thick mass of sediment may represent a part of the Cement-stone group of other districts, but even if some deduction is made on this account there remain many hundred feet of stratified deposits, for which there does not appear to be any stratigraphical equivalent elsewhere in Scotland. The distinguishing features of this series of strata are the thick zones of white sandstone, with occasional bands of fine conglomerate, the abundant seams of dark shale, often highly carbonaceous (oil-shales), the cyprid limestones and the seams of coal. Such an association of deposits may indicate a more humid climate and more varied conditions of denudation and deposition than are presented by the typical Cement-stones. The muddy floor of the shallow water must, in many places, have supported a luxuriant growth of vegetation, which is preserved in occasional seams and

[1] See Maclaren's "Geology of Fife and the Lothians," the *Memoirs of the Geological Survey of Scotland*, on Sheets 31 and 32, and my Memoir, already cited, *Trans. Roy. Soc. Edin.* vol. xxix. (1879) p. 437.

streaks of coal. Numerous epiphytic ferns grew on the subaerial stems and branches of the lycopodiaceous trees. Large coniferae clothed the higher grounds, from which the streams brought down copious supplies of sediment, and whence a flood now and then transported huge prostrate trunks of pine. In the lagoons animal life abounded. Cyprids swarmed to such a degree as to form by their accumulated remains bands of limestone, which in the well-known Burdiehouse seam sometimes attain a thickness of 70 feet. Fishes of many genera haunted the waters, for their scales, bones and coprolites are found in profusion among the shales and limestones.

When the puys began their activity, this district was gradually dotted over with little volcanic cones. At the same time it was affected by the general movement of slow subsidence which embraced all Central Scotland. Cone after cone, more or less effaced by the waters which closed over it, was carried down and buried under the growing accumulation of sediment. New vents, however, continued to be opened elsewhere, throwing out for a time their showers of dust and stones, and then lapsing into quiescence as they sank into the lagoon. Two groups of volcanoes emitted streams of lava and built up two long volcanic ridges—those of Fife and West Lothian.

The occasional presence of the sea over the area is well shown by the occurrence of thin bands of limestone or shale, containing such fossils as species of *Orthoceras*, *Bellerophon* and *Discina*, which suffice to prove the strata to be stratigraphical equivalents of the Lower Limestone shale, and part of the Carboniferous Limestone of England (Fig. 170). Yet the general estuarine or freshwater character of the accumulations seems satisfactorily established, not only by the absence of undoubtedly marine forms from most of the strata, but by the abundance of cyprids and small ganoids, the profusion of vegetable remains, and the occasional seams of coal.

The portion of the Forth basin within which the puys are displayed extends from near Leven in Fife, on the north, to Crosswood Burn near the borders of Lanarkshire, on the south, a distance of about 36 miles, and from near Culross in Fife and the line of the Almond River between Stirlingshire and Linlithgowshire, on the west, to the island of Inchkeith on the east, a distance of about 16 miles (Map IV.). But these limits do not precisely mark the original boundaries of the eruptions. To the north and south, indeed, we can trace the gradual dying out of the volcanic intercalations, until we reach ground over which no trace of either lavas or tuffs can be detected. To the east, the waters of the Firth conceal the geology of a considerable area, the island of Inchkeith with its bedded lavas and tuffs showing that these rocks extend some way farther eastwards than the position of that island. But in Midlothian there is no evidence that any of the puy-eruptions took place to the east of the line of the Pentland Hills. On the west side, the volcanic rocks dip under the Millstone Grit and Coal-measures, so that we do not know how far they extend in that direction. But as the Carboniferous Limestone series, when it rises again to the surface on the west side of the Stirlingshire coal-field, is destitute of included lavas and tuffs, the westward limit of the eruptions cannot lie much beyond the line of the River

FIG. 170.—Junction of amygdaloidal basalt with shales and limestone, Shore, half a mile east from Kinghorn, Fife. (From a photograph by Mr. R. Lunn.)

Almond. We shall probably be within the mark if we set down the original area over which puys broke out and spread abroad their lavas and tuffs as covering about 300 square miles of the lagoons and jungles of Central Scotland.

I have already shown that the range in geological time of the puy-eruptions in this district extends from a low horizon among the Calciferous Sandstones through the Carboniferous Limestone series, up to nearly the level of the Calmy Limestone, which lies not far from the top of that series. The vertical thickness of strata between these two stratigraphical limits, when there are no intercalated volcanic rocks, is probably more than 4000 feet.

The vents from which the volcanic materials were ejected, so far as they are now to be observed at the surface, may be divided into two groups, one lying to the north, the other to the south of the Firth of Forth. The northern or Fife group may be followed over an area 15 miles long, and about three miles broad. Some fifteen separate vents may be recognized in it, distributed chiefly at the two ends of the belt, a cluster of about six rising around Burntisland, while another of nearly as many appears at Saline. The characters of some of these necks have been already given in the foregoing pages.

The southern or West Lothian group includes about a dozen vents which are scattered over an area of some 60 square miles, extending from the coast-line between Borrowstounness and Queensferry southwards to Bathgate and Uphall. In this group Binns Hill, a mile long by almost half a mile broad, and rising to a height of nearly 300 feet above the sea, forms the most prominent individual. But the vents are generally smaller in the southern than in the northern group.

The manner in which the vents have been left filled with volcanic material has been described in previous pages. Most of them are occupied by tuff or agglomerate. In many cases the neck consists entirely of pyroclastic detritus, as in most of the vents of eastern Linlithgowshire and many of those in Fife. Not infrequently a column of basalt has risen in the funnel and solidified there, as exemplified by Binns Hill and Saline Hill, or the basalt has filled rents in the tuff and now appears in dykes like those on the Binn of Burntisland (Figs. 148, 149, 159, 166, 168).

But it is possible that in some cases vents may be represented by bosses of basalt or dolerite, unaccompanied by any agglomerate or tuff. Perhaps the best example of this suggested origin is supplied by Galabraes Hill, which rises through the Hurlet Limestone and the volcanic series of the Bathgate Hills, about a mile north-east from the town of Bathgate. This eminence rises to a height of 940 feet above the sea, and consists of a rudely elliptical boss of basalt, measuring 3500 feet in its greater and 3000 feet in its minor axis. It disrupts the sedimentary and volcanic series, which can be traced up to it on all sides. Some of the smaller circular or elliptical bosses in eastern Linlithgowshire and western Fife may perhaps belong to the same category. But undoubtedly most of the intrusive basalts and dolerites of this volcanic region are sills.

Over the greater part of the district, only fine tuffs were ejected. These occur as interstratifications among the ordinary sediments, and vary from mere thin partings, marking the feeblest and briefest explosions, up to continuous accumulations several hundred feet thick. As an example of the least vigorous emission of tuff I may refer to the sections already given on pp. 437, 438. The thicker bands are well illustrated by that which lies some way above the Houston Coal between Drumcross and West Broadlaw in Linlithgowshire, and by the great mass of tuff which occurs immediately below the Calmy Limestone on the River Avon near Linlithgow Bridge, and which may be 300 feet thick.

It is a striking characteristic of the tuffs that they may be met with in their solitary beds intercalated in the midst of ordinary sediments at a distance from any other trace of volcanic activity, their parent vents not being visible. I may cite in illustration an interesting case in the Swear Burn, near the southern end of the volcanic district. A band of tuff about ten feet thick lies there intercalated in a group of dark shales and thin coals. Below it there is a seam of coal four inches thick, and among the blue shales overlying it is another coal ten inches thick. The tuff is pale green, almost white in colour, fine in texture, like a volcanic mud, while some of its component beds, one foot in thickness, are made up of fine laminæ and are even false-bedded. We might infer that the volcanic vent lay at some distance, so that only the finest dust fell over the swamps in which the coal-vegetation was accumulating, but for the presence of occasional blocks of basalt one foot in diameter scattered through the tuff. When the eruptions ceased, the deposition of ordinary mud and the accumulation of plant-remains went on as before, and animal life crowded in again over the spot, for between the partings of the coal above the tuff, abundant fragments of eurypterids and scorpions may be found.

One of the most extensive volcanic discharges of fragmentary material was that which produced the "Houston marls" already referred to. These strata appear to mark a peculiar phase in the volcanic history of the Lower Carboniferous rocks of the Firth of Forth, when exceedingly fine ash, or perhaps even volcanic mud, was erupted in considerable quantity. The "marls" attain in some places a thickness of nearly 200 feet, and can be traced through most of the eastern part of Linlithgowshire, over an area of perhaps more than 50 square miles. This volcanic platform, which has been followed in mining for oil-shale, is one of the most extensive among the puy-eruptions. The material, which probably came from one or more vents among the Bathgate Hills, is not always of equal fineness, but passes into and even alternates with ordinary granular tuff. Thus in the Niddry Burn, above Ecclesmachan, the dull sage-green and brownish red Houston marls contain a few inconstant layers of green tuff, in which may be noticed pieces of black shale and lapilli of the usual basic pumice. Not far to the west, beyond Wester Ochiltree, and thus probably nearer to the active vents that ejected the dust and ashes, the Houston marls are replaced by or include a bedded granular tuff or basalt-agglomerate, which lies above the 2-feet coal of the district. The matrix of this rock is in part a dull green granular mudstone, wrapping round the lapilli and ejected stones, which, when they fall out under the action of the weather, leave casts of their forms behind them. The enclosed fragments vary in size up to blocks three feet in diameter, and consist in great measure of a compact volcanic grit, composed of a fine mud mixed with minute fragments of black shale, grains of sand and flakes of mica. There are likewise blocks of cement-stone and shale. Thin courses of black shale interlaminated with the tuff show its bedding.

The thickest and most continuous accumulations of tuff occur round some of the larger tuff cones, particularly round the Saline Hills, and in the Burntisland district. In the first-named area the copious eruptions of fragmentary material brought the volcanic history there to an end; but around Burntisland they were only the prelude to a prolonged and varied series of discharges.

I have already remarked that in the area of the puys of the Forth-basin, while the majority of the vents were tuff-cones, and emitted only fragmentary discharges, there were two well-marked tracts where lavas were poured out extensively and during a long geological interval. One of those lies in the southern, the other in the northern area.

The southern or Linlithgowshire lava-ridge forms now what are known as the Bathgate and Linlithgow Hills. The lavas extend for about 14 miles from north to south, dying out in both directions, while their present visible breadth is about three miles at its widest part. The highest summit reaches a height of about 1000 feet above the sea. The structure of this long ridge reveals an interesting record of volcanic eruptions. It consists mainly of sheets of basalt, sometimes separated by layers of tuff (Fig. 155). But on one or two horizons the volcanic rocks cease, and ordinary sedimentary deposits take their place. As has been already stated, the Main or Hurlet Limestone can be traced through the heart of the volcanic masses. This seam attains there an exceptional thickness of as much as 70 to 80 feet, and is nowhere more abundantly fossiliferous. During its deposition there seems to have been a subsidence of the area, together with a cessation of volcanic activity for a time. The crinoids, corals, brachiopods, bryozoa, lamellibranchs, gasteropods, cephalopods and fishes, which swarmed in the clear water, built up a thick calcareous layer above the lavas and tuffs of the sea-bottom.

Among the sandstones and shales that cover the limestone, bands of tuff make their appearance, indicating the renewal of volcanic activity. These are immediately surmounted by another thick pile of basalt-sheets. Subsequently, during pauses in the eruptions, while the general subsidence continued, renewed deposits of sediment spread over the submerged volcanic bank. One of the longest periods of quiescence was that during which the coals and even the Index Limestone of Bathgate crept northwards over the sunken lavas and tuffs. But the whole of the ridge does not seem to have disappeared at that time under water, at least these intercalated strata have not been traced across the thick pile of volcanic material near Linlithgow.

During the final period of eruption, the outpouring of lava and discharge of ashes, neither in united thickness nor in horizontal extent, equalled those which had preceded them. When the volcanoes ceased their activity, the area continued to sink, and over the submerged lavas marine organisms crowded the sea-floor, so as to build up the Calmy Limestone. After that time volcanic action seems to have become extinct in this region, for no trace of any intercalated lava or tuff has been detected either in the overlying Millstone Grit or in the Coal-measures. The total thickness of rock in the Linlithgowshire volcanic ridge is about 2200 feet. It will probably not be an exaggeration to place the proportion of lava and tuff in that depth of material at nearly 2000 feet.

The northern or Fifeshire district over which lavas were abundantly erupted stretches along the coast from Aberdour to Kirkcaldy and inland to near Lochgelly, as well as seawards to Inchkeith. It may comprise an area of about 30 square miles. In many respects this is the most important locality in Britain for the study of Carboniferous volcanic history. The sea has cut an admirable coast-section in which the structures of the rocks are laid bare. The bottom and top of the whole volcanic series can be seen. The vents and their relation to the lavas and tuffs that were emitted from them may easily be made out; while the interstratification of well-known seams of rock in the Scottish Carboniferous system permits the sequence and chronology of the whole volcanic series to be traced with great clearness.

Most of these features have already been described in foregoing pages, for the district is a typical one for the study of Carboniferous volcanic phenomena. Thus the group of vents about Burntisland has been illustrated by the Binn of Burntisland rising among the bedded lavas and tuffs. The characters of the Carboniferous basalt-sheets have been enumerated, together with their intercalated layers of tuff and bole, and their fine partings of ashy material that was thrown out over the lagoons during the intervals between two outbursts of lava. But it may be of service if I insert here a detailed section of the whole volcanic series as it is displayed along the coast-section between Burntisland and Kinghorn. The lowest intercalated lavas of that section lie a little above the horizon of the Burdiehouse Limestone, and are thus probably rather earlier than those of Linlithgowshire. The highest reach up to the base of the Hurlet Limestone. The volcanic energy manifestly died out here long before it ceased on the south side of the Firth. Yet so vigorous was its activity while it continued, that it piled up one of the thickest masses of volcanic material anywhere to be seen among the puy-eruptions of the British Isles. The following tabular statement of the alternations of material in this great mass in descending order, was drawn up by me on the ground many years ago, before the construction of fortifications and other changes partly concealed the rocks.

Fig. 171.—Columnar basalt, Pettycur, Kinghorn, Fife. (From a photograph taken for the Geological Survey by Mr. R. Lunn.)

SECTION OF THE VOLCANIC SERIES BELOW THE HURLET OR MAIN LIMESTONE ON THE COAST OF FIFE, WEST OF KINGHORN, IN DESCENDING ORDER [1]

75. Reddish and white sandstones.
74. Shale with hard ribs of limestone and ironstone nodules. Fossils abundant.
73. Limestone, crinoidal, 8 or 9 feet.
72. Blue shale, becoming calcareous towards the top, where shells are plentiful.
71. Reddish false-bedded sandstones, with bands of reddish and blue shale.
70. Basalt in two sills separated by 2 or 3 feet of sandstone and shale.
69. Dark fissile sandy shale, passing up into white shaly sandstone, and including a thin parting of impure coal.
68. Limestone (HURLET or MAIN SEAM) in a number of bands having a united thickness of 25 feet. Abundant fossils.
67. Black shale becoming calcareous at top, and then enclosing abundant *Productus*, etc., 8 or 10 feet.
66. Red and green tufaceous marl and tuff. About 30 feet.
65. Basalt, the lower part strongly amygdaloidal.
64. Tufaceous red marl and tuff; comparatively coarse below, becoming finer above, 3 or 4 feet.
63. Basalt, earthy and amygdaloidal, with an irregular bottom involving masses of the shales below.
62. Dark calcareous shale and dull green tufaceous marly shale, 2 or 3 feet.
61. Crinoidal limestone in several bands with a united thickness of 10 feet.
60. Shale, 1 foot.
59. Fine green sandy tuffs in a number of bands of varying coarseness, about 6 feet.
58. Dark shale with abundance of *Aviculopecten* immediately under the tuffs above, 1½ feet.
57. Soft, light, marly shale with fragmentary plants, 1½ feet.
56. Dark fissile shale, full of fish-scales, plants, etc., 3 feet.
55. Basalt, rudely columnar, dark fine-grained in centre, becoming highly amygdaloidal and scoriaceous at bottom and top.
54. Basalt, like the sheet above, vesicular at top and bottom, with a parting of red clay on top.
53. Fissile rippled sandy shale, with plants, having a red and green marly parting at the top, 12 or 14 feet.
52. Basalts; a group of beds, probably in part sills, involving three bands of sandstone or quartzite.
51. Quartzite—a hard white altered sandstone, 2 to 3 feet.
50. Basalt, light green, earthy, amygdaloidal.
49. Sandstones and shales with plants, 25 feet.
48. Basalt, with a highly amygdaloidal central band. There may be several sheets here.
47. Green tufaceous shale and marl, 1 foot.
46. Basalt, dark, firm and amygdaloidal.
45. Sandstones and shales with plants.
44. Basalt forming west side of Kinghorn Bay, and including more than one sheet. The rock is very black, compact, irregularly columnar, with the usual amygdaloidal earthy band at the base, and forms the crag called the Carlinehead Rocks. An irregular and inconstant band of dull green tufaceous shale, sometimes 2 feet thick, serves to separate two of the basalt-sheets. Below it lies a remarkable scoriaceous almost brecciated basalt, which has been broken up on cooling in such a manner that at first it might be mistaken for a volcanic conglomerate.

[1] The succession of rocks in this interesting coast-section was briefly given by Dr. P. Neill in his translation of Daubuisson's *Basalts of Saxony*, Edinburgh, 1814, note *f*, p. 215. He was secretary of the Wernerian Society, and in his enumeration the Wernerian terminology is used without a hint that any single band in the whole series is of volcanic origin.

43. Basalt, a compact black solid rock, with a vesicular and amygdaloidal bottom, about 40 feet. This sheet runs out into the promontory of Kinghorn Ness.
42. Basalt, firm, compact and highly amygdaloidal throughout, 15 feet.
41. Basalt, earthy, amygdaloidal and scoriaceous in the upper part, compact below.
40. Red tufaceous marl, clay or bole, a few inches thick.
39. Basalt: one of the most compact sheets of the whole series, about 40 feet. The top is formed of a thick zone of scoriaceous and brecciated material, the bottom is singularly uneven owing to the very irregular surface of the underlying bed.
38. Basalt more or less scoriaceous throughout, especially at the bottom, the vesicles being drawn out round the slag-like blocks.
37. Green tufaceous shales with bands of fine green tuff, 7 to 8 feet. The lower bands consist of a gravelly tuff passing up into a fine volcanic mudstone, with scattered lapilli of basalt an inch or more in diameter.
36. Basalt, with an upper, earthy and highly amygdaloidal portion, 30 feet.
35. Tufaceous sandstone and shale, 6 to 8 feet.
34. Basalt, in a thick bed, having an earthy, slaggy top and a scoriaceous bottom.
33. Basalt, very slaggy below with a compact centre.
32. Basalt, like that below it.
31. Basalt, firm, compact, black rock, with a rough, green earthy band, from 6 inches to a foot, at the bottom, and becoming again very slaggy at the top.
30. Green shale like that below the underlying limestone, a few inches in thickness.
29. Coarse, green, sandy tufaceous limestone, averaging 1 foot in thickness.
28. Black shale with plants, 12 or 14 feet, becoming green and tufaceous at the top.
27. Basalt—the most striking of the whole section—a fine compact black olivine-bearing rock, beautifully columnar, 30 feet. The columns reach to within a foot of the bottom of the bed and cease about 10 feet from the top, the upper portion of the bed being massive, with vesicles which are drawn out parallel to the bedding of the series. The lowest part of the bed is a broken brecciated band, 3 or 4 inches thick. (See Fig. 171.)
26. Black shale with fragmentary plants, 3 feet.
25. Basalt, with plentiful olivine, 12 to 16 feet. The base is not highly scoriaceous, but finely vesicular. Towards the top it becomes green, earthy and roughly brecciated. It partly cuts out the tuff underneath.
24. Tuff, green, fine-grained and well-stratified, consisting chiefly of fine volcanic dust, but becoming coarser towards the top, where it contains lapilli and occasional bombs of highly vesicular lavas.
23. Black carbonaceous shale, 3 feet; approaching to the character of an impure coal in the lower part, and becoming more argillaceous above with some thin nodular calcareous bands.
22. Green tuff, 12 feet, well stratified and fine-grained, with minute lapilli of highly vesicular basic lavas; becomes shaly at the bottom.
21. Basalt, compact, amygdaloidal, with highly vesicular upper surface, 20 feet.
20. Basalt, hard, black and full of olivine; an irregular bed 3 to 6 feet thick.
19. Basalt, dull brownish-green to black, full of kernels and strings of calcite, and showing harder and softer bands parallel with upper and under surfaces, which give it a stratified appearance.
18. Basalt, some parts irregularly compact, others earthy and scoriaceous. The distinguishing feature of this bed is the abundance of its enclosed fragments of shale, ironstone and limestone, which here and there form half of its bulk. The roughly scoriaceous upper portion is especially full of these fragments. In the ironstone balls coprolites may be detected, and occasional pieces of plant-stems are embedded in the basalt. This lava has evidently broken up and involved some of the underlying strata over which it flowed. This rock overhangs Pettycur Harbour.
17. Shales and limestone bands more or less tufaceous, 8 to 10 feet, with plants, cyprids, etc. The intercalation of fine partings of tuff in this band has been already cited on p. 438, as an illustration of the feeble intermittent character of many of the volcanic explosions between successive outflowings of lava.

Owing to a change in the direction of strike the rocks now wheel round and for a time run nearly parallel with the coast-line, while they are partly concealed by blown sand and herbage. The shales and limestones just mentioned are not constant, and are soon lost, but about a quarter of a mile westward a band of tuff begins on the same horizon or near it, and increases in thickness towards the west, where it is associated with other sediments. The shore ceases to furnish a continuous section, so that recourse must be had to the craggy slopes immediately to the north, where the rocks can be examined on a cliff face (Fig. 153). There the tuff just referred to, together with some overlying bands of sandstone, is seen to pass under the group of basalts last enumerated. It is a green, stratified rock, perhaps 60 feet thick at its maximum, but dying out rapidly to north-west and south-east. It encloses balls of basalt and subangular and rounded fragments of sandstone, limestone and shale. A mass of coarse volcanic agglomerate which is connected with it and cuts across the ends of some of the basalts below, probably marks the position of the vent from which the tuff was ejected (Fig. 152).

16. Black and grey shales forming a thin band at the summit of King Alexander's Crag.
15. Basalt, dark compact rock, with an upper and lower highly scoriaceous and amygdaloidal band, 15 feet.
14. Black shales, tufaceous green shales, sandstone, and 6 inches of coal, forming a group of strata about 12 feet thick between two basalts; plants and cyprids abundant. (The coal seam is shown in Fig. 151.)
13. Basalt, dull, earthy and highly amygdaloidal, with abundant calcite in kernels and veins; about 15 feet, but varying in thickness.
12. Basalt, forming a well-marked bed from 12 to 25 feet thick. It is a compact black olivine-bearing rock, sparingly amygdaloidal, but showing the usual dull green, earthy scoriform base. The upper surface is singularly irregular, having, in flowing, broken up into large clinker-like blocks, which are involved in the immediately overlying basalt. The bottom also is very uneven, for the basalt has in some places cut out the underlying shales, so as to rest directly upon the basalt below.
11. Black shale, varying up to 6 inches, but sometimes entirely removed by the overlying lava-stream.
10. Basalt, containing large irregularly spheroidal masses of hard black finely vesicular material enclosed in more earthy and coarsely vesicular rock. The vesicles are sometimes elongated parallel to the bedding, but have often been drawn out round a spheroid; some of them measure nearly a foot in length by 2 or 3 inches in breadth. The upper surface is uneven and coarsely amygdaloidal.
9. Basalt, hard black, with abundant olivine, and a columnar structure.
8. Green shale, 6 inches to 1 foot, much baked and involved in the overlying basalt.
7. Basalt, dull-green, earthy, amygdaloidal, varying from 10 to 40 feet in thickness.
6. Blue shale, disappearing where the basalt above it unites with that below.
5. Basalt with olivine, forming a thick irregular bed, which in some places is black and compact, in others green, earthy and amygdaloidal. The upper part is particularly cellular.
4. Sandstones forming a thick group of beds which have long been quarried for building-stone at the Grange and elsewhere.
3. Black shales.
2. Limestone (Burdiehouse).
1. Sandstones, shales and thin limestones forming the strata at Burntisland through which the sills of that district have been injected (Fig. 159).

The phenomena of sills are abundantly developed among the Carboniferous rocks of the basin of the Firth of Forth, and some of the more remarkable examples in this district have been already cited. Taking now a general survey of this part of the volcanic history, I may observe that if the sills are for a moment considered simply as they appear at the surface, and apart from the geological horizons on which they lie, they form a wide ring surrounding the Falkirk and Stirlingshire coal-field.

Beginning at the Abbey Craig, near Stirling, we may trace this ring as a continuous belt of high ground from Stirling to the River Carron. Thence it splits up into minor masses in different portions of the Carboniferous system, and doubtless belonging to different periods of volcanic disturbance, but yet sweeping as a whole across the north-eastern part of the Clyde coal-field, and then circling round into Stirlingshire and Linlithgowshire. There are no visible masses to fill up the portion of the ring back to Abbey Craig. But through Linlithgowshire and the west of Edinburghshire a number of intrusive sheets form an eastward prolongation of the ring. Large as some of these sheets are at the surface, for they sometimes exceed two or three square miles in area, a much larger portion of their mass is generally concealed below ground. Mining operations, for example, have proved that in the south-east of Linlithgowshire areas of intrusive rock which appear as detached bosses or bands at the surface are connected underneath as portions of one continuous sill, which must be several square miles in extent.

But it is in Fife that the sills reach their greatest development among the Carboniferous rocks of Scotland (Fig. 172). A nearly continuous belt of them runs from the Cult Hill near Saline on the west, to near St. Andrews

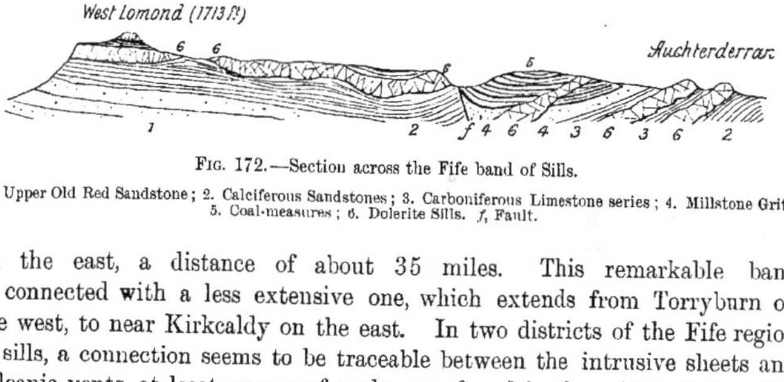

FIG. 172.—Section across the Fife band of Sills.

1. Upper Old Red Sandstone; 2. Calciferous Sandstones; 3. Carboniferous Limestone series; 4. Millstone Grit; 5. Coal-measures; 6. Dolerite Sills. *f*, Fault.

on the east, a distance of about 35 miles. This remarkable band is connected with a less extensive one, which extends from Torryburn on the west, to near Kirkcaldy on the east. In two districts of the Fife region of sills, a connection seems to be traceable between the intrusive sheets and volcanic vents, at least groups of necks are found in the midst of the sills. One of these districts is that of the Saline Hills already described, the other is that of Burntisland. In the latter case the evidence is especially striking, for the vents are connected above with bedded lavas and tuffs, while below lie three well-marked sills (Fig. 159).

It is certainly worthy of remark that sills are generally absent from those areas where no traces of contemporaneous volcanic activity are to be found. No contrast in this respect can be stronger than that between the ground to the east and west of the old axis of the Pentland Hills. In the western district, where the puys are so well displayed, sills abound, but in the eastern tract both disappear.

Another question of importance in dealing with the history of these sills is their stratigraphical position. By far the larger proportion of them lies in the Carboniferous Limestone series. Thus the great sill between Stirling and Kilsyth keeps among the lower parts of that series. On the same general horizon are the vast sheets of dolerite which stretch through Fife in the chain of the Cult, Cleish, and Lomond Hills on the one side, and in the eminences from Torryburn to Kinghorn on the other, though the intrusive material sometimes descends almost to the Old Red Sandstone. In Linlithgowshire and Edinburghshire, as well as in the south of Fife, the sills traverse the Calciferous Sandstone groups.

If the horizons of the sills furnished any reliable clue to their age, it might be inferred that the rocks were all intruded during the Carboniferous period, and as most of them lie beneath the upper stratigraphical limit of the puy-eruptions, the further deduction might be drawn that they are connected with these eruptions. I have little doubt that in a general sense both conclusions are well-founded. But that there are exceptions to the generalization must be frankly conceded. On close examination it will be observed that the same intrusive mass sometimes extends from the lower into the upper parts of the Carboniferous groups. Thus, in the west of Linlithgowshire, a large protrusion which lies upon the Upper Limestones, crosses most of the Millstone Grit, and reaches up almost as high as the Coal-measures. Again, in Fife, to the east of Loch Leven, a spur of the great Lomond sill, crossing the Carboniferous limestone, advances southward into the coal-field of Kinglassie. In Stirlingshire and Lanarkshire numerous large dolerite sheets have invaded the Millstone Grit and Coal-measures, including even the upper red sandstones, which form the top of the Carboniferous system in this region. It is thus obvious that if the puy-eruptions in the basin of the Forth ceased towards the close of the deposition of the Carboniferous Limestone series, there must have been a subsequent injection of basic lava on a gigantic scale in central Scotland. I shall recur to this subject in Chapter xxxi.

2. NORTH OF AYRSHIRE

In this part of the country another group of puys and their associated tuffs and lavas may be traced from near Dalry on the west, to near Galston

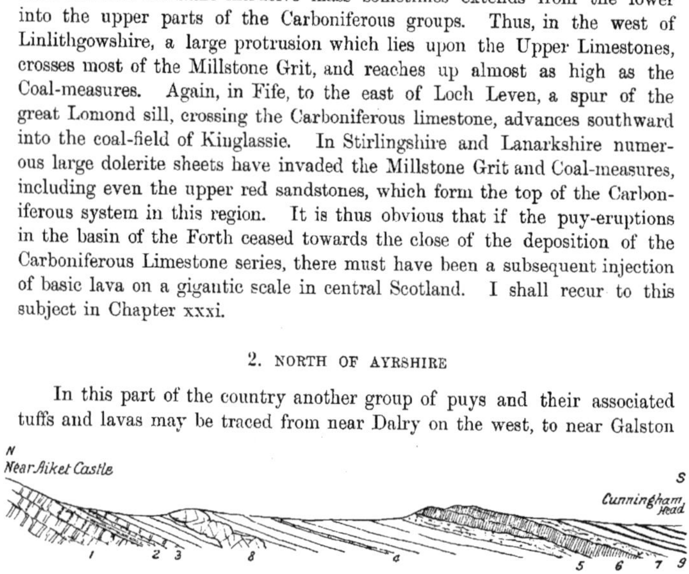

FIG. 173.—Section across the Upper Volcanic Band of north Ayrshire. Length about four miles.

1. Andesite lavas of the Clyde Plateau; 2. Tuffs closing the Plateau volcanic series; 3. Hurlet Limestone; 4. Carboniferous Limestone series with coal-seams; 5. Lower tuff zone of the Upper volcanic band; 6. Basic lavas; 7. Upper tuff zone; 8. Basic sill; 9. Coal-measures.

on the east (Map IV.). The length of the tract is about sixteen miles, while its breadth varies from about a furlong to nearly a mile and a half. I have had occasion to allude to this marked band of volcanic materials which here intervenes between the Carboniferous Limestone and the Coal-measures, and from its position appears to mark the latest Carboniferous volcanoes. Its component rocks reach a thickness of sometimes 600 feet, and as they dip southwards under the Coal-measures, they may extend for some distance in that direction. They have been met with in borings sunk through the northern part of the Irvine coalfield. Even what of them can be seen at the surface, in spite of the effects of faults and denudation, shows that they form one of the most persistent platforms of volcanic rock among the puy-eruptions of Scotland.

Where best developed this volcanic band has a zone of tuff at the bottom, a central and much thicker zone of bedded basalts, and an upper group of tuffs, on which the Coal-measures rest conformably. A few vents, probably connected with it, are to be seen at the surface between Fenwick and Ardrossan. But others have been buried under the Carboniferous sedimentary rocks, and, as already described, have been discovered in the underground workings for coal and ironstone (p. 434). These mining operations have, indeed, revealed the presence of far more volcanic material below ground than would be surmized from what can be seen at the surface. Here and there, thin layers of tuff appear in brook-sections, indicating what might be conjectured to have been trifling discharges of volcanic material. But the prosecution of the ironstone-mining has proved that, at the time when the seam of Blackband Ironstone of that district was accumulated, the floor of the shallow sea or lagoon where this deposition took place was dotted over with cones of tuff, in the hollows between which the ferruginous and other sediments gathered into layers. That seam is in one place thick and of good quality; yet only a short distance off it is found to be so mixed with fine tuff as to be worthless, and even to die out altogether.[1]

3. LIDDESDALE

A remarkable development of puys lies in that little-visited tract of country which stretches from the valleys of the Teviot and Rule Water south-westwards across the high moorland watershed, and down Liddesdale. Through this district a zone of bedded olivine-basalts and associated tuffs runs in a broken band which, owing to numerous faults and extensive denudation, covers now only a few scattered patches of the site over which it once spread. The geological horizon of this zone lies in the Calciferous Sandstones, many hundred feet above the position of the top of the plateau-lavas (Map IV.).

So great an amount of material has been here removed by denudation that not only has the volcanic zone been bared away, but the vents which supplied its materials have been revealed in the most remarkable manner over an area some twenty miles long and eight miles broad. Upwards of forty necks of agglomerate may be seen in this district, rising through the Silurian, Old Red Sandstone, and lowest Carboniferous rocks. It fills the

[1] See Explanation of Sheet 22, *Geol. Surv. of Scotland*, pars. 29, 33, 45.

geologist with wonder to meet with those stumps of old volcanoes far to the west among the Silurian lowlands, sometimes fully ten miles away from the nearest relic of the bedded lavas connected with them.[1] That these vents, though they rose through ground which at the time of their activity was covered with the volcanic series of the plateaux, do not belong to that series, but are of younger date, has been proved in several cases by Mr. Peach.

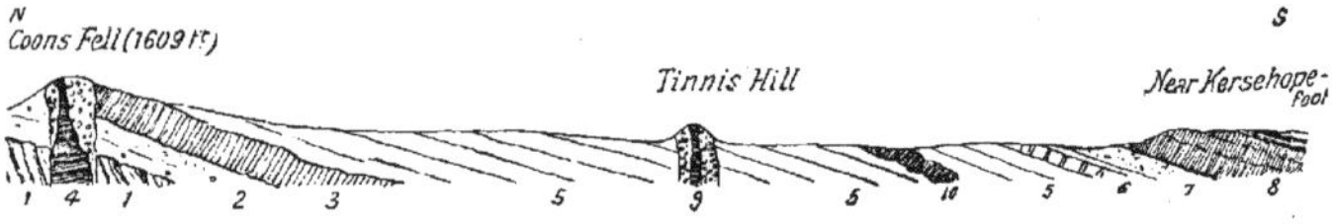

FIG. 174.—Section showing the connection of the two volcanic bands in Liddesdale.

1. Upper Silurian strata; 2. Upper Old Red Sandstone; 3. The lavas of the Solway plateau; 4. Agglomerate neck with lava plug, belonging to the plateau system; 5. Calciferous Sandstone series; 6. Thick Carboniferous Limestones; 7. Tuff, and 8. Lavas, of the upper volcanic band, connected with the puys; 9. Agglomerate neck with lava plug belonging to the puy-system; 10. Basic sill.

He has found that among the blocks composing their agglomerates, pieces of the sandstones, fossiliferous limestones and shales of the Cement-stone group, overlying the plateau-lavas, are to be recognized. These vents were therefore drilled through some part at least of the Calciferous Sandstones, which are thus shown to have extended across the tract dotted with vents. After the volcanic activity ceased, fragments of these strata tumbled down from the sides into the funnels. Denudation has since stripped off the Calciferous Sandstones, but the pieces detached from them, and sealed up at

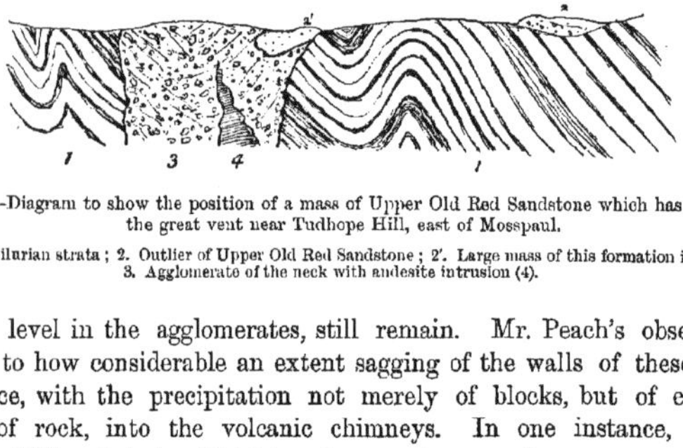

FIG. 175.—Diagram to show the position of a mass of Upper Old Red Sandstone which has fallen into the great vent near Tudhope Hill, east of Mosspaul.

1. Upper Silurian strata; 2. Outlier of Upper Old Red Sandstone; 2'. Large mass of this formation in the vent; 3. Agglomerate of the neck with andesite intrusion (4).

a lower level in the agglomerates, still remain. Mr. Peach's observations indicate to how considerable an extent sagging of the walls of these orifices took place, with the precipitation not merely of blocks, but of enormous masses of rock, into the volcanic chimneys. In one instance, between Tudhope Hill and Anton Heights, a long neck, or perhaps group of necks, which crosses the watershed, shows a mass of the red sandstone many acres in extent, and large enough to be inserted on the one-inch map, which has fallen into the vent (Fig. 175).

The materials ejected from the Liddesdale vents include both basaltic lavas and tuffs. The former are sometimes highly vesicular, especially

[1] They have been recognized and mapped by Mr. B. N. Peach for the Geological Survey. See Sheets 11 and 17, *Geol. Surv. Scotland.*

TO ACCOMPANY SIR ARCHIBALD GEIKIE'S "ANCIENT VOLCANOES OF BRITAIN"

Map IV

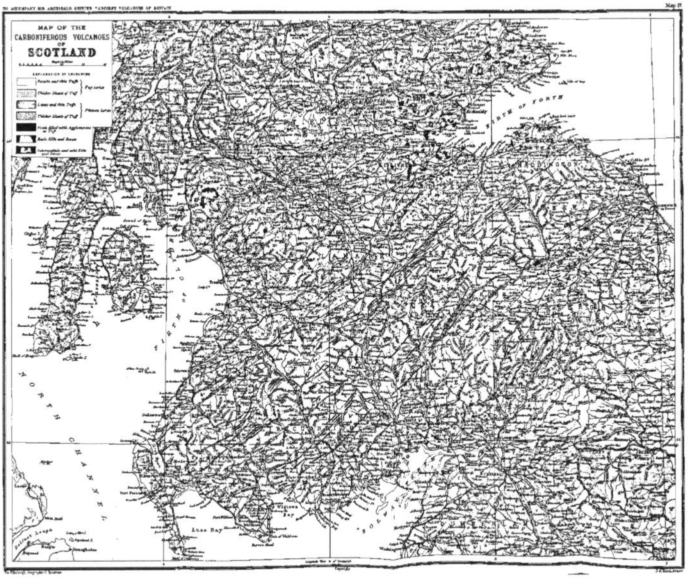

along the upper parts of the flows. They are thickest towards the north, and in Windburgh Hill attain a depth of at least 300 or 400 feet. In that part of the district they form the lower and main part of the volcanic series, being there covered by a group of tuffs. But a few miles southwards, not far to the west of Kershopefoot, they die out. The tuffs then form the whole of the volcanic band which, intercalated in a well-marked group of limestones, can be followed across the moors for some six miles into the valley of the Esk, where an interesting section of them and of the associated limestone and shales is exposed (Fig. 174). At Kershopefoot, a higher band of basic lava overlies the Kershopefoot limestone, and can be traced in scattered patches both on the Scottish and English side of the Border.

END OF VOL. I.

Printed by R. & R. CLARK, LIMITED, *Edinburgh.*

MAP OF THE
VOLCANIC DISTRICTS
OF THE
BRITISH ISLES
BY
Sir ARCHIBALD GEIKIE, D.C.L., F.R.S.
English Miles
Tertiary
(Probably Tertiary)
Permian
Carboniferous
Old Red Sandstone & Devonian
Silurian
Cambrian
Bristol Channel
Cardigan Bay
Isle of Man

BY THE SAME AUTHOR.

TEXT-BOOK OF GEOLOGY. With Illustrations. Third Edition. Revised and Enlarged. Medium 8vo. 28s.

CLASS-BOOK OF GEOLOGY. Illustrated with woodcuts. Second Edition. Crown 8vo. 4s. 6d.

ELEMENTARY LESSONS IN PHYSICAL GEOGRAPHY. Illustrated with woodcuts and ten plates. Fcap. 8vo. 4s. 6d. QUESTIONS FOR THE USE OF SCHOOLS. Fcap. 8vo. 1s. 6d.

PHYSICAL GEOGRAPHY. Illustrated. Pott 8vo. 1s. [*Science Primers.*

THE SCENERY OF SCOTLAND VIEWED IN CONNEXION WITH ITS PHYSICAL GEOLOGY. With a Geological Map and Illustrations. Second Edition. Crown 8vo. 12s. 6d.

GEOLOGY. With Illustrations. Pott 8vo. 1s. [*Science Primers.*

OUTLINES OF FIELD GEOLOGY. New and revised Edition. Extra Fcap. 8vo. 3s. 6d.

GEOLOGICAL SKETCHES AT HOME AND ABROAD. With Illustrations. 8vo. 10s. 6d.

THE TEACHING OF GEOGRAPHY. Suggestions regarding principles and methods for the use of teachers. Globe 8vo. 2s. [*Geographical Series.*

GEOGRAPHY OF THE BRITISH ISLES. Pott 8vo. 1s. [*Geographical Series.*

Box of Geological Specimens to illustrate Geikie's Primer of Geology. 10s. 6d.

MACMILLAN AND CO., LTD., LONDON.

MACMILLAN AND CO.'S BOOKS FOR STUDENTS

OF

PHYSICAL GEOGRAPHY, GEOLOGY, AND MINERALOGY.

PHYSICS OF THE EARTH'S CRUST. By Rev. OSMOND FISHER, M.A., F.G.S., Hon. Fellow of King's College, London. 2nd Ed., enlarged. 8vo. 12s.

TABLES FOR THE DETERMINATION OF THE ROCK-FORMING MINERALS. Compiled by F. L. LOEWINSON-LESSING. Trans. by J. W. GREGORY, B.Sc., F.G.S. Glossary by Prof. G. A. J. COLE, F.G.S. 8vo. 4s. 6d. net.

MICROSCOPICAL PHYSIOGRAPHY OF THE ROCK-MAKING MINERALS. By H. ROSENBUSCH. Trans. by J. P. IDDINGS. 8vo. 24s.

METEOROLOGY. By T. RUSSELL. 8vo. 16s. net.

A POPULAR TREATISE ON THE WINDS. By W. FERREL, M.A., Member of the American National Academy of Sciences. 8vo. 17s. net.

ECONOMIC GEOLOGY OF THE U.S. By R. S. TARR, B.S. 8vo. 16s. net.

ELEMENTARY PHYSICAL GEOGRAPHY. By R. S. TARR, B.S. Crown 8vo. 7s. 6d. net.

ELEMENTS OF CRYSTALLOGRAPHY, for Students of Chemistry, Physics, and Mineralogy. By G. H. WILLIAMS, Ph.D. Crown 8vo. 6s.

POPULAR LECTURES AND ADDRESSES. By Lord KELVIN, P.R.S., Vol. II. GEOLOGY AND GENERAL PHYSICS. Crown 8vo. 7s. 6d.

THE PLANET EARTH. By R. A. GREGORY, F.R.A.S. Crown 8vo. 2s.

PHYSIOGRAPHY. An Introduction to the Study of Nature. By the Right Hon. T. H. HUXLEY, F.R.S. Illustrated. Crown 8vo. 6s.

SHORT STUDIES IN EARTH KNOWLEDGE. Introduction to Physiography. By WILLIAM GEE. Illustrated. Globe 8vo. 8s. 6d.

OUTLINES OF PHYSIOGRAPHY—THE MOVEMENTS OF THE EARTH. By J. NORMAN LOCKYER, F.R.S. Illustrated. Crown 8vo. Sewed, 1s. 6d.

PHYSIOGRAPHY FOR BEGINNERS. By A. T. SIMMONS, B.Sc. Globe 8vo. 2s. 6d.

MACMILLAN AND CO. LTD., LONDON.

Spotted something that isn't quite right?

Let us know and we'll fix it for future reprints. And as a thank you for your diligence, we'll send you a free Sothis book of your choice.

publisher@sothis.press

This page intentionally left blank

Except of course for that line above. And this one.
And whilst we have your attention, perhaps you'd
like to follow us on Instagram @sothis.press. Go on.

Made in United States
Orlando, FL
07 November 2026

90337541R00300